To Xie Yun and the memory of Guo Lining

Allan Kulikoff

From British Peasants to Colonial American Farmers

THE UNIVERSITY OF NORTH CAROLINA PRESS

CHAPEL HILL AND LONDON

Manufactured in the United States of America
Designed by April Leidig-Higgins
Set in Minion by Keystone Typesetting, Inc.

The paper in this book meets the guidelines for permanence and durability of the Committee on Production Guidelines for Book Longevity of the Council on Library Resources.

Library of Congress Cataloging-in-Publication Data
Kulikoff, Allan.
From British peasants to colonial American farmers / by Allan Kulikoff.
p. cm.
Includes bibliographical references (p.) and index.
ISBN 0-8078-2569-7 (cloth: alk. paper)
ISBN 0-8078-4882-4 (pbk.: alk. paper)
1. United States—Economic conditions—To 1865. 2. Great Britain—Economic conditions—18th century. 3. Great Britain—Economic conditions—17th century. 4. Europe—Economic conditions—18th century. 5. Europe—Economic conditions—17th century. 6. Agriculture—Economic aspects—United States—History. 7. Agriculture—Economic aspects—Great Britain—History. 8. Agriculture—Economic aspects—Europe—History. 9. Farm tenancy—United States—History. 10. Farm tenancy—Great Britain—History. 11. Farm tenancy—Europe—History. 12. United States—Emigration and immigration—History. I. Title.
HC104.K85 2000
338.1′0973—dc21 00-029904

04 03 02 01 00 5 4 3 2 1

contents

Acknowledgments xi

INTRODUCTION

From Peasants to Farmers 1

PROLOGUE

The Remaking of Rural England 7

The Medieval Legacy 8

The English Road to Agrarian Capitalism 16

Households and the Transformation of Gender Relations 27

CHAPTER ONE

From England to America 39

Colonial Promotion and Immigration 42

Motivations for Migration 53

Servants and Farmers, Mechanics and Merchants 61

From Emigrant to Immigrant 71

CHAPTER TWO

This Newfound Land 73

This Land Is Not Ours 75

This Land Should Not Be Theirs 85

This Land Is Ours 106

CHAPTER THREE

New Lands, New Peoples 125

Hugging the Coast: The Land System of Tidewater Areas 127

Conquest of the Borderlands 138

Moving to the Empty Land 145
Making a Home in the Hills and Valleys: Colonizing the Borderlands 150

CHAPTER FOUR

Deprivation, Desire, and Emigration 165
A Peasant's and Worker's Paradise? 167
Economic Development and British Emigration 170
Poverty and Poor Law in Britain 174
Decaying Feudalism in German Lands 180
Migration and Emigration 184
The Process of Emigration 190

CHAPTER FIVE

Markets and Farm Households in Early America 203
To Market, To Market 205
Both a Borrower and a Lender Be? 216
Household Formation 226
Family Labor and Hired Hands 242

EPILOGUE

The Farmers' War and Its Aftermath 255
Disruption of the Agrarian Economy 256
Who Plowed the Fields and Made the Corn? 261
The Ravages of War 268
The Balance of Terror 275
Refugees and Émigrés 277
Rebuilding Farm Markets 280
New Peoples and New Lands 283

AFTERWORD

Toward a Farmers' Nation 289

Notes 293
Bibliography 367
Index 471

maps

1. European Population in Mainland British North America, ca. 1675 107

2. British Mainland Colonies, ca. 1675 108

3. British Mainland Colonies, ca. 1770 128

4. European Population in Mainland British North America, ca. 1770 146

acknowledgments

DURING 1985 I began thinking about farm families in early America. My interest started innocently enough. Until the end of the nineteenth century most Americans lived on small farms, but I could tell students in my rural history class at Princeton very little about their history, their aspirations, or their relations to slaves, Indians, or rich men, much less the dynamics of their households. I thought I would use a month or two of leave to review the literature, develop an interpretation of small farmers, and write a short article. The essay (modestly titled "Class, Gender, and Race in Early America, 1650–1900") quickly mushroomed, turning into the longest paper ever imposed upon the good people who attended seminars at the Philadelphia Center for Early American Studies. The response at that meeting in late spring 1985 was encouraging. I was peppered with questions, queries, and disagreements about my arguments, but most in attendance thought the project worthy and agreed I should try my hand at a short book on the topic. I then revised the paper (now titled, more modestly, "The Rise and Destruction of the American Yeoman Classes"), shortening it for presentation at the American Historical Association. At that meeting Richard D. Brown and especially Elizabeth Fox-Genovese tore my argument and narrative to shreds, just what I needed to revise my thinking.

During an NEH fellowship in 1986–87 I returned to small farmers and rewrote the manuscript, adding details and making it even longer. But it was so tightly and densely written that reader misunderstanding was inevitable. To clarify my thoughts (for myself as much as for my readers), I began writing interpretive and theoretical essays about capitalist transformation in America. I completed writing (and revising) these essays at the Newberry Library in 1990 (where I enjoyed an NEH–Lloyd Lewis fellowship) and at Northern Illinois University in 1991, using a fellowship from ACLS. In 1992, while I continued to work on a big history of small farmers, I published *The Agrarian Origins of*

American Capitalism, a collection of these essays. (I have borrowed phrases, sentences, and small segments from that book, reshaping them to fit the concerns of the present work.)

As my work progressed (financed, in part, by summer stipends from the history department and the graduate school at Northern Illinois University), it grew longer and longer. A projected short volume on small farmers, 1600–1900, became a very long one. I knew I either had to cut way back (impossible, given the richness of the story) or write more than one volume. My long-suffering editor, Lewis Bateman, eagerly agreed that I could write several volumes, if I would disgorge a book (any book) faster. So in 1991 I began to compose a colonial volume. At first the chapters seemed of reasonable length. Supported by a John Simon Guggenheim Foundation Fellowship, my writing progressed especially rapidly during the 1996–97 academic year. But chapters begat new chapters (two projected chapters each split three ways!), so I divided them into two volumes, one emphasizing demographic and economic structures; the other, small farmer (yeoman) identity. During spring and summer 1999, I thoroughly revised the entire manuscript, taking criticism into account and incorporating the rich literature of the preceding few years into my story.

During my years of research and writing, I have incurred many academic debts. The document retrieval department at the Northern Illinois University library has ordered hundreds of books and articles. The history department at Northern Illinois University has been a fine home. Three chairs—Otto Olsen, George Spencer, and Elaine Spencer—lent their support by allowing me to double up in the fall semester to free up writing time during the spring. The department granted me two years and the graduate school awarded me three years of summer money. Fellowships from the NEH, the ACLS, the Newberry Library, and the Guggenheim Foundation played an essential role, as did leaves and sabbaticals at Princeton University and Northern Illinois University.

Two eminent scholars have provided absolutely essential help. Stanley Engerman has supported this project since it began, reading drafts, making comments, sending encouragement, and expressing enthusiasm even as the project grew bigger and its completion more distant. Elizabeth Fox-Genovese lent an ear and made astute suggestions for structural changes as I elaborated endlessly on each writing crisis. When I most despaired about the length of the book manuscript, she read my proposals for splitting the volume into two parts and advised me on the best way to proceed. And she read the entire manuscript, making masterful comments about its structure and argument.

I have presented pieces of the book at Northern Illinois University, the Newberry Library, Indiana University, and the Capital Historical Society. Scholars too numerous to mention have commented on this project over the past decade; they include my current and former colleagues at Northern Illinois University, especially William Beik, Mary Furner, Simon Newman, Marvin Rosen, and Alfred Young. Many of my graduate students—among them Michelle Gillespie, Marcia Sawyer, Michael Smuska, Susan Branson, Lucy Murphy, Thomas Humphrey, Terry Sheahan, Robert Hagaman, and Sean Taylor—have shared my interest in agrarian history, and their scholarship continues to inspire my work. I am grateful to Alison Games and Lisa Bywaters, who each read and commented on a chapter of the manuscript; to Sean Shesgreen and Sarah Maza, who provided intellectual stimulation throughout my research; and, in particular, to David Bywaters, who read the entire manuscript with the keen eye of an editor, pointing out inconsistencies in argument as well as stylistic blunders.

This volume is dedicated to my father-in-law, Xie Yun, and the memory of my mother-in-law, Guo Lining, two extraordinary Chinese individuals who impressed me with their integrity and humanity. I cherish the times they spent with us in our home, cooking for us, taking care of our young daughter, and arguing with me about American and Chinese politics, as we tried to make sense of the new world order. I thank them for their interest in my work and their emotional support.

Although my now six-year-old daughter Xie Rachel Kulikoff frequently chased me off the computer so she could play "Reader Rabbit" and kept me from writing with her myriad demands, I must credit her for the completion of this work. In her own unique way, she goaded me to finish, calling me "history man," explaining to her friends what "yeomen" means, making me a birthday card with a picture of my book on it, and asking me a million times, "Baba, when are you going to finish the chapter?" At long last my book is done, and I hope that someday Rachel (who loves to see her name in print) will read it.

Finally, I thank my wife, Lihong Xie, for her enduring love, understanding, and support. As always, she is my first reader and editor, my best companion and friend.

DeKalb, Illinois
September 1999

FROM BRITISH PEASANTS TO COLONIAL AMERICAN FARMERS

introduction

From Peasants to Farmers

"Those who labour in the earth," Thomas Jefferson wrote in a famous passage in his *Notes on the State of Virginia*, "are the chosen people of God, if ever he had a chosen people, whose breasts he has made his peculiar deposit for substantial and genuine virtue." These men—who looked "to their own soil and industry . . . for their subsistance"—had achieved independence, unlike those who "depend for it on the casualties and caprice of customers. Dependance begets subservience and venality, suffocates the germ of virtue, and prepares fit tools for the designs of ambition." Jefferson knew that a majority of white Americans belonged to families of these virtuous small farmers. But everyone realized that most families in Britain and Europe, the ancestral (or recent) homeland of nearly all free Americans, suffered the greatest privation, debasement, and subservience.[1]

Who were the small farmers Jefferson so praised? Their farms varied in size but generally ranged from 25 to 200 acres, large by British standards. Unlike their counterparts in Britain or Germany, they owned or expected to own their land. Landownership buttressed their economic independence. Farm owners and their families grew much of their own food, took small surpluses to market, exchanged goods with neighbors, and nurtured future generations in the soil. Small farms like theirs had disappeared from Britain and Germany, yet from the early seventeenth century a substantial majority of colonists lived on them.

This book tells how British and German peasants became small American farmers. Our story takes place on two continents over more than two centuries. It has a vast cast of characters: English landlords, peasants, and merchants; American Indians and settlers bent on taking their land; colonial small farmers and their gentleman-farmer patrons or protagonists; immigrant labor recruiters and land speculators; the children of colonial farmers and newly arrived British and German peasants and laborers; and farmers—with the critical help of wives, children, hired hands, and the occasional servant or slave—struggling to build farms for themselves and their children.

We begin in England, where peasants had enjoyed secure land tenure for centuries. But in the sixteenth and seventeenth centuries, landlords and improving farmers (agrarian capitalists) threw peasants off the land, leaving them and their descendants with a great yearning for land. The same men who evicted peasants financed colonial ventures that promised land to former peasants. Persuaded by labor recruiters, family, or friends, hundreds of thousands of English, Scot, Irish, and German immigrants came to America. Most had to work in temporary bondage to pay their passage, but all expected to get land after completing their terms. Two-thirds of the emigrants (and an equal portion of their descendants) did acquire land.

Capitalist transformation, then, stands at the center of our story. In popular language, the words "capitalist" and "capitalism" evoke the power of captains of industry and finance, or a set of entrepreneurial and market ideals most Americans shared. But "capitalism," as used in this work, refers to a society dominated by two classes: capitalists who own the means of production (banks, factories, tools, and productive land) and workers who have only their labor to sell. By this definition, capitalism had not yet reached our shores as late as the American Revolution. But because Britain had turned capitalist, colonists swam in a capitalist sea, selling goods to English capitalists, buying goods made by wage workers, and seeking the aid of capitalists in financing colonies.

Capitalist transformation began in England a century before colonization and by the mid-eighteenth century had spread to Scotland, Ulster, and (to a lesser degree) German states. Capitalists stole peasant land, consolidated the land they took, and leased land to improving tenants. With their profits, they financed commerce, manufacturing, banks, and colonial ventures. The old relation between lord and peasant, which had guaranteed peasant use of land, crumbled, and an army of surplus people tramped the countryside, looking for work. Displaced peasants harvested crops, spun thread and wove cloth, mined coal, and moved to cities. When capitalists—eager to import corn, tobacco, and sugar—founded colonies, former peasants were eager to move to get land. Living on the periphery of the capitalist world, emigrants and their descendants made farms and got the land capitalists had denied them at home.[2]

During early modern times, hundreds of thousands of Britains and Europeans sold their possessions, left farm cottages or city hovels, and moved hundreds or thousands of miles to Ireland, Prussia, eastern Europe, and the Americas. Fleeing persecution or seeking opportunities, they settled thousands of square miles, setting off violent conflicts with native peoples. Far more seventeenth-century Britons moved to Ireland and the West Indies than the mainland; more eighteenth-century Europeans moved east than west. Seventeenth-century English immigrants nonetheless secured a foothold on the coast of the New World and gradually forced Indians to leave; eighteenth-century Europeans repeopled a vast backcountry, stretching through the foothills and across the mountains, making new farms wherever they went. When land became scarce, their descendants moved inland. A constant churning thus engulfed the colonies, as youths and families formed new neighborhoods from northern New England to Georgia.[3]

Traveling to America was an expensive undertaking, which the poorest folk, hardest hit by economic change, could rarely afford on their own. Because of high transportation costs and exit fines, for instance, poor eighteenth-century Germans rarely crossed the ocean. Although the market in indentured servants provided some opportunities for poor seventeenth-century Britons, immigrants usually began by selling some land or other productive property. They spent much of the proceeds traveling to a port, and many of them still had to indent themselves for service in the colonies; the rest often arrived with few reserves. Only after accumulating property in older regions could they move to a frontier and procure a farm substantial enough to provide for them and their children.[4]

At first, immigrants merely wanted land, but—seeing unimproved land spread endlessly before them—they came to expect to own it. Land prices, the costs of surveying land, governmental willingness to give land away, and treatment of squatters structured opportunity for landownership. While small farmers wanted cheap or free land, land speculators wanted an unfettered market where supply and demand set land prices. Colonial governments, caught in the middle, vacillated, sometimes giving land to farmers to contain the Indian threat, other times granting thousands of acres to land speculators. Whenever they could, small farmers got title to their land, but if land became too expensive or rich men refused to sell, they squatted and refused to move. Despite these difficulties, an astonishingly high percentage—two-thirds of colonial families—owned land.[5]

Once farmers got land, they worked to improve it. They sought a competency, which they defined as the ability to grow (or trade for) most of their food, make most of their own clothing, and cut down most of the firewood needed for winter. Within the constraints of the search for competency and the marketplace, farm husbands and wives negotiated the division of labor and figured out what goods they would trade with neighbors or send to market. While farm men (helped by wives and children) cleared land, farm women worked to feed and clothe the family. Vegetables from women's gardens fed small farmer families; women made the clothes the family wore and sold eggs and butter. Relations between farm women and men were thus central to the formation and preservation of small farm households. Farm men, who legally controlled their wives' property, achieved economic independence by exploiting their labor, but neither state coercion nor cultural norms prevented conflicts between spouses.[6]

No family, no matter how rich, could completely feed or clothe themselves. To pay taxes or buy imported sugar or coffee, farmers took surplus tobacco, rice, or wheat to market. Farm families used most of their cash to pay taxes or buy land. Such trade financed intensive barter between families. When a farm wife needed butter, she borrowed it from a neighbor, eventually giving thread in return; farm men traded corn for pork or cider for lumber. Neighbors helped neighbors build cabins, raise barns, or harvest corn, knowing they could count on the same help. Seeking always to sustain their families and provide for the next generation, these farmers thus made "composite" farms, combining household production, local exchange, and market sales.[7]

I have framed this book around economic development, migration, land acquisition, and the relation farm families forged with the market. I begin by examining how English peasants organized their households; how rich Englishmen got capital to finance colonies; and how others lost their land, tramped the countryside, and became eager to emigrate. Chapter 1 details immigrant recruitment in seventeenth-century England and patterns of migration to the colonies. Once they arrived, Chapter 2 shows, colonists faced hostile Indians, deep forests, and a climate far more extreme than England's. Despite the struggle with Indians, most families did get land and made it their own. As Chapter 3 relates, during the eighteenth century, after coastal lands filled with settlers, colonists moved to new frontiers, chasing Indians away and improving more land. Chapter 4 shows who left eighteenth-century Britain and Europe and explains why so many peasants moved east rather than west. Turning from economic and demographic issues to the process of farm making, Chapter 5 describes the gender division of labor on the farm, exchange between farm families, and the relation between market and household. The American Revolution, the Epilogue argues, temporarily stopped migration, ended international trade, thrust families into subsistence production, and ignited vicious partisan and Indian warfare; after the war, internal migration and farm making resumed and intensified.

Building such an interpretation has become more difficult in recent years. Historians, who knew little about farm families a quarter-century ago, now swim upstream against an ocean of data—community studies, demographic and immigration histories, economic analysis, works about farm rebellion and rural faith—covering every corner of North America, Britain, and Europe. And newly published primary sources appear every year.

To interpret these materials fully would take a book far longer than this one. I have, therefore, not included here the class identity of small farmers and have played down elements of farmer identity—kinship, community, ethnicity, and faith—central to class formation. I will address these issues in a companion volume, showing how the yeoman class of Jefferson's day—despite its origin deep in the history of medieval England—had made itself a new class, nurtured in a rich and abundant new land. That book will show how small farmers found a voice after revivals that swept through the colonies in the eighteenth century, in the new ethnic communities British and German immigrants made, and in repeated struggles with landlords, merchants, and rulers bent on depriving them of their farms.

Recent historians have turned from master narratives such as this one to study small events intensively. These microhistories put a human face on big events, processes, and social changes; they revel at the contingency of events; and they draw out cultural meaning absent from structural arguments. The drama of Paul Revere's ride, the story of unredeemed captive Eunice Williams (of Deerfield, Massachusetts), or the report of the botched abortion that killed Connecticut teenager Sarah Grosvenor in 1742 reveal much about politics, communal norms, and local culture. But one cannot assume that every farmer (much less farmer's wife, servant, or slave) understood this ride, or captivity, or abortion in the same way. Nor can one easily jump from an abortion or night ride to the dreams of people who saw neither Revere nor Grosvenor, much less use these events to explicate an entire culture.[8]

Stories such as these should be embedded in an analysis of material conditions. The reaction of Grosvenor's village to her abortion, the way Deerfield villagers coped with the loss of children to Indian captivity (and adoption), or the reasons villagers along Revere's route rallied to his cause grew out of the way trade and local exchange, migration and settlement, Indians and climate, and inheritance and class relations had framed choices farmers could make. Long ago Karl Marx captured the ambiguity of human agency in a world of structural determination: "Men make their own history," he wrote, "but they do not make it as they please; they do not make it under circumstances chosen by themselves, but under circumstances directly encountered, given and transmitted from the past."[9]

Big economic and demographic structures, moreover, tell a story all their own, one as compelling as narratives of Indian captivity, biographies of common folk, or tales built from court cases. What could be more dramatic than the dispossession of thousands of families by English capitalists; the struggles of immigrants with Indians, bad weather, and deep forests; the rhythm of thousands of hoes and plows clearing virgin land; or the struggle for subsistence the violence of the Revolution brought?

With the profusion of specialized studies and new fields—women's history, African American history, Indian history, and rural history—the center of historical understanding has collapsed. American historians have struggled to incorporate the findings of new fields and frame a new "master narrative," replacing older political narratives; but such attempts are inevitably incomplete, whether they focus on society, economy, politics, class, or gender. Each must include some voices but exclude others, emphasize some themes but minimize others. Like horses on a carousel, each surrounded by others but never touching, a carousel of master narratives best interprets colonial history. One horse on that carousel, this volume stands as a master narrative among others. Its focus on small farm families slights slaves, Indians, planters, land speculators, and merchants; its emphasis on population and economics puts religion, ideology, and class formation to one side.[10]

This book is remarkably different from four interpretations—each master narratives—published in the 1980s. Unlike the export-based model of John J. McCusker and Russell R. Menard's *Economy of British America*, it emphasizes farm activity. Beginning in England, as David Hackett Fischer's *Albion's Seed* does, it rejects Fischer's insistence on cultural stasis, underscoring instead how changes in England influenced migration and colonial development and how the colonial environment transformed customs migrants brought with them. Less comprehensive than Jack P. Greene's *Pursuits of Happiness* (ignoring the West Indian colonies he incorporates), it emphasizes similarities among the colonies over Greene's regional interpretation. Spatial organization, the metaphor behind D. W. Meinig's *Shaping of America*, is less important here than the productive relations that reshaped the environment. The small farmers who serve as the central actors in this volume play a major role in none of the previous studies. Instead, McCusker and Menard emphasize the economy; Fischer, immutable folkways; Greene, regional societies; and Meinig, human geography.

Our story—of immigration and migration, of struggles with nature and Indians, and of farm making and remaking—recovers the experiences of thousands of colonists. It connects the remote past to nineteenth-century immigration (much of it to farms) and settlement of the trans-Appalachian and trans-Mississippi wests. And in this era of corporate farms and agribusinesses, it reminds us of a lost world of small farms, home production, and neighborly exchange.

› *prologue* ‹

The Remaking of Rural England

A sense of crisis pervaded early modern England. In 1549 Sir Thomas Smith, Elizabethan diplomat and intellectual, tried to explain the "manifold complaints of men touching the decay of this Commonweal," grievances shared, he thought, by rich and poor: "Dearth [high prices] of all things though there be scarcity of nothing, desolation of counties by enclosures [of crop land into pasture], desolation of counties towns for lack of occupation and crafts, and division of opinions in matters of religion hale [drag] men to and fro and make them contend one against another." A century and a half of economic development intensified the problem. Crisis bred rebellion; enclosure, Arthur Standish wrote at the time of the 1607 Midlands Revolt, had made wheat "too deare a rate for the poor Artificer [artisan] and labouring man; by which dearth, too oft ariseth discontentments, and mutinies among the common sort." In 1688 Gregory King calculated that just over half the families in England—seamen, laborers, cottagers (who rented a small house and garden), paupers, and soldiers—earned less money than needed for bare subsistence, thereby requiring public aid and decreasing "the wealth of the Kingdom."[1]

Peasants and rural laborers agreed. Some Hampshire peasants complained in 1579 that "a piece of our common and heathe" had been "ditched and hedged and enclosed and planted with willows," displacing "our cattle, which have hitherto many yeares past prospered verie well . . . ;—wherefore we desire it [the hedge] be pulled down again and levelled" or "in short time yt will be taken from our common to some particular man's use, which were lamentable and pitiable and not sufferable. For as our ancestors of their great care and travail have provided that and like other many benefits for their successors, so we thinke it our dutie . . . to keepe, uphold, and maintain" our right "for our posteritie to come, without diminishing any parcel or part from yt." These peasants captured the sources of the discontent of common people. They had enjoyed rights to the commons since time immemorial, inheriting it from their ancestors. Clearly rejecting individual ownership of communal property, they viewed the commons as a collective entitlement. By reducing resources needed for subsistence, enclosure raised an unstated (but real) fear of dispossession.[2]

People in all ages have protested poverty and inequity, but these complaints reflect great economic change. All around them, peasants like those in Hampshire saw the dissolution of their world. Rich men had enclosed their lands, forcing many of them into wage labor. The rapid inflation of prices but slower growth of wages had reduced many a laboring family to poverty. Transients tramped the countryside, working when they could and stealing for their bread when they could not work. Dispossessed peasants flocked to towns, where they struggled to earn a living; husbands sometimes deserted their wives, leaving them to support themselves and their children with the meager wages they could get from spinning.[3]

At the same time, trade and manufacturing grew, busy market centers developed, London's population skyrocketed, and improved transportation integrated markets. Per capita gross domestic product grew as much as 1.6 percent a year, nearly as high as during the industrial revolution. Despite intermittent depressions, by the mid-seventeenth century, if not earlier, England produced surpluses of food and clothing. Landlords and improving tenants developed their lands, draining swamps and growing new crops, thereby increasing farm productivity. As networks for distributing goods improved, subsistence crises diminished and then disappeared, for even in times of dearth, food could be imported from elsewhere. Not surprisingly, while poverty increased, many families prospered.[4]

Such stories of poverty amid prosperity permeate the history of early modern England. We will plumb these stories, paying special attention to the dispossession of the peasantry and changes in the relations of men and women in rural households. This will not only uncover reasons families might have wanted to leave England but suggest their expectations once they arrived in the New World. As we shall see, the kinds of households English migrants to North America made, the markets they developed, and their responses to Indians and the physical environment in America all grew out of their English experiences.[5]

The Medieval Legacy

The Hampshire peasants remembered the prosperity of late medieval England after the Black Death—the plague and subsequent epidemics during the mid-fourteenth century—had reduced population and serfdom had withered away.

They perhaps had heard the "Ballad of a Tyrannical Husband," well known by the late fifteenth century. The song laid bare the abundance of the time. "The goodman and his lad to the plough are gone," the balladeer sang, while "the goodwife had much to do, and servant she had none." When he came home "early in the day" and demanded his dinner, his wife complained about her duties. She lay "all night awake with our child," rose early, and found "our house chaotic." While her husband slept, she milked their "cows and turn[ed] them out in the field." Later she made butter and cheese, fed the chickens, tended the hens and ducks, baked and brewed, carded wool and spun flax, fed "our beasts," and made the midday meal. Belittling her labor as unnecessary, her husband "began to chide," groaning about his own labors. "Damn you! I wish you would go all day to plough with me, / To walk the clods that are wet and boggy, / Then you would know what it is to be a ploughman."

Passed from generation to generation, folk memories like these told of high wages, widespread landholding, and peasant independence. The tyrannical husband was a lucky man. He not only held enough land to raise crops but had rights to pasture animals on the commons. The joint labor of wife, husband, and servant—the cultivation of grain and flax and the keeping of all kinds of fowl and livestock, including milk cows—made nearly enough for subsistence. Peasants strove to reduce their dependence on barter and sale. The goodwife insisted she had to spin, card, and weave. "Either I make a piece of linen and woolen cloth once a year / In order to clothe ourselves and our children," claimed she, "Or else we should go to the market and buy it very dear." Tellingly, the balladeer neglected the relations between lord and peasant. Although feudal lords had controlled the political order, taking surpluses from peasants, long-remembered custom enforced in manorial courts limited the fees such lords could impose.[6]

To understand these folk memories, it is crucial to delve into medieval times. We will emphasize elements of those times—the characteristics of peasant households and the economics of the peasant farm—most critical in shaping folk memory and the ways English families understood the changes of the sixteenth and seventeenth centuries. Remarkably, colonists replicated many characteristics of medieval peasant farms. Knowing how much recent events had debased their lives, seventeenth-century English emigrants looked back on a supposedly more prosperous time and sought to remake it in America.[7]

The folk memories of later emigrants began with changes in the late fourteenth century. Peasant poverty had characterized earlier times. For several centuries before the Black Death, the population of England had been rising. As land grew scarce, peasants divided it into smaller and smaller plots; two-fifths or more of them often held less than 7.5 acres. Close to half the men held no land but worked as day laborers for lords or landed peasants. At the same time, landlords raised rents and demanded labor from peasant families (many of whom were still serfs) to work the land (the demesne), and the cost of both food and clothing doubled. Peasant resistance to increased appropriation of their labor and surpluses was common but unavailing, given the dense population and the power of the lords. Profits flowed from peasant to lord. Unable to challenge the lords' growing power, peasants survived only by turning pasture into arable land and farming more intensively the small plots they inherited or could procure. Richer peasants took advantage of the improved markets population growth brought, buying or leasing land, hiring poor peasants to work their land, and selling surpluses at the growing number of market towns and agricultural fairs. Most peasants, however, owning too little land for subsistence and unable to increase farm production or pasture additional animals, turned to wage labor, learned a craft, and postponed or avoided marriage.[8]

The Black Death reduced the English population by at least a third and probably half. Some places lost three-fifths of their populace to the disease. Even by the early sixteenth century, the population had not fully recovered. Because the population declined so much, the quantity of land under cultivation dropped by a quarter or more, and many lords turned arable land into pasture, at times doubling the number of animals, especially sheep and cattle rather than milk cows, which required greater labor. On some estates, tens of thousands of sheep replaced tenants who had died of plague. Areas like Norfolk, where arable farming intensified, nonetheless emphasized dairy cattle, sheep, and swine far more than they had before the plague. The great reduction in population changed relations between lords and peasants, giving those peasants who survived the upper hand in struggles over peasant surpluses. Profits stayed within peasant households. Notwithstanding attempts by national authorities to control wages and maintain rents, wages rose and rents fell during the century after the plague.[9]

Before the plague, at least half, and probably more, of the peasants had been serfs, an unfree state that tied them to the land and the lord. Serfdom withered away after the plague and disappeared entirely by the end of the fifteenth century. Most peasants, whether serf or free, were smallholders; but free men, unlike serfs, could move without the lord's permission, sell and buy land at will, gain secure access to common pastures, and bargain for lower rents and reduced fees at times of family need or crisis, such as marriage, land buying, and death. Insisting that they were free men, serfs regularly revolted, and the economic openness of the post-plague era completed their emancipation. As the holdings of richer peasants grew, they competed with the lord for labor, forcing wages up and encouraging serfs to abandon land constrained by feudal dues. To entice new tenants or get men to work their fields, lords had to reduce rents, eliminate servile payments, and pay wages. With so many opportunities for land and higher wages, free laborers preferred to contract for work by the day. To coerce wage laborers, lords tried—with little success—to enforce long-term contracts with threats of imprisonment. They were forced, however, either to pay day laborers the prevailing wage or to add food and drink to the wage and give annual workers land, pasture rights, and clothing, thus making them independent producers.[10]

John Gower, a Kent manor lord, captured the changed relations between peasant and lord in an essay written in the 1370s. Peasant workers, formerly pliant and agreeable, had become "sluggish," "scarce," and "grasping. For the very little they do they demand the highest pay. . . . Yet a short time ago one performed more [labor] service than three do now." Not only did they refuse to work as servants or sign a yearly contract; they wanted "the leisures of great men." When he hired workers "for even a day's pay," they labored "now here, now there, now for myself, now for you." Because such a man was hired as a member of the household, "he scorns all ordinary food. . . . He grumbles . . . and he will not return tomorrow unless you provide something better."[11]

What kind of farms did peasants cultivate in the fifteenth century? They held ten to about thirty acres, much more than earlier, under secure land tenure, paying rent to the lord, buying more land when needed, and bequeathing land to their children. Like the tyrannical husband, they mostly used family labor, perhaps hiring laborers at planting and harvest. As population declined, marketing diminished, but peasants still took surpluses to the remaining markets. Even the poorest landholders enjoyed rights to commons, which allowed them to pasture livestock.

Only access to land could make possible the degree of self-sufficiency the tyrannical husband achieved. Under feudal land tenure, which persisted through the medieval era, no one owned land as private property (except, perhaps, the owners of freehold land, which was available only in a few places). Neither peasants nor lords could fully alienate land or keep the other from it. Rather, peasants and lords had differing use rights to land. In return for the payment of rent, fees, or (less often) labor service, peasants could grow crops and graze livestock on land held collectively by the village. Upon payment of fees, peasants could bequeath these rights and use the land as collateral for loans. Lords not only owned the demesne, which they worked with peasants whose labor services they demanded or, more commonly, leased to richer peasants, but also had the right to demand rent from peasants for using other manorial land. They could not, however, evict peasants and hire wage laborers to work their holdings.[12]

The feudal land system, combined with a vigorous land market, allowed more and more peasants to get land, to consolidate their holdings, or to accumulate land for their heirs. Land markets had developed long before the

Black Death. Peasants had often transferred small parcels of land (one or two acres) to kindred or close neighbors, keeping the rest for the family. Since so many close relations died during the plague, the proportion of land transferred to neighbors, more remote kin (such as in-laws), and strangers increased in the late fourteenth century. Because so many fewer families sought land, tenants often consolidated their holdings, and the proportion of large holders (thirty or more acres) often increased. Fathers on occasion ordered land sold (and the proceeds distributed to heirs) rather than given to a son; others bequeathed land to the church, bringing more land onto the market. Since land remained widely available but potential tenants scarce, peasant sellers sometimes granted mortgages, thus allowing poorer families to buy it. To be sure, land sales and inheritance did not change ownership of land (which stayed with the lord), but they transferred use rights the landlord had to acknowledge after the proper fees were paid.[13]

The huge quantities of vacant land that came on the market provided new opportunities for landless peasants. Complex exchanges of land among villagers ensured that families held enough for subsistence, but no more land than they could farm with their own labor. When sons came of age and needed land, fathers repurchased family land sold earlier; alternatively, sons took up land deceased kindred had farmed in nearby villages. Landless artisans bought plots peasants placed on the market; some cottagers and poorer peasants accumulated land, becoming middling peasants; middling peasants acquired more lands; and richer peasants increased, consolidated, and enclosed their holdings and specialized in sheepherding. Inequality thereby increased, and differences in wealth among peasant families appeared; but even so, the differences among peasant holdings were small, ranging from thirty to fifty acres for rich peasants to fifteen acres for middling ones.[14]

Soon after a young couple married, they probably acquired land and began farming. Some took over (or inherited) land, chattels, and personal property. Families strove to accumulate land or rights to commons and livestock sufficient to ensure the survival of the household and pay feudal dues. Even the poorest cottager could afford to build or rent the small houses (450–675 square feet) found in most villages, and better-off peasants built larger, permanent homes and small bake houses, barns, and granaries. Although each farm family had its own house, land, and livestock, it thrived only by cooperating with the peasant community and the manorial lord. The success of the farm thus depended not only on the particular labor of each member of the farm family but on decisions about crops and livestock the peasants made collectively and on the price their goods received in local markets. We will examine each of these issues in turn.[15]

Peasant men and women alike labored to make food, clothing, and shelter.

Women's tasks centered near the house, while men worked mostly in the fields. Men plowed, planted, and harvested grain crops; herded livestock; and collected wood, sometimes taking up craft work in slack times. They handled family finances and represented the family at the manorial court. Richer men dominated village baking, butchery, and brewing. Women fetched water, washed, cooked, milked cows, made butter and cheese, fed poultry, sheared sheep, spun thread, wove cloth, and cared for children. Although the farmyard—containing ovens, wells, barns, vegetable gardens, and brewing vats—was a center of female economic activity, peasant wives had important roles in cultivating grain. They occasionally drove plow oxen; they harvested grain with sickles as often as men and gleaned harvested fields more often. Like their husbands, women had to help gather the lord's crops. On occasion women, especially widows, appeared at the manorial court in civil suits or land transactions, and a few worked as brewers, bakers, petty retailers, and creditors.[16]

Family members (often just parents and two small children) provided nearly all the labor needed on the farm; once the children of poorer peasants reached adolescence, many left home to work as servants in towns or in husbandry, reducing the farm's permanent labor force to husband and wife. Although only the richest peasants (15 percent in 1377) could afford to employ live-in servants-in-husbandry, peasant families readily hired harvest labor. Attracted by rapidly rising wages, as many as a third of late-fourteenth- and fifteenth-century peasants—women as often as men—worked for wages. But few of them, except near London, resorted to permanent wage labor, since a mere laborer, lacking even a cottage garden, could fall into abject poverty and face hunger. Most wage workers, embedded in local communities, took whatever jobs were available at home or in nearby villages, working by day in carpentry, plowing, brewing, or harvesting, depending on the season. Cottagers and even middling peasants sometimes worked for wages to supplement what they could make on their land. Even landed peasants who hired labor often worked for hire at harvest time, getting cash to pay rents, taxes, and feudal dues. Landholding husbands and wives engaged in harvest labor, sometimes together, sometimes moving about the countryside in different circuits.[17]

Within the peasant household, the father and husband held the preeminent position, but neither women nor men could escape the lord's power. This double bond constrained the behavior of peasant women. A father often arranged his daughter's marriage, thwarting her voluntary choice, and then asked the lord's permission for her to marry. Before the marriage, families of bride and groom arranged property transfers (the woman's dowry; the man's land). Understanding that a peasant farm could succeed only if husband and wife cooperated, wives followed the dictates of their husbands and in return received food and clothing. Only 1 or 2 percent of peasant marriages ended in recriminations, adultery, or separation. Lords demanded labor from them and collected merchet, the demeaning marriage tax, from every landholding serf. To protect his labor supply and control inheritance of manorial land, the lord charged less when a woman married a manor resident than when she wed an outsider or landed freeman.

The condition of peasant women may have improved after the plague. Widows could hold land, and when land availability increased, the pressure on heiresses to remarry diminished, leaving widows with more land. Wives, moreover, gradually gained greater control over land through devices such as joint ownership. As the labor supply grew scarce, unmarried women gained new opportunities to work as agricultural laborers and as servants to town artisans or petty retailers. Notwithstanding their exclusion from highly paid work, they earned as much as men did for identical labor. Since they could accumulate savings, more women postponed marriage until their mid-twenties and paid merchet themselves, thereby gaining some control over their marriage decisions.[18]

Although peasants farmed independently, most had to cooperate with neighbors. Peasants who lived in open-field regions—over half of England's land in the early sixteenth-century and nearly all good crop land in the midlands and the south—held strips of land scattered over a manor or village. (Pastoral lands in the north and much of East Anglia had always been enclosed in compact farms.) Since they owned adjacent strips, peasants plowed and harvested land communally, sharing oxen and agreeing about the grazing of livestock, the cultivation of grain, the timing and organization of harvest labor, the rules for gleaning the stubble, and the time fields were to be left fallow. They enjoyed rights to the commons, where they could graze their livestock and collect berries and nuts, peat or wood for heating or construction, and bracken for bedding or thatch. In surrounding forests, common rights extended to building materials, firewood, and livestock pasturage (grass and acorns). Even cottagers, who had legal rights only to a house and a garden, often used the commons. Richer and middling peasants in open-field villages enforced agreements they made about land use in the manorial court or village meeting—institutions they controlled—and publicly humiliated villagers who refused to conform. This system developed slowly, but by the fifteenth century all the elements—the strip system, commons, field rotations, and communal assemblies—were firmly established in open-field regions.[19]

Fifteenth-century peasant communities extended to nearby market towns. Peasants needed cash to pay rent, fees, and feudal dues and to buy what they could neither make nor get through barter. Unsurprisingly, market towns, organized for the sale of farm surpluses, were ubiquitous in medieval England. Although marginal markets (as many as half the total) atrophied after the Black

Death, others expanded their business, taking advantage of the increased sale of livestock. Peasants traded in a five- to ten-mile circuit; itinerant traders tramped from market to market on a regular route. Poor peasants walked to town markets to hawk dairy products, and better-off peasants, who owned horses, rode there to sell produce and buy consumer goods. Women sold produce, butter, eggs, hens, and ale; men bought and sold pigs, leather, beef, wheat, and wool and cloth, often incurring small debts in the process. Loaning and borrowing money and labor to finance purchases, neighboring villagers established complex credit networks, sustained more often by mutuality than by class patronage; richer and middling peasants traveled beyond the village to sell and buy goods. Such growth of markets does not mean that a market economy, where supply and demand determined price, permeated medieval society. Only a small portion of the output entered regional commodity markets. Most peasant women and men engaged in small-scale exchange within their own village, trading butter for bread or swapping household goods.[20]

As more land became available and labor grew more scarce after the Black Death, peasants moved more often, usually over short distances (twenty miles or less), searching for land or work on the best terms. When lords attempted to pay laborers the rates mandated by law, the workers would "take flight and suddenly leave their employment and district" and "go from master to master as soon as they are displeased about any matter." Peasants thereby abandoned a few villages and left others with a tiny fraction of their people. But migrants usually replaced departing peasants, thereby maintaining village populations, albeit at a much lower level than before the plague. As surviving kindred moved around the countryside, extensive kinship networks, covering adjacent villages, developed. Marriage between people from different villages became common, and newlyweds settled on lands with the most advantageous terms. Over a lifetime, middling and rich peasants often held land and established ties to families in several villages. Poorer peasants and youths moved as well, not only to find land but to take advantage of rising wages and growing opportunities to work as servants-in-husbandry or replace artisans who had died. These migrants, many of whom may have owned a bit of land in their home villages, tramped a circuit of several miles, planting and harvesting crops or working as artisans.[21]

After a downturn in the wake of the plague, per capita income and gross domestic product grew, at a low rate, between 1300 and 1470. This greater income allowed fifteenth-century peasants and laborers to improve their standard of living.[22] Improvements in diet illuminate the trend. Before the Black Death, most peasants drank water and ate bread (made of beans and barley) and pottage (a thick vegetable soup) along with a little meat and game they poached from forests and game reserves. With greater access to land, more peasants and cottagers grew food, especially grains and vegetables, and laborers demanded better food from employers as part of their wage. Catering to more refined peasant tastes, big producers retained wheat acreage but reduced acreage for cheaper grains. As landlords turned arable land into pasture, more mutton, pork, and beef reached village markets. Village bakers and butchers thereby supplemented what peasants made. Since wages rose faster than the price of grain, cheese, and meat, laborers improved their diet as well. Rural folk ate less bread, and nearly everyone had a varied diet of white (wheaten) bread, milk, cheese, eggs, ale, beans, vegetables (cabbage, onions), fish, and meat.[23]

By 1500, peasants had successfully challenged the economic basis of feudalism: serfdom had disintegrated, rents had declined, and wages had jumped. At the same time, poverty among peasants had increased, some land had been enclosed, and geographic movement and wage labor may have risen. But feudalism as a legal system remained largely intact: lords still collected payments from peasants; peasants, who held the vast majority of England's lands, still participated in manorial courts. The regulation of common fields strengthened in many open-field villages; traditional village hierarchies, with their reciprocal if unequal class relations, remained in place. However, such feudal classes and the peasant communities they supported collapsed as capitalism spread through the countryside. By the seventeenth century, only folk memories of secure peasant life and a craving to somehow re-create that world remained.[24]

The English Road to Agrarian Capitalism

Midlands peasants, furious at enclosure, explained their 1607 revolt in a moving petition. "Wee . . . doe feele the smart of these incroaching Tirants [enclosing landlords] which would grind our flesh upon the whetstone of poverty" for the benefit of their "Hearde of fatt whethers." The landlords had "depopulated and overthrown whole Townes and made thereof sheep pastures nothing profitable for our commonwealth. For the common Fields being layd open," they insisted, "would yield as much commodity, besides the increase of corne, on which stands our life." Despite protests, enclosures stayed. Commoners thrown off the land could have recourse only to wage labor. In the late 1620s, when Spain prohibited the import of some kinds of cloth, 40,000 to 50,000 Essex County cloth workers faced destitution. Those "who live by those manufactures" could not "subsist unlesse they bee continually sett on worke and weekly paied," they insisted in a 1629 petition to the king; many "cannot support themselves and their miserable families unless they receive their wages everie night." With neither work nor credit, they urged public aid "to give some order and direction by which many thousand of your poor and faithful subjects may be preserved from utter perishing."[25]

The plight of the Midland peasants and the Essex County cloth workers illustrates the speed of England's economic transformation. In 1579, when the Hampshire peasants had urged that enclosed land on their manors be thrown open, most peasants still held land. Less than a century later, landlords had evicted growing numbers of peasants, who were then forced to work for wages, and those who still held land feared a similar fate. For more and more Englishmen, the reality of landholding receded into folk memory. But their yearning for land remained strong. This history of the dispossession of English peasants is essential to our story, for it illuminates the desire for communal rights, familial self-sufficiency, and independence shared by nearly all those who crossed the Atlantic.

The process that began with landlord expropriation of peasant land constituted a "transition from feudalism to capitalism." Landlords (and the improving farmers to whom they leased big farms) made enough food to permit towns to grow and to support an increasing population with little or no access to land. Former peasants worked as rural wage laborers or moved from farm to town. Landlords and merchants took up manufacturing, hiring the dispossessed peasants to spin, weave, or mine coal. Rural (and city) industry thereby increased. These changes turned peasants into workers, allowing them to control their work but not the fruits of their labor.[26]

Capitalism began with enclosure of open fields and the elimination of commons. After the Black Death, rich peasants had consolidated and enclosed their plots, but the perpetuation of common rights assured subsistence for all. In contrast, sixteenth- and seventeenth-century enclosures extinguished common rights by distributing commons to landholders. By "giving the Earth to some, and the denying the Earth to others," in Digger Gerrard Winstanley's 1649 words, enclosers created private property. "Fence well," William Lawson wrote in 1618, "therefore let your plot be wholly in your own power." Land became a commodity rather than a bundle of use-rights. Where feudal landlords only had the right to a revenue from their lands, capitalist landlords, unconstrained by communal demands or the cries of the evicted, surveyed and mapped their land to determine its bounds, sold their property, adopted specialized agriculture, charged market rents, and reduced long-term tenants to short-term leaseholders or evicted cottagers. The extinction of common right and the spread of private property discomforted traditional gentlemen. After enclosures, Bishop Hugh Latimer asked in a 1548 sermon, "what man will let go or diminish his private commodity, for a common wealth?" He added, "Who will susteine any damage for the respect of a publique commodity?"[27]

Landlords and rich tenants used enclosures to consolidate their economic power. When land was enclosed, landholders (but not cottagers) received small, consolidated farms, but everyone lost their rights to commons. Between 1541 and 1600, the population of England rose by nearly half, from 2.8 to 4.1 million, and it grew by another quarter to 5.3 million by 1651. This population explosion provided ample opportunity for landlords to reverse the good fortune peasants enjoyed after the Black Death and to take advantage of growing urban markets for food. When the population had grown in the thirteenth century, smallholders had cultivated marginal land and subdivided their plots ever more minutely. Now landlords enclosed common fields, pastures, and wasteland, sometimes evicting peasants into the winter cold and on occasion laying whole villages waste. Landlords usually sought the approval of the larger landholders but rarely the smallholders, much less the growing numbers of cottagers.[28]

Although only 4 percent of land in open fields was enclosed in the sixteenth century, any enclosure, by eliminating common rights, reduced opportunities for smallholders and threw cottagers off the land. In the Midlands, in particular, landlords enclosed a fifth of the cultivated land between 1485 and 1607, dispossessing innumerable peasants, often replacing them with thousands of sheep. In the seventeenth century, enclosures covered a quarter of all the land in England—almost half of the land held in open fields in 1600—and further reduced opportunities for peasants. Despite enclosures, however, open fields still predominated in nearly a third of England at century's end. Like their ancestors, poor families in these places continued to rely on access to commons and wasteland to guarantee their subsistence and on local government to sustain their rights to land.[29]

Landlords not only enclosed land but reduced traditional land tenure to short-term leases. Freeholders and tenants in open-field villages had paid a fixed, customary rent to the lord. In the inflation of the century after 1540 (when prices more than doubled), peasants with customary rents—especially the big farmers who produced for the market—gained at the expense of landlords and poorer peasants. Entrepreneurial landlords, seeking to regain lost profit, reduced the security of those still on the land, imposing short-term leases or much higher rents or, if raising rents proved impossible, demanding high fees or payment in commodities rather than cash.[30]

Enclosure erased all rights to commons and wasteland. Communal rights to the commons (pasture, forest, and wasteland) had guaranteed small landholders and cottagers alike subsistence beyond the grain they made. The commons had supplemented wages cottagers and small landholders earned and the produce of peasants' fields and cottagers' gardens. Grazing rights, guaranteed to tenants and bought with a pittance by cottagers, gave the family milk, butter, wool, dung (for fertilizer), and lighting (meadow rushes). Common rights to forests yielded game animals, fish, fruit, berries, green vegetables, edible roots, fodder, medicinal herbs, building materials (timber, clay, and thatch), wood (for furniture and kitchen implements), and fuel (firewood and peat). In addition, peasants and cottagers sold surplus meat and wool from commons pastures and some of the forest products they gathered. Peasant communities jealously guarded rights to commons, limiting them to manorial residents. These rights became even more essential in the sixteenth and seventeenth centuries, when the price of food grains skyrocketed and the size of farms plummeted, leading villages to limit access to residents. The continuation of a part of the commons preserved some of these rights.[31]

Peasants feared the loss of their rights to commons more than enclosure of open fields, for many did receive small plots of land after enclosure. Men who farmed fewer than twenty acres (two-thirds of those with land) had too little to come close to self-sufficiency and could feed their families only when they had rights to commons.[32] Cottagers, in contrast, could lose all access to land, including their gardens, after enclosure. Although early-sixteenth-century cottagers and wage laborers retained some land, their rights to commons diminished thereafter. When cottagers and smallholders lost these rights, they had to buy more (if not all) of their food and clothing at a time of rising food prices and stagnant wages. To subsist, they fell into wage labor, harvesting grain or working in mines or forests. Given the low wages men received and the unemployment (often lasting half a year) they suffered, women had to work as well, joining husbands and brothers in grainfields and working at home as spinners for clothiers, on occasion even doing construction or casual labor in town, weeding gardens, cleaning streets, or working as domestic servants.[33]

As wage levels fell in the late sixteenth and early seventeenth centuries, the standard of living for laborers and cottagers plummeted. They had to use most of their cash, garden crops, and milk to buy bread and clothing. Laborers sometimes found it nearly impossible to subsist, even if the whole family worked. Some laborers (but not their families) received food and drink as part of their wages; a few probably kept fowl or a pig, and cottagers, of course, produced much of their own food. Nonetheless, poor landless families ate bread and porridge, on occasion supplemented by milk, ale, cheese, eggs, or cheap meat—a far less bountiful diet than that of medieval peasants or contemporary landed families, which included puddings, butter, cheese, fish, and meat of all kinds. In good times, laborers suffered malnutrition; in times of dearth, many died, and others stole food or joined in food riots.[34]

Understanding the critical importance of land in a time of higher prices and dispossession, peasants increasingly kept land in their families rather than selling it to strangers, as a case study of Earls Colne (in East Anglian Essex) shows. From the mid-sixteenth to the mid-seventeenth century, village land typically stayed in the family, nine-tenths of the time passed from husband to wife, father to child, or grandfather to grandchild. Families typically held on to the land for three generations (roughly eighty years). Those with larger holdings kept land within the family far more successfully than cottagers and smallholders, who regularly lost family land they had mortgaged to stay afloat; at best, they sold their traditional rights to land and then sublet it.[35]

A minority of rural English families benefited from enclosures. Enterprising peasants rented large farms and turned to specialized market agriculture; gentlemen and other rich peasants (and even some smallholders) leased much of their land to other peasants. Enclosure—along with the draining of fens, greater manuring and the reduction of fallows, and perhaps the rising productivity of day labor—led to a modest increase in cereal yields; turning more acreage over to fodder crops, large-scale farmers raised far more cattle, swine, and (probably) sheep. Such agricultural improvement allowed landlords and their major tenants to enjoy higher income and an improved standard of living, but at the same time it led to more enclosures. After enclosing the land, landlords consolidated their estates, imposed higher rents, leased farms to tenants, and hired wage laborers. They fenced in their holdings and expropriated wasteland, keeping animals of other farmers out and evicting land-poor cottagers. When evicted families squatted on wasteland, improving landlords often claimed these lands and forced the squatters into full-time wage labor.[36]

Yeomen benefited the most from enclosures. These commercially oriented landholders farmed 50 to 100 acres. They constituted a small group—a fifteenth of Gloucestershire's male household heads in 1608, at most a tenth of the late-seventeenth-century population. Only in farm villages with little rural industry did yeomen head as many as a fifth of the households.[37] The richest yeomen, chief tenants of gentlemen and the direct employers of cottagers and farm laborers, became well-off commercial farmers. They exchanged parcels vigorously, accumulated land, produced large surpluses for expanding markets, borrowed large sums of money from gentlemen or rich kindred to improve their estates, extended credit to one another, augmented their income through artisan trades, invested in industry, and apprenticed their sons to merchants or wealthy craftsmen.[38]

The ability of rich yeomen to get property and emulate gentlemen excited hostile comment. In 1577 William Harrison remarked that yeomen had "a certain preheminence . . . and commonlie live wealthilie, keepe good houses, and trauell [work] to get riches." By "grasing, frequenting of markets, and keeping of servants [who] get both their owne and part of their masters living," they came to great "welth," which they used to "buie the lands of unthriftie gentlemen." By educating their sons or "leaving them sufficient land whereupon they may live without labour," they turned their progeny into gentlemen. "Not contented . . . to be counted yeomen," Thomas Wilson added in 1600, they "skipp into his velvett breches and silken dublett" and thereafter "skorne to be called any other than gentleman." Most yeomen acquired less wealth; nonetheless, they lived more comfortably than rich medieval peasants or contemporary artisans, building estates worth £100–£200 and earning £50 or more a year, twice that of independent artisans.[39]

The significance of the yeoman class extended beyond its small size. Many failed to buy land, preferring to rent from gentlemen. They spent much of what they earned from specialized production of market crops on rebuilding their houses and furnishing them with an ever increasing quantity of consumer goods. As their demand for consumer goods rose, the variety and quantity of cloth and stockings and glassware and ceramics sold by village craftsmen, local storekeepers, or peddlers rose as well. Yeomen thereby sustained rural manufacturing, leading to the employment of men and women thrown off the land as textile workers, rural artisans, and peddlers.[40]

If yeomen prospered, husbandmen (who held five to at most fifty acres) and cottagers struggled to maintain a foothold on the land. Sustained by the improvement of wasteland, partible inheritance, and the perpetuation of open fields in many places, smallholders and cottagers proliferated between 1560 and 1640. Families improved vast quantities of wasteland and forest, especially in the thinly populated North; others in pastoral regions rented ten acres of cropland and ten to twenty acres of pasture and raised cattle, sheep, and horses. Evicted cottagers often squatted on unclaimed wasteland. Fathers with land divided their holdings among their sons. Although rich yeomen might give whole farms to several sons, land division by husbandmen inevitably led to a multiplication of tiny holdings often no larger than a cottage and an acre. Even when the eldest or youngest son received the bulk of his father's land, other siblings inherited a cottage or profits from the estate, divided the personal property with the widow, or received their portions from the inheriting son, an act that sometimes forced him to sell the land. Sons of cottagers and small landholders in most places bought or rented a tiny sliver of land; most, lacking a landed patrimony, rented a cottage and combined vegetable gardening and herding with an artisanal craft.[41]

Landed families took advantage of higher prices and improved markets for grain. Grain prices multiplied more than four times between 1550 and 1640, 1.6 times faster than the wages landholders paid. Not only did some 760 market towns still survive, but most of them attained substantial size, becoming centers for the distribution and sale of wheat, or specializing in particular commodities—butter or cheese or cattle or sheep—some brought in from great distances. Private marketing expanded greatly as well, providing additional opportunities for gentlemen and yeomen (but rarely small-holding husbandmen) to sell their wares. Buttressed by an extensive and growing network of roads, London markets (both public and private) attracted innumerable gentry and yeoman sellers to feed the city's multiplying populace.[42]

By the mid-seventeenth century the peasantry had disappeared, and a new class structure of improving landlords, capitalist tenants, and wage laborers had spread over much of southern England. As early as the 1520s, appreciable numbers of cottagers and landless laborers, numbering at least a tenth of the householders, populated the countryside and especially the industrial villages.[43] Thereafter English population and enclosures grew too fast for land division and settlement on wasteland to provide farms for many sons of husbandmen. Small landholders, evicted by landlords or unable to repay loans from larger operators, lost their land at a high rate. The proportion of cottagers and wage laborers among household heads grew from one-quarter before 1560 to two-fifths by 1620, and at the same time the proportion of small-holding husbandmen dropped from about two-fifths to less than one-third.[44] During the 1650s only half of the men in three Lancashire villages worked in agriculture, while two-fifths worked in the textile industry. By 1688 only a quarter of rural families—gentlemen, yeomen, husbandmen, clergymen, and some shopkeepers—leased or owned land. Half were cottagers, landless farm laborers, or vagrants, and one-seventh worked exclusively in textiles, mining, and other village industries. At least a third of late-seventeenth-century rural families lacked even a cottager's garden and survived exclusively on the wages of family members, including children and wives.[45]

No matter how poor, smallholder and cottager families with access to land, especially those with kinfolk nearby, remained members of a rural community and fully participated in its social and religious life. Smallholders and landless families in Willingham (two-thirds of the total by 1603) not only witnessed one another's wills but those of their economic superiors. Rich folk sometimes showered their less fortunate cousins or neighbors with jobs and loans; some poor youths leased a sliver of land from kinfolk or got a tiny inheritance. But increasing numbers of families moved about so much that no well-off kindred lived nearby, and they could establish only relations of profound dependence with superiors bent on controlling their behavior.[46]

Internal migration, fed by the dispossession of peasants and the growth of wage labor, was increasingly common in early modern England. Only a quarter of the people stayed in their parish of birth; as many as two-fifths of the people (a quarter of the families) of early-seventeenth-century villages left each decade. Almost no one, except propertied religious dissenters, stayed in the same village for two or more generations. Servants stayed in a parish for only a year or two before moving on. Most rural folk usually moved short distances (up to five or at most ten miles), searching for land or work, but an important minority moved to large towns and ultimately reached London. These migrants—many of them poor transients—went great distances, often over 100 miles, from depressed farm areas to towns. As wages declined and dispossession quickened, migrants forced to move for subsistence overwhelmed those who sought land. Many moved each year, working as farm or town laborers, until marriage; then they persisted for longer periods.[47]

The English migration system was related to family labor but responded to labor markets. Youths first moved during adolescence, leaving home for nearby villages or big towns to work as apprentices or servants, at times with parents' (or siblings') help, at times despite their wishes. Most apprentices in cities such as Bristol or Southampton moved from nearby villages where their parents had farmed. The boys came to learn craft skills; the girls, housewifery. Most left town before the end of their terms. Familiar with searching for work through the neighborhood, they became more deeply involved in the labor market, and the migration it required, once they left apprenticeship or service. Even skilled journeymen rarely stayed in one place for more than a year, much less found permanent jobs or set up craft shops; at best they squatted on fen land, took up a cottage, or worked by the day in a nearby town. Often settling for seasonal work, they moved across the countryside, harvesting grain or performing specialized labor, as individuals or in work crews organized by petty capitalists. Cottagers labored near home, but most moved about, often migrating to textile towns to find work. Lacking access to credit, they worked for capitalists as spinners or weavers.[48]

The experience of servants-in-husbandry—unmarried youths hired by yeomen, husbandmen, or craftsmen on annual contracts—illustrates the ambiguous status of laboring people. At least three-fifths of rural English youths—mostly children of laborers and cottagers—worked as servants-in-husbandry during the seventeenth century. Starting in their mid-teens, nearly all boys became servants; fewer girls entered agricultural service, but many served town craftsmen and merchants. Servants moved a few miles from home, choosing masters who lived nearby, but sometimes returned home after serving a stint. Although servants saved most of their wages, their savings rarely permitted them to get a tenancy, and most became laborers or cottagers.

Opportunities for youths to become servants varied with population growth and prices. Population increases glutted the servant market and reduced wages, encouraging farmers to replace day laborers with servants. High prices for wheat, relative to livestock, led farmers to limit their herds (which required constant labor) and increase grain output (which required harvest labor). Low population and high beef prices in the mid-sixteenth century suggest substantial opportunities, but servitude declined rapidly from the late sixteenth to the mid-seventeenth century, when population growth combined with high grain prices to reduce demand for servants but increase demand for planting and harvest labor.[49]

Enclosure, the growth of wage labor, and declining wages led to what officials considered an alarming epidemic of vagrancy in the late sixteenth and early seventeenth centuries. Called "masterless men" or unruly women, vagabonds were able-bodied migrants who could obtain only casual labor. In the 1630s at least 25,000 vagabonds—runaway apprentices, servants searching for new masters, journeymen looking for work, seasonal wage laborers, sailors and soldiers returning home, deserted wives, unmarried pregnant women, prostitutes, abandoned children, Gypsies, Irish, disabled people seeking cures, and unlicensed beggars—tramped the countryside. Most were young and unmarried men, but half were over thirty, and a fifth to a third were women. Upland areas and towns, toward which most transients moved, had neither jobs nor housing for them. The "Beggar Boy of the North," protagonist of a contemporary ballad, resembled many vagrants. "From parish to parish," he sang, "I roam," crying everywhere, " 'Good your worship, one token!' " Born "in the North Country" to a poor family of beggars, he went "throughout all Christendome." His travels eventually took him "naked unto London City," where he continued to beg. Like the beggar boy, most vagabonds probably traveled alone, but on occasion they joined together for mutual protection, sometimes stealing food or clothing for their subsistence. Tramping the roads, looking for work, and thieving as they went became a way of life.[50]

Transients built stable lives with great difficulty. Mary Bond, born about 1605, was still an unmarried servant in 1635 when she testified in a slander case. Bond claimed to have lived in Brampford Parish, Devon, for a decade, but in reality she moved in and out of the parish, struggling to find work, "always coming and going to other places in that time," villager Edward Panie related, "and never lived in the parish more than one and a half years at one time." She had worked for Katherine Mogridge on two occasions for less than a year, and between times, Mogridge related, "she went away." About two months before testifying, she returned home "unto her father, and remains there." Her transience and poverty led Panie to dismiss her curtly, saying, "She is a poor woman, and such a one as little credit can be given to her sayings and depositions."[51]

The growth of poverty and vagrancy horrified England's ruling class. Although legislative attempts to control labor and enforce long-term contracts failed, local courts set maximum wages for female spinners and farm laborers lower than the cost of living, hoping to force independent women into subservience. Insisting that underemployed laborers disrupted social order, landholders tore down the cottages the poor built on wasteland or evicted them from town. Farmers needed workers at planting and harvest, and clothiers needed them when demand for yarn peaked; both groups wanted them to disappear thereafter. The mid-sixteenth-century English state experimented with harsh penal measures (including servitude or death for convicted vagrants); laborers who took to the road during planting and harvest seasons but failed to get work risked punishment for vagrancy. When punitive measures failed to prevent the spread of poverty, local officials turned to relieving the condition of the poor. To guarantee food to the poor in times of dearth and to prevent food riots, they prohibited the sale or export of grain, regulated its price, and forced the sale of hoarded grain cheaply to the poor. Poor laws, codified at the end of the century, reconciled labor needs of farmers and clothiers with local demands for control by collecting taxes from landholders to pay the unemployed to spin or weave. They also gave a stipend to children, widows, the elderly, and the lame and punished able-bodied persons who refused to work. By the mid-seventeenth century this system had spread through the realm, thereby insuring much of the populace against dearth, sickness, and old age.[52]

This reserve army of the dispossessed without access to land supplied workers for agriculture and for rural industries set up to meet yeoman and gentry demand for consumer goods and laborer demand for food and cheap clothing. During much of the sixteenth century, agricultural labor and cottage industries employed most of the excess population. But by the 1570s, rural areas had become saturated, and migration to cities accelerated, completely severing dispossessed peasants from the means of subsistence.[53]

An unprecedented surge of cottage industries employed ever increasing numbers of rural women and men in the sixteenth and early seventeenth centuries. Not only did the English textile industry expand into less expensive fabrics, but a host of new or nearly new industries—coal mining, ore smelting, glassmaking, iron making, metalware manufacturing, and shipbuilding—flourished. Manufacturing villages, mostly located in pastoral areas or forests where poor peasants had moved to build cottages on wasteland, proliferated. Half or more of the inhabitants of these villages worked more or less full time in industry, joined part time by many cottager and smallholder families. Even in agricultural villages a fifth to a third of the populace participated in industry.[54]

Seventeenth-century capitalists failed to sever all laborers from the land, but as the textile industry shows, the independence of workers slowly eroded. Carding, spinning, weaving, and often finishing took place at home, allowing men to direct their wives and children. In some areas cottagers supplemented textile work with subsistence activities or agricultural labor. Male weavers and female spinners often bought raw material and then sold their output. The vast majority of women in the Kent weald (along with innumerable children), for instance, spun wool fiber at home for capitalist clothiers; spinners worked beside their weaver husbands, and most combined that labor with farming. But more and more people relied on cloth making for their subsistence, despite occasional ownership of cottage gardens. A putting-out system, in which capitalists hired cottagers to spin or weave raw materials they owned, soon predominated. Single women, widows, and deserted wives in Kent worked for below-subsistence wages. In the 1590s or 1620s, when hard times struck, cloth workers unable to fall back on farm labor faced destitution.[55]

Driven by the scarcity of firewood in the London region and the growing uses of coal in smithing, ore smelting, glassmaking, and brick making, annual coal output exploded during the century after 1530, increasing from 200,000 to 1.5 million tons. Mines employed thousands of workers to dig shafts, scoop up coal, carry ore to the surface, and take it to market. By the 1630s rich capitalists operated the biggest mines, which employed 100 or more laborers. Capitalists engaged overmen to hire workers and direct day-to-day operations. Overmen recruited in poor pastoral areas near the pits, in Scotland, and among transients and youths seeking work. A few local men prospered, working as overmen and carters or providing food to workers, but the rest—men, boys, and a few women—lived on tiny wages and risked injury or death from accidents. Able to gain but seasonal employment, most mine workers were poor transients. Coal mines transformed the surrounding countryside. The population of mining villages increased rapidly, and pollution destroyed wells, reduced pasturage, and ruined horticulture.[56]

Urban growth developed because of rural pressures. Early-sixteenth-century English towns remained small, losing industry and population to their hinterlands, where petty entrepreneurs could make cloth more cheaply. In 1520 just one of twenty people in England lived in towns with a population of over 5,000; London, the largest city in the realm, counted only 55,000 people. Economic depression continued in most towns until late in the century, but London started to grow, attracting thousands of unemployed migrants. By 1600 about 200,000 people, one-twentieth of England's populace, lived in London and its suburbs; an additional twelfth of the population lived in places with over 5,000 inhabitants.[57]

As dispossession in the countryside grew, rural-to-urban movement accelerated. With neither land nor prospects in the depressed textile industry, thousands of workers tried their luck in cities. The percentage of England's people living in cities with populations over 5,000 rose during the seventeenth century from 8 to 17. London became the great magnet, its population multiplying five times between 1560 and 1670 and reaching more than a half-million by 1700. A tenth of England's people lived there during the mid-seventeenth century, but nearly twice as many passed through sometime in their lives. Cities attracted workers from a vast hinterland. Migrants to Norwich and York came from an area within twenty miles of town; to Bristol, within sixty miles; and to London, within 125 miles. Migrants arrived in their late teens or early twenties after working several years in the countryside. Half the immigrants came from villages, and most of the rest were from small towns. Men and women, former servants-in-husbandry and would-be female domestic servants, and skilled workers and vagrants moved to cities seeking work. But without craft skills, most urban migrants fell into wage labor or domestic service, wandered in and out of town, or became vagrants.[58]

Urban growth reverberated through the countryside. Seeking to take advantage of the London food market, landlords improved their farms, displacing peasants, who moved to cities. Rural workers made food and cloth for a growing urban market. The city attracted poor and respectable alike. Having lost any possibility of independent subsistence, city migrants had to rely on wages for survival. Rarely serving as apprentices but concentrated in proletarian neighborhoods, they evaded the class control still possible in the industrial countryside and created new institutions, such as alehouses, that served as hostels, employment agencies, and trade clubs.[59]

By the early seventeenth century, crisis permeated English society. Enclosures multiplied, peasants lost land, the textile industry faced depression, wage labor spread, and vagabonds tramped every country lane. But rural folk considered permanent wage labor (much less vagabondage) a debased status unworthy of freeborn people, because it made them absolutely dependent on others for their survival. As the next section reveals, wage labor turned the relation of husband and wife upside down.[60]

Households and the Transformation of Gender Relations

Farmer Richard Inckpen, a 1638 account relates, "laboureth in husbandry ordinarily with his own hands, holdeth the plough, maketh hay, selleth corn at market himself, and keeps no man or attendant upon him but such as are employed in labouring and husbandry." Englishmen like Inckpen knew that husbands and wives shared the productive tasks of the farm. Later in the century William Stout's mother, who had worked in the fields and marketed corn, sent a trusted female servant to train her daughter-in-law in "housekeeping" and, presumably, farmwork. In 1647 farmer Adam Eyre not only built a henhouse for his wife's chickens but rode six miles to find her a brewing pan. Such expectations of productive households, of course, presumed that families had land for crops and livestock.[61]

Rural English families thus envisioned households much like those of their ancestors. Thomas Tusser, an Eton- and Cambridge-educated East Anglian gentleman, related these expectations in *Five Hundred Points of Good Husbandry*, a practical guide to farming. Aimed at gentlemen, yeomen, and substantial husbandmen like Inckpen, Stout, and Eyre, the book was popular, going through fourteen editions from 1573 to 1600 and another six by 1638. Tusser imagined a diversified and nearly self-sufficient farm where one could make "a competent living" practicing thrift and agricultural improvement. Such a farm had to be large enough for a garden, fields with several types of grain, and pastures to graze milk cows, cattle, swine, and sheep. Hard work permeated the lives of Tusser's idealized farm family, which counted "no travell slavery that brings in penie saverlie" and shunned "the path to beggery."[62]

Every farm needed a wife. The word "husbandman" presumed that the landholder had married; the word "yeoman" was at first limited to married landowners. Tusser imagined a subservient farmwife who kept house, washed linens, carded and spun fiber, sewed and mended clothes, nursed children, and cooked and served meals for husband, children, and servants. Successful farming nonetheless depended on the mutual affection and joint labor of husband in grainfields and wife in house, garden, and dairy. But cooperation, Tusser insisted, went deeper. For instance, "Good milch-cow and pasture, good husbands provide / the res'due, good huswives know best how to guide." When a husband left on business, his wife had to know how to manage the farm. Any attempt to overthrow this natural order bred disaster, as the popular ballad "The Woman to the Plow and the Man to the Hen-Roost" attested.[63]

Yeoman households had a servant or two, and husbandmen often employed one as well. Servants provided essential labor for farmers whose children were too young to work or had left home and, at the same time, allowed poor families to reduce the number of mouths they had to feed. Tusser urged masters and mistresses to discipline yet behave fairly toward the servants and wage laborers they hired. Farmers should "keepe servant in awe"; the mistress should "shew servant his labour, and shew him no more," not even his food, and teach the maid "to stirre, when hir mistress doth speake." If master and mistress failed to supervise carefully, the servants would surely "loiter," leave the barn dirty, and pilfer and "carry home corn." But if workers "earneth their meate," their employers had an obligation to feed them "husbandlie fare" and pay them promptly. At the harvest farmers should not only pay "harvest-folke, servants and all" "more, by a penie or twoo / To call on his fellows the better to doo" but "make all togither, good cheere in the hall."[64]

Tusser detailed the work of husband and wife month by month, starting after the harvest. In September, while the husband stored grain, sowed winter rye and wheat, and threshed seed, the wife planted berries in part of her garden. In late fall and winter, husbands (and servants) plowed fields for new crops; threshed barley and hay; sowed barley, oats, peas, and beans; planted orchards; bred stock; hedged and ditched fields; cleared fields of bushes and roots; slaughtered pigs and dried fish; put dung in a rack; and cleaned privies and chimneys. Wives weeded their gardens, fed cows and cared for sows, and took care of calves. In the spring the pace of work quickened. Husbands made fences, manured and plowed fields, and planted hops and sowed barley. Wives planted, weeded, and watered their vegetable gardens; milked cows; made cheese, cream, and butter; and sowed flax and hemp. The harvest began in early summer. While husbands and servants sheared sheep, weeded fields, mowed meadows, carted hay, and began harvesting beans, wives gathered hemp and flax. During August, husbands mowed or raked barley and wheat, manured newly harvested fields, cleaned seed corn, and gathered fuel for the winter. Wives harvested the garden, saving seeds for next year.[65]

This division of labor in farm households persisted among landed families through the seventeenth century. As Tusser showed, all family members—husbands, wives, children, and servants—contributed to subsistence and made goods that could be sold to pay rent or buy manufactures. Wives as well as husbands marketed what they made. Despite this interdependence, men made all crucial economic decisions. Passed from father to son, land sustained paternal authority and helped guarantee subsistence. Common law mandated the subjection of wife to husband, granting him the right to her body and her property. Husbands thus controlled all land, including their wives' dowries, determining its use and selling its surpluses. When the husband neglected familial responsibilities or the wife asserted autonomy, neighbors or public authorities stepped in and enforced communal norms of appropriate behavior.[66]

Tusser missed significant elements in the organization of household labor. Households typically contained one or two children. Children played a small role in Tusser's story, but by age seven most had joined their parents in sex-appropriate farm tasks, running errands such as carrying food to harvest workers. Children worked full time by early adolescence. They washed or cooked, spun, herded animals, milked cows, weeded crops, and plowed, gaining competency in these skills by age sixteen or seventeen. Tusser assumed that yeomen and husbandmen hired mostly adults or older adolescents. Yet prosperous farmers and craftsmen apprenticed orphaned boys and girls or children of paupers, some as young as eight, but usually about age twelve. At ten Joanne Redwood, for instance, "was then able to Milck the Cows, Make Beds, Attend Children, or any other ordinary worke about the house." Other youths, barely teenagers and many not orphans, worked as servants-in-husbandry or as day laborers on the farm or in the textile industry.[67]

Gentlemen like Tusser insisted that enclosure, the growth of private property, and the rise of wage labor enhanced good household order. He imagined a world of productive and acquisitive peasants living on enclosed farms. Where common fields and pastures prevailed, livestock wandered over crops and destroyed wheat, peasants pastured too many animals and planted without regard to soil fertility, and men "theevishlie loiter and lurke." Enclosure, in contrast, would bring "more plentie of mutton and biefe, corne, butter, and cheese of the best, / More wealth any where (to be briefe)." The greater productivity of enclosed farms would give "more worke for the labouring man, as well in the towne as the feeld," saving him from destitution and vagrancy.[68]

Land consolidation, the rise of regional markets, and the growing rural textile industry enhanced paternal control, just as Tusser predicted. Husbands took advantage of improved markets to consolidate their power. They rebuilt their homes, adding rooms with domestic functions they allocated to their wives. Children often remained at home under parental discipline until their early to mid-teens: teenage boys worked in the fields under the master's eyes; teenage girls labored in the house, dairy, and garden under the mistress's authority. Even poor husbandmen retained sufficient rights to commons to allow fathers to direct wives and children in farm labor. Families of craft workers, moreover, cooperated in production under the husband-father. Landed weavers, for instance, maintained authority in the household, directing their wives' spinning while they wove cloth from yarn their wives made.[69]

Supporters of enclosure like Tusser misunderstood the ways commons helped peasants attain a degree of self-sufficiency and maintain a community. Tusser replaced the universal peasant community, with its control over resources, with a more restricted neighborliness. He encouraged neighborhood borrowing: "lending to neighbour, in time of his need," wins his support "and credit doth breed." Without such credit, farming would collapse, and "buying and selling must lie in the dust." After trading with neighbors, a farmer could sell his crop and buy what the family needed with the proceeds. The idea of neighborly trade is best illustrated in ballads, for instance, in the lamentation of a country man who slaughtered his cow before her time. After killing the cow, the country man sold every part of her: her hide to the tanner, her tallow to the candle maker, her horn to the hunter, her feet to the tripe-woman, and her meat to the butcher. Such individual trades, however, hardly resembled the control over arable farming and herding in peasant communities, nor could the growing number of dispossessed peasants who had no rights to commons engage in them.[70]

Cottager families could hardly make ends meet, much less lend and borrow. A "careful wife," in a ballad from the 1620s or 1630s, laments this "hard yeare" when they had barely enough to pay "the brewer and the landlord his rent, / The butcher, the baker, and the collier his score." Apparently with only a garden, the wife had little to trade and could add to family income only by "spinning and reeling." Once they lost their rights to commons, cottager families no longer had anything to swap beyond their labor and thus could not participate in this new, privatized farm community.[71]

Enclosure thus transformed familial relations among the poor. When peasants lost their land or rights to commons, an important buttress of paternal authority—the father's ability to direct family labor—was severed. Robert Payne, a rich estate steward, laid out the logic of wage labor in a petition to the Privy

Council in 1586. He sought to reduce restrictions on the cultivation of woad, a blue dye plant. Woad cultivation would displace peasants, yet it would, Payne argued, provide jobs. Forty acres, enough for three husbandman farms, "will kepe in worke one hundred and sixtie persons, the most parte weomen and children, one theerde parte of the yere." He would pay tiny wages of 4d. per day. Underemployed husbands and sons need not apply. "Yet see what greate good it do the poor laboringe man," Payne insisted, for "before this tyme [he] kept his wiffe and familie with his owne bare wages" of 3s. 4d. a week; "now his wiffe and two children by this meanes bringe in more for a good space together of 5 or 6s. a weeke." (Payne underestimated labor demands; a late-sixteenth-century woad harvest employed 387 people.)[72]

Husbands and wives earned wages independently, each contributing to the family budget. Even when wives worked with husbands as farm laborers, landholders directed their work. While wives earned low wages spinning yarn for clothiers or sold manure they had collected to farmers, husbands searched for farmwork or odd jobs. Wives of sailors and tramping artisans had to fend for themselves. When work was scarce, poor-law authorities put husband and wife to work at different tasks. At harvest time they might work on the same farms, but masters directed their labor. Widows and deserted wives earned wages and received relief insufficient to feed themselves and their children, but they remained masterless women.[73]

Women who worked separately from men created societies that included unattached women or men who lived independently or boarded with local families. Coal mining villages were heavily male; cloth villages and cities housed surpluses of women. Some of these laborers defied communal norms, living outside patriarchal households. Men had few opportunities, for instance, in pastoral textile villages, but single women could find work in spinning or dairying. Widows of craftsmen in small cities enjoyed similar opportunities, for they could control family property and open alehouses to support themselves rather than remarry.[74]

The severing of families from the land, the growth of wage labor, and the rising independence of working-class women disrupted rural households. Evidence of unsettling changes could be found everywhere, from slowly rising rates of bastardy to the increasing numbers of masterless adults tramping the roads, from the epidemic of abandoned children to the rise of infanticide. Poor women suffered the most, since they had neither the protection of a landed husband nor the pleasures of a companionate marriage. Since lonely men wanted female companionship, prostitution flourished. Unemployed husbands beat their wives, deserted their families, or turned to adultery. When women asserted independence—scolds slandered neighbors or berated husbands, reputed witches put a curse on neighbors, or unhappy or battered women deserted their husbands—they sometimes suffered judicial punishment. On occasion they suffered mental depression: two-thirds of a group of late-sixteenth- and early-seventeenth-century rural mental patients were women, and troubled courtships, marital problems, or bereavements explained three-fifths of their illnesses.[75]

With little property to protect, poor unmarried men indulged their passions, sometimes violently, in fornication before marriage. Most brides were pregnant when they married. Unable to support a family by their wages alone, poor men seduced and deserted their betrothed; masters seduced or raped their servants. A growing minority of children—one in ten by the early seventeenth century—were born out of wedlock. Deserted by would-be husbands, jilted servant girls gave birth in secret. Mother and child had to fend for themselves or rely on poor relief. Fearful for their own survival, on occasion mothers killed their newborns.[76]

The search for work far from home separated youths from their families. Parents, even those with little property, had insisted that children gain their consent before marriage, and landed parents often helped arrange the courtship and wedding. Seeking a dowry or portion so they could set up a household, children of smallholder families craved that consent, though some resisted parental attempts to choose their spouse. But when youths left home, neither masters nor the village community took over parental oversight. Servants-in-husbandry regularly courted fellow workers and made private, legally binding marriage contracts (sealed by sexual relations) far from home and without the consent of parents. Servant girls courted boyfriends at work, without the master's knowledge. Friends, neighbors, and kinfolk influenced courtship, but they had a smaller role than parents of children living at home. Parents and kin had even less oversight over children who went to London. Daughters of yeomen or husbandmen who worked as servants-in-husbandry and then moved to London to take up domestic service or the needle trades conducted their own courtship and married men they chose. Not only were they separated from parents and village, but few had kin resident in London to turn to for advice.[77]

Laboring families looked outside the home for companionship. Servants met and courted at festivals, weddings, wakes, dances, alehouses, fairs, markets, or parks, far from the eyes of their masters. The village alehouse became an alternative household, attracting transients, servants, laborers, textile workers, and poor craftsmen. Although men predominated, transient women took lodgings, wives came with husbands, and unattached women attended baptisms or weddings. Not only did alehouse keepers provide lodging and serve food and drink to travelers, but they pointed out employment opportunities to newcomers. Poor villagers came there to drink—and to buy bread or have meat cooked, borrow money or pawn property, attend weddings or wakes, play games or go to festivals.[78]

Sustained by wage labor and nurtured by the alehouse culture, egalitarian plebeian households emerged by the late sixteenth century. Headed by cottagers, laborers, poor widows, wives separated from husbands, or unmarried mothers, these households required the income of all family members to assure the meanest survival. The most secure lived on a tiny holding; others had neither permanent home nor regular work. Knowing that survival depended on the wages of wives and children, plebeian wives expected equality within marriage. Husband and wife together directed children in industrial outwork or farm labor or left them to care for younger siblings or fend for themselves. When unemployed men tramped the roads or left home to work in the mines or go to sea, wives had to support the family and often needed poor relief. At the end of the century, social commentator Judith Drake concluded that in laborer families, "though not so equal as that of Brutes, yet the condition of the two Sexes is more level, than amongst Gentlemen, City Traders, or rich Yeomen."[79]

Unwilling to accept egalitarian marriages, some poor men turned on women, blaming them for the decline in male familial authority. Men took independent and gossiping women to court on slander or witchcraft charges. One-pence ballads, plays, and pamphlets—hawked throughout the countryside by peddlers and sung at taverns and in fields—praised affectionate spouses and chaste and obedient wives but berated independent, disorderly, and rebellious women. Ballads described the cuckolded husband, the adulteress, the gossip-hungry (and alcoholic) wife, the lazy wife but hardworked husband, the shrewish and scolding wife, the domineering or violent wife, and the prostitute. Between 1557 and 1709 at least 3,000 ballads appeared. One observer lamented in 1595 that "in the shops of Artificers [artisans] and cottages of poor husbandmen," one would "sooner see one of these new Ballades" than "any of the Psalms." Misogynist proverbs attacked woman as scold, shrew, gossip, seducer, lecher, parasite, or adulteress: "A wife brings but two merrie daies to her husband, the one when she is married, the other when she is buried"; "A woman without a tongue, is as a soludier without his weapon"; "He that a wife hath strive hath"; "An undutifull wife is a house-traitor"; "A woman can do more than the devil"; "Women have but two faults, they can neither do well nor say well."[80]

Viewing female striving for equality as insubordination and unable to control their wives, some plebeian husbands beat them, inflicting what one woman called "the unspeakable tyrannies of an hard-hearted Yoak-Fellow." A popular proverb—"A woman, an asse, and a walnut tree, Bring the more fruit, the more beaten they be"—captured their belief that violence would cure a scolding wife and instill obedience. *Wife Lapped in Morels Skin*, a sixteenth-century chapbook, portrayed a sadistic beating. After his temperamental wife hit him and called him "whoreson," a husband "gave her than so many a great cloute, / That on the grounde the bloud was seene." She apologized and thereafter, the ballad-writer related, behaved well. The wife's mother condemned the punishment, but—tellingly—her father forgave "the yongman if he did sin," insisting "he did nothing amisse" and "did all thing even for the best." These ballads and proverbs document ready acceptance of wife beating, and court records show that husbands killed wives twice as often as wives killed husbands.

Most women in the ballads, however, continued to resist even after a beating, thereby defeating its purpose. In the "Patient Man's Wife" the husband sought that his wife "be still." "She will not grant me what I require," the ballad continued, "but sweares she'le have her will / Then if I chance to heave my hand, / straight-way she'le 'murder' cry." Other wives responded to beatings with equal violence. A man who beat his wife, one ballad related, "caught two blowes upon his head / For every one he lent." Husband killing, usually motivated by adultery or marital violence, was the subject of at least fifteen ballads, some based on actual cases. Such stories of violent women placed blame for male violence on their victims, thereby justifying yet more violence.[81]

Propertied men, especially Puritans, rejected violence, as did a few ballads. However, they blamed the ills of worker families on their independence. By supporting the poor independently, wage labor subverted conceptions of household order. Without the discipline of a good master, servants and laborers, Puritan cleric Richard Sibbes argued, became "wild creatures, ruffians, vagabonds, Cains." The poor lived under a master's discipline only briefly before forming their own households. Finding no all-powerful patriarch heading plebeian families, gentlemen denied that they were households at all. They searched for ways to reassert control over unruly men and independent women. Neither lordship nor the medieval ideal of a great chain of being (setting God over man, king over lord, lord over commoner, and husband over wife) could sustain male control in plebeian homes. Since all subjects had to be under the authority of a master—father, husband, magistrate, lord, and king—masterless men and women challenged both the state and the communal order. Only communal control, vigorously enforced, might suppress the lewdness of the poor and unpropertied.[82]

Changes in household governance reached the yeoman and gentry classes, further complicating the job of defenders of conventional morality. Taking the egalitarian promise of Protestantism seriously, some women demanded more control over their property, and a few went entirely outside their sphere by beginning to preach. To keep control of their property, a small but perhaps growing minority of women, aided by bequests payable at majority but before marriage, did not marry. Landed women—a declining minority among landholders—held land jointly with husbands or informally controlled the money, household goods, and livestock (typically worth £40–£50) they brought into marriage. A widow received (and controlled) far more than the third of her husband's personal estate guaranteed her, and on remarriage she made settlements to protect her children's property. On occasion women marrying for the first time kept their property separate, circumventing the law of coverture that gave husbands control. Women regularly went to court or testified in controversies, most of which arose out of disputes over land. Widows of gentlemen, yeomen, and prosperous husbandmen living in southern England participated most frequently, but wives (suing with husbands) came as well, taking an active interest in the management of assets they brought into marriage.[83]

While yeoman and gentry women jostled for more control over their households, men pushed village women out of every craft (including brewing and baking) a few had previously practiced. Prosperity further reduced women's productive roles in well-off households and accentuated the importance of a wife's domestic tasks, nurture of children, and loving obedience to her husband. In return the husband had to cherish his wife and provide for his family. While the husband ruled the whole family, his wife joined in governing children and servants. William Gouge, a prominent early-seventeenth-century London minister of Puritan leanings, emphasized these themes of mutual love, joint paternal governance, female subservience, and male familial support in his well-known work, *Of Domesticall Duties*. "A Woman's Work is never done," a ballad dating from the 1650s, also captures the change. The ballad's protagonist works all day cleaning house, making beds, tending the fire, dressing and feeding her children, cooking for her husband, washing, sewing, and knitting but makes nothing for market or house. Only the wives of country gentlemen or city merchants and professional men could possibly exchange *all* productive labor for these domestic tasks.[84]

Such mixed messages may have led to struggles over the proper role of women in the households of merchants, rural manufacturers, and improving tenants. Rich women had the leisure to pursue learning and the urge to control their own property. William Gouge found his congregation full of such independent-minded women. After preaching, in conventional terms, about the duties of wives while railing against wife beating, he found that "much exception was taken against the application of a wives subjection to the restraining of her from disposing of the common goods of the family without, or against her husband's consent." Some wives even asserted that "husband and wife are equall," that wives were in "no way inferiour" to husbands. Such women went against God, for they assumed that "*their* will must be done, *they* must rule and over-rule all, *they* must command not only children and servants, but husbands also, if at least the husband will be at peace."[85]

Seeking to make middling women chaste, silent, and obedient again, in the pulpit and in pamphlets sixteenth- and early-seventeenth-century commentators assailed "the insolence and impudence of women." They published over 100 treatises on women. The stereotypes found in them—woman as shrewish and promiscuous, chaste and pious—flowed from medieval or classical images and resembled those in plebeian ballads and proverbs. These circulated widely among bourgeois families.[86]

These works espoused a form of patriarchalism, a theory popular in the early seventeenth century, which linked monarchical authority in the state to male authority in the home. Husbands and fathers, patriarchalists argued, held absolute sway over their families, like that of the king, the nation's father, over his subjects. In a Protestant world, with neither priest nor feudal lord, a father stood between the king and his dependent wife, children, and servants. Because God had given this power to fathers and kings after the fall from Eden made government necessary, it neither required consent nor could it be limited by contract. To instill subservience, political authorities mandated that children study a patriarchal catechism, and adults heard patriarchal interpretations of the Bible in sermons and read patriarchal maxims in domestic advice literature.[87]

Images of the good wife and the good husband were firmly embedded in patriarchal literature. A virtuous wife should be obedient to her husband, industrious in her household labor, and devoted to increasing the family fortunes. An honorable husband should share household management with his wife, work assiduously to support his family, and remain faithful to his marriage vows. These stereotypes, which ignore the productive activities of poor rural women, point to the separation of the male public sphere and the female domestic sphere already under way in gentry and yeoman households.[88]

Patriarchal theorists urged husbands to share authority with their submissive wives. Drawing on older marriage ideals, Protestant commentators insisted that marriage be based upon companionship, mutuality, love, and the shared labor of providing for the family. Husbands, pamphleteers agreed, should make economic decisions; wives should run the household and care for the children. In the new spiritualized household of the Protestant bourgeoisie, the father took on the role of priest, responsible for the moral discipline of dependents. But parents together led family prayers and catechizing, teaching children obedience to parents, church, and state.[89]

This patriarchal ideal influenced expectations in gentry and yeoman households and, given the large number of servants in them, affected much of the rural populace. Some households approximated the headship, good order, and cooperation patriarchalism mandated, but the contradiction between paternal supremacy and reciprocity in marriage as often led to conflict between spouses. Husbands often compromised, making joint decisions with wives; when husbands were absent or had died, women acted as household heads. Nor was everyone a member of a respectable household. Only a wage, not household discipline, connected farm laborers and city workers to their masters.[90]

The behavior of servants-in-husbandry uncovers the contradictions of patriarchal family government. Since the state delegated paternal authority over servants to masters, masters considered servants subordinate members of the household. Masters housed, fed, and clothed servants and in return put boys to work in their fields and girls to work in their homes, dairies, and pastures. Both male and female servants helped harvest and thresh grain. Nonetheless, masters were temporary guardians who hired servants for one-year terms and paid them a stipend. Frequent changing of masters disrupted patriarchal authority and reduced the master-servant relation to a short, contractual agreement. Masters considered wages heavy expenses and occasionally refused to pay. Unwilling to recognize the master's paternal authority, servants often shirked work, sabotaged crops, ran away, married without permission, or—worse—tried to live independently.[91]

Under the pressure of the revolution of the 1640s, patriarchal order broke down. Revolutionary-era policies reached toward turning the family into a private institution bound by an implicit contract between husband and wife. Advanced thinkers not only denied that marriage was an immutable sacrament but challenged male authority, insisting that decisions within marriage required the wife's informed consent. Further removing marriage from the public sphere, Parliament debated divorce and instituted civil marriage, making religious ceremonies illegal. Although divorce remained illegal and opposition to civil marriage was so great that it was soon abandoned, the debate pushed couples into more conscious private decisions.[92]

The revolution severed the link between absolute monarch and authoritarian father. Contractual theories of the origins of government and market interpretations of the economy soon predominated. If contract underlay the origins of the state, the little commonwealth of the family had to be based on consent as well. Consensual theories of the family justified the retreat of bourgeois women into the home and the new division between public and private spheres in gentry and yeoman families.[93]

Once the state no longer legitimated itself through theories of family discipline, plebeian families gained more space to divide responsibilities as they pleased. With improved economic conditions, diminished population growth, and the mass migration of young men to the colonies after mid-century, laborers often chose leisure over additional work. Employers discovered that the labor of the poor was an asset that could be harnessed without regulating their families. Consequently the state slowly lost interest in enforcing family discipline, as the decline in prosecutions between 1630 and 1660 for alehouse keeping and failure to attend church attests. The discipline of the workplace, not the family, would soon ensure the productivity of wage laborers.[94]

COLONIAL EMIGRANTS left a rapidly changing England where the rich and the poor multiplied. As one balladeer sang, they dreamed of social justice in a place where "farmers consider'd the price of graine," where "honest poore tradesmen" charged fair prices, where "poor tenants' landlords / would not rack their rents," and where men helped their neighbors. Above all they dreamed of land. Having witnessed venal men take their land (or that of their parents) or fearing the loss of land they still had, they wanted to go where they could live independently, grow most of their own food, and run their families as they saw fit. As Chapter 1 relates, some of these families came to England's new American colonies.

The laborers, husbandmen, and (to a lesser degree) yeomen who moved to America pictured an idealized household where husband, wife, and children worked together, much as Thomas Tusser imagined. But poor and middling emigrants viewed familial relations differently. Middling men wanted to maintain patriarchal households; the poor (and especially the few women among them) sought greater domestic equality. Whatever their desires, however, hostile Indians and deep forests mandated familial cooperation. Only after conquering Indians and the forests, Chapters 2 and 5 relate, could farm men reassert authority and farm men and women together create markets and neighborhood borrowing networks that could sustain growing communal self-sufficiency.[95]

chapter one

From England to America

As John Winthrop agonized in 1629 over the reasons to found an "intended Plantation in New England," he reflected on England's economic condition. He put economic problems into a religious context; a colony would "carry the Gospell" and "raise a Bulworke against the kingdom of Ante-Christ." As a justice of the peace in a depressed textile county, Winthrop saw England's devastation firsthand and thought colonization would provide an economic refuge for suffering families. England, he insisted, "grows weary of her Inhabitantes, soe a man whoe is the most praetious of all creatures, is her more vile and base than the [earth?] we treade upon, and of lesse prise among us than a horse or a sheepe." The poor were everywhere, unable to make a living; the rest of the people were "growne to that height of Intemperance in all excesse of Riott, as noe mans estate allmost will suffice to keepe saile with his aequalls." How much worse it would be, Winthrop reasoned, to stay in England, where "many men" spend "as much labour and coste to . . . keepe sometimes an acre or twoe of Land," than go where they "would procure many C [hundred] as good or better."

Winthrop circulated his "Reasons to be Considered" widely, eliciting comments from potential investors and colonists. As he heard objections, Winthrop honed his text. The faithful were not abandoning England, taking "awaye the good people," as some believed. "The number wilbe nothing in respecte of those that are lefte," for "many that live are no use heere" and "are likely to doe more good there than here." Nor should the "many and great difficulties" of colonization or the "ill successe of other Plantations" dissuade prospective colonists. Immorality and irreligion explained the failure of Virginia's colonizers, whose "mayne end was Carnall and not Religious," and who transported "a multitude of rude and misgouernd persons the very scumme of the Land." A virtuous colony based on true religion would succeed where Virginia had failed. Using these arguments, Winthrop and his friends recruited 700 London and East Anglian Puritans to sail to Massachusetts.[1]

Winthrop's arguments had a fifty-year history. Commentators examining economic dislocation had often argued that only colonization could relieve England of her surplus people and thereby reduce English poverty. In 1584 Richard Hakluyt had insisted that "western planting" would employ "nombers of idle men" who "for want of . . . honest employmente" grew into "multitudes of loyterers and vagabondes." If colonies were founded, petty thieves could be condemned to colonial service, "felling of timbers for masts"; working in "mynes of golde, silver, copper, leade or iron"; or cultivating crops. As settlements grew, men at home could be put to work making "a thousande triflinge thinges" for colonists. Promoters of every seventeenth-century mainland colony saw them as a refuge for England's dispossessed. Poet John Donne, dean of St. Paul's in London, repeated the argument when he preached before the Virginia Company in 1622. Transported to Virginia, "many a wretch would be saved from the . . . hands of the Executioner." Virginia, he insisted, "shall sweep your streets, and wash your dores, from idle persons, and children of idle persons, and imploy them." Unlike poor Englishmen, even the "meanest" servants in Virginia or Maryland, John Hammond wrote in 1656, did not spend "wearisom lives in reliance of other mens charities," nor did they "make hard shift to subsist from hand to mouth, until age or sicknesse takes them off from labour and directs them the way to beggerie." He pitied England's poor, for rather than "remove themselves," they lived in England "a base, slavish, penurious life."[2]

Seventeenth-century Atlantic migration grew out of economic conditions—population growth, enclosure, dispossession, and the development of capitalism. Capitalist transformation provided the income necessary to establish colonies as well as migrants with so little to lose (and so much to gain) that they risked a perilous voyage across the ocean to join an invasion of unknown worlds. Before a single ship could leave, merchants or the national government had to build and provision it, recruit immigrants, finance the voyage of those unable to pay, and provision the infant colony. Since the crown refused to finance emigration, colonization began only after English capitalists had accumulated sufficient surpluses to permit speculation in colonization.[3]

The New World thus began as an extension of the old. The history of the seventeenth-century English colonies must be seen as a subordinate part of a greater Britain that encompassed not only England but Wales, Scotland, Ireland, and the North Atlantic basin. Migration to Ireland or Maryland was part of a process that began in the smallest English village. English emigrants did not cease being English (or East Anglians or Londoners) just because they crossed the Atlantic; rather, they carried their class relations and culture with them. English adventurers wrested Irish and American colonies from native inhabitants, and colonial histories thereafter became so intertwined with that of England as to make a single whole. Much like the sixteenth-century settlements among the Celtic Irish, the North American colonies developed as economic outposts of England, dependent on the homeland for sustenance, defense from hostile natives, and a supply of colonists. Ireland, Virginia, Massachusetts, and Barbados became the economic periphery of England, obligated to take England's surplus people, make corn or sugar, and buy English pots or cloth.[4]

A small minority of the English came to the infant colonies, and even fewer arrived from Ireland, Scotland, or the continent. Nearly all the English emigrants had already left their native villages to work as servants-in-husbandry, to marry, to find work or land to lease, or to escape poverty. Some, like the East Anglians who went to New England, had moved but a few miles (if at all) and stayed in close contact with their families. But an important minority had reached large towns or London, where recruiters most often solicited colonists. Urban migrants, already severed from their villages and families, listened to recruiters, and some of them, usually unmarried men, signed on for colonial service or settlement in Ireland, Virginia, the West Indies, or New England. For those people, colonial emigration became an extension of their search for work at home.[5]

We will never know precisely why these folk left family and friends. Many, especially those who came to New England, explained emigration in religious terms. But diminishing opportunities for economic security at home and the search for land in America stood behind many migration decisions, even of the most religious emigrants. Indentured servants bound for the Chesapeake or the West Indies, a majority of seventeenth-century immigrants, left because of their desperate straights at home. Once immigrants got a foothold in Massachusetts or Virginia, they wrote home and encouraged others to join them in exploiting the land's bounty and making godly societies. The decision to emigrate was thus complex. Potential immigrants weighed conditions at home against the possibilities in America, considering family and friends and the godly work of colonization.

This chapter examines early emigration. We will begin by examining the financing of the first colonies and of the servant and immigrant trade. Then we will turn to the complicated issue of the motivations of emigrants. Finally, we will document the social composition of immigrants to various colonies and speculate about what they expected after they arrived. As we shall see in Chapter 2, the particular and different composition of immigrant populations and the discrete reasons men and women came to each new colony helped form distinctive patterns of settlement.

Although the English crown refused to invest in colonial ventures, much less finance emigration, the monopolies it granted colonizers permitted them to begin risky enterprises without fear of competition. England, moreover, did allow men—even dissatisfied citizens—to leave. Travelers needed passports to depart, and Puritans leaving for Massachusetts had to present evidence that they were in good standing in their parish. But immigrants could evade these requirements or smuggle themselves aboard ships. Prominent Puritan clergy, for example, called themselves servants or changed their names. A growing flood of cheap chapbooks, plays, ballads, fiction, and pamphlets—averaging forty-two a year between 1621 and 1650—narrated the heroic actions of early voyagers, described America, and extolled or satirized emigration. How successfully English colonial promoters competed in this feverish marketplace for information on America and how well they enticed emigrants is the subject of this section.[6]

Colonial enterprises had to be financed by private individuals or companies, but neither the nobility nor merchants invested heavily in sixteenth-century missions. The reasons are not hard to find. The first explorers were "adventurers"—gamblers, speculators, and soldiers—who risked their lives in foreign conquests or their pocketbooks in chancy ventures. Lusting for gold and silver and aiming at plunder, conquest, and military honor, they took soldiers, not colonists, with them. Since explorers could only hope to plunder Spanish ships laden with gold, men with capital had little reason to believe that their adventures would yield a profit. Mission organizers often had to provide most of the money themselves. Their ventures usually failed, but their exploits, recounted at length, familiarized merchant and gentleman alike with the *possibility* of colonization.[7]

Early-seventeenth-century colonizers knew how often colonial adventurers had failed. The Roanoke fiasco, where a colony had disappeared without a trace, intensified their difficulty in persuading potential investors and colonists. To attract capital, they engaged in promotional campaigns, circulating manuscripts extolling colonies, writing innumerable pamphlets, and responding to every criticism about colonial conditions. Promoters fought bad publicity by extolling the high profits adventurers could make from fish and shellfish, skins and furs, tobacco and sugar, olives and silk, grapes and wine, cherries and strawberries, fruit trees and grains, herbs and spices, and lumber and firewood.[8]

The long—and implausible—lists of commodities that graced promotional pamphlets suggest the dire need of colonizers for investors and their realization that only greed could lure rich men to make such risky investments. But the lists also point to a major change in the purpose of colonial ventures from the search for gold, honor, and plunder to relieving English poverty, transporting immigrants, exporting farm products, and increasing commerce—goals meant to appeal to merchant and gentry investors. They painted a picture of a new England that replicated, even improved upon, hierarchical society at home. Hard work (under strong leadership and severe laws) would turn vagrants into productive citizens who could afford to buy what merchants exported. By planting a garden "in soiles most sweet, most pleasant, most strong and most fertile," in the words of Richard Hakluyt the Elder, investors would establish colonies, win glory, and attract the families the colonies needed to prosper.[9]

Colonial promoters sought investors to finance flagging operations far more than they sought emigrant farmers. Rich speculators, but hardly farmers, might be enticed by the supposed big profits to be made from wine or olive oil or silk. The promoters' ambiguous, even hostile description of tobacco reveals their aim. Praised for its reputed medicinal qualities, tobacco (some domestically grown) had been smoked in England since the late 1560s. High prices prevented the diffusion of smoking beyond the rich, but in the late 1590s a German traveler found that taverns sold small quantities to patrons for a penny. Educated Englishmen like Sir John Beaumont, who wrote an ode to tobacco in 1602, knew Virginia Indians grew the plant. Despite the possibilities of big profits, promoters at best mentioned the leaf in their long lists. Thomas Hariot wrote two paragraphs on tobacco in his 1588 *Briefe and True Report* but said nothing about its commercial prospects, while he praised the marketable qualities of woad, the dye plant. John Smith gave tobacco a few lines in his early work. By 1620, after Virginians had turned to tobacco, promoters found the crop an embarrassment. In Smith's words men were "rooting in the ground about Tobacco like Swine." Trying to reverse flagging interest, promoters such as Smith reprised earlier lists, adding the agricultural improvements Virginians had already made.[10]

Merchants and rural gentlemen resisted mightily entreaties from colonial promoters. They easily found safer investments. Gentlemen could buy land and live off timber and rising rents and grain prices; they could lease land to a tenant who would increase production; they could rebuild their houses and buy luxuries. London merchants and tradesmen, who controlled most of England's capital, found more opportunity in foreign trade, fishing, or new domestic businesses such as coal mining or provisioning cities than in colonization. Expeditions to North Atlantic fishing banks proved profitable for London and West Country merchants. Foreign commerce guaranteed safe profits: although the cloth trade floundered in depression during the early seventeenth century, rich foreign traders—members of the Turkey, Levant, and East India companies—enjoyed great success in trading to Greece, Turkey, and the Far East. These merchants exchanged commodities and carried goods but were not involved in risky production; colonial ventures could return a profit only to those willing to invest in plantation agriculture over many years. Adventurous men, finally, might help finance privateering vessels, an undertaking that cost about £2 million but returned £2.7 million between 1585 and 1603 alone.[11]

Ireland gave the North American colonies strong competition. Potential investors and immigrants from the English west and northwest, south Wales, and southwestern Scotland viewed that nearby island—almost as thinly peopled as North America—as the most desirable area to colonize. Small numbers of Britons had lived there for centuries. The crown considered Ireland a part of the realm and, beginning in the 1560s, repeatedly tried to conquer it. Over 50,000 troops worked to suppress rebellions between 1585 and 1602, an expenditure far greater than any indirect investments the crown made in colonizing activities. From the 1580s to the 1640s English and Scot investors established many colonies in Ulster and Munster, sometimes with direct royal participation. Gentlemen eagerly invested, expecting to reap big profits as landlords or cattle barons. By 1640 colonizers had brought over as many as 100,000 immigrants; some were former soldiers, but many were middling farmers and craftsmen unwilling to move to North America. One New England promoter, believing that potential immigrants found Ireland "a fitter place to receive them than New-England, Being Nearer, Our owne, Void in some parts, [and] Fruitfull," insisted that it was "well-nigh sufficiently peopled already."[12]

Since the richest merchants and nobles refused to subsidize colonial ventures, colonizers had to look elsewhere for money. The first colonizers—granted a monopoly by the crown to settle particular parts of North America—formed joint stock companies. In this innovative enterprise, new in the mid-sixteenth century, investors bought shares and entrusted management to a governor and a small board of directors, almost always dominated by rich London merchants. Early companies rarely made a profit, but they did start colonies and transport thousands of immigrants. The survival of the first colonies in Virginia and New England proved critical in enticing other colonial investors. After 1630 the richest Englishmen clamored for grants of proprietary colonies. Once colonies began to thrive, especially after 1640, smaller private investors took over, bringing tens of thousands of servants and many fewer free immigrants to the colonies and, in turn, hauling away crops the colonists made.

Colonial joint stock companies opened their membership to a far wider range of men than had invested in earlier trading companies. The joint stock form allowed gentlemen and merchants to invest sums as small as £10 or £15. Company members sacrificed immediate returns for potential profit once the colony became productive, but if the enterprise collapsed, they forfeited small sums they could afford to lose. When members of the Levant or East India companies refused to invest in colonizing, organizers recruited members of other London trading companies, lesser London merchants not involved in foreign trade, and the rich provincial gentry and peers. Merchants predominated among stockholders, constituting three-fifths of the members of twelve colonial joint stock companies, but nearly two-fifths were gentry or peers. Two-thirds of the money invested in colonial joint stock companies came from merchants. Gentry investors, as interested in bringing honor to England as in profits, provided a third of the capital the companies collected and gave them critical protection at court from state interference.[13]

These companies rarely returned a profit to the subscribers. Such profit required continuous investments in agriculture over many years, but investors—especially the rich merchants who dominated the cloth trade and Far Eastern commerce—refused to wait that long. The companies, as a result, were far less capitalized than trading companies. The East India Company raised £2.9 million, more than eight times the total (£464,000) invested in North American exploration and settlement between the 1580s and 1630s. The Virginia Company, the largest colonial company, collected £76,500 from all sources, more than one-fifth of the total raised by all the colonial companies. In contrast the Plymouth adventurers invested just £7,000, and the Massachusetts Bay Company merely £5,500. Members fell behind or refused to make payments on their stock, and company leaders could rarely recruit enough new members to make up the shortfall. Nor did inducements such as lotteries, promises of land, or the sale of fishing or trade monopolies to merchants bring in enough money to make companies profitable. Failure or revocation of charters was inevitable. The failure of the chartered companies opened the way for small-scale investment in the colonies; private investors—company members among them—spent £100,000 trading tobacco, transporting servants, and building plantations even before the Virginia Company's charter was revoked.[14]

The Virginia Company was both the greatest success and the greatest failure among joint stock companies. It raised £37,000 (£30,000 by 1612), nearly three times as much as the Plymouth adventurers and the Massachusetts Bay Company combined. By far the largest company with 1,684 members, it had a higher than average proportion of gentry, most of whom, unlike the merchants, joined in the 1610s, after the colony's problems had become a scandal. But after 1613, stockholders—even gentry—refused to pour more money down the Virginia sinkhole. From 1612 to 1621 the company relied on lotteries rather than stock to pay its debts. It tried with some success to extort money for the great London lotteries from the same trading and livery companies that adamantly refused to invest in more stock. Sometimes plagued with scandal and often unable to sell enough tickets to hold a drawing, the lotteries nonetheless raised £29,000, three-quarters of the company's income after 1613. Such creative financing did not save the company, racked with internal dissension between gentry and merchant members, from dissolution in 1624.[15]

Without joint stock companies, the English could not have begun the first North American colonies. Despite their collapse, companies backed exploration of the North Atlantic coastline; financed the first voyages to Virginia, Plymouth, and Massachusetts Bay; and supplied the colonies during their critical first years. These costs were heavy. Investors had to build or buy ships, hire seamen, stock the vessels with food and water, pay servants' travel costs, and provision the new colony until crops could be harvested. The £5–£6 free immigrants paid for their passage barely began to cover these expenses. The more limited the goal, the more likely revenue would meet expenses. The Massachusetts Bay Company, financed mostly by London's Puritan merchants, most of whom had already contributed to other Puritan ventures, used their funds to send ships to Salem in 1629 and the Winthrop fleet to Massachusetts in 1630, but thereafter it made few commercial investments. In contrast, the Virginia Company not only sent many shiploads of immigrants (to replace those who had died) and provisions to Virginia but financed expensive experiments in producing iron, making silk, planting grapes, and educating Indians, all of which collapsed.[16]

Seventeenth-century colonial companies faced a daunting problem in attracting immigrants. As travelers returned from the colonies, stories of the starvation, poverty, and hostile Indians immigrants faced filled the taverns, where merchants recruited poor men to sign on as colonial indentured servants. Pamphlet writers lamented the "sulphurious breath of every balladmonger," the "slander of the Countrey," the "abuse" of dramatists. Tantalizing examples survive of this oral tradition. Ben Jonson, George Chapman, and John Marston ridiculed the yen for riches of early adventurers in their 1605 (reprised in 1613–14) play *Eastward Ho!* Seagull, an adventurer bound for Virginia, entices Scapethrift to join him with tales of riches: "I tell thee, gold is more plentiful there than copper is with us," and when you arrive, "you shall live freely there, without sergeants, or courtiers, or lawyers, or intelligencers." For the scene to work, the audience (gentlemen, merchants, or the court) had to be aware of the history of colonial adventures.[17]

Anticolonial rumor and truth traveled far down the social order. Colonial promoters had to contend with a flood of intelligence from immigrants that reached their families back home. Letters to kinfolk, carried across the ocean by merchants, mariners, and travelers, reported American wonders or dangers and complained about American conditions. Highly literate colonists—a small minority of the whole—wrote most often, but illiterates could hire a scribe to write home for them. John Winthrop swapped letters with English correspondents, but so did Richard Frethorne, a miserable servant employed by the Virginia Company. Frethorne wrote home in 1623 about "sickness and death," lack of clothing, petty thievery, poor diet, and massacres by Indians. His plea that his parents "redeem" him, "release" him "from this bondage and save [his] life," must have circulated through his old neighborhood. In London in early 1620 a number of poor children, swept off the streets and held at the Bridewell Hospital for transport to Virginia, made clear "their unwillingness to go to Virginia," fearing work "under severe masters."[18]

Popular ballads, printed in cheap, single-sheet broadsides, were the most potent anticolonial media. Set to popular tunes and passed from hand to hand, early-seventeenth-century ballads circulated widely and probably reached illiterate workmen. Usually topical responses to a current event or rumor, anticolonial ballads satirized colonial conditions, the practice of kidnapping boys to labor as servants in Virginia, and Puritan intolerance in early New England. Three examples suggest the power of these ballads. "The Maydens of London" probably parodied the expectations of the 147 women, mostly middling Londoners, who traveled to Virginia on two voyages in 1620 and 1621, financed by the Virginia Company, to be wives of settlers. "Like loving Rogues together," the "London girls, that are disposed to travel" signed on for the voyage. They would be well fed on the trip and would find "Gold and Silver Mines, and Treasures much abounding, as plenty as New-Castle coals" when they arrived. But alas they were put to hard labor, as another ballad—possibly from the same period—relates: "the Axe and the Hoe have wrought my overthrow." As their clothes grow threadbare, the Maydens lamented: "thinne cloathing then may serve your turn, when as you do come thither, for there the Sun is hot enough and very warm the weather." The "West-Country Man's Voyage to New England" (1633) lampoons early Massachusetts. It relates the tale of an immigrant to Dorchester, Massachusetts. "Before she went o'er Lord how Voke did tell how vishes did grow, and how birds did dwell, all one among t'other, in the wood and the water." But he found neither a natural paradise nor a civil society. "The strong buildings" were all like the tents "at Bartholmew Fair"; no bells called him to prayer "because they had never a Bell in the Town"; and strange customs, like adult baptism and civil marriage, abounded. When a ship left for England, he "got leave to depart" and "bid farewell to those Fowlers and Fishers" in New England, where "they shal catch me go thither no more."[19]

Colonial promoters, hoping to persuade hardworking folk jaded by untrue visions of riches and an easy life, had to respond to such effective challenges to colonization. They resorted to printing pamphlets and publishing ballads that extolled the colonial virtues or played down the colonial problems. They had to deceive potential immigrants, yet they could not evade well-reported starving times and Indian massacres. To answer anticolonial sentiments, ballad writers, pamphleteers (like John Smith), and promoters of later proprietary colonies (Pennsylvania and Maryland) pursued three strategies: they attacked men who spread rumors about the colonies, admitted colonial imperfections and promised that problems had been resolved, and argued that with hard labor (avoided by the first colonists) any immigrant could become a landholder and live in comfort. In a land-poor society, where most adults struggled mightily and yet suffered great underemployment, such appeals to hard work and land acquisition must have been very powerful.[20]

In 1615 the Virginia Company commissioned a ballad to advertise a lottery to raise funds for the colony. In 1623 Virginia faced the more daunting task of soliciting immigration after the Indian massacre of 1622. That year Edward Waterhouse published a pamphlet exonerating the Virginians and making clear that (with better defense) a prosperous life awaited immigrants, and a Virginia planter published a ballad in London to defend the colony. After lamenting how by "sauage trecheries" the Indians murdered "full many a mothers sonne," the ballad tells how the English "slew those enemies, that massacred our men," burned their villages, and destroyed "both tempell, Botes, houses, and weres for fishing." Since the Indians had been killed or chased away, it was safe for men to immigrate; "of late from England safe ariu'd," the ballad continues, "a thousand people." The ballad ends with a list of opportunities that awaited the settler, from "Iron Workes and silk workes" to "'ndico seed, and Sugar Canes, and Figtrees."[21]

New England publicists pursued similar strategies. The author of *New-Englands Plantation* (1630) admitted the cold, heat, rattlesnakes, and wild Indians the settlers would find but immediately added that men could readily get land and find vast supplies of wood. "New England's Annoyances," written in 1643 by Edward Johnson, admitted every slander about New England, except heresy, but painted the region's deficiencies as virtues. Far from being a natural paradise, from December through February "the ground is all frozen as hard as a stone." Nor was farming easy, for "our corn being planted and seed being sown, the worms destroy much before it is grown." Since the settlers lacked money to import cloth, "our garments begin to grow thin." But Massachusetts was a godly, equal society. "We have a Cov'nant with one another," Johnson wrote, "which makes a division 'twixt brother and brother." If "some are rejected" for church membership, the others are saints, "equal in virtues and wants." English Puritans "who the Lord intends hither to bring" should "forsake not the honey for fear of the sting"; if one could "bring both a quiet and contented mind," he concluded, "all needful blessings you surely shall find."[22]

Pamphlets aimed at investors and labor recruiters told a similar story. By 1609 London merchant Robert Johnson had substituted the "Excellent fruites by Planting in Virginia" for the search for riches. He urged that the "swarmes of idle persons" with "no meanes of labour to releieve their misery" be sent, but more than that, he wanted industrious workers—"Carpenters, Ship-wrights, Sawuers, Brickemakers, Bricklayers, Plowmen, Sowers, Planters, Fishermen"—to come to grow sugar, oranges, or lemons, "to make . . . all necessaries, for comfort or use of the colony." All who paid their own passage would receive a share in the Virginia Company and, eventually, land. In 1616 John Smith, who knew firsthand the refusal of Virginia immigrants to work, proposed a different kind of colony in New England. His Virginia experience suggested that no "other motive than wealth, will erect there a Commonweale." Good fishing banks and farmlands abounded, waiting for the "imployment" of the "idle." Laborers could farm "good ground" that "cost nothing but labour; it seems strange," he concluded, that "any such there should grow poore." The author of *The Planters Plea* (1630), an English Puritan apparently involved in settling Massachusetts, agreed. Everyone should "serve one another through love, in some profitable and useful calling," a task difficult in England, where "handy-craftsmen, as Shoomakers, Taylors, nay Masons, Carpenters" were "overcharged [oppressed]." These men, the "able Ministers, Physitians, Souldiers, Schoolemasters, Mariners, and Mechanicks," were "fit to be employed in . . . planting a Colony." The author tried to overcome objections to planting a colony: immigrants were not abandoning England but doing God's work; plentiful fuel would keep families warm through the cold winter; and Virginia may have failed to prosper, but Massachusetts—unlike Virginia—was undertaken for "the advancement of the Gospell."[23]

The king granted later-seventeenth-century colonies—Carolina, Maryland, Pennsylvania, the Jerseys, and New York—to proprietors. Able to attract settlers from nearby colonies, they had far fewer survival problems than Virginia or Plymouth. Moreover, while proprietors did not use the joint stock form, they borrowed its substance, making large land grants (the equivalent of stock) in return for money that could be used to finance emigration. Calvert (Maryland) relied on rich Catholic gentlemen; the eight Carolina proprietors jointly funded operations, giving £500 each. Penn (Pennsylvania) not only set up German and Welsh enclaves (leading the Welsh to set up several land companies) but encouraged rich Quaker merchants in London and Bristol to form a joint stock company, the Free Society of Traders. Unlike earlier companies it was never dominant, but it did send servants to Pennsylvania, organize fishing expeditions, and operate a gristmill and brick kiln. Colonists and later investors benefited from the unintended gifts—ships, supplies, and investments in production—of merchant and gentry company members.[24]

William Penn perfected earlier strategies of joint stock companies in his highly effective 1680s campaign to gain investors and entice settlers to Pennsylvania. As population growth diminished after mid-century, economic conditions in England improved, and more men could find work at home. Penn therefore aimed his campaign, in part, at fellow persecuted Quakers. Since he knew that too few Quakers would migrate, he promoted his colony widely to non-Quakers and foreigners, appointing agents in England, Scotland, Ireland, and Holland to recruit immigrants. To aid recruiters and attract immigrants, he published seven tracts between 1681 and 1686 in English (twenty editions), Dutch (five editions of four pamphlets), German (four editions of three pamphlets), and French (a compilation of several pamphlets). The pamphlets refuted arguments against colonization and recounted the great opportunities to be had on Pennsylvania's bountiful lands if one worked assiduously. Far from weakening England, colonies had already "inrich'd and so strengthened her." The labor of the industrious "is worth more than if they stay'd at home," he added, "the Product of their Labour being in Commodities of a superiour Nature to those of this Country." Penn sought "industrious Husbandmen and Day-Labourers," "Laborious Handicrafts," "younger Brothers of small Inheritances"—anyone who wanted to improve his station. The poor would not get rich, but men who could "hardly live and allow themselves Cloaths" in England "do marry there, and bestow thrice more in all Necessaries and Conveniences" for their families. Later, less defensive pamphlets recounted the colony's resources in prosaic detail, specified how immigrants could get land, portrayed the public order and dense settlement already achieved, described the peaceful Indians, and printed a map of the colony that showed thick habitation by whites, but no Indian villages.[25]

Intensive promotional efforts were most needed during the planning stages before the first settlers arrived and at the outset of colonization. Pamphlets extolling New England first appeared in the 1610s and continued from the early 1620s to the early 1630s; Penn published his pamphlets in the 1680s. The campaign to populate Virginia continued for several decades after 1607 because the colony's survival remained in doubt. Once colonists established households, markets, and local institutions, investors published much less promotional literature. Large-scale, organized expeditions—best financed by wealthy companies—were no longer needed. With colonies well established, rich men could move about (from Barbados to Virginia, Carolina, or New England, and from New Netherland or New England to Virginia) and bring servants and slaves with them. English immigrants, however, were still necessary to hold land conquered from Indians or to work as farm laborers on recently cleared lands. Literature became more focused on them: "New England's Annoyances" was circulated in the 1640s, when more people returned to England than left for New England, and John Hammond wrote *Leah and Rachel* in 1656 to attract indentured servants to Chesapeake tobacco fields.[26]

Once colonies appeared secure, entrepreneurs entered the colonial servant and commodity trades. Englishmen had to find people to work the Indian land they conquered. Labor shortages appeared everywhere. In New England, family labor sometimes proved insufficient for local subsistence, and in the southern colonies, householders could not meet demands for tobacco in England or meat in the West Indies. Poor and youthful wage laborers in London and the countryside, the most promising potential recruits, could ill afford to pay £5–£6 for passage across the ocean, a sum that equaled more than half a year's wages. At first the Virginia Company transported laborers and set them to work on company plantations; but workers ran away to Indian villages, and by 1619 the company had sold the servants to planters. After the company's demise, independent London and outport merchants met colonial labor demands. They sent shiploads of servants and free immigrants, bought tobacco, and hired agents to engage in colonial business. Even before the dissolution of the Virginia Company, over half the investment in the colony came from independent merchants. In 1619–21, for instance, four prominent Gloucestershire gentlemen recruited 100 people to settle on the upper James River. By 1635, 175 men traded tobacco, and many of them were involved in the servant trade. Thousands of small-time merchants, mariners, and planters sent a few servants to the colonies, and nearly 3,000 participated in the trade in Bristol alone from 1654 to 1686. Small entrepreneurs, who paid the way of four or fewer people, transported to the Chesapeake region over half the servants who left Bristol from 1654 to 1660 and London from 1682 to 1686.[27]

This recruitment system worked efficiently, for at least two-thirds of seventeenth-century British emigrants to North America came as indentured servants. Indentured servitude maintained the form of service-in-husbandry but destroyed the constraints that bound master to servant. While servants-in-husbandry chose their master and worked a year at a time, indentured servants over twenty-one signed on for four years to repay their passage, and those without indentures worked even longer. Indentured servants could not choose their masters but had to work for and accept the authority of whoever bought their contracts; they could be bought or sold during their terms. Infrequently paid a wage, indentured servants received freedom dues at the end of their term. The longer terms and frequent sales they suffered suggest that they were less likely than servants-in-husbandry to be treated as family members.[28]

Given the difficulties in persuading anyone to emigrate, the success of servant recruiters was undeniable. While recruiters occasionally kidnapped boys or women or spirited men aboard ships bound for the colonies, nearly all servant immigrants had to be persuaded. Immigrant recruitment became a major enterprise in London and the outports, where recruiters set up offices, posted notices at alehouses, hired men to sing glowing ballads, and sent men to comb city streets and village byways. Recruiters promised high wages and land

and tried to keep the most damning facts from potential recruits (especially those with skills), who soon learned each colony's reputation. Before signing indentures, youths negotiated with the merchant who paid their travel costs; those willing to go to the West Indies demanded terms six months shorter than the five years served by those bound for Virginia. Skilled servants, literate servants, and older youths (who had gained strength or agricultural skills) negotiated shorter terms. The few servant girls who emigrated to colonies short of women could demand a premium of a year and a half off their terms. The poorest and least skilled, however, may have had little choice of either destination or term of service.[29]

John Hammond's *Leah and Rachel* (1656) mimics arguments servant recruiters probably made. Writing in the middle of the heaviest English migration to the Chesapeake region, Hammond posed as an honest broker. He admitted Virginia had been "an unhealthy place, a nest of Rogues, whores, desolute and rooking [cheating] persons; a place of intolerable labour, bad usage and hard Diet" but attributed these failures to the venality of the first immigrants, the lack of civil justice in the colony, and a "curell Massacre committed by the Natives." Virginia's "honest and vertuous inhabitants" herded cattle and grew corn and tobacco, making the country a "wholesome, healthy and fruitfull" place. To lure immigrants, Hammond offered practical advice. Those unable to "defray their own charges" should beware of "mercinary spirits" and sign an indenture. Such servants would receive "corne, dubble apparrell, tooles necessary," but not land, which went to the master. But Hammond did promise land, the great inducement, even to servants. Some "industrious" servants might "gain a competent estate before their Freedomes" on income made from the garden plot masters granted them. Once freed, an industrious man could get land cheaply, make "his owne corne and neede take no care for bread," grow tobacco for "a good maintenance," and "by diligence" become a "great merchant." Hammond's pose of honesty, his willingness to admit problems and offer practical advice, made his arguments about colonial opportunity persuasive when repeated to ill-informed potential servants. By the time servants learned of the high mortality immigrants suffered and the long years they had to work before getting land, it was too late.[30]

This private system of labor recruitment was remarkably effective in bringing Europeans to North America. The English migration rate was higher than those of continental European powers. About 160,000 to 164,000 English men and women immigrated to the mainland English colonies during the seventeenth century, most between 1630 and 1670; 116,000 to 120,000 went to the Chesapeake colonies. New England, which attracted few servants, received 21,000; about 23,500 English, but many other Europeans, came to the mid-Atlantic colonies at the end of the century. If one includes migration to the West

Indies, the count rises to 350,000. In comparison, between 650,000 and 750,000 Europeans, nearly all Spaniards, reached Spain's colonies between 1492 and 1700. Only 30,000, almost all French, came to New France, but over two-thirds of them returned home. In other words, 2,300 English immigrants reached the mainland colonies annually from 1630 to 1700 (5,000, including the West Indies), while only 3,400 came to Spain's colonies in the century after 1492, and 200 arrived in Quebec (and just 65 stayed).[31]

Motivations for Migration

About 350,000 Britons stepped into the unknown by traveling to North America (a voyage of five to nine perilous weeks), where they risked Indian attack, disease, and death. What motivated them to take this step? So few immigrants set down their reasons for emigration that we will never precisely understand motivations, but enough information survives to allow us to make educated guesses. We know that promoters explained away the dangers, but potential colonists could decode their lies. Statistical data on the timing of the migration suggest that immigrants made more or less informed decisions, comparing prospects at home to opportunities in America.

Historians have vigorously debated *individual* motivation, implicitly assuming that each individual or family made a separate decision to come to America. Yet migration to America involved communal decisions. Religious communities might encourage or discourage migration by members. Economic conditions—population growth, underemployment in London, the depressed cloth industry—provided an essential backdrop to migration. The 1619 commission to seat a town at Berkeley Plantation, Virginia, captures the ambiguities of emigrant motivations: the founders would "settle and plant our men" there "to the honour of Almighty God, the enlarging of the Christian religion, and the augmentation and revenue of the general plantation . . . , to the particular good and profit of ourselves, men, and servants."[32]

The reasons why families and individuals emigrated thus varied by time of migration, region of origin, class, and religion. The timing of colonial charters and the migration of free families point to a strong economic motivation. Every seventeenth-century colony was founded during a depression. The Massachusetts Bay Company received its charter in 1629 at the beginning of a depression; in 1633, when the first ships reached Maryland, the depression deepened. Joint stock companies and proprietors made countercyclical investments, hoping that colonizing would make money at times when few investments were profitable. Families more willingly sold their assets and emigrated during depressions when profits from farms and businesses had reached a low point. A depression in the East Anglia cloth industry stood behind the great migration to New

England of the 1630s; free families migrated to the Chesapeake colonies most commonly during the depression and war years of the 1650s and 1660s; Pennsylvania was founded in 1681, at the beginning of a depression, and heavy migration continued throughout that depressed decade.[33]

Nearly all seventeenth-century English men and women stayed in England. They were, Richard Eburne wrote, "wedded to their native soil like a snail to his shell" and would "rather even starve at home than seek store abroad." Those who expected to own land or practiced skilled trades had few reasons to emigrate, but even laboring people had other choices: remain in one's home village, tramp the countryside searching for work, migrate to a market town, or move to London. Fear of the unknown, inability to accumulate the money necessary for passage, unwillingness to trade freedom for four or more years of servitude, and the desire to stay near one's family kept the vast majority of English families at home. Those in groups most prone to migrate often stayed. Samuel Rogers, a young Puritan, remained at home to please his father, despite friendship with clerical émigrés, exposure to promotional efforts, and his own desire to "one day see New England and the beauty of . . . thy lively and pure ordinances." The half-million people who did emigrate from England between 1630 and 1680 represent no more than 4 percent of the people of the country, and the mainland colonies had to compete with Ireland and the West Indies even for these people.[34]

Some immigrants, moreover, considered their stay in America a temporary refuge from religious turmoil or a poor economy, expecting to return home once English authorities tolerated their faith or they had made enough money. Only richer men—especially gentlemen, merchants, or Puritan clergymen—could afford to return to England. Many gentlemen who came to Virginia before 1620 returned to England when they discovered that sickness, death, and Indians lurked everywhere. Other Virginians wanted only "a present crop, and their hasty return." As early as the 1630s, discontented New Englanders, unwilling to abide the rule of the saints, returned to England. Their numbers increased in the 1640s when a depression sent some back home to seek better opportunities and when civil war beckoned the faithful back to fight for Protestant liberty. As many as a sixth of immigrants who arrived in New England from London in 1635 moved to the West Indies or returned to England in the 1640s and 1650s, as did a twelfth of Watertown's early settlers. Most returnees were middling or educated men and their families. One-quarter of the university-educated immigrant clergymen returned; lacking stable positions in New England but enjoying good reputations in England, they moved back to help fellow Puritans and to gain employment in parishes seized from followers of Archbishop William Laud.[35]

Britons who moved to the colonies considered an array of economic, cultural, religious, familial, and personal factors difficult to disentangle. Migrants were pushed out of England by grinding poverty, fear of becoming poor, religious persecution, and the enthusiasm of neighbors or family. They were pulled to North America by the excitement of adventure, promise of riches, opportunity to own land, and desire to build a godly community. We can rank these possibilities by looking at the groups who made up a majority of colonists: the earliest adventurers, servants and free immigrants bound for the Chesapeake, ministers and laymen who went to New England, and free families who repeopled Pennsylvania. We will examine three dimensions of migration: the presence of kin in the colonies, the costs of passage and setting up households in America, and relative opportunities in England and America.[36]

A yearning for riches and glory motivated late-sixteenth-century adventurers and some of their seventeenth-century successors. The gentlemen among them knew the risks of a colonial enterprise and hardly needed to migrate to earn their daily bread. Rank-and-file adventurers crossed the ocean to plunder native peoples, to mine gold and silver they believed abundant, and to live without labor in a tropical paradise. As late as 1612 William Simmonds lamented that some emigrants deemed Virginia "not delightful because not stuffed with . . . heaps of gold and silver, nor such rare commodities as the Portugals and Spaniards found." Such men lusted for adventure—fighting the Spanish or the Indians, converting Indians, and planting English sovereignty on foreign soil—as much as gold. They had neither knowledge of the hard work necessary to make a settlement nor the desire to labor to accomplish that end.[37]

Unemployment and destitution at home, combined with colonial opportunities, propelled servants bound for Virginia or Maryland. As the population and the labor force grew and the number of enclosures rose during the early seventeenth century, agricultural and industrial wages declined. Servants immigrated during times of low English wages but high tobacco prices, enticed by the possibility of setting up their own tobacco farms once their terms expired. The Chesapeake region attracted the English whose futures were most bleak. Servant emigrants from Bristol during 1654–86, for instance, mostly came from depressed textile towns, poor pastoral regions, or other marginal areas. Those who left London could find only intermittent work and faced the unpalatable choices of destitution, petty thievery, or colonial emigration. Merchants or sea captains paid their passage, and they would lose only their meager and irregular wages. In return they were promised, at a minimum, subsistence during their terms.

Yet only a small minority of young, dispossessed city laborers signed on for the colonies. What distinguished them from those who stayed in London or Bristol? Servant emigrants may have been more isolated than most of the poor. They received neither subsistence nor succor from their families. Youths who shipped out as servants had left home years before to find work in farming or industry. Most had lived in depressed cloth villages where their families scraped by on the wasteland. As youths they had moved to London, Bristol, or Liverpool. Usually illiterate, they maintained sporadic contact with their families. Many were orphans, tenuously tied to siblings and other kin. Those without indentures were so young that they may have been abandoned by their families. Without families, they could expect no inheritance; without an inheritance, they could not marry. Even Virginia, with its remote promise of family, farm, and community, must have seemed a better choice than the certainty of poverty and loneliness in London.[38]

Many New England and Pennsylvania servants—those who worked in the fishing banks or the half of early Pennsylvania servants brought over by merchants in groups of seven or more—shared the experience of men bound for the Chesapeake. But others, like most of those on the *Mayflower* or in the Winthrop fleet, had been hired by emigrating families and came over as part of a household. Similarly, at least a third of early Pennsylvania servants—those held by owners of one to three workers—came as family workers. Unlike Virginia servants, they *chose* their masters and probably enjoyed more paternal treatment. They had to endure longer terms than servants-in-husbandry, but if Pennsylvania indentures are representative, they served four years, less time than most plantation-bound servants. Rather than desperation, a desire to be with people who treated them as kindred may have led them to emigrate; some may have shared the master's religion, whether Puritan or Quaker, and like him, they left to escape possible persecution.[39]

Women constituted a small minority—one in four in the second half of the century—of servants. The need for men to clear land and grow tobacco reduced demand for female servants; the unwillingness of women to move away from home, kin, and civility reduced the supply. What induced women to leave? The fact that they were in their late teens and early twenties, younger than male servants or female servants-in-husbandry, suggests that many were orphans, abandoned children, or daughters of the very poor—women with few prospects of marrying or making a living. If they came to Virginia or Maryland and survived servitude, they might find husbands and become planters' wives, a status much higher than they could achieve in England. Planters who wanted wives bid down the term of service by nearly a year and a half. Former servant women married in their mid-twenties, the same age as the daughters of propertied men in England.[40]

Free immigrants financed their own passage. Given the expenses of the voyage and of making a new household, they had to have substantial means. Before 1640, until local artisans appeared and livestock reproduced, emigrants expecting to set up households had to carry everything with them. The animals and fodder, flour and peas, butter and cheese, dried meat and fruit, clothing and bedding, pots and pans, axes and hoes, and guns and fishing nets, along with transportation, cost a frugal couple around £25. A couple with several children paid £40; a gentry household, £80. Few families had that much money. Gentlemen might finance emigration from current income, but yeomen and master artisans had to sell possessions. With property worth £50 or less, cottagers, urban laborers, and husbandmen—three-quarters or more of the populace—had no hope of financing emigration, even if they sold all they owned. In order to emigrate, many artisan and husbandman families scrimped to save and took less than what they needed to survive after they arrived. After 1640 middling immigrants could buy what they needed in the colonies, but they still required money or credit to buy land, furniture, tools, food, kitchen goods, and clothing to last them their first year.[41]

The gentry, yeomen, merchants, and craftsmen who paid their way to the Chesapeake region did forgo income and comfort in the hope of greater opportunities. Propertied men secure in their livelihoods or unwilling to take financial risks to make a fortune stayed at home, but younger sons of gentlemen—who could not expect to inherit land—emigrated to escape burdensome debts or to build their estates, become landed gentlemen, and provide for their sons. Merchants, some of gentry lineage, came to the region to exploit the tobacco and servant trades. The middling majority probably hoped to acquire land and an independent subsistence denied them in England. Fear of losing out in the struggles over enclosure and dispossession pushed them out of England; hopes of making a comfortable living growing tobacco lured them to the Chesapeake. A small inheritance of little use in England might buy passage to the Chesapeake and with it a headright, giving a husbandman or craftsman a right to survey and patent his own land.[42]

For gentlemen and merchants, migration to the Chesapeake entailed complex family negotiations. Some pairs of brothers migrated together; immigrants encouraged relatives to join them in Virginia; younger sons moved to relieve pressure on their older siblings as well as to establish an independent fortune; and gentry or mercantile families set up colonial businesses in London and sent sons or brothers to the Chesapeake to be their agents. When rich Barbadian merchants migrated to the Chesapeake (with family, servants, and slaves), they not only set up plantations and mercantile businesses but kept in touch with kin and neighbors at home, encouraging them to come as well. Other emigrants clashed with their families, even to the point of being disinherited. Those without family help could call on friends in London to help them emigrate; the Chesapeake allowed these men to start over in a new land.[43]

If economic incentives pushed most families to migrate to the Chesapeake, religious and familial reasons were behind the decisions of others. The Catholics, Protestant dissenters, and Quakers who moved to Maryland sought a refuge where they could practice their faith and work their plantations in peace. One-fourth of the first immigrants to Maryland were Catholic, mainly gentlemen and small property holders; most of the Protestants were servants. Immigrant Catholics built a plantation community in St. Mary's County, where as late as 1708 they constituted a third of the populace. So many radical Protestants, seeking toleration, flocked to the colony that Charles Calvert complained in 1677 that "the greatest part of the Inhabitants . . . doe consist of Presbiterians, Independents, Anabaptists, and Quakers." During the 1650s Puritans and Presbyterians emigrated from England. Quaker missionaries converted many immigrant Puritans; by the end of the century, one in eight Marylanders may have been Quaker. Puritans and Quakers immigrated to Virginia as well. As early as the 1610s Calvinists regularly preached at Virginia churches, and a congregation of Brownists—radical independents—arrived in 1618. By mid-century the lower James basin and Eastern Shore had become Puritan strongholds and had attracted Quaker missionaries. When Virginia authorities suppressed Protestant radicalism, many Quakers and Puritans moved to Maryland.[44]

Economic conditions combined with religious persecution drove migration to Massachusetts. The most influential promoters—Puritan clergymen and lay gentry—looked at New England as a refuge from oppression and disaster. They saw evidence of catastrophe everywhere in the 1630s, from vagrancy to public encouragement of sport and drink. Poverty, hunger, poor harvests, and a depressed cloth industry stalked their villages, and religious persecution tipped the balance in favor of leaving. When Archbishop Laud began to remove Puritan clergy from their parishes, forbidding them to preach, more than a hundred college-educated Puritan clergy and laymen left. Laud had suspended the clerical rights of at least forty-seven of the seventy-six emigrant Puritan clergy. In England they refused to give in to Laud's demands; in New England they could make a church of their own and freely preach salvation. These men had an importance that far outweighed their small numbers, for about half persuaded followers to emigrate with them.[45]

Religious motivations inspired artisan and farm families to migrate as well. Bitter conflicts between Puritan preachers and episcopal officials occurred in 53 of 165 East Anglian towns that sent emigrants to New England. Sympathetic to their ministers, parishioners listened attentively when they preached about emigration. A few people from each beleaguered village subsequently emigrated. Roger Clap, born into a religious family of five children in Devonshire (with an annual income "not above eighty pounds"), was one such emigrant. At about age twenty he left home to be near a "famous preacher" in Exon and lived as a servant in "as famous a [gentry] family for religion as ever I knew." He "never so much as heard of New England until [he] heard many godly persons that were going there," including preacher John Warham. After some struggle with his father, he boarded a ship with "many godly families," including Warham's, convinced that "God brought [him] out of Plymouth" and "landed [him] in health at Nantasket." Like Clap, a substantial minority of emigrants had developed such great piety and extensive knowledge of the Bible and Calvinist theology that they reported detailed spiritual autobiographies soon after they arrived. A third of those admitted to the Cambridge church linked immigration to their striving for salvation.[46]

Religious concerns pushed families toward emigration, but other incentives were usually necessary to make them leave England. An emigrant might, like defrocked Puritan cleric Thomas Shepherd, leave England to protest ungodly practices of the church but, at the same time, hope to find a "way of subsistence in peace and comfort to me and my family" impossible in England. Or an emigrant could follow a charismatic preacher, go with friends or kindred, be persuaded by husbands, or join family members already in New England. Others wished to live in a holy community uncontaminated by religious error; when they got to New England, they formed towns with like-minded countrymen. Even the faithful disposed to move rarely left without inner turmoil. Mary Angier Sparhawk, who emigrated to Cambridge with her prosperous husband, listened to John Rogers, "one of the most awakening preachers of the age." When she heard of "New England, she thought if any good here it was." But when her husband resolved to emigrate, she feared if "God should not help all would rise to greater condemnation." This "fear of no blessing" stalked her until she thought that if "her children might get good it would be worth [her] journey." William Manning, who moved first to Roxbury and then to Cambridge, had more personal doubts. "To come out of" his sinful state, "having at last thoughts of . . . New England, my wife and I hearing some certainty of things here, I desired to come hither." But "when the Lord brought me to sea, I was overcome with a discontented mind, meeting there with hard and sad trials as fear of losing my wife."[47]

Even as he hoped in 1630 that "the most sincere and godly" immigrants would "have the advancement of the *Gospel* as their maine scope," Puritan minister John White knew that "necessities," "noveltie," or "hopes of gaine in time to come may prevaile with" others. Fear of impoverishment at home and hope for success in New England combined powerfully with religious, familial, and communal motivations to propel families to emigrate. Poor harvests in East Anglia during the early 1630s raised food prices, heightening the distress. Textile workers who emigrated from East Anglia and the West Country probably sought escape from that depressed industry. Even if they were not directly affected, the devastation around them served as a potent reminder of what might happen to them.

John Dane's experience shows how economic and religious motivations were linked in the decision to migrate. Dane, a tailor's son, followed his father's trade and moved from place to place in search of work. He grew up in a religious family but led a dissolute life dancing and attending cock fights until he became a servant of a pious Puritan, married, and began to listen to sermons. He explained his decision in religious terms; it was a way to free himself from temptation. But immigration to New England was also part of his search for stable employment. Like John Dane, every emigrant expected to make a decent living once he arrived. Defrocked clergymen hoped to find a new parish; artisans either wanted to reestablish their trades or find land to farm; farmers wanted land for themselves and their children.[48]

Economic motives were undoubtedly paramount for some New England immigrants. The single young men (and many fewer families) who emigrated from Wiltshire (in the West Country) mainly left towns and pasture-forest areas to escape the consequences of rapid population growth—pressure on land, rising rents, growing numbers of cottagers, postponed marriages, and constricted opportunities. The religious radicalism of the area further encouraged emigration. Many of these migrants landed on the Maine, New Hampshire, or northern Massachusetts coast, where they joined servants and youths from other parts of England to exploit fisheries or forests. Economic motivations predominated among some Connecticut River Valley pioneers as well. The settlers of Springfield, whether servants or free people, came to trade with Indians or work for the Pynchons, the premier family in town.[49]

The origins of emigration to Pennsylvania can be traced to William Penn's (and other Quakers') urge to develop a refuge for that persecuted sect, but economic and religious motivations for migration were entangled here, too. Quaker theology linked prosperity to bourgeois values of domesticity, self-sufficiency, and self-denial. In Restoration England, Quaker meetings were illegal, and Quaker missionaries could be prosecuted as vagrants. Hundreds of Friends suffered large fines and imprisonment and some faced banishment to the colonies. Most Quaker emigrants had suffered greater disruptions of their households, the center of their value system, than those who stayed in Britain and paid larger fines for infractions. They could ill afford the fines, for they lived in the impoverished North Midlands and Wales, adjacent areas nearly untouched by the prosperity and improving agriculture characteristic of regions that sent emigrants to the Chesapeake and New England. Families had responded to population growth by subdividing estates, rather than the consolidation common elsewhere in England. Landholders, whatever the form of their tenancy, thereby accumulated less wealth than those who lived in more advanced parts of the realm. Emigrants hoped that Pennsylvania would bring a flowering of prosperity, radical faith, and bourgeois social mores. Once the promise of Pennsylvania became known, numerous middling Quakers departed, sometimes hiring neighbors' children as servants. Meetings of Friends sent other families too poor to afford passage across the ocean. Pushed by persecution and pulled by opportunity, nearly entire Quaker meetings on occasion picked up and moved to Pennsylvania.[50]

Servants and Farmers, Mechanics and Merchants

Like all migration, emigration from England to North America was highly selective. Each colonial venture attracted or repulsed men and women from different segments of English society: Londoners or villagers, single men or families, orthodox Protestants or dissenters, and farmers or craftsmen. Prior migration and British and colonial labor markets structured immigrant recruitment by colonizers, shippers, and land speculators. Recruiters, in turn, sent immigrants to places where demand for labor was high or opportunities great. As colonial labor markets and opportunities grew or declined, immigrant destinations changed. One can distinguish immigration to the Chesapeake colonies from that to New England, early-seventeenth-century from mid- and late-seventeenth-century immigration, and migration to newer colonies such as the Carolinas and Pennsylvania from movement to Virginia or Massachusetts. Each of these migration streams points to differences in the timing of immigration; the sexual, age, and occupational distribution of migrants; and the contrasting mixes of free people and servants among immigrants.[51]

English emigrants fell into two groups: servants and free people. Servants were young (between ages fifteen and twenty-five), mostly male, and commonly from the laboring classes. Usually recruited in cities such as London or Bristol, many had worked as servants-in-husbandry before they moved to town. Free emigrants, in contrast, were more evenly divided by sex, included families as well as unmarried men, and typically came from the smallholder, artisan, or less often, gentry classes. Recruited from countryside as well as city, they often moved with neighbors or kinfolk. During the first decades of white habitation, the character of colonial society depended in large measure on the mix of free and servant immigrants. The rapid development of commercial agriculture in the southern colonies, along with the failure of New Englanders to find an exportable staple, led recruiters to send most servants to the southern colonies. As a result, the proportion of servants in the immigrant population rose as one traveled south from New England.[52]

Heavy emigration to North America began in the 1630s. Less than 10,000 people moved to Virginia or New England between 1607 and 1630, and most of them died of hunger or disease. Large numbers of English emigrants landed in the Chesapeake colonies between 1630 and 1680 and in New England in the

1630s. Immigration averaged 20,000 a decade in the 1630s and 1640s and 25,000 a decade between 1650 and 1690, almost all to the southern and middle colonies. By the last third of the century, immigration and a declining English rate of natural increase had lowered population growth by a third, reducing the number of new workers entering the labor force, raising wages, and decreasing both internal migration and colonial emigration. At the same time, opportunities diminished in the Chesapeake colonies, where as many as two-fifths of British immigrants had gone. English emigration therefore plunged to just 10,000 a decade in the 1690s and 1700s.[53]

The Roanoke colonists, the first permanent white residents of the mainland, acted as soldiers out to conquer and hold territory for English adventurers and the crown. At best they created a barracks society, not a settled farm community. At least a ninth of the adventurers were gentlemen, scholars, and officers. All of the 108 setters in Roanoke in 1585–86 were men. A few gentleman adventurers out to make a fortune in precious metals migrated, but the large majority were Londoners or West Country men hired by Sir Walter Raleigh as servants; perhaps half were soldiers. Although many had experience as farmworkers, they got their food from supplies they carried with them or from the Indians. The 1587 Roanoke colony (the "lost" colony) was similar: 100 of the 117 immigrants were men or boys, mostly servants; only 17 were women. Some, especially the 32 who migrated in family units (15 couples, 2 of which had one child) might have expected to establish a farming community.[54]

Between 1607 and 1633, 9,000 English migrants, from much of southern England, came to Virginia. Like Roanoke, Virginia was organized on military principles. The first three shiploads of colonists—293 men and 2 women, including 110 gentlemen, 51 laborers (most of whom were gentlemen's personal servants), and 27 city craftsmen—were unprepared to feed, clothe, or house themselves. Most died, and their replacements included fewer gentlemen. Through the 1610s the colony attracted unruly men who could hunt or fight Indians but few people with the skills necessary to feed the colony. The Virginia Company and independent merchants plucked laborers from London streets, sold them to planters, and set them to work farming tobacco. The company, moreover, made strong efforts to recruit men accustomed to farm labor. It sent over about 100 in 1619, in a ship filled with 89 company tenants including 19 husbandmen and 43 artisans, 29 of whom followed trades needed in a farm community. Three years later another ship, with at least 103 passengers, carried private adventurers and their servants. In 1625, when Virginia authorities took a census of the survivors of famine, disease, and massacre, three-fourths of the white men in the colony were servants; most were between ages fifteen and thirty. The gentry (an eleventh of white men) and other big planters bought most of the servants. As late as 1635 over two-thirds of the Virginia immigrants were between fifteen and twenty-four years of age; presumably nearly all were servants.[55]

Neither the company nor Virginia's planters attracted many women. By 1610 a few women had migrated, but most were too ill to work. Even in 1625, after the Virginia Company sent several shiploads of "younge, handsome, and honestly educated Maides" to marry single planters, there were more than six men for every woman in the colony. Planters, seeking the highest possible tobacco profits, continued to buy male servants but avoided purchasing women: in 1635 they still transported six men for every woman. Planters keenly felt the absence of women in the colony. A former servant might find a wife, but only if he survived for years after his term of service had ended. Women, moreover, were essential to the colony's survival, performing tasks men considered beneath them, such as cultivating vegetable gardens, milking cows, rearing children, and keeping house. Most important, women would bring stability to a violent society, "tye and root the Planters myndes to Virginia by the bonds of wives and children."[56]

The servant trade to the Chesapeake colonies peaked between 1630 and 1680, when nearly 100,000 English men and women, the vast majority of whom were servants, immigrated. Servant immigration grew from the 1630s to the 1670s and was heaviest during the third quarter of the century. English wages and the tobacco trade regulated English emigration. Servants indentured themselves to Chesapeake planters when English wages declined and tobacco prices rose; they refused to go when English wages rose and tobacco prices dropped. This convergence of English wages and Chesapeake tobacco prices suggests that servants migrated expecting greater opportunities and that planters used servants they bought exclusively to produce tobacco. After 1680, as fewer Englishmen came of age and wages rose, English youths refused to go to the region, preferring opportunities at home (especially in times of war, when men joined the army or privateers) or in newer colonies. As a result, servant prices rose, and Chesapeake planters seeking workers turned to slavery.[57]

Chesapeake planters most wanted young and strong male servants who could tolerate the backbreaking work of cultivating tobacco. During the middle decades of the seventeenth century they attracted poor youths, many of them vagrants, but fewer farmworkers than they wanted. A quarter of the servants were female, a substantial increase from early in the century. Nearly all servants were young and unmarried. Those who came with indenture contracts were usually poor, unskilled youths in their late teens or early twenties; the two-fifths without indentures who served with the custom of the country were usually in their mid- to late teens. Most (three-fifths) had been born in the countryside, had migrated ten to forty miles to London, Bristol, or Liverpool, and had worked in the city for several years before migrating to the Chesapeake region as servants.[58] Men who signed indentures during the mid-seventeenth century included laborers, unskilled youths, craftsmen, and husbandmen in roughly equal measure; by the late seventeenth century, the proportion of skilled workers, laborers, and the poor among indentured servants had risen.[59]

Servants who arrived in the Chesapeake colonies after 1640 entered an environment far different from that of their predecessors. Servants were integrated into an ongoing agricultural society. They lived with families, not in barracks. Planters set all servants, male and female, skilled and unskilled, to work growing tobacco and corn, thereby guaranteeing that both subsistence and an exportable staple would be made. Despite the continuing high ratio of men to women among servants, there were enough female servants to ensure that many male servants who survived their term could marry, begin households, and start plantations.[60]

Passengers who paid for their own and their family's voyage from England were always a small minority, amounting to 25,000, or one-fifth, of Chesapeake immigrants. Most arrived in two spurts, 1646–52 and 1658–67, when they constituted one-quarter of arrivals. Like indentured servants, they migrated from the south and west of England, especially from the London and Bristol regions; an important minority arrived from Barbados, bringing servants and slaves with them. Unlike servants, three-fifths were twenty-five or older. Despite their small numbers, free immigrants played a key role in the region's society. They could take advantage of headright land grants, which were given to anyone who paid a person's passage. Men outnumbered women two to one, but over two-thirds emigrated in family units. Over two-fifths of the men were in their thirties and forties, prime years for capital accumulation. Many gentlemen and merchants—a quarter of the free immigrants—came to the region. Most of the rest were skilled craftsmen, but few husbandmen or yeomen risked the ocean voyage. Given the heavy mortality and low fertility of earlier migrants, merchant and gentry immigrants were in a strong position to seize and retain economic and political power. They financed the emigration of thousands of indentured servants, bought slaves, and exported more and more tobacco. One group of English gentlemen, Royalist in politics and conservative and hierarchical by inclination, swept to the top of Virginia society and dominated provincial and some county governments; others—middling merchants and craftsmen—filled in the void on other county courts, taking on the title "gentleman" as soon as they arrived in the region.[61]

Far more women and families traveled to New England than to the Chesapeake region, but at first unmarried men dominated New England. During the 1610s fishermen set up bases on the Massachusetts and New Hampshire coast, plying the rich waters nearby for three months and trading furs during the off season. For decades coastal communities such as Salem, Gloucester, and Marblehead attracted predominantly unmarried and transient sailors and fishermen. West Country fishermen colonized Salem in 1626, and while the town distributed land and attracted some farmers, close to a third of those who arrived by 1637 were mariners, fishermen, or marine craftsmen, most of whom were unmarried. Gloucester, first seated in the 1640s, developed into a fishing village populated by contentious West Country and Welsh fishermen; half the men in mid-seventeenth-century Marblehead provisioned a fishing fleet manned by an equal number of transient fishermen from every part of Britain.[62]

Plymouth colony shared these disruptions. Two-thirds of the immigrants on four ships to the colony during 1620–23 were men or boys, most between fifteen and thirty. Three-fifths of the 104 *Mayflower* passengers, most originally from East Anglia, came in family units, but twenty-nine unmarried men (including fourteen servants and four seamen) and two female servants were onboard. Few servants shared the religious commitment of most of the men and therefore failed to sign the Mayflower Covenant that symbolized order in a strange land. No more prepared to make their own food than men in Jamestown, immigrants traded fur and fished rather than farmed; they, too, relied on Indians for subsistence and suffered starving times in their first years. Plymouth's problems were heightened in 1622, when two new ships arrived. The passengers, sixty-seven men recruited from the streets of London, were brought to Plymouth by Thomas Weston, a speculator who had helped finance the colony. They stayed several months, ate precious food, and then left to begin a colony near the site of Boston, where they refused to grow corn, preferring to steal from Indians and setting off warfare with them.[63]

Winthrop and other promoters of the Massachusetts Bay colony concentrated their activity in Puritan strongholds in East Anglia, the West Country, and London. Three-fifths of the passengers in the Winthrop fleet—founders of most early Massachusetts towns—came from East Anglia (Norfolk, Suffolk, and Essex). East Anglia accounted for two-fifths or more of the 1630s migrants to Massachusetts, and at least another tenth came from adjacent areas socially similar to East Anglia. Other large groups, a third of the total, came from London and the West Country. East Anglians dominated Salem, Boston, Cambridge, and other large towns in the province, sometimes forcing families from other regions to peripheral villages. The culture of eastern England, combined with the Puritans' Calvinist theology, structured social life in the province and, at the same time, incited religious and social conflicts.[64]

More than any previous group of English immigrants, the 13,000 to 20,000 people who moved to Massachusetts Bay during the 1630s came in family groups. Although more than a quarter of the emigrants were single (mostly male) youths between fifteen and twenty-four, about 50 percent higher than in England, almost two-fifths, the same proportion as in England, were between twenty-five and fifty, the prime ages for child rearing. The 1630 migration of 700 people with John Winthrop set a pattern. Nearly two-fifths were female, more than twice as many as Virginia-bound emigrants. Very few children had come to Virginia, but more than a fifth of those in the Winthrop fleet were children traveling with their parents; in all, half the emigrants were under twenty. Three-quarters in total arrived as family members. Similarly, two-thirds of the passengers boarding ships to New England from West Country and East Anglian ports from 1635 to 1638 traveled in family groups (most were headed by young parents with one to three children), a number that jumps to nine-tenths if servants are counted as family members. More than two-fifths were children, a proportion as high as in England's population. Even on the ships that departed from London, where more single men were recruited, two-fifths of the passengers were women and nearly a third were children under fifteen. About 200 of these 700 immigrants to Massachusetts Bay died soon after they arrived, but the population thereafter grew rapidly from immigration and natural increase.[65]

East Anglians who migrated to Massachusetts in the 1630s were remarkably close-knit. Two-thirds traveled with a prominent gentleman, local clergyman, or kindred. Promoters organized intensive neighborhood meetings. In 1627 "some friends beeing together in Linconsheire, fell into some discourse about New England and the plantinge of the gospell there." After "deliberations," they "imparted our reasons" to Puritans in the West Country and London. Similar groups of emigrants, organized by Puritan gentlemen or clergymen, formed all over East Anglia. Even more—some two-fifths of East Anglian emigrants—came with large (often twenty or more), less formal groups of extended kin (including siblings, cousins, and in-laws). In other cases several family members or neighbors established a beachhead, and others followed. A substantial portion of some village populations thereby abandoned their homes for Massachusetts. For instance, 143 people (a third of the town) left St. Andrew Parish in Hingham with their pastors. Not surprisingly, they formed markedly cohesive communities after they arrived; the Hingham, England, group founded Hingham, Massachusetts.[66]

A representative sampling of the middling classes of England could be found among the Massachusetts immigrants of the 1630s. No more than one in four persons came from the growing class of day laborers; perhaps one in twenty was a gentleman, merchant, or substantial clothier. Most of these migrants had little property, and a few were barely able to get the money needed for their passage. Artisans were heavily overrepresented among the emigrants. Only one in four had been a husbandman or yeoman, nearly an equal number had worked in the depressed textile industry of East Anglia or the West Country, and another quarter had engaged in other crafts. The rest were laborers or servants. Two-thirds had lived in London or large towns before migrating. Most were ill prepared for the full-time farmwork they would face in Massachusetts.[67]

Although a big majority of migrants to New England in the 1630s came in family units, as many as a fifth of emigrants and a third of those who boarded in London were servants. Given the large number of children, these servants constituted close to half of the male labor force of early New England. More than a third of the families who emigrated between 1635 and 1638 brought servants, though all but the richest families employed only one or two. Some of these young men (and a few women), most of whom were between fifteen and twenty-five, labored in the fields and workshops of the colony's gentry, yeomen, and craftsmen. Other young, male servants, like most of the 180 the Massachusetts Bay Company sent to Salem in 1628, worked in the fishing industry, where they regularly defied notions of good order Puritan magistrates brought with them from England. Nearly a seventh of Anne Hutchinson's Antinomian followers during her conflict with Massachusetts authorities (1636–38) were servants, probably of her gentry and merchant supporters. When English emigration dried up after 1640, the servant supply dropped, and only the richest gentlemen could recruit them. In the 1650s Cromwell's government sent Irish and Scot war prisoners to New England to work as servants, and a few servants (including 165 from Bristol between 1654 and 1686) came to the region thereafter. These men failed to meet local demand, and most families had to rely exclusively on family labor.[68]

English immigrants migrated directly to Virginia, Plymouth, and Massachusetts Bay but reached other mainland colonies from adjacent colonies as well as England. Virginia immigrants moved within the Eastern Shore to Maryland and sailed up the Chesapeake Bay to Kent Island or to Potomac River settlements; others colonized North Carolina's Albemarle Sound region. Massachusetts immigrants founded colonies in Connecticut, New Haven, and Rhode Island. Groups of New Englanders settled Long Island and, later, East Jersey, and the first whites in South Carolina came equally from England and the West Indies. These colonies, however, were not just offshoots of earlier settlements. Religious dissenters fleeing banishment or religious persecution founded outlying communities. Others, like Connecticut, New Haven, and South Carolina, attracted young and unattached men from older colonies.[69]

English immigrants to Rhode Island, Connecticut, New Haven, and New Hampshire usually landed in Boston, stayed a few years in Massachusetts, and then moved on to the new colony. Only 23 of the 939 immigrants who left London in 1635, for instance, moved to Connecticut or New Hampshire on arrival, but within a few years another 164 left for Connecticut, New Haven, or Rhode Island. Settled families typically stayed in Massachusetts Bay, but the

New Haven and Connecticut migrants, most of whom were male, single, and young, resembled early Virginia immigrants more than the main body of immigrants to Massachusetts. New Haven's founders arrived in Massachusetts in 1637 in two large companies from London and East Anglia and founded a colony on Long Island Sound the next year. Men outnumbered women 3.5 to 1 among the first 450 people to reach New Haven; three-fifths of the men were still in their twenties. Similarly, half the men who founded Connecticut were unmarried, and two-fifths were under thirty. A crucial core of settlers migrated in families. Many founders of Windsor, Connecticut, had emigrated from a small area in western Dorset and Southwest Somerset and stopped in Dorchester, Massachusetts, before moving to Windsor. Orthodox Puritans, seeking fresh lands along the coast or on the Connecticut River, predominated among settlers of New Haven and Connecticut. Rhode Island's pioneers, however, were malcontents, such as Roger Williams or Anne Hutchinson and her followers, who had been banished from Massachusetts Bay. Richer than the Connecticut settlers, two-thirds of the seventy-eight Antinomians banished from Massachusetts to Rhode Island or New Hampshire were married, and many had children. Some failed to gain a foothold in their new home and moved on to Dutch lands on Long Island or New Jersey.[70] (See Map 1.)

The first English inhabitants reached South Carolina in the 1670s. Roughly equal numbers came from the British West Indies and from England; two-thirds were men, but a quarter migrated in family units. As sugar replaced tobacco on West Indian plantations after 1650, thousands of white men who lacked the land, slaves, and expensive equipment to make sugar left the islands. Rich Barbadians dominated the adventurers who tried to establish Carolina colonies in the 1660s, but during the 1670s most returned to England or migrated to New England or Virginia. Other West Indians—mostly freed servants, smaller planters, and artisans searching for greater opportunities—came to established Carolina settlements in the 1670s. The immigrants included some of the islands' richest gentlemen, who escaped the cramped conditions of the islands and found fresh acres for their children. Islanders had a great impact in South Carolina, where they dominated the new colony's political leadership. Merchants, gentlemen, and middling planters from the islands brought an equal number of servants and slaves with them. The West Indians were joined by English immigrants. Some younger sons of gentlemen, yeomen, and craftsmen were among these migrants, but close to three-fifths were indentured servants, nearly all men, purchased by wealthy planters.[71]

If the demography of early South Carolina resembled Virginia's, the initial immigrants focused sharply on commercial agriculture rather than the search for riches that had consumed the first white Virginians. They aimed at provisioning West Indian sugar plantations with meat and grain. Located closer to the islands than any other colony, they were well positioned to dominate that trade. But they needed labor. During the 1680s and 1690s, South Carolina planters attracted a small stream of English indentured servants, and a handful of Englishmen brought servants with them to Carolina; but improving conditions in England kept most of them at home. Planters quickly turned to West Indian and then African slaves to herd cattle, cultivate grain, and (later) make rice.[72]

Some 15,000 immigrants and migrants from other colonies came to the Delaware Valley (New Jersey and Pennsylvania) between 1670 and 1700; 8,000 arrived between 1681 and 1685, the peak of this migration. Despite William Penn's promotional efforts in the Netherlands and Germany, more than three-quarters came from England and Wales. Half left Quaker strongholds in the English Midlands, but more than a quarter came from other parts of England, including half of the nearly 600 rich "first purchasers" (investors), many of them from London and Bristol. Some were Quakers, but others sympathized with the movement and joined a Friends meeting after they arrived. Nonetheless, the diversity characteristic of eighteenth-century Pennsylvania was already apparent. There were Baptists, Anglicans, and Lutherans among the immigrants. About 800 Dutch and Scandinavians lived in the region in 1680, and close to a quarter of the late-seventeenth-century immigrants were German, Irish, or Dutch. English, Welsh, and German immigrants settled in their own communities, where they could maintain their language and culture.[73]

Immigrants to late-seventeenth-century Pennsylvania combined the servant-dominated movement to the Chesapeake and the family-based emigration to New England, as lists of immigrants from Philadelphia and Bucks County attest. Three of every five passengers came with kin; a third were children traveling with parents, a higher proportion than in the Winthrop fleet. But two-thirds of the adults were men, close to three-fifths of them indentured servants. Almost two-fifths of the women were servants. Most servants had probably been servants-in-husbandry or wage laborers, but others were skilled craftsmen brought over as salaried workers. The rest of the men were householders of modest fortunes, evenly divided between artisans and husbandmen, but a smattering of gentlemen and merchants emigrated, including half of the first purchasers, who quickly rose to the top of Philadelphia society. These immigrants were ideally situated to make farms. While wives and their female domestic servants made homes and started gardens, the men and their male servants cleared land; built houses, barns, and shops; planted grain; and made goods for sale.[74]

During the 1620s and 1630s the Dutch, Swedes, and Finns established small colonies in the Delaware Valley, but because Swedish and Dutch colonial joint stock companies were more interested in the fur trade than in farming, they sent few emigrants. The low population densities of these colonies enticed land-hungry English inhabitants of adjacent colonies. The Swedes set up outposts along the Delaware River in the 1630s and 1640s, but when the Dutch conquered the colony in 1655, just 300 Swedes and Finns lived there, along with 100 Dutch and English colonists. In the 1650s and 1660s, after large numbers of Dutch arrived to farm, the English demanded sovereignty and in 1664 conquered the region. After the conquest, English settlers enveloped the 800 inhabitants who lived in Dutch, Swedish, and Finnish enclaves. These colonists, mostly men, were survivors and descendants of 2,000 Walloon, Finish, Swedish, and English migrants who had arrived since 1624.

The far more prosperous and populous Dutch colony in New Netherland (New York) succumbed as well to English settlements and conquest. From 1624 to 1655 just 2,800 people came from the Netherlands, mostly male servants to the patroon (large landowner) Van Renselaer or to the Dutch West India Company. Since few Dutch wanted to leave their prosperous country, servants were paid a wage and served no more than four years, far better terms than those of English servants. Desperate for colonists to fend off Indians and secure land for the Dutch Republic, the company granted town sites with full privileges to settlers from New England. Recruiting throughout Europe between 1656 and 1664, the company sent nearly 3,000 people to New Netherland. Although half the immigrants came from the Netherlands and others had lived in adjacent regions, about two-fifths—mostly Protestants fleeing persecution—were French, German, or Scandinavian. Like New Englanders, over two-thirds of them migrated in family units, many with small children. They were ready to operate family farms but too weak to forestall the onslaught of English settlers. The Dutch left their cultural imprint on the Hudson River frontier, where they retained their majority, but by 1664, more than half of Long Island's people could trace New England origins.[75]

The patterns of immigration to North America had lasting consequences in the colonies. Migration streams did overlap. Indentured servants reached every colony, and at first they predominated everywhere but Massachusetts. Puritans migrated to Virginia and Maryland as well as to New England. Quakers could be found in New England and the Chesapeake colonies. But colonial promoters recruited from such disparate social classes and in such different parts of England and America that the demographic, social, and religious composition of each settlement was, in some ways, unique. With a higher proportion of family members, the first English settlers of the northern and mid-Atlantic colonies were well situated to begin farms and found communities. Lacking a product in high demand in Britain, however, they had to rely on family labor. The men, who predominated among immigrants to the southern colonies, set up households with difficulty, and international demand for tobacco kept servants coming in, perpetuating the sexual imbalance. The southern colonies had to incorporate proportionately much larger numbers of footloose young men and devise institutions such as orphans' courts to regulate precarious households.[76]

From Emigrant to Immigrant

What did immigrants expect once they arrived in North America? The question is difficult to answer, for rich men wrote nearly all surviving documents. A lay sermon by John Winthrop or a pamphlet by John Smith reveals more about the expectations of colonial promoters than about the aspirations of the immigrants they led. Yet the success of promoters suggests that their appeals were popular and that what they wrote can provide clues to popular consciousness. When one adds the written word to immigration patterns and to what we know of the life experiences of emigrants before they left, what immigrants anticipated after they arrived becomes clear.

Nearly all immigrants expected access to the land colonial promoters promised. Told repeatedly that America was empty and that land was waiting to be farmed, they wanted their share. Laborers signed indentures in the hope that once they completed their terms, they could leave the ranks of debased laborers and become landholders. Middling families, fearful that they or their children would lose the land they had in England, expected to build landed fortunes in America. The system of land tenure they would enjoy was probably not clear to them, but they likely did not expect to own land outright. Private property in land (with the absolute right of alienation that went with it) was new in seventeenth-century England. All male migrants did expect some kind of secure land tenure with rent and taxes low enough to allow them continued occupation of the land. Propertied men expected to hold and cultivate more land more quickly than servants, but land hunger gripped everyone. Knowing that poor Englishmen had not even a cottage, in 1610 the Virginia Company expected to lure servants with the promise of "day wages for the laborer, and fir his more content, a house and garden plot!" In this the company was mightily mistaken and soon had to promise more.[77]

Immigrants expected to marry and create the independent households so many had lost in a century of land expropriation. After the disasters at Roanoke and Jamestown, few anticipated great profits, and those who did soon left. By the 1620s or 1630s the ridicule that anticolonial ballads, plays, and pamphlets had heaped on the search for gold and silver had permeated London and probably much of the realm. Even after constant disaster and disappointment, however, a few men still craved gold. Sebastian Brandt wrote a friend from Virginia in early 1622 that as soon as he recovered from illness, he expected to search "up and downe the hills and dales," for "good Mineralls here is both golde, silver, and copper." Nearly all migrants anticipated hard work. Still, men expected to feed and clothe their families, and their wives planned to rear children, cultivate gardens, and make clothes.[78]

Immigrants expected to farm, perhaps as servants or wage laborers for a while, but on their own farms eventually. Most newcomers had experience as servants-in-husbandry or day laborers, but few had been full-time farmers or farm laborers. Some had been born in London; others had worked for years in London, Bristol, or some other town. Many village emigrants, moreover, had labored in the textile trade or spent much of their time in craft labor. Artisans and urban folk among immigrants anticipated that they would return to the household patterns and work regime of their peasant ancestors. The London peddler understood he could not hawk goods on a city street, and the Essex weaver did not expect to find extensive clothing manufacturing; both knew they would have to learn farming skills or relearn old ones long forgotten. Those who attempted to make cloth or tan hides soon discovered their folly and turned to farming.[79]

As Chapter 2 will relate, immigrants were often surprised, even shocked, when they landed. The woods were thicker and living conditions more uninviting than they had imagined. They discovered Indians, whose demise had been reported to them, still on the land, and were forced to conquer land they believed should be theirs. The hot summers and cold winters amazed a people used to a temperate island climate, and their surprise was sharpened by the deliberately false reports of colonial promoters. At first, before the truth began to seep back to England, many may have been surprised at the illness, incapacity, and death that followed immigrants on the islands and the southern mainland.

chapter two

This Newfound Land

Bitter cold and scorching heat, unbearable humidity and searing drought, thick forests and raging rivers, the bloody flux and scurvy, and hostile Indians and wild animals greeted the first English immigrants. In 1623 Virginia immigrant Richard Frethorne reported that "there is nothing to be gotten here but sickness and death." Indians killed colonists and destroyed corn. Frethorne and his compatriots lived in "fear of the enemy every hour," danger made worse by the colonists' infirmity. Unable to hunt forest fowl and never seeing "deer or venison," they would surely starve without English supplies. Nor was early New England any better. John Winthrop reported "bitter frost and cold" in December 1630, frozen rivers in January 1631, and "many snows and sharp frost" in late February. Settler John Pond, who lived through the same cold, lamented to his parents in March 1631 that the "rocky and hilly" terrain was good for nothing unless fertilized, and complained that the Indians were "a crafty people" who would "cozen and cheat." Timber, fish, and game abounded; but the settlers had few boats, and "wild fowl" were "hard to come by." No wonder supplies were scarce and the colonists "died of the scurvy and of the burning fever."[1]

The society Frethorne and Pond expected did not magically arrive with the first ships. Immigrants landed in what they saw as a void, a world with neither towns nor villages, houses nor barns, churches nor courts, and cows nor pigs. They had to bring everything with them. What they left behind they had to make themselves, get from Indians, or do without. This void was especially apparent to the first settlers, who came to Virginia before 1630, to Plymouth in the 1620s, and to Massachusetts Bay and Maryland in the 1630s. As families built homes and made farms, a landscape resembling England's took shape. Their material lives improved, but their homes and possessions lagged behind those of similar classes in England. Although later immigrants and whites who moved to frontier lands bought surplus food from earlier arrivals, they shared, to a degree, the poverty of the first immigrants.[2]

Settlers accepted such primitive conditions more or less willingly because they craved land. Land meant everything to immigrants (and to their children). It was the bedrock of their prosperity. The basis of their economic independence in England, landed property symbolized their political and class position; it sustained their cultural and spiritual life. The original inhabitants used and understood land quite differently, but they, too, used it to gain an independent subsistence and made the land (and its flora and fauna) the font of their spiritual lives. No wonder colonist and Indian contended so fiercely over control of the land.

However wild the land might have seemed to immigrants, colonists knew that it could be improved. Land, whether in forests or in meadows the Indians had cleared, was the raw material that would sustain them. Land hunger consumed immigrants, who realized that without it they could neither marry nor begin households. But first the English had to acquire land by buying it, stealing it, or conquering the Indians to get it. Once they controlled the land, they made their living from it by growing food for subsistence and producing surpluses to sell. They changed the face of their land, clearing forests, cultivating fields, and building houses and barns.

The original inhabitants, from the perspective of the settlers, were part of the new environment they faced. Ordinary colonists' relations with Indians were ambiguous. Indians knew far better than the interlopers how to live in the forest. Understanding how essential Indian trade, agriculture, and trails were to their survival, settlers traded manufactures for furs and food. But Indian villagers occupied, cultivated, and hunted on the land settlers wanted, and they refused to relinquish sovereignty over it for any price. Bloody wars between settlers and Indians invariably ended in conquest of Indian lands. English leaders, reflecting the ambitions of settlers, justified conquest by arguing that settlers' "civil" rights to farmland superseded the "natural" rights of Indians to hunt game and gather food—claims colonists, who had relied on the Indians' corn, knew to be false.[3]

The settlers' encounters with nature and with the Indians played a critical role in the kinds of farms they made. Climate and soil determined what crops would grow, but as settlements expanded, colonists attempted to shape the physical environment. Although their first contacts with nature and the Indians led to disaster, disease, and death, they slowly learned to live with the climate and endemic diseases. After they conquered the Indians, they sent them into exile and expropriated their villages. As they made farms, colonists so transformed the landscape around them that they believed that their actions had remade nature, turning the soil more fertile and the climate more temperate. The beliefs about nature that colonists brought with them structured how they interacted with the environment, but the environment, in turn, transformed the meaning of English customs and habits.[4]

This Land Is Not Ours

Immigrants had experienced nature (weather, soil, topography, forests, and animals), habitat (fields, pastures, woods, houses, and highways), and order (farms, villages, and cities) far differently than they would in America. England's climate was temperate; its soil, fertile. The English lived in a crowded landscape full of wheat fields, market gardens, sheep pastures, villages, mills, and houses. At the end of the seventeenth century, cultivated land, pastures, and meadows covered over half of England and Wales and far more of the East Anglian, West Country, and London regions that sent nearly all early emigrants to America. Much land was still divided into strips of open fields, creating a pattern of cultivated land as far as the eye could see. But even enclosed areas, with their hedgerows and fences, were packed with sheep or wheat and studded with farmhouses. Villagers saw the invisible boundaries that separated fields and manors; they knew where their land ended and their neighbors' began. They acknowledged the growing pervasiveness of landed private property that enclosures represented and that the rising number of estate plans and county maps symbolized. Most rural folk in open-field areas lived in compact villages within ten miles of one of southern England's 600 market towns. Highways capable of handling wagons in good weather linked village to market town, market town to regional center, and all of England to London.[5]

The North American landscape shocked colonists, especially those who had read glowing reports in the pamphlets of colonial promoters or listened to their enthusiastic agents. Distinguishing between proper and improper landscapes, they knew America was disordered because it was so different from England. The climate was too cold and too hot; the land was too empty, but the Indians were too hostile; the forests were too deep, but the clearings were too full of stumps. What colonists saw was so alien that they had no words to describe it. They could see forests, creeks, inlets, and hills but could make little sense of the whole. New Englanders who survived into the 1640s came to believe that they had lived in a "desart wildernesse," a "scant wilderness in respect to the English," "an unsubdued wilderness yielding little food," or "a hideous and desolate wilderness, full of wild beasts and wild men." The land they invaded was not a wilderness, if that term meant an uninhabited, uncultivated place, yet they knew it was wild in comparison to England.[6]

As late as the 1690s, two-fifths of England was covered by "Woods & Coppices," "Forests Parks & Commons," and "Heaths Moors Mountains & barren Land." But less than a sixth of England's acres were forested, and most emigrants had not seen these forests. Many heaths, mountains, and forests were located in the north of England, far from the center of emigration. In southern England, home of most early emigrants, forests had been cut down and over half the villages had no woods. Royal parks, with as much pasture as forest and as many decayed as healthy trees, and heavily inhabited wood-pasture regions, where woodland industry sustained cottagers and squatters, contained most of southern England's woodlands. By the mid-seventeenth century, so much timber had been cut down for iron making, heating, house building, and furniture that England faced a severe wood shortage and turned to imports or substituted coal for wood. Deep forests had disappeared, swamps had been drained, and thousands of cottagers had moved into fens and woodlands, forever changing the landscape.[7]

The vast forests of eastern North America dwarfed anything immigrants had seen. Between 80 and 95 percent of the land along the eastern seaboard was covered by woods, and the trees were much taller (nearly 100 feet) than those in England. Colonists found the land "wilde and overgrowne with woods." Forests comprised 97 percent of Virginia's territory, 95 percent of Maryland and Connecticut, 90 percent of Massachusetts, and 87 percent of New Jersey and South Carolina.[8] Swamps (some forested) and sandy beaches covered part of the rest, especially on the coast. A few sandy pine barrens or grass prairies covering thousands of acres encompassed a small part of the land. Colonists avoided these areas, fearing that they were not habitable. The woods alternated between thick, impenetrable masses of trees, foliage, and forest and parklike areas of well-spaced trees crossed by many paths. The more open woodlands were located near the coast or in hunting grounds adjacent to Indian villages; thicker forests could be found in northern and western regions, where few settlers ventured.

The only completely open acres were those Indians had cleared for their villages and cornfields. As thousands of Indians died in the early seventeenth century from European diseases, survivors abandoned many villages. Colonists gravitated to lands Indians had cleared. "The ground" at Plymouth, Edward Winslow wrote in 1621, "is very good on both sides, it being for the most part cleared. Thousands of men," he added, "have lived there, which died in a great plague not long since." Settlers had a moral obligation to reseat the land, he concluded: "Pity it was and is to see so many goodly fields, and so well seated, without men to dress and manure the same." When those lands ran short, they bought Indian land or conquered villages still inhabited by Indians. A group of mid-seventeenth-century settlers along the Potomac River, for instance, bought land from the Chicacoan (an Algonquian group) and made their first plantations at their village sites, on fields the Indians had already cleared.[9]

Hundreds of natural harbors, great bays, wide rivers, and large lakes, as well as thousands of small streams and creeks, drained into the Atlantic. Forty-eight rivers alone drained into the Chesapeake Bay, the greatest estuary on the east coast. This network of waterways expedited Europeans' exploration of the interior (after they learned how to use Indian canoes to navigate narrow streams) and later fostered international and intercolonial trade. At the same time, however, it created barriers almost as great as the forests. The widest rivers—the James in Virginia, the Potomac in Maryland and Virginia, the Susquehanna in Maryland and Pennsylvania, the Hudson in New Netherland, and the Connecticut in New England—were so difficult to cross in the seventeenth century, an era before ferries and bridges, that they served as borders between colonies, counties, or towns.[10]

Rivers and forests teemed with wondrous fish and wild animals. Some were sources of food or fur (beaver, elk, and deer), others were predators (crows, foxes, bears, and gray wolves), and still others seemed monstrous (two-headed snakes or the female opossum, with her two vaginas and two uteruses). Colonial promoters had listed many of these creatures as commercial assets, along with exotic vegetation and forest products. John Smith saw huge cod, salmon, and sturgeon and a multitude of shellfish off New England's coast; early settlers everywhere hunted the small game—duck, geese, and turkey—that abounded in the woods. Sixty million beaver and 24–34 million deer roamed North America's woods on the eve of colonization. Beaver and deer skins, hunted (and overhunted) by Indians, of course, quickly became a staple of trade. Settlers hunted predators and the food and fur animals Indians left behind, paying no attention to poorly enforced game laws.[11]

If the land had been as desolate of human society as colonial promoters imagined, all the colonists would have died, for they had no idea how to survive in the forests. Hundreds of Indian settlements interrupted the woodlands of early-seventeenth-century coastal America, even after epidemics had killed so many Indians. Late-sixteenth-century Roanoke adventurers found twenty-seven villages near their settlement. When the first colonists arrived in Virginia, there were about 150 villages associated with the Powhatan confederacy, and 40 more could be found in Maryland at the time of that colony's founding. Indian habitation was nearly as thick in New England. The Massachusetts coast and Cape Cod each contained 27 Indian villages; early English immigrants along the Connecticut River settled among 8 to 13 villages.[12]

Far from a wilderness, the first colonists found a densely occupied and exploited forest sufficient to feed and clothe a big population. Long before the

English arrived, Indians had remade the landscape to fit their needs, burning forests, growing corn, and clearing paths. Everywhere the English went, from southern New England to the Carolinas, Indians had burned the forest floor near their villages to make travel easier, facilitate hunting, provide forage for animals, allow wild berries or fruit trees to grow, or destroy older trees to open meadows where they could plant corn. They burned their fields annually, but they fired the surrounding forest every three to ten years. Between fires, women collected twigs and cut small trees along the paths for firewood; men cut larger trees to make canoes.[13] Indian villages filled with semipermanent houses took up considerable space. Indian women cultivated corn outside their villages (in hills to protect the plants), alternating corn with beans and squash and burning the fields to maintain fertility. Some of these fields contained 100 or more acres. After crop yields dropped, Indians cleared new acreage, and their old cornfields slowly returned to forest. They created a network of paths that covered the coastal plain and piedmont and connected village to hunting ground, village to village, tribe to tribe, creek to river, Powhatans to the New York Iroquois, the Carolina Cherokee to the Powhatans, and northeastern Indians to those in the Great Lakes and Gulf regions. Winding along or between streams or around hills or connecting well-trod navigable waterways, these paths—only a foot or two wide—allowed Indians to walk through the woods single file, going long distances to visit, forage, hunt, trade, or wage war. English settlers used and widened these paths.[14]

The colonists nonetheless insisted that the forests were wild and uninhabited and identified them with the "wild" Indians who lived there. The woods contained no English-style towns, their churches and marketplaces emblems of civility. During the hunting season, Indian villagers moved to hunting grounds, deserting the cornfields and leaving the appearance of an unoccupied land. In the face of epidemics, Indians had abandoned many villages, leaving only old meadows and corn hills as evidence of their residence, and those became overgrown with weeds, bushes, and pine trees. Whites knew that Indians lived in the woods, some within a few hundred yards of their farms. But few colonists except traders or officials visited Indian villages, and those who did were not reassured. Indian villages differed markedly from English streets lined with wood or stone houses or the two-room cottages of the poor. Indians often lived scattered over the landscape, not in nucleated villages; cornfields sometimes surrounded their villages, but Indian women cultivated other areas far from the village. Indian round houses or wigwams, 13–16 feet wide by 18–20 feet long, were temporary sapling structures supported by poles and covered with bark or mats; the houses rotted after three to five years and had to be rebuilt. A central fire or fires kept the house warm, and several families slept on bunks that ran the length of the structure. Similarly constructed long houses, 40–400 feet long and 20–40 feet wide, housed an entire matrilineal clan. Indian villages, in English eyes, were unfit for permanent habitation, for they lacked the amenities—chimneys, family privacy, and furniture—even the poorest English family expected.[15]

While Indians saw themselves as part of nature, the English wanted to conquer and exploit it. Like Indians, settlers girded trees to kill them and planted crops around the dead trunks. Since they lived in more heavily populated settlements and used more wood than Indians, colonists cut down forests at a greater pace, using trees to build houses, make farms, and produce marketable surpluses. As soon as colonists occupied a place, they widened Indian paths to accommodate horses and carts, chopping down acres of trees for each mile of road they built. To make fields for crops, fences to protect crops from animals, pastures for livestock, and firewood for themselves, farmers clear-cut two or three acres each year, far more than an Indian family would destroy. Farmers either exploited their own woods or took from the English-style commons, communal woodlots, or swampy wastelands New England villages established. At first colonists lived in Indian-style wigwams (in Maryland and Plymouth) or dug caves for homes (in Pennsylvania), but they soon built small clapboard or wooden houses. As the colonies prospered, middling and wealthy farmers replaced their crude dwellings with larger houses and built barns for cattle and hogs, using even more wood. Colonists heated their homes in inefficient fireplaces, each requiring fifteen to twenty cords of wood a year. As the immigrant population grew, farmers cleared more land and created more and larger open and cultivated spaces.[16]

Colonial farmers harvested trees for commercial purposes as well. By the last quarter of the seventeenth century, settlers were using idle time during winter to make shingles from timber and supply firewood to towns, such as Boston and New York, that ran low on wood for housing and heating. Wharves, churches, and courthouses also used timber bought from area farmers. Forests not only supplied lumber; they were sources for potash and lye, needed in the textile and glass industries, and pitch, tar, and turpentine, used in shipbuilding. On occasion, northern farmers made potash and lye from wood ashes and sometimes collected them from fires farmers set to clear land for crops. Chesapeake and especially Carolina farmers and their slaves made pitch and tar from local forests in the off-season. Forest industries grew in the eighteenth century and spread to Georgia. The seventeenth-century timber industry, which supplied masts and lumber to English shipbuilders, lumber products to the West Indies, and pipe staves for Madeira wine makers, was centered in northern New England. Although herdsmen and farmers participated, merchants and sawmill operators organized it. Some timber was cut everywhere, however, and farmers may well have delivered raw timber to local sawmills, often built soon after Europeans arrived.[17]

We should not exaggerate this destruction of forests. Seventeenth-century farmers consumed no more than 650,000 acres for firewood and a smaller amount for other uses. Even in 1760, no more than 5 to 10 million acres had been cut down. Mid- to late-seventeenth-century regulations on cutting timber in a few New England towns suggest that some trees had been unnecessarily destroyed. But the practice of setting swine and cattle to forage in the woods continued everywhere, and forests remained, even on the coast. In 1700, 70 to 80 percent of Connecticut and Massachusetts remained in forests, as did half of Rhode Island in 1767 and two-fifths of central Delaware in 1797. Farmers in Concord, Massachusetts, cut only 0.4 percent of their timber each year. Abandoned Indian land often returned to forests before white habitation began; farmers—at least in the South—also abandoned land after intensive use and allowed it to lie fallow for twenty years and become overgrown with pine trees. Most farmers would have agreed with Hugh Jones, a Maryland clergyman, who complained in 1699 that "tho we are pretty closely seated, yett we cannot see our next neighbours house for trees." He expected otherwise "in a few years" because tobacco "destroyes abundance of timber." But twenty-five years later another Hugh Jones reported that in Virginia "the whole Country is perfect Forest, except where the Woods are cleared for Plantations, and Old Fields, and where there have been formerly Indian Towns." No serious local shortages of timber or firewood appeared in the colonies until the mid-eighteenth century, and then only in the northern cities, southern towns, and the oldest settled regions, like the Tidewater Chesapeake.[18]

The colonial climate, with its bitterly cold winters and feverishly hot summers, shocked immigrants as much as did the vast forests. In contrast, England's climate, a visitor wrote in 1598, was "most temperate at all times," and other observers commented that its climate was milder than that of France or Italy, cool in winter and summer. Although farmers in the hills of Scotland and northern England abandoned some land during the little ice age of the sixteenth and seventeenth centuries, moderation persisted in southern England. Temperatures did drop about 1°C in winter and ½°C in summer from the medieval era. Nonetheless, during the early 1660s, January temperatures in central England stayed consistently above 0°C, averaging 3°C (37°F), and July temperatures were cool, averaging 15–16°C (59–61°F)—similar to temperatures earlier in the century. Rain fell regularly throughout the year, and the country experienced one or two weeks of snowy days each year; but floods and droughts were both infrequent.[19]

England's pleasant climate set high expectations for weather in the colonies.

A moderate climate like England's brought immense benefits, nourishing "mechanicall and politike Arts" and sustaining health and productivity. Believing that climate was invariable across the same latitude, colonial promoters thought that New England, located south of London, should have enjoyed a warmer climate, milder in winter and warmer in summer than England's. They feared the heat of the southern and island colonies yet expected that the Chesapeake colonies, lying parallel to Spain and Italy, would produce tropical crops without suffering extreme heat. If the winters were mild, then ice would never clog rivers. The climate rarely prevented a bountiful harvest in England, but a longer growing season in the colonies would ensure plenty for everyone—and a profusion of exotic produce. Virginia settlers, acting on this climatic theory, tried (but failed) to make decent wine and silk.[20]

Any prospective emigrant who believed promoters' accounts of colonial climate would have been shocked. Regulated by weather systems that flowed over land from west to east, the North American east coast had hotter summers and colder winters than Britain's. During the first half of the century all of eastern North America suffered from the little ice age. Winters were 1°C colder than in the twentieth century, and winter snow may have been greater. Summers, in contrast, resembled those of the twentieth century. Winter cold persisted during the second half of the century in the Northeast and intensified in the 1680s and 1690s, but it moderated in the South, where winter temperatures averaged just ½°C less than in the twentieth century. In the Philadelphia region, halfway between northern and southern settlements, winter temperatures (December–February) averaged about 2°C (36°F), compared with 4°C (39°F) in central England. Philadelphia summers (June–August) averaged about 21°C (70°F), compared with 15°C (59°F) in central England. Such climatic differences required substantial adjustments by English immigrants.[21]

Extremes of temperature and weather frightened the immigrants, who knew that either cold or heat, drought or flood, could be dangerous. Storms in America, moreover, were more violent and destructive than any immigrants had known. Newfoundland's climate, George Calvert discovered, had "ayre so intolderable colde" that it turned into a "hospitall" for his men, leading him to abandon the place for supposedly more temperate Maryland. Droughts may have been as infrequent as in England, but Roanoke settlers arrived during the most severe two-year drought (1587–88) in 800 years, and the first Jamestown settlers immigrated in the midst of the harshest seven-year drought in 770 years (1606–12). New droughts hit the Chesapeake region in 1615–17 and 1632. High temperatures and humidity accompanied the droughts; crops withered, and ill-prepared men died in droves from heat and malnutrition. Wet years followed drought, and as far south as Virginia, winter snow could be much deeper than England's. Similarly, annual spring rains in South Carolina and Georgia ranged from 200 to 700 millimeters (8 to 28 inches) during the early and mid-seventeenth century.[22]

Immigrants soon recognized that the climate of colonies varied markedly. As one traveled from north to south, winter temperatures and length of growing season both rose. New England winters and springs were so bitterly cold in the 1630s and 1640s that Massachusetts Bay on occasion froze over. Such cold winters—accompanied by heavy snow—continued with regularity through the end of the century. January temperatures rarely rose above freezing, averaging −1.4°C (30°F). Snow stayed on the ground for two months in the mildest winters; in the worst years, cold and snow reduced personal mobility, retarded planting until April, and decreased the growing season to 160 days, 50 fewer than in southern England. As Edward Johnson lamented in the 1640s in the satiric ballad "New England's Annoyances," "from the end of November till three months are gone, the ground is all frozen as hard as a stone, . . . our mountains and hills and vallies below, being commonly covered with ice and with snow." July temperatures averaged 18°C (65°F), cool by continental American standards but sufficiently warmer than England's to provoke warnings about the heat.

Summers were hotter and winters milder in the Chesapeake colonies than in New England. The Chesapeake rarely had as many as "thirty days of unpleasant weather," but severe cold occasionally struck the region, leaving mounds of snow and frozen rivers. January temperatures in milder years averaged as high as 4°C (40°F), warmer than central England, but with greater variations. Great heat, high humidity, and the strong sun provoked fear of the deleterious effects of summer farmwork. The Chesapeake region was much hotter than England, with average July temperatures reaching 26°C (79°F). With its cold winters and cool springs, the growing season in the Chesapeake region averaged 180–200 days, similar to England's and insufficient for tropical plants. Coastal South Carolina, in contrast, enjoyed warm Januaries with temperatures averaging 7°C (45°F) but suffered tropical Julys, with high temperatures of 33 to 38°C (90 to 101°F) and average temperatures of 27°C (81°F). Mild winters increased the coastal South Carolina growing season to about 280–300 days, sufficient for experiments with tropical crops.[23]

Settlers slowly adjusted to North America's climate. The first immigrants to reach Virginia, Plymouth, and (to a lesser degree) Massachusetts failed to plant crops in time to harvest or build weather-resistant houses. The starving times, with their high death rates, followed. At first, colonists subsisted on corn and other American crops or learned to eat "strange beasts." As soon as they could, they cultivated European crops, gradually learning what would grow and when to plant. Adjustments to cold weather in New England came easily. Families survived the winter by staying indoors, building weather-tight houses, burning

"Christmas fires all winter," and dressing more warmly than they had in England. Crossing frozen rivers by sled could speed travel. Newly arrived immigrants to the South avoided intense work in the summer heat and humidity. Bathing, the use of herbs (guided by Indians), and the construction of open, airy houses mitigated the effects of heat. Cool clothing helped, but some southern immigrants, Robert Beverley of Virginia bitingly wrote in 1705, "wisely go sweltering about in their thick Cloaths all the Summer, because they used to do so in their *Northern Climate*," much to their discomfort and ill health. Destruction of the forests and cultivation of crops, settlers believed, would moderate the climate by allowing the sun's rays greater access to the land, rendering further adaptations to cold and heat unnecessary. The mild climate of the mid-seventeenth century seemed to prove their point. But the bitterly cold New England weather at the end of the century reduced crop yields and disappointed Puritans, who could explain the reversal only in terms of their sinfulness.[24]

Epidemics and death intensified the strangeness of North America. Although life expectancy in England fell throughout the late sixteenth and seventeenth centuries, most seventeenth-century immigrants had led healthy lives. Famine had vanished from southern England, though it persisted in the north. Even when epidemics or dearth hit an area, births exceeded deaths in England as a whole. About 133 of every 1,000 infants died before they reached their first birthday—high by twentieth-century standards but lower than the rate of most colonial populations. Those who survived to age twenty lived another thirty-seven years. Nonetheless, many potential emigrants, who lived in less healthy parts of England, may have been willing to risk debility and death in the colonies, believing it no worse than their current situation. The poor, who dominated immigration to the Chesapeake and the West Indies, had suffered greater exposure to disease than the middling classes. Londoners (and to a lesser degree persons who lived in low-lying areas) suffered from the plague, smallpox, and fevers. Malaria, which weakened resistance to disease, was prevalent in badly drained marshy areas. Cold winters increased the severity of respiratory diseases; hot summers led to more waterborne ailments.[25]

Nothing prepared immigrants for life in the colonies. Weakened from the ocean voyage and exposed to disease in the close quarters of the ship, immigrants faced illness as soon as they arrived. Unable to produce or trade for enough food, the first colonists at Jamestown, Plymouth, and Boston suffered from malnutrition that reduced their resistance to disease. A high proportion—more than half in Plymouth and Virginia and an eighth in Massachusetts—of the first settlers died from dysentery, typhoid fever, influenza, or other ailments before they made their own food and became acclimated to the new disease environment of America. High mortality levels disappeared within several years in New England but persisted in Virginia. Sixty-nine of the first 104 English immigrants to land at Jamestown in 1607 died. Newly arrived Jamestown immigrants continued to die at high rates throughout the 1620s, probably from dysentery and typhoid found in the brackish water they drank. As immigrants moved away from brackish water the death rate declined, but during the 1630s one in seven still died from disease.[26]

Even after Virginians learned to avoid brackish water, disease and mortality remained a great threat. During the second half of the century every immigrant underwent a process of seasoning that lasted one or two summers. Colonists suffered from malaria brought to the region by infected immigrants who passed the disease on to mosquitoes that infected other immigrants. Low-lying swampy areas were the unhealthiest, but malaria spread elsewhere as well. Victims slowly built immunity to the sweating, chills, and fevers that accompanied malaria, but a more virulent strain of the disease was brought to the region from Africa in the 1680s, causing greater distress and mortality. Even persons who built up immunity to malaria never fully recovered and were therefore more susceptible to other diseases. Mortality, as a result, was much higher among Chesapeake immigrants than those who stayed in England. One- to two-fifths of immigrants died during their seasoning. Survivors lived only another twenty years, into their early forties. Women, who had lowered immunities during pregnancy, died even younger. The disease environment in coastal South Carolina was worse. The hotter and more humid climate, combined with the propensity of white immigrants to settle on swampy, mosquito-infested land, sustained a malaria pandemic more severe than that in the Chesapeake. Here, too, malaria lowered resistance to other diseases and led to higher infant mortality and perhaps lower adult life expectancy than in Virginia.[27]

New Englanders suffered much less from disease. Once they learned to feed themselves, death rates plummeted. Cold winters, cool summers, dispersed farms, and well-drained land kept villagers healthy through the seventeenth century. To be sure, occasional smallpox, diphtheria, and influenza epidemics and even malaria killed many New Englanders. But infant mortality in seventeenth-century New England resembled England's, while adult life expectancy was, by English standards, astonishingly high in farming villages. Men at age 20 could expect to live another 45 to 48 years; women, 42 to 46 years. Even in unhealthy Salem, men at age 20 lived another 36 years. As population density increased, the rural populace became more susceptible to epidemics; the throat distemper of the 1730s, in particular, probably increased infant mortality and reduced adult life expectancy. New England, however, remained far healthier than the southern colonies.[28]

In every place they conquered and settled, colonists reshaped the landscape to make it habitable for Europeans. They cut down forests, built homes and barns, and constructed churches and courthouses. They brought swine, sheep, and cattle and turned them loose to forage in the forest, where they multiplied rapidly, displacing wild game. To regain the taste of home, settlers planted European grains and vegetables. In so doing, they tried to manage nature and to create an artificial environment. They sharply distinguished the world they made from that of the Indians among whom they lived. They cleared land for farms and "built fair towns of the land's own materials." But beyond European settlements lived the Indians and colonists could not make a European landscape until they had vanquished that foe.[29]

This Land Should Not Be Theirs

Friday, March 22, 1622, dawned as any other day in the infant Virginia colony. "As in other dayes," Edward Waterhouse reported, Indians "came unarmed into our houses, without Bowes or arrowes, or other weapons, with Deere, Turkies, Fish, Furres, and other provisions to sell, and trucke with us, for glasse, beades and other trifles." Some "sate down at Breakfast with our people at their tables." This peaceful scene soon turned murderous. Without provocation, the Indians took up the "tooles and weapons" of their hosts and "basely and barbarously murthered" them, "not sparing either age or sexe, man, woman, or childe." They found other settlers in the fields, "planting Corne and Tobacco, some in gardening, some in making Bricke, building, sawing, and other kinds of husbandry," knowing well where "each of our men were, in regard of their daily familiarity, and resort to us for trading."[30]

Friendship and hate, trade and war permeated relations between Virginians and Powhatans. Unfamiliar with how to survive in the woods and too few to conquer Indian lands, immigrants like the Virginians had to cooperate with Indians. At first the Powhatans controlled communication, demanding that it be on their terms using their language and adopting their forms of gift exchange. As the white population grew and Indians died in epidemics or war, whites forced Indians to adopt their ways of doing business. Settlers came to see them as impediments to their control over land. Despite the benefits of trade, this struggle over land set immigrants and Indians against one another. The accommodation of the first years of settlement broke down into acrimony and warfare, and by the end of the century, settlers and Indians had become strangers, separated by hundreds of miles.[31]

The first adventurers—at Roanoke, Jamestown, and northern New England—hungered for gold or fish. Unwilling to plant corn, they relied on Indians for their bread, bartering trade goods, sometimes at the point of a gun. Indians made but small surpluses, and conflict was inevitable, especially in the spring and summer months, before crops matured. At these times Indians survived on fish, shellfish, dried nuts, and tuckahoe (a tuber pounded into flour). Roanoke Indians, mindful of their rules of hospitality, planted corn for the intruders, but English demands soon exceeded these meager provisions. When their provisions were exhausted in the following spring, the Roanokes deserted the settlement, leaving the English hungry.[32]

The Powhatans may have made corn surpluses as small as those of the Roanokes, but in addition they had to give as much as four-fifths of their corn as tribute to Powhatan. He redistributed some of this food as gifts, but the Powhatans lived at the edge of subsistence each spring. To gain the allegiance of the English, Powhatan gave them corn from his surplus, much as he would any tributary, but the English dismissed his leadership and refused to give him the guns he wanted. At first, adventurers and Indians feverishly traded corn for goods and furs for exotic English foods. But Indians, who spent much time getting alternative food, could not make more corn for the English. When hungry Indians refused to feed Jamestown settlers in 1609–10, Virginians raided their villages and extorted corn from them; when that failed to turn up enough corn, settlers turned to cannibalism. To avoid such problems, the Virginia Company urged colonists to grow their own corn, their demands growing more insistent after Indians nearly destroyed the colony in 1622. But as John Smith related two years later, the planters still neglected corn to make money growing tobacco. Eventually settlers learned to feed themselves, and relations with Indians turned on issues of land.[33]

The first settlers lived in a sea of Indians, and their widely scattered farms were easy targets. In 1613 Virginians farmed tiny settlements at the mouth of the James River, at Jamestown, and on the upper James near present-day Petersburg. By 1622 their settlements covered both banks of the James River but hugged the shore. A decade later they spread to the southern end of the peninsula between the James and the York Rivers, and the southern Eastern Shore as well. Similarly small colonies could be found in New England in the 1620s and 1630s, where vast forests separated the Plymouth, Massachusetts, and Connecticut settlements.[34]

Far more Indians than Europeans inhabited the east coast during the early seventeenth century. Before contact, two-fifths of eastern Indians had died in infancy or childhood, but those alive at age twenty lived another two decades, ensuring a low level of natural increase. Many Indians had died from European microbes brought by explorers and traders. But epidemics skipped communities distant from coasts or rivers, and thousands of Indians remained after contact. A half-million Indians lived east of the Mississippi River, most far from the first colonists—at least 25,000 in Virginia and Maryland, 6,000 on Long Island, 12,000 in the Hudson Valley and adjacent Jersey, 11,000–12,000 in the Delaware Valley, and 60,000 in southeastern New England. As soon as settlers reached an area, Indians succumbed to the diseases they brought (more than four-fifths of the Pequots, for instance, died in an epidemic in 1616–19). Indian populations recovered between epidemics, however, leaving many alive to confront the invaders. Some 1,500 Powhatan warriors faced a few hundred English of all ages (and their 600 Indian warrior allies) during the early 1610s; in 1624, when Virginia's white population had reached 2,000, Indians amassed 800 soldiers. In the early 1640s, after smallpox killed thousands of southern New England Indians, 30,000 remained. As late as 1676 New England Indians mobilized several thousand warriors. Settlers quickly learned that a larger population guaranteed neither easy conquest nor safety for outlying settlements. Only at the end of the century did Virginia and Massachusetts achieve the overwhelming superiority supremacy required.[35]

Although they could not speak one another's language, Indians and English and Dutch immigrants communicated using gestures (whites pointed to animal skins, Indians to cloth to exchange goods), facial expressions, or signs. But gestures could easily be misinterpreted. When an Englishman who had come to negotiate with the Pequots in 1636 entered a wigwam, seeking horses, all those inside immediately left except the Sachem's wife. She "made signs he should be gone," but her gestures were misinterpreted as hostile signs that the Pequots intended to cut off the Englishman's head. Soon whites and Indians learned a few words of one another's language, or enough of a third language (such as French), to communicate. When John Dane, on the road in Massachusetts from Roxbury to Ipswich in the mid-1630s, encountered "forty or fifty indiens," he warily greeted them with "What chere," and they—apparently knowing a bit of English—"with a lug voise, laughing," "cryd out What chere, What chere." In Charles County, Maryland, indentured servant Elizabeth Brumley could speak a bit of an Algonquian language. When four Indians (two of whom she knew) came to her mistress's door in 1665 and tomahawked one of her mistress's sons, she yelled "*Kaquince machissno Chippone* why he did soe?" One of the Indians answered in his own language, "Because hee would." Others used trade languages understood by both groups. These pidgins mixed Indian and English phrases and grammatical forms.[36]

Most settlers learned enough to carry on simple trade; more complex negotiations required interpreters, a few of whom lived in each community. Maryland hired two St. Mary's County planters, both of whom had arrived as servants and built small fortunes, to interpret, a service they might have carried on in their own neighborhoods. Thomas Savage and Robert Poole, who spent their adolescence and youth as interpreters in Virginia, could provide similar service when they settled in the 1620s on the Eastern Shore and James River. At the same time, Plymouth trader Isaac Allerton learned the language of Indians among whom he lived. Later in the century Hilletie van Olinda, daughter of a Dutch trader and a Mohawk woman, and two Dutch captives, who learned

Iroquois during their captivity, became provincial interpreters. Religious activities increased the number of Indians and whites fluent in one another's language. Missionaries learned Indian languages, and they taught English to a few Indians. Other colonists without nearby experts might have resorted to word books, such as Roger Williams's, to aid communication.[37]

During peacetime Indians and whites enjoyed many opportunities to learn about one another. Colonial leaders, military officers, fur traders, and merchants interacted most often with Indians, but no farmer or farmwife could avoid contact. At first they watched one another at a distance. Indians observed white farmers herding livestock, while white farmers examined Indian markings, attire, and corn cultivation. Observation turned to trade. A few Virginia artisans lived temporarily in Indian villages, making implements and perhaps learning the language, and a handful of Indians worked for or lived in English settlements (a 1619 Virginia muster counted four). The Piscataways saw Maryland settlers at court, and the two groups often lodged with each other when on hunting expeditions. Similar contact occurred in New England. William Wood learned enough about New England Indians during a four-year stay (1629–33) to write about their culture. Indians lived in Springfield, trading and drinking with the English, for decades. During the late seventeenth century, Goodwife Osborne of Hartford, Connecticut, and Daniel Denton, an English resident of Long Island, visited an Indian village and saw a green corn ceremony.[38]

Exchange of ideas and goods between Indians and colonists was common. Many Algonquian words for foods, plants, and animals unique to America passed into English, among them *maize*, *skunk*, *raccoon*, *hominy*, and *pone*, along with words describing Indian life, such as *sachem*, *squaw*, *moccasins*, and *wigwam*. Colonists and Indians often satisfied one another's tastes. Before colonization began, Indians had acquired a few European manufactures from traders or sailors. But soon thereafter they owned so many European goods—blankets, cloth, glass beads and bottles, copper kettles, ceramic bowls, axes, hoes, knives, tobacco pipes, and rifles—that they used them in cooking and farming, traded them with distant Indians, and filled their graves with them. Indian leaders owned more European goods than ordinary Indians, who had little more than a gun, a kettle, and a blanket. Indian artisans made bowls for farm families, hunted fowl, and skinned beavers; farmers traded metal tools, knives, beads, cloth, or clothing for furs or fowl. Increased incidence of dental caries suggests that Indians ate sugar and milled flour, acquired in trade. Indians hunted the colonists' livestock and sold the meat to them as venison. When merchants visited Indian villages, they traded with Indian women and swapped European goods for corn, baskets, bowls, and mats the women made, along with furs provided by the men.[39]

Indians frequented European settlements, traveling at times 100 miles or farther to trade, as visits of the northern New England Abenaki to Boston, the North Carolina Tuskaroras to Virginia, and Maryland Indians to the St. Mary's settlement show. John Smith reported in 1616 that Indians "daily frequented us with what provisions they could get, and would guide our men on hunting, and oft hunt for us themselves." Indian women exchanged corn and beans for shovels and hoes useful in corn cultivation. Such daily trade between Indians and the English in Virginia remained common, setting the stage for war in 1622. Trade and hunting continued after the war, but the assembly repeatedly prohibited both practices and urged planters to be self-sufficient in corn production.[40]

Such contact was as common farther north. So many Indians lived in settler homes that Massachusetts authorities prohibited that practice without a license; in 1632 they ordered "a trucking howse" be set up in each village "whither the Indians may resorte to trade to avoid there coming to several howses." By the 1640s, Indians living at Nonatum (now Boston suburb Newton) sold fruit, fish, brooms, and baskets at markets. At the same time they supplied villagers in Springfield, Massachusetts, with so much corn and venison and so many beans, nuts, and baskets that the town fathers tried to limit settler corn purchases by taxing corn bought from Indians; other colonists got baskets, bowls, snowshoes, and canoes. In 1677, immediately after King Philip's War, Indians returned to Boston's marketplace. The first Dutch settlers exchanged dairy products and imported glass beads for fresh food (maize, fish, and venison); Indians who came to truck on New Netherland farms sometimes ate an elegant dinner at the buyer's cabin and stayed the night, sleeping on the floor beside the family's bed. Indians on occasion stole pots or blankets from Dutch houses and then resold them to other, unsuspecting Dutch settlers.[41]

Colonial authorities, who wanted to keep guns out of Indian hands, prohibited the gun trade and mandated severe punishment (death in Virginia and New Netherland) for such sales. But runaway laborers in Jamestown sold rifles to Indians and taught them how to shoot, and a deadly arms race began; Virginia settlers sold guns to Indians, who used them with great accuracy. A similar illegal gun trade developed in New England in the 1620s. New Englanders wanted Indian furs so badly that they sold firearms and gunpowder to Indians or repaired guns they owned that could be used against settlers. Commoners sold or lent guns to the Indians. Widow Horton of Springfield, Massachusetts, for one, lent her husband's gun "to an Indian for it lay spoilinge in her seller." If English colonists refused to sell guns, powder, or flintlocks to Indians, the French and Dutch settlers would provide them with these items, or Indians would steal what they wanted from colonists' cabins. To try to control the trade, in the 1660s New England governments licensed traders to sell weapons. But by mid-century—with so many traders willing to sell firearms—nearly every tribe had acquired a supply.[42]

Officials feared the liquor trade as much as the arms race. Indians who got alcohol drank until inebriated. Drunken Indians who frequented English and Dutch settlements, officials believed, got angry with farmers and merchants who cheated them and pilfered their homes, killed their cattle, and incited violence that could lead to war. In the mid-1680s, when a Pennsylvania settler (and rum trader) refused to sell liquor to Indians, they attacked him and stole his rum; in Maine in 1677, settlers gave Indians rum and killed them when they became drunk. Massachusetts banned the trade in 1633, but it persisted and led to intermittent prosecutions. Despite similar bans throughout New England and in New Netherland and New York, Pennsylvania, New Jersey, and the Chesapeake colonies, colonists sold liquor to Indians throughout the colonial era. Traders and tavern keepers dominated the liquor trade, but during the seventeenth and early eighteenth centuries, Indians got rum from Dutch and English farmers as well.[43]

Long before Europeans arrived, northeastern tribes had exchanged tubular, blue clamshell beads. Used as ornaments and in ceremonies such as marriages, they were in short supply. The Dutch and New Englanders, realizing that cheap beads could be traded for furs, showed Indians how to use metal tools to make standardized beads that could be woven into belts, necklaces, or collars (called wampum). Indians on both sides of Long Island Sound rapidly increased wampum production and established permanent villages (abandoning long hunts) to take advantage of European and Indian demand. As early as 1608, Virginia settlers traded copper and glass beads, perhaps made locally. To facilitate the beaver trade, Dutch and New England colonists imported glass beads (used by Indians for jewelry) and set up wampum factories, where farmers worked; fur traders, such as the Pynchons of Springfield, paid farmers to string wampum for use in trade. Sometimes traders acquired wampum from Long Island Indian producers, giving cloth or other goods in exchange, and then used the wampum to buy furs from distant Iroquois. As the quantity of wampum rose, the Iroquois wanted more of it, thereby raising the wampum price for furs Europeans craved. Needing a medium of exchange, colonial legislatures latched onto wampum as money. Wampum circulated among white farmers in the cash-poor New England and Chesapeake colonies, as the wampum payments defeated Indians had to pay to the English as tribute attest. Long Island Indians alone paid New Englanders 6.9 million beads, worth £4,800, between 1634 and 1664.[44]

White merchants carried on nearly all the fur trade, buying beaver pelts and deerskins. Yet seventeenth-century white farm families did participate. The Chesapeake beaver trade, started in the 1620s, engaged men who traded and planted tobacco, as the widespread planter use of pelts as money shows. Planters provided merchants with corn and tobacco they traded with the Susquehannocks of the interior; men in outlying districts continued to exchange guns for beaver pelts into the 1660s. Ordinary New Englanders participated even more. Thomas Morton established a short-lived colony near Plymouth in 1625 to grow corn and trade for furs. From the 1630s through the early 1660s, William and John Pynchon held a legal monopoly of the Connecticut Valley fur trade, but they hired agents to barter with the Indians and other residents to transport furs to market. Pynchon's agents were rich farmers who used the trade to build landed estates; poorer farmers or artisans who acted as agents bought few furs.[45]

New Netherland, closest to northern fur supplies, established a large fur trade, open at first to farmers. The first settlers both farmed land and sold furs they acquired from Indians to the Dutch West India Company, but by 1625 the company had tightened its grip on the trade and tried to eliminate the farmers as middlemen. Dutch farmers defied the law and continued to buy beaver pelts throughout the century, often "walking in the woods" to meet Indians with furs, acting as brokers for the fur traders, or buying furs in their own houses. Dutch authorities, seeking immigrants, often failed to enforce or even rescinded the laws, opening the trade to everyone. So many settlers tried to persuade Indians to sell furs to them, each seeking to outbid the next, that they complained of being assaulted and held prisoner in farmers' houses until they sold their furs.[46]

Indians, who knew how to survive in the woods, passed forest lore on to the English. Not only did they guide settlers through the forests, but they drew maps in the sand for them, which the English reproduced on paper. Indians showed immigrants how to make wigwams, which the newcomers built and used until they could make English-style houses. Observing Indian success in hunting and traveling through the woods, immigrants wore moccasins, snowshoes, and dull brown or green clothing. Indians taught settlers how to stalk game and efficiently catch and prepare fish. Observing old fields filled with girded trees, the English soon learned to burn the brush and plant beneath dead trees. Immigrants remembered and taught their children forest lore useful in making new settlements in the woods.[47]

The agricultural practices of the colonists owed much to Indian tutelage. Maize was not grown in Europe but was crucial to survival in America, and Indian beans, peas, squash, and melons supplemented the settlers' diet. Indian women showed Europeans when to plant; how to make hills for corn; how to keep weeds down; how to interplant corn, beans, and pumpkins; when to harvest; and how to husk the ears. The English, at first skeptical about the nutritional value of maize, came to see it, in the words of John Winthrop Jr., as "wholesome and pleasant" food "of which great Variety may be made." Like Indians, the English cultivated fields extensively, moving on when nutrients in the soil had become exhausted rather than manuring the land to extend its use. The Dutch in New Netherland burned the underbrush to clear land, taking care to prevent the fire from consuming their houses.[48]

Indian training was essential for survival in the first years of European habitation, as an incident in New Netherland in the 1640s indicates. An Iroquois Indian, seeing Andriaen van der Donck clearing land, told him, "It is very good soil and bears corn abundantly—which I well know, because it is only 25 or 26 years ago that we planted corn there and now it has become wooded again" and thus ready to cultivate again. Indians must have repeated farming lessons endlessly as immigrants arrived and tried to hack farms out of the woods; later white frontier families, already knowledgeable about Indian techniques, may have temporarily returned to Indian ways.[49]

At first colonists used techniques Indians taught them. But seeking markets and missing the labor-saving techniques of English agriculture, they soon made their farms resemble those of England. Despite their preference for English grains, they continued to plant corn. But they cleared more land and planted continuous fields (rather than the corn rows interspersed with beans, peas, and squash); they used metal hoes, axes, scythes, sickles, and plows rather than the granite and wooden hoes, spades, and corn planters of Indian women. On occasion New Englanders plowed cornfields at planting and weeding times and fertilized their fields with fish (perhaps an Indian custom as well), rather than plant a new field. Moreover, they brought seeds of familiar English grains and fruits with them or ordered them as soon as they could, making English-style fields, gardens, vineyards, and orchards. Fearful that their cattle would not thrive on native wild rice and broom straw, they planted English varieties for pasturage and for hay.[50]

The accommodation of settler to Indian had distinct limits. Unlike Frenchmen, who married Indians until French women arrived, Dutch and English men refused to marry Indian women. Despite a scarcity of European women, no Dutchman wed an Indian, though lonely men often procured Indian prostitutes, and several cohabited and sired children with Indian women. In New England, too, men outnumbered women. Epidemics that depleted male populations caused some Indian women to try to win settler husbands and their trade goods. But despite sexual attraction, few immigrant men married Indian women, for the men feared they would become savages through contact with Indian women. In the early 1620s Thomas Weston's men at Wessagusset kept Indian women; men in Thomas Morton's renegade Massachusetts town danced and cavorted with Indians, but waited for English women to arrive to choose marriage partners. By 1640 plenty of English women had come to New England, rendering intermarriage unnecessary. A few frontiersmen fled to Indian villages and cohabited, but a white man found guilty of "intiseing an Indian woman to lye with him" was "severely whipped."[51]

Because so few European women migrated to the southern colonies, men in the South had more reason to wed Indians. Sexual attraction, sexual intercourse, and concubinage between Indian women and Virginia men may have been common during the 1610s, when forty to fifty colonists ran away and cohabited with Indians. But few Anglo-Virginians wed Indian women, despite the scarcity of English women and the example of John Rolfe's marriage to Pocahontas. Not only did clergymen warn against racial mixing, but Indian women refused to wed men so poorly prepared to hunt, fish, and support families. After the 1622 massacre, English settlers, no matter their rank, sought the extermination or at least exile of Indians; sex-hungry men might have turned to sodomy. After peace returned, a tiny number of Chesapeake whites (most of whom were indentured servants) slept with or married Indian women, but by the end of the century enough Englishwomen had immigrated to make such action unnecessary. English Indian traders and eighteenth-century backwoodsmen frequented Indian prostitutes and took Indian mistresses; eager to take advantage of kin ties to develop trade, a few traders married Indian women.[52]

A few isolated settlements of the offspring of interracial marriages appeared in the colonies during the eighteenth century. A tiny number of servants and poor whites did marry Indians, but most residents traced their ancestry to runaway slaves and free blacks who had married Indians or to sexual unions between free black women and white men. The Maryland Brandywine people, for instance, probably began with the intermarriage of several Catholic servant women and Indian men or free blacks and grew through intensive intermarriage within the group. Unable to find Indian men to marry because so many pursued dangerous work in the militia or on whaling ships, Christian Indian women in New England increasingly married free African American men; descendants could choose to belong to either community. Groups similar to these could be found on the Eastern Shore and in North Carolina.[53]

Indians took war captives, both soldiers and outlying settler families, to their villages. We know the most about the 1,641 New Englanders captured between 1675 and 1763. Two-thirds were male, half were adults, but the other half were impressionable children. These captives—two-thirds or more of whom were repatriated—learned much about Indian society, which they shared with their neighbors after they came home. Mary Rowlandson, a captive in King Philip's War of 1675 who wrote the first famous captivity narrative, told her readers about her captors' strange diet (ground nuts, corn pancakes, and wild game) and clothing, wigwams, forms of exchange (she traded shirtmaking for food), and war councils. A few years later ten-year-old John Gyles of Maine saw his captors torturing prisoners and almost suffered the same fate. He hunted with the men, planted corn with the women, and learned their language so well that he could report their fables and marriage customs. Such experiences became part of the oral tradition of frontier communities, and the Indian skills captives learned could be silently incorporated into white society.[54]

Even in peaceful encounters between settlers and Indians, misunderstanding lay just below the surface. The Dutch dubbed Indians "Wilden," wild people, and the English agreed. Indians' dress and painted bodies (dubbed nakedness by the English), their strange foods and supposedly homeless lives in forest and swamp, and their gender roles, religion, land use, and trade made Indians appear to Europeans as witches, minions of Satan. Imagine the horror of Englishmen who saw Indian women work unsupervised in the cornfields and (unlike European women) build houses, make wampum, and hold the land themselves—a horror magnified when the men had to learn farming from these women. Imagine Indian incomprehension of men who wore tight clothing (Narragansetts called Europeans *Wautacondug*, coatmen) and grew unkempt beards, worked in the fields like women, valued goods strictly in monetary terms, and sent their pigs and cows into the woods to eat forage meant for deer.[55]

When the English traded, they saw a contract, an economic nexus governed by price and profit; when they granted credit, they expected goods to be delivered or collateral to be taken. When Indians traded, they viewed the transaction as an exchange of gifts in which each side should eventually receive fair value. If the English failed to reciprocate a gift of land or corn, Indians expected the English to recognize them as superiors; the English, viewing the gift as a sale, refused. Seeking the civility of their own kinds of gift exchange, Indians viewed the colonists' behavior as selfish. At first Indians wondered why Europeans wanted furs, which were so easy to obtain through hunting, while Europeans deemed the beads and trade goods the Indians craved as trinkets and were amazed that "for a copper kettle and a few toyes," in John Smith's view, Indians "will sell you a whole Countrey." Wonder soon turned into conflict. Indians saw the settlers' demand for food as extortion rather than a gift exchange; settlers later viewed hungry Indians asking for food as beggars.[56]

As soon as Indians and Virginia settlers learned each other's forms of trade, they tried to gain advantage. In 1607 the Powhatans invited the English into their villages, proffering hospitality and hoping that they would become Indian; misunderstanding the nature of the gift, the English turned to economic negotiation based on contract, which the Indians did not understand. Since they received no gifts, the Indians took trade goods—and the English accused them of stealing. By the fall of 1607 the Indians understood the nature of English trade, turned to barter, and demanded weapons for their corn and turkeys. When the English extorted corn and burned canoes, Powhatan mandated a trade embargo that lasted nearly half a year. When he reopened trade, Powhatan abandoned gift exchange and sought to trade guns for food, which the English refused, leading to renewed extortion of corn and war.[57]

Long contact barely lessened misunderstanding. On a Sunday in 1675, Nathaniel Saltonstall reported, "Seven or Eight of King *Philip's* Men came to *Swansea*" seeking "to Grind a Hatchet" at a local farm. But "the Master told them, it was the Sabbath Day, and their God would be very angry if he should let them do it." Enraged that he had turned down their simple request, they answered that "they knew not who his God was, and that they would do it for all him, or his God either." After leaving his house, they "went to another House and took away some Victuals, but hurt no man." Encountering an Englishman on the road, they detained him a "short time, then dismist him quietly," after ridiculing Puritan religion and culture by "giving him this Caution, that he should not Work on his God's day."[58]

Peace became conflict in a moment. Whites and Indians insulted and struck one another for wrongs real or imagined. New England and New Netherland settlers built large herds and allowed their hogs and cows to meander in Indian cornfields and orchards and the hogs to forage on their shellfish. To pasture their herds, colonists demanded Indians sell more land. Indians hunted on unimproved lands Europeans bought, and they not only killed livestock in retaliation for the destruction by colonists' animals but took food or farm tools they believed were theirs by right, inciting violent responses from farmers. Virginia Indians killed so many hogs in the aftermath of the 1622 war that few were left. Rich Rhode Island farmers wanted to hire Indians or poor whites to tend hogs; poor whites—who made money from foraging hogs—often refused, inciting repeated conflicts. Such tensions persisted after the English won superiority. The town leaders of Dedham, Massachusetts, complained in 1681 that Christian Indians from nearby Natick "rob us of our corn & other provisions out of our fields, That cattle in ye woods have been torn by their dogs." Only careful negotiations could avert disaster. In 1648, when English settlers at Southampton, Long Island, complained that Indians planted on land they had sold and that colonists' livestock fell into Indian food-storage pits, Montank leader Wyandanch negotiated a compromise whereby the Indians filled in their pits and the English built fences. But similar disputes broke out in 1655 and 1657 requiring yet more delicate arbitration.[59]

From the day they arrived, settlers struggled with Indians for control of the land. Indians and settlers held strikingly different conceptions of land use. Indian villages claimed sovereignty over their cornfields and hunting grounds. Bounded by barriers such as rivers, the village held these territories as common lands that outsiders could use if village leaders granted permission. Leaders customarily granted use of cropland and the food it produced to village women, but no Indian *owned* the land. Hunting or trapping parties had exclusive rights to hunt in their territories and to keep animals they killed, but the village collectively owned their hunting grounds. English law, in contrast, separated sovereignty and ownership. The king or proprietor held sovereignty over colonial territory, but individuals owned land and had absolute rights to sell it. When colonial officials signed land treaties with a sachem, they believed that he had transferred sovereignty, and with it the right to sell or give land to settlers. Only settled habitation—which settlers denied Indians had—legitimated ownership once sovereignty changed hands. Indians, in contrast, thought that they had granted settlers the right to use land that remained under their sovereignty.[60]

At times settlers acquired land on Indian terms as part of a gift exchange, and sovereignty did not change hands (even by English lights); but they soon demanded deeds that transferred ownership. Roger Williams, for instance, understood that the Rhode Island plantation given him in 1636 remained under Indian political control. The next year Williams had the Indians sign a deed that did transfer ownership of some land to Providence, but it merely allowed use of the rest and neglected to give precise boundaries. By 1645, deeds in the colony followed English form, handing over land unencumbered and listing bounds.[61]

Different ideas about land use led to hostility between settlers and Indians. Noting how few women (so crucial in Indian society) came with the first settlers, the Powhatans assumed the white men had come for temporary trade and would soon leave. Settlers, especially land speculators, repeatedly bought or seized Indian land for a fraction of its value, despite the bans colonial governments—fearful of war—placed on such sales. One Long Island Indian signed ninety-one deeds between 1655 and 1703, most for small tracts (15 to 100 acres) in the Oyster Bay area. Merchants granted credit to Indians for manufactures, taking land as collateral and making new land sales inevitable. A few buyers had their farms destroyed and the land returned to Indians, but more often they held on to purchases by making added payments. Others squatted on Indian land, prompting complaints to English authorities. Indians, for instance, complained to the Virginia Assembly in 1652 that farmers had taken their lands, "forceinge them into such narrow Streights, and places That they Cannot Subsist, Either by plantinge, or huntinge." Fearing war, the assembly prohibited settlers from planting on Indian lands. Settlers nonetheless pushed their leaders into demanding ever more land from Indians. Colonial governments or New England towns then bought land from Indians.

Indians signed land cession treaties in Plymouth, Massachusetts, Maryland, Connecticut, Rhode Island, and Virginia before 1650, but they resented English claims of ownership once they understood that it entailed permanent loss of land. As soon as Indians learned what signing a deed meant, they tried to bend

English law to their own benefit, retaining hunting rights on lands they had sold. This strategy worked where few English farmers settled. Throughout the seventeenth century, Maine Indians sold use rights to their land but retained ownership because whites craved their furs. When colonists wanted the land, such tactics failed. At best Indians gradually sold off parts of their territory and received a fair price for it, as did the Lenape of southeastern Pennsylvania and the upper Delaware Basin. English law—with its doctrine of free alienation of land—accelerated Indian losses. If Indians permitted land guaranteed to them to be surveyed (as did the Lenape) or let officials grant them a township (as in Massachusetts), they implicitly accepted English sovereignty. Their land could then be sold to colonists, no matter what protections were built into treaties.[62]

Once white farm families moved onto Indian land, they changed the landscape, setting cattle and hogs to forage in meadows and forests used by animals Indians hunted and in the unfenced Indian cornfields. To protect livestock, colonists killed animals Indians needed for subsistence. White occupation of the land, combined with European demand for beaver and deer skins, depleted coastal forests of game, depriving Indians of part of their subsistence. When Indians killed hogs in the forests for their meat—not recognizing them as private property—immigrant farmers became furious. They viewed Indian use of land as unproductive and therefore felt free to appropriate it, sometimes in advance of treaties. New England and Virginia writers argued that Indians had but a natural right to land that the English could end merely by cultivating it; Virginia pamphleteers insisted that the Indians' savagery justified wars of expiation and land confiscation.[63]

Struggles over corn and land repeatedly led to violence and war. Colonists and Indians assaulted one another and stole each other's property. Usually caused by encroachment on Indian land, wars were triggered by the murder of an Indian or a white by the other group. Through the 1640s the two sides were equal, and Indians had reason to believe that they might dominate. Once begun, wars were destructive and bloody. Indians and colonists had firearms and knew how to use them; both sides burned villages, houses, and fields and, on occasion, massacred women and children, scalping their victims. Colonists and Indians fought such destructive wars because Indians sought to hunt on the same land colonists wanted to farm. Seeing the Indians' growing cornfields and prosperous villages, soldiers craved these rich farms, igniting more wars.[64]

The English and the Powhatans began attacking each other in 1609. The English extorted corn, burned Indian villages, or murdered Indians, and the Indians responded in kind, killing a fifth to a quarter of the colonists. Virginians searching for food risked death. One foraging group was found "slain, with their mouths stopped full of bread, being done . . . in contempt and scorn, that others might expect the like when they should come to seek for bread and relief." In 1609–10 Powhatan laid siege to Jamestown. The Powhatans monopolized corn, causing the deaths of two-thirds of the colonists. In 1611 the English stole enough corn to prevent a recurrence. Warfare continued until 1614, when fresh English forces forced the Powhatans to sue for peace. After spending eight years gathering weapons, they struck in 1622, massacring 347 colonists (a quarter to a third of the total) and capturing at least 19 women. The attack took place in March, when corn supplies were low and the Powhatans felt particularly incensed at English demands for bread. The English retaliated viciously, killing Indians, burning their villages, destroying their fishnets, and burning or stealing their corn in a decade-long campaign. Instead of negotiating for the release of their women, they poisoned 200 Powhatans who came to parley and shot 50 more. Intermittent warfare in both Chesapeake colonies, along with alliances between the English and distant Indians, continued until 1644, when the Powhatans struck again, killing about 400 settlers, or 4 percent of the colonists. English retaliation ended the Indian threat in Tidewater Virginia.[65]

Similar violence permeated New England and New Netherland. Skirmishes began in New England as soon as the English landed, and full-scale warfare broke out in 1637 when the Pequots—unwilling to pay heavy tribute for a trade treaty or turn over men who had murdered a drunken white trader—raided an English settlement. The war (which pitted the English and their Indian allies against the Pequots) escalated, with both sides burning fields and killing civilians. When the war ended, half of the Pequots were dead; about 600–700 were soldiers, women, and children the English burned alive at Fort Mystic or shot when they tried to escape the inferno. Minor warfare broke out in the 1640s, but New England enjoyed peace until the 1670s. Hog killing by Indians led the Dutch on Staten Island to murder several Indians in 1639, and Indians retaliated by slaying several whites. In 1643 the Dutch massacred eighty local (and peaceful) Indians who had come to them for protection against invading Mohawks; in 1655 Indians attacked New Amsterdam, killing fifty settlers, burning farms, and stealing 600 cattle.[66]

The climax of Indian-settler conflict in New England came in 1675–76 when Indians, "provoked . . . to anger and wrath," as one Nipmuck explained, went to war. As Roger Williams watched his home burn, he angrily asked "Why they assaulted us With burning and Killing." Having lost land and subsistence, the Indians were desperate: "You have driven us out of our own Countrie and pursued us to our Great Miserie, and Your own, and we are Forced to live upon you," one responded. In retribution for this dispossession and dependence, they aimed to destroy everything that made settlements English. As the Nipmuck explained, "We have nothing but our lives to lose but thou hath many fair houses cattell & much good things." One English survivor lamented, "They burnt our milles, brake the stones, ye our grinding stones," forcing settlers to make flour like Indians. And "what was hid in the erth they found, corne & fowles, kild catel & tooke the hind quarters & left the rest. . . . They burnt cartes wheeles, drive away our catel, shipe, horses."[67]

Led by King Philip, Indians attacked Plymouth and Narragansett Bay villages and laid waste the Connecticut River and Maine frontiers, burning barns and houses, killing cattle, capturing some settlers, and murdering others, including pregnant women, grandparents, and babes in arms. Only coastal settlements north of Boston went unscathed, and whites living there were apprehensive, for Philip's men raided villages ten miles from Boston. About 2,500 English died, a twentieth of the region's people. Indians pushed the frontier back toward the coast, sending countless families fleeing, and destroying towns, such as Deerfield, Massachusetts, that were not to be fully repeopled by whites for forty years. The English and their Indian allies, with their greater numbers, defeated the rebel Indians, burning their villages and food stores and killing 4,000–5,000 natives, two-fifths of the populace. In one massacre the English torched a Narragansett settlement, killing at least "three hundred old men, women and children burnt in the wigwams." The victors captured another 2,000, including dependents, executing a few and enslaving hundreds. Intermittent fighting between Indians (often joined by the French) and settlers continued on New England's northern and western frontiers throughout the early eighteenth century. Colonists and Indians raided each other's settlements from 1689 to 1702, and Deerfield suffered its second burning in 1704.[68]

Virginia warfare, too, began in 1675–76 with a conflict over trade and escalated into mutual violence with Indians and frontiersmen killing one another. When Governor William Berkeley refused to support colonists' desire to clear out all Indians, frontier planters formed an irregular army led by the rebel Nathaniel Bacon and attacked peaceful tributary Indians (who had gone to war against Susquehannocks hostile to Virginia) rather than hostile Indians. When colonists murdered an unarmed band of Susquehannocks who had come to parley for peace, the survivors went on a rampage and attacked outlying families throughout settled Virginia. At the end of this bloodletting, no Virginia Indians, hostile or friendly, stood in the way of the expansion of English plantations. They had been pushed to the piedmont or mountain regions, where no settlers yet lived. Any Indians who came to frontier settlements faced death, as did the Iroquois who passed through backcountry Virginia in 1679 searching for Indian enemies and seeking food.[69]

Warfare between Indians and settlers in the Carolinas began and ended later than in the older colonies. Until the early eighteenth century, trade—in deerskins and Indian slaves the coastal tribes had captured in the interior—kept relations between the English and coastal Indians peaceful. But as settlers occupied Indian lands, Indians feared dispossession. The Tuscaroras struck first, killing over a hundred German and English colonists. In response North Carolina's government asked its neighbors to suppress the Indians, a task South Carolina whites and their Yamasee allies readily undertook in order to enslave more captives. The Yamasee soon turned against their allies. By eating the forage that game animals relied on, the settlers' swine and cows had displaced the deer. At the same time, the supply of Indians the Yamasee could sell into slavery plummeted, and Yamasee debts mounted. Fearful that they would be forced into slavery to pay debts, Yamasees allied themselves with the Creeks and went to war in 1715, killing traders and other colonists, burning outlying settlements, and pushing survivors back toward Charleston. With the help of Cherokee allies, the English put down the rebellion and drove Creeks and Yamasees alike from their territories.[70]

During the seventeenth century, freed Chesapeake servants or sons of New England settlers repeatedly fought Indians. Virginia authorities armed every plantation after the 1622 massacre; once he had firepower, nearly every white man went to war, killing Indians, burning villages, destroying cornfields, and taking captives. A smaller army of the best militiamen responded to the 1644 attack. Clashes in Virginia during the 1670s involved local militias defending their fields and expeditions of frontier small planters and ex-servants, out for revenge and plunder, against peaceful Indians. Connecticut mobilized so many farm men during the Pequot War that the grain crops were threatened; thinly peopled Plymouth colony sent three or four dozen men, a quarter of their youths. In 1675 Connecticut drafted a quarter of the militia, mostly unmarried farmers' sons needed at home to plant and harvest crops. Nearly all Plymouth men capable of using rifles participated; most took part in home defense, but a tenth were involved on the battlefield. Frontier Deerfield, Massachusetts, lost fourteen men, more than a third of the community's men. Before the end of the century, however, fighting Indians had become far less common for farm youths. Draft resistance was common in New England during King Philip's War, and many Boston servants and apprentices replaced farm boys. Not only did the Indian menace recede in older settlements after 1700, but expeditionary forces (a tenth or less of adults) and Indian allies, rather than farm youths drafted from the militia, defended frontier New England after 1680 and the Carolinas in the 1710s.[71]

Once colonists had conquered the land, Indians lost their autonomy, and some who remained fell into abject dependence. Eastern bands hunted scarce fur-bearing animals or traded with Indians farther west, hoping to get corn, pork, cloth, hoes, and gunpowder. They fell short, however, and had to get some supplies from the English on credit. Survival required complete acquiescence to the English legal system, as Awashunkes, the "Squaw Sachem" of Saconet, discovered in 1683 when colonial authorities charged her with infanticide, a common practice among her people. The Maryland Piscataways temporarily avoided this fate. Never defeated in war, they held on to their land and maintained autonomy through the seventeenth century, choosing their own leaders, burying their dead by Indian custom, and rebuffing missionaries. But their circumstances were special. They lived far from the center of Maryland's English population, and they not only allied with the English but supplied them with corn and deerskins.[72]

The English frequently set up reservations, on occasion at the request of Indians seeking to keep land. But reservations excluded Indians from land occupied by whites while allowing colonists to encroach on Indian territory. Isolated on small reservations, dependent on colonists for survival, often forced to speak English and accept English ways, and posing no danger to whites, who could squat on their land, these Indians lived peaceably among colonists. Virginia followed a policy of separation after the 1622 massacre, set up its first reservations after defeating Indians in 1644, and formed new reservations after 1676. After 1660 Maryland fur traders, hoping to maintain supplies, patented several small reservations on Maryland's Eastern Shore, but planters soon surrounded them and eliminated the fur trade; much of the land soon came into English hands. Maryland Indians maintained some autonomy, living in wigwams and practicing their language, on larger Eastern Shore reservations (located on swampy land settlers avoided). As game disappeared and settlers squatted on their lands (on occasion stealing their property or burning their cabins and fences), Indians had to lease (for low, unpaid rents) and then sell much of their remaining lands, and their autonomy disintegrated.[73]

New England expanded its reservation system after King Philip's War. During the war, Massachusetts praying Indians—unfairly accused of arson and treason—were forced to abandon their livestock and crops and go to barren Deer Island in Boston harbor, where some were illegally sold into slavery and the rest went hungry. After the war they had to live in the remaining praying villages, where they endured white commissioners and guardians, laws regulating their behavior, prohibitions on their traveling or hunting without permission, and encroachments of white squatters. Although they vigorously defended their collective property rights, eventually those, too, dissolved. Christian Indians in Natick, Marlborough, or Martha's Vineyard lived in separate villages until settlers intruded and forced them off the land. They first lost their hunting territories and then their villages through sales to speculators and settlers. When Indians instituted private property, which allowed them to sell their own land without tribal approval, colonists bought yet more of their territory, mostly in small parcels. Such sales gradually destroyed any corporate Indian community. During the 1720s and 1730s, Natick Indians sold a third of their land, nearly half to whites.[74]

Warfare turned farmers and their leaders against any Indian presence. Not long after the Indian rising of 1622, Virginia governor Sir Francis Wyatt captured the colonists' thinking: "Our first worke," he insisted "is expulsion of the Salvages to gain the free range of the countrey for encrease of Cattle, swine, &c which will more than restore us, for it is infinitely better to have no heathen among us, who at best were but thornes in our sides, than to be at peace and league with them." Even though treaties guaranteed hunting rights on reservations, Indians (such as those on Virginia's Eastern Shore) retained guns with great difficulty.[75]

Surrounded by hostile settlers, often defeated in war, and with their game chased away and their lands fenced in, some Indians in English areas adjusted to white ways to retain some land while continuing to hunt if they could. New England Indians gradually abandoned wigwams for frame houses and acquired beds, tables, chairs, and kitchenware. Indian women began to spin and to cultivate one- to two-acre, English-style vegetable gardens. The 2,000 Massachusetts praying Indians, who adopted Puritan religion, were the best example of such groups. While trying to retain a semblance of their culture, many spoke English, while a few learned to read their own language (less than a third) or read English (a seventh). Others adopted English farming, clothing, and housing. By the late seventeenth to early to mid-eighteenth century, Indians at Nantucket, Natick, and Stockbridge had adopted English farming techniques, using English tools, building mills, plowing fields, cultivating apple orchards, and (in Natick) distributing land to individuals.

When hunting diminished, Indian men owned livestock and thus added herding to their regime of hunting, gathering, farming, and fishing. By the mid- to late seventeenth century, Indians living amidst settlers in New England, Long Island, and the Hudson Valley often owned pigs, sheep, cattle, and chickens. They particularly liked hogs because the animals needed no care and could be hunted, adding to their store of protein during the spring dearth. King Philip, for one, raised hogs in the 1660s, setting them to graze on Hog Island, which was used as a commons by Portsmouth, Rhode Island. Indians not only ate pork but sold it at Boston's markets, getting firearms in return. Indians who remained in the Tidewater Chesapeake also took up European farming, fencing in cornfields and raising pigs and horses. During the 1710s William Tapp, the last leader of the Wicomoco, continued to hunt, fish, and clear land, much like his ancestors, but he also raised sheep, milk cows, and hogs. He also slept on a feather bed and even read English books.[76]

If settlers had left Indians to build a new life, they might have prospered, keeping their language, religion, and gender roles. But no matter how much the Indians adapted to white ways, settlers still thought them savages who held land the newcomers wanted. Insisting that the hogs Indians raised had been stolen, settlers in Warwick, Rhode Island, tried to prevent Indians from branding or keeping them. Tensions over livestock lay behind the war King Philip led. Portsmouth town fathers demanded Philip take his hogs off the island. As the praying Indians of Natick learned in the early eighteenth century, European farming required purchase of tools and animals that could be financed only by selling land. Indians at Warwick similarly fell into debt and sold themselves into servitude.

The adoption of Christianity by Indians did little to help their plight. Distrusted by Indians who wanted nothing to do with alien Christian ideas of sin, heaven, and hell and by whites who forced them to live under the authority of white missionaries and magistrates, the praying Indians were unable to sustain their cultural balancing act. Soon after Natick was founded in 1650, the town fathers of neighboring Dedham began a protracted court battle to recover the entire new town, claiming that the land had been granted to them. Some praying Indians joined King Philip's campaign; others tried to remain loyal to the English, hoping for secure land in return. English soldiers not only forced loyal Indians to exile on Deer Island, but four soldiers murdered three Christian Indian women and three children in the Concord woods for the crime of picking berries. After the war the number of Christian villages declined from forty-six to thirty-two. With their farm tools destroyed, the praying Indians returned for a while to hunting, fishing, gathering, and fur trading. Only on Martha's Vineyard, where few whites settled, did praying Indians succeed for a while in combining Christianity with Indian customs.[77]

War made it impossible for any Christian Indian to remain true to Christianity and maintain an Indian identity. Accused of disloyalty by the English, Joseph Tukapewillin asserted his Christian belief and recounted how both the English and Philip's men had mocked his faith and ruined his farm, where he labored like an Englishman at the plow. "The English have taken away some of my estate, my corn . . . my plough, cart, chaine & other goods." At the same time, "the enemy Indians have also taken a part of what I had, & the richest Indians mock & scoff at me, saying now what has become of your praying to God. The english also censure me, & say I am a hypocrite."[78]

The victorious English tolerated only Indians who became servants or wage laborers. Increasing poverty forced some Indians into indentured servitude. During the 1660s, Eastern Shore Indian families, unable to feed their children, indentured numerous seven- to fourteen-year-old sons; after King Philip's War, at least thirty-six children of enemy Indians were forced into servitude until age twenty, and by the 1680s, other Massachusetts Indian children faced servitude to pay their parents' debts. More worked for wages. By the end of the century, Indians were joining Long Island and New England whaling voyages, sometimes under coercion. Some Rhode Island Indians lived in white households, where they worked as servants, herded livestock, killed wolves, and built stone fences. Connecticut Indians hunted wolves, built fences, and herded cattle, receiving cloth or other trade goods as wages. Virginia Eastern Shore Indians built canoes and mended fishnets for planters. In the 1640s, Massachusetts praying Indians worked as laborers for merchants; two decades later, they marketed shingles to the English and worked on construction projects. After King Philip's War, many men worked for farmers, making stone walls, cutting wood, and harvesting and weeding—work reserved for women in their societies. (Edward Hutchinson had hired several of the local "*Sachems* Men, in Tilling and Plowing his Ground" even before the war.) Women joined their men, and some of them learned to spin. Much to the colonists' disgust, temporary seasonal farm labor allowed Indians to return home after the harvest and hunt deer.[79]

Captured by slave hunters, taken in war, or forced into bondage to pay debts, a small minority of Indians became slaves. Slave-hunting expeditions were common in South Carolina, where coastal tribes raided the interior and sold their captives, mostly women and children, to slave traders and planters. Some of these captives were exported, but most remained. In 1703, 300 Indian slaves worked in South Carolina, but by 1708—after extensive raids in north Florida and sales by Indian slave catchers—South Carolina planters held 1,400 Indian slaves, nearly a third of the unfree labor force. Unlike African slaves, they were predominantly women (43 percent) and children (21 percent). The number of South Carolina Indian slaves peaked around 1720 at 2,000 but declined rapidly thereafter, to no more than 500 by 1730. Whites enslaved Indians captured in war in other colonies. But—except on Rhode Island cattle ranches, where as many as one-seventh of the people and two-fifths of the slaves were Indians—they were few in number. New Englanders on occasion bought Indian slaves from Carolina traders. Some New England Indians captured during the Pequot and King Philip's Wars were enslaved locally, but most were transported to the West Indies or Spain. In late 1676 the English sent at least 180 captives—including women, children, and even men who fought for the English—into West Indian or (when the islands refused to accept them) North African slavery. Indian slaves posed no threat to white occupation of Indian land, but far fewer Indians than Africans became slaves because Indians ran away with greater success and died from European diseases at a much higher rate than Africans.[80]

Refusing to accept dependence, servitude, or slavery or to live in a land settlers made uninhabitable, the vast majority of Indians moved west. During the late seventeenth and early eighteenth centuries New England Mahicans and Narragansetts, mid-Atlantic Lenapes, Carolina Tuscaroras, and Chesapeake Nanticokes emigrated to the upper Susquehanna and Great Lakes regions, some taking refuge in Iroquois villages. At the same time, Abenakis moved farther east. After English planters invaded the Potomac basin late in the seventeenth century, Piscataways moved northwest along the Potomac and then left for Pennsylvania. Similarly, the Lenapes sold their lands and left for western Pennsylvania in the early eighteenth century, and Powhatans moved toward the remote piedmont southside after whites encroached on their lands. By the 1770s, Indians still inhabited most of northern New England, western New York and Pennsylvania, Georgia, and the entire territory west of the Appalachians—lands with few, if any, settlers.[81]

The separation of settlers and Indians was especially apparent in the South. More Indians (130,600) than whites and blacks combined (79,600) lived there in 1700. Nonetheless, just 1,300 Indians lived in Tidewater and Piedmont Virginia, compared with over 60,000 whites and blacks. Indians remained in the Carolinas for decades. In 1700, 7,600 Indians lived in North Carolina east of the mountains (45 percent of the population), but by 1715 the Indian populace had plummeted to 3,000 (15 percent). Similarly, South Carolina's Indian population declined from 7,500 (53 percent) in 1700 to 5,100 in 1715 (27 percent). Indians continued to leave settled parts of the South. Epidemic disease reduced the Indian population in 1760 to 53,600, an eleventh of the South's population. But the 2,400 Indians who still lived in settled parts of Virginia and Carolina were surrounded by more than a half-million whites and black slaves. Because most Indians in the South lived far from European settlements, the trade between European and Indian farmers nearly disappeared. After the Yamasee War, for instance, the Cherokees rarely visited English settlements, and when they did, colonists showed a great deal of hostility. Cherokees thus only had contact with resident traders, not with settlers.[82]

Until the late seventeenth century, Indians had been a preeminent part of the life of English farm families. Indians lived in their midst, trading or going to war with them. At first nearly every farm was on the front line, as the destruction in Virginia in 1622 shows; later, families in outlying areas might see their farms burned and husbands, wives, or children taken captive or killed. On occasion Indians had challenged colonial survival or had pushed settlement back toward the coast. With landownership so insecure, land scarcity hemmed in English settlers. So much land—but much of it a howling wilderness in the colonists' eyes; so much land—but Indians controlled most of it and threatened to force them off the rest. By the 1720s, however, the Indians had been pushed to the fringe of white settlement near or beyond the Appalachian Mountains, and only unmarried traders and the roughest frontiersmen knew Indians who lived independently. Land scarcity turned into land plenty. As Indians disappeared and folk traditions of cooperation with them faded, colonists invading the West saw them as impeding their own settlements. Not surprisingly, warfare among English settlers, French habitants, and Indians repeatedly broke out. Although

Indians fought valiantly to keep their land, neither English control nor the ultimate victory of settlers would ever be in doubt again.[83]

As confrontations with Indians receded into memory, the Indian became an abstract symbol representing savagery or independence, dignity or struggle. Indians had often captured colonists in wartime, but the captivity narrative, a new and popular genre, arose in New England (and to a lesser degree in the South) by the late seventeenth century. Readers could vicariously experience danger, exotic peoples, strange foods and harsh living conditions, Indian bravery and kindness toward women, and violent killing and torturing of men they would never encounter themselves. Set at first in the context of the Calvinist story of sin and redemption but later in more secular language, New England narratives followed captives from the time they were taken from their farms, through the march into the interior, and to their return to colonial society. More dispassionate southern writers described Indian customs and emphasized Indian cleanliness, bravery, and loyalty. Generations of settlers brought up on these tales came to see Indians as the independent denizens of the woods, outside the bounds of corrupt English society. To symbolize revolt against tyranny, patriotic Bostonians donned Indian garb at the Tea Party; backwoods Virginians fought the British in Indian hunting shirts, moccasins, and leggings; and Maine pioneers made themselves into Indians when rebelling against land speculators. Such activities had been inconceivable as long as Indians and whites were implacable foes.[84]

This Land Is Ours

Making the land English involved more than expropriating Indian territory. English law increasingly viewed land as private property, not as commons or a communal resource. If land was private property, it had to be owned by individuals. So even before settlers could begin to farm, colonial governments had to devise ways to distribute or sell ownership rights and to determine the boundaries between acreage individuals owned. Once land was distributed, land markets regulated its allocation everywhere. Distributing land was only the first step in a long process by which colonists made the land their own. Farmers changed the face of the land and gave it familiar English names; they built English houses and put English hogs and cattle on it; and they cultivated it for English ends of commerce.

Colonization began with the legal fiction that the king could dispose of Indian land—most of which no English person had seen—because he held sovereignty over it. But only occupation of the soil (building houses, planting gardens, and putting up fences) and not discovery alone could justify English sovereignty. To hasten occupation of the soil, the English adopted developmen-

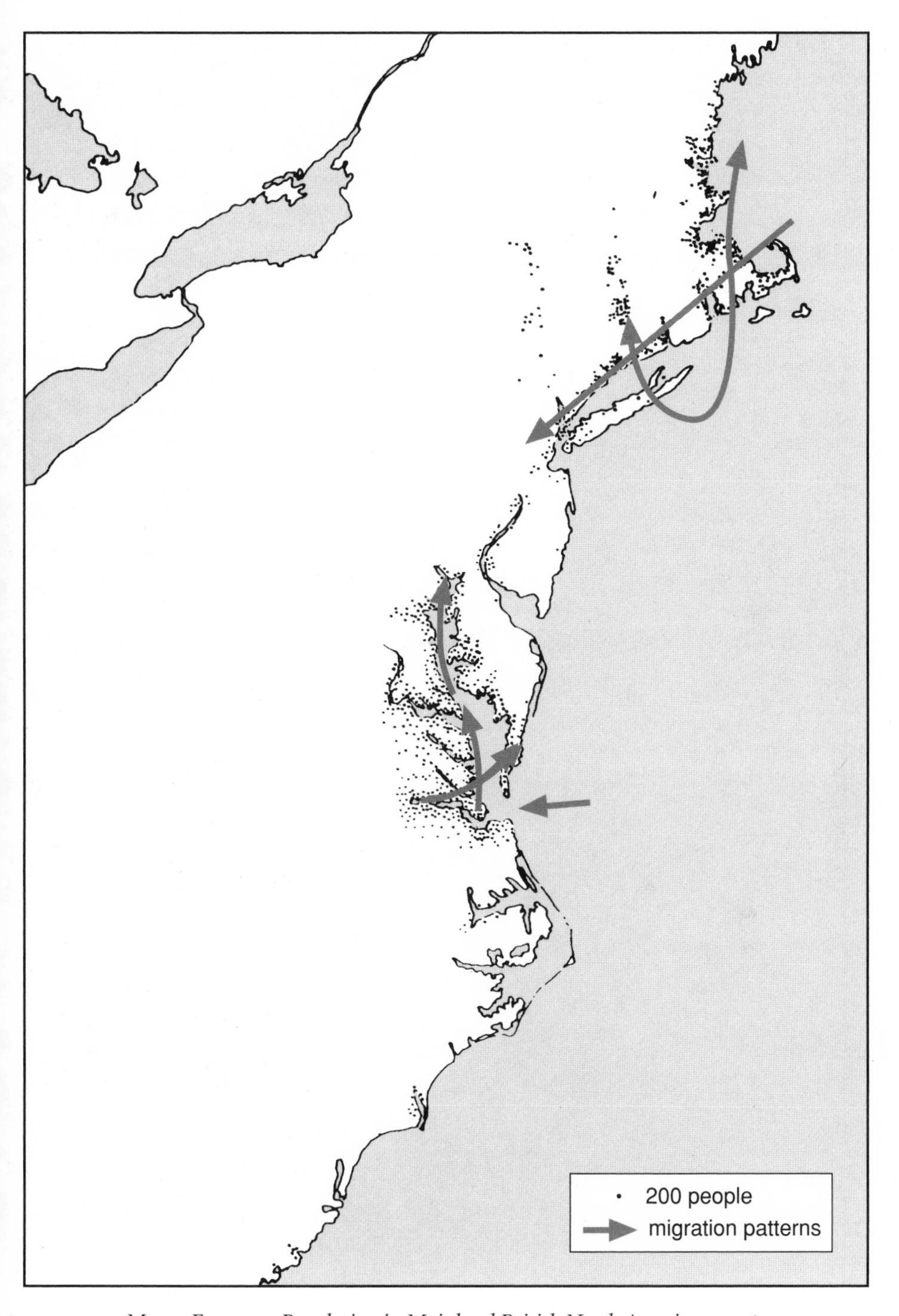

Map 1. European Population in Mainland British North America, ca. 1675

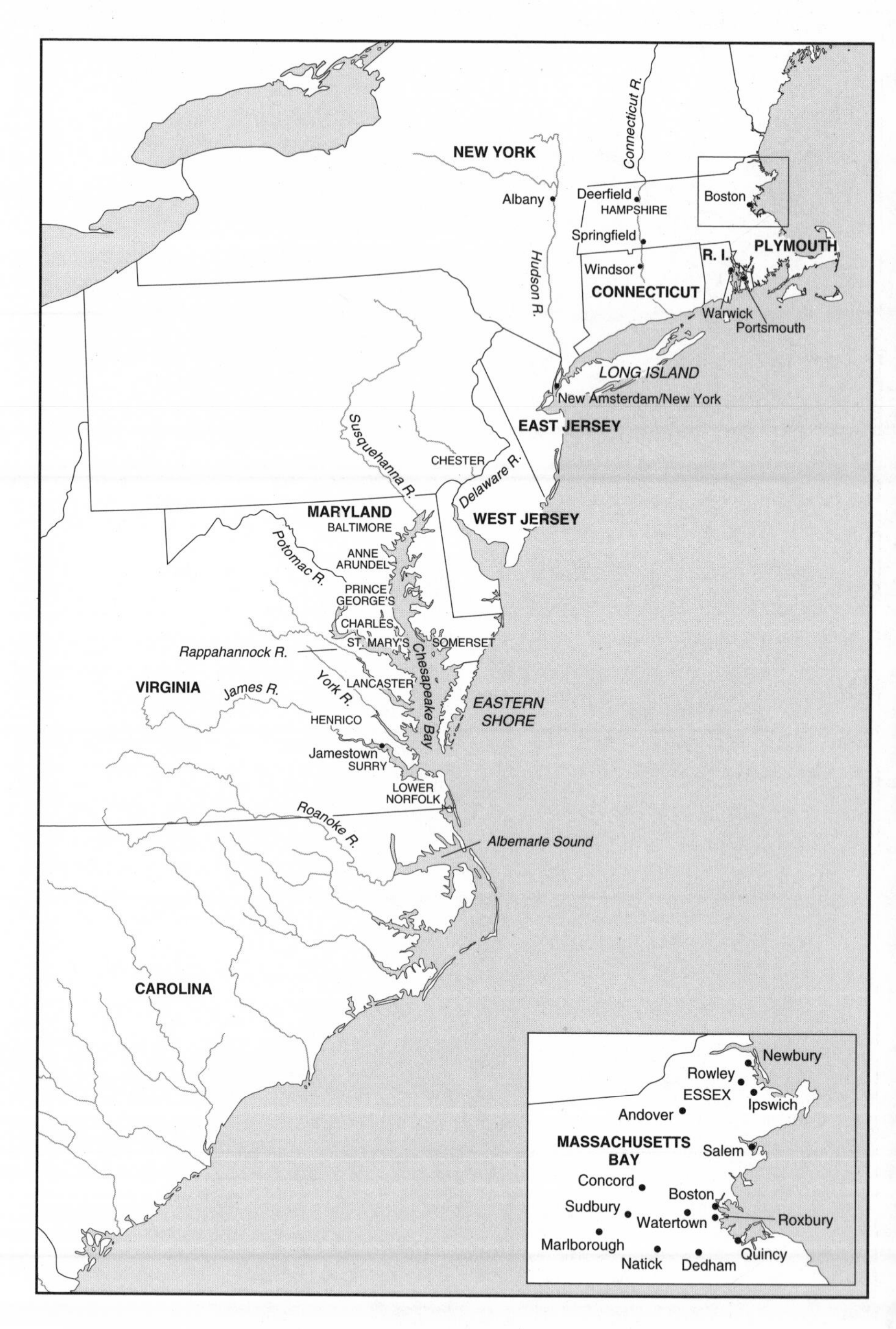

Map 2. *British Mainland Colonies, ca. 1675*

tal grants, commonly given to those who drained land or mined coal in England, to colonial needs. Joint stock companies or proprietors petitioned the crown for a charter giving investors the right to allocate or sell land (in fee simple or freehold tenure). Such tenure gave owners private property rights. Proprietors could form governments, appoint judges, collect taxes, and coin or print money—rights associated with sovereignty. Even in the seven colonies under royal control, local officials retained sovereignty rights, constrained only by the possibility of a royal veto. Charters, then, gave colonies a legal structure that justified stealing Indian land and defending it with force. In return proprietors had to plant and defend a colony loyal to the crown and pay the monarch a nominal fee and a fifth of the gold and silver found in the colony.[85]

With one eye on profits and the other often on creating an idealized utopia, adventurers and proprietors devised schemes to distribute land long before immigrants arrived. Uniformly seeking to repopulate their holdings rapidly, they also wanted to profit from their investments. Despite the capitalist nature of their enterprise some devised schemes to re-create feudal societies complete with manors. But only the promise of landownership would entice English emigration and sustain profits, and only the allocation of thousands of acres to investors would persuade them to risk their money. Manorial societies never materialized, except in parts of New York and Maryland, and colonial proprietors soon realized that land had to seem easy to acquire.[86]

After attempts to keep land in the hands of a few investors failed, the Virginia Company devised a system of land distribution designed to resolve the labor shortage and provide land to middling and rich immigrants. The headright system guaranteed anyone financing immigration the right to land. An immigrant who paid his own passage received land in return, as he did for each person—servant, slave, wife, or child—he brought over. Once an individual got this headright, it had to be registered. Then land had to be chosen, surveyed, and patented and the tract improved with a house, crops, or livestock. Headrights were larger at the outset of English occupation, when proprietors had to establish a beachhead against the Indians, but the system continued as long as land obtained from the Indians remained unoccupied. In 1619 the Virginia Company—which had previously held all land—granted 100 acres to settlers who had lived three years in the colony and 50 acres to the rest. To encourage immigration, the company gave 50 acres to each immigrant (and an additional 50 to anyone he transported) who stayed three years. This system continued after the dissolution of the company, with the added proviso that the holder improve the land within three years of registering the headright. Servants received no headright unless it was specified in the indenture, but a few men who brought over many servants and, later, slaves received rights to thousands of acres. The headright system thus distributed Virginia's land widely but unequally, and landless men had to buy land or headrights. These certificates, by the 1650s sold regularly to land speculators, were then purchased by small planters.[87]

New Netherland, Maryland, the Carolinas, Pennsylvania, and New Jersey copied the headright system. Desperate for immigrants, in 1638 the Dutch in New Netherland offered them land; two years later, they paid the passage of immigrants and gave them free land. During the 1630s, seeking to create big manors and rapidly people his infant Maryland colony, Lord Baltimore gave a 1,000-acre headright for each five persons transported, but only 100 acres per person to those who brought in fewer people and 50 acres for each child. By the late 1640s he had reduced the headright to 100 acres per person, no matter the age or number transported. Masters or the proprietor (after 1648) had to give each freed servant a 50-acre headright, but lacking money to register and survey the land, the new owners often waited until someone bought their headright before registering it. Seven-tenths of servants freed between 1669 and 1680 who registered their claims immediately sold their rights. The earliest Carolina settlers received 75–150 acres for transporting themselves and 30–50 acres for each servant and child, but thereafter the acreage was reduced to 50 per person. Headright grants were less common in the mid-Atlantic region. East and West Jersey established 20- to 70-acre headrights; William Penn granted a 50-acre headright for each servant brought to the colony but distributed nearly all the land in larger blocks to investors.[88] (See Map 2 for places.)

On New England and English Long Island, land was distributed differently. A group of proprietors petitioned the legislature or Dutch authorities to grant them a town. Led by prominent men capable of influencing the legislature and enticing landless families to move to the wilderness, proprietors included most men seeking to farm in the town. Once a town site had been granted, proprietors bought Indian land and had the area surveyed. Proprietors or the town meeting set up commons (grazing land that all families could use) and distributed land, keeping much of it in reserve for future allocations to householders or their sons. Principles of land distribution varied. In half the New England towns, each family received a share proportional to its contribution to buying and surveying the land; in most other towns, the family's wealth determined its share. A few people in half the towns did not participate in allocations but either received a small gift of land or leased their farms. The proportion of town lands in New England allocated in each division and the acreage given each family varied according to land distribution systems in the settlers' English homeland. Newcomers rarely shared in land divisions; a few original families may have been landless tenants; and as many as half the rest held less than fifty acres. Town proprietors nonetheless allocated land more evenly than the headright system.[89]

Colonial proprietors often engaged in fantasies of establishing a manorial system, complete with feudal dues and a status-bound feudal hierarchy. Swept away by the failure of gentlemen to migrate permanently, labor shortages, and abundant land, such delusions rarely materialized. Colonists wanted their own land, unconstrained by dependence on a manor lord. New York and Maryland appear to have been exceptions, however. The Dutch created a few large patroonships in New Netherland, but only one—Rensselaerwyck—succeeded, and that only because the proprietor paid for transportation and equipped farms, taking a share of the crop as rent. During the 1680s the English created new manors in the Hudson Valley, but to attract tenants landlords had to finance migration, build mills, plant orchards, provide livestock, and give good terms (long nominal rents or leases for life, with rights to sell improvements). Maryland manors were similar. Seeking to create a manorial system, Lord Baltimore gave anyone bringing in enough settlers to warrant a 1,000- or 3,000-acre grant the right to establish a manor. Six manors were created; half were owned by the proprietor's brother. Although planters initially settled near the manors, the system collapsed, bedeviled by labor shortages and the unwillingness of ex-servants to remain tenants. As population grew, the Calverts formed new manors, which eventually numbered twenty-three. But like New York landlords, they peopled manors only by offering inexpensive developmental leases (10 shillings per 100 acres) for three lives or 99 years.[90]

Soon after English occupation, land markets developed; by the late seventeenth century, direct sales had become the norm everywhere. Beset by corruption, the headright system ended in Maryland in 1683 and was replaced with direct sales from the proprietor; about the same time, Virginia governors began selling unpatented land. Nonresident landowners, found in two-thirds of New England towns, sold land to newcomers. Elsewhere land speculators, colonial investors, or governors' favorites—who acquired land through grants from proprietors or by sponsoring immigration of many individuals—sold surplus land. William Penn, for one, sold Pennsylvania proprietary land to the "first purchasers," who resold it to farmers, and his land office later sold smaller parcels, varying the price by location and soil quality. Government land sales (mostly to rich men) coexisted alongside headright grants in late seventeenth-century South Carolina. Once colonists gained ownership, they sold land to accumulate capital or to pay debts. Although the sons of original landowners might inherit property, those who immigrated after an area's land had been allocated or sold had no choice but to buy from current owners or move to a frontier.[91]

In an English land system, where individuals owned specific acres within known bounds, establishing the borders of tracts was essential. Before landownership could be recorded, the property had to be surveyed; persons who farmed land before a survey was completed were squatters. Officials wanted compact farms, each with its share of the best and worst lands; seeking the best land, farmers wanted as much water frontage as possible. Surveyors divided the land into a grid system with rectangular plots (in New England and parts of Pennsylvania) or skinny strips as long as three miles (in other parts of New England) or used natural boundaries such as streams (in the South). Surveying in forests, through "THICK WOODS our COURSE to know," surveyors divided one farmer's land from the next. To mark the corners or bounds of tracts, they selected natural barriers such as creeks or streams, notched unusual trees, or linked the tract to farms already surveyed. Virginia owners could patent as much water frontage as they desired, but surveyors devised a method that gave each owner an equal opportunity to get it. To encourage thicker settlement, West Jersey, Maryland, and Carolina officials limited the length of shoreline a patent might include. In the Chesapeake colonies, once the survey was completed, a map and a description of the tract were copied into patent books, the owner paid a surveying and patenting fee, and the land was his. A similar but less efficient system, where colonists could gain titles without patenting land, could be found in South Carolina, New Jersey, and Pennsylvania. In New England, town land was surveyed before residents arrived, and then a town official resurveyed it as he distributed it to village heads of household, limiting access to waterfront as he proceeded.[92]

Farmers showed boundary trees to their children, wives, and neighbors. When parcels of land overlapped or bounds were challenged—common occurrences given the crude tools surveyors used, the inexactness of surveys, and the burden of rapidly surveying hundreds of tracts—witnesses remembered the boundary trees. Maryland landowners, who could petition for resurveys when controversies occurred, counted on people testifying about boundary trees. In 1729 John Middleton (age fifty-five) of Prince George's County reported that his father had shown him the boundary tree of their land thirty-two or thirty-three years previously. Similarly, Edward Heneberry (age forty-six) testified in 1732 that sixteen or seventeen years earlier his master Christopher Beanes, while "Walking in the Woods" with him, showed him a boundary tree so "that he might be of Service to his Children after his death." The boundaries of Virginia land had become so confused by the 1660s that the assembly mandated that inhabitants in each parish inspect and note boundaries every four years. In 1634 the Massachusetts Court of Assistants ordered a similar procedure to ensure that the town book accurately recorded land.[93]

Colonial systems of land allocation were remarkably successful. By the end of the seventeenth century, most householders owned land. A few landless men lived in most New England towns, but all proprietors received arable and pasture in town distributions. All but 4 of the first 238 inhabitants of Salem,

Massachusetts, got land, and later arrivals fared nearly as well, eleven-twelfths (134 of 146) getting land. New England land continued to be widely distributed. In three towns in Essex County, Massachusetts, in the late seventeenth century, half the men owned land before they were thirty, as did 95 percent of men over thirty-six. Before 1660 two-fifths of Connecticut settlers, most of them young men, had no land, but by the 1690s six-sevenths of all farmers owned land. Similarly high levels of landownership could be found in the Chesapeake colonies. In 1660 four-fifths of the white men in Charles County, Maryland, were landowners; as the opportunity for former servants to get land plummeted, the proportion of owners among taxable men declined to seven-tenths in 1675 and six-tenths in 1690. Most landless men either moved from the county or died young, before they could acquire land. In both 1687 and 1704 nearly two-thirds of the household heads in Surry County, Virginia, held land, as did three-quarters of householders in Talbot County, on Maryland's Eastern Shore, in 1704. Landownership, moreover, might have been nearly universal in early Pennsylvania; during the 1690s eight-ninths of the householders in one Chester County township owned land.[94]

Most seventeenth-century farmers had enough land to plant grain, cultivate a garden, and graze cows and pigs. To grow tobacco, Chesapeake planters had to own 100 acres. Maryland planters in the 1660s typically owned 250- to 300-acre farms; by the end of the century, farm sizes had declined to 150 to 200 acres. Between 1635 and 1655 settlers on Virginia's Eastern Shore held 300 to 400 acres; in the 1650s planters in Lower Norfolk County owned more than 300. From 1650 to 1675 Virginians patented over 2 million acres, thereby maintaining big farms. In 1705, after this land boom ended, the average Virginia planter owned a 225-acre farm. Nearly all had viable tobacco farms; a quarter owned 100 acres or less, but three-fifths of this group had 100 acres. Significant differences in landholding had begun to appear. In frontier Henrico County, farms averaged 300 acres; but only a third of the county's land had yet been seated, and speculators and merchant-planters with over 1,000 acres owned three-fifths of the patented land. As more families moved in, population density grew, big holders sold their surplus, land became more scarce, and farm sizes dropped. By the end of the century that process had been completed in counties at the southern tip of the peninsula between the James and York Rivers. Landowners there had patented more than four-fifths of the available land; typical farms averaged 150 acres, and most men got land through inheritance.[95]

Even though seventeenth-century New England, Long Island, and East Jersey proprietors allocated land by social status, granting more to ministers, elders, merchants, or rich men who had financed purchasing the town, nearly everyone received land. In Swansea, Plymouth, during the 1680s the town meeting devised a three-rank system, allocating the top group three acres to the middle group's two and the low group's one. Since proprietors held land in reserve, farms were much smaller than those in Virginia or Maryland but sufficient to grow corn and make gardens, and residents grazed livestock on the town common. Families initially received less than 20 acres in Rowley, 20 acres in Salem, 33 in Newbury, 34 in Marlborough, and 64 in Watertown. These averages changed little over the century. Later allocations increased holdings; sales and bequests reduced them. Men who stayed in town ultimately received more, perhaps as much as 125 to 150 acres in both Dedham and Watertown. Similarly, New Englanders who moved to East Jersey late in the century patented 50- to 150-acre farms.[96]

To get land, immigrants left older towns and moved more often and over longer distances than they had in England. New England immigrants typically moved once, but between one-third and two-fifths stayed put. Young men moved most often. Three-fifths of those under forty but only two-fifths of older men left their first American residence. Those with little land moved most often. While two-fifths of the first-generation families in Ipswich, Massachusetts, with allotments of ten or more acres moved during their lifetime, at least seven-tenths of those with less than ten acres (about half of the early settlers) left town. Landed men with large families moved as well, fearing they would not be able to accumulate enough land for their children.[97]

The possibility of acquiring more land clearly motivated the movers. Sudbury and Andover, Massachusetts, settled by men leaving Watertown and Newbury, respectively, are good examples. At first Sudbury proprietors allotted an average of just 21 acres, but in 1658 they divided most of the commons, giving proprietors and their sons enough land to increase average holdings to 130 acres. Andover's proprietors gave settlers house lots of 4 to 20 acres, but in 1662, after four new divisions, the original forty householders owned on average 153 acres. Newcomers or landless sons—forty by 1686—got no land but could buy 20-acre plots. Early residents of Windsor, Connecticut, received over 150 acres; by the 1670s Connecticut landholders age forty to sixty had 80 acres. Long Islanders were less successful. Farmers in five Long Island Dutch towns held on average 39 acres of improved land in 1683, but those in the English towns owned just 16 acres, no more than in Salem or Newbury. As many as two-fifths of the English farmers on Long Island owned 10 or fewer acres, enough to plant some grain and make a vegetable garden, but even with access to commons, they had to work as artisans or fishermen or whalers to support their families.[98]

Immigrants repeatedly moved short distances to take up land. After they arrived on the coast, they fanned out across the countryside, following the courses of rivers and the contours of the land and seldom traveling more than two days from the ocean or venturing onto Indian lands. Most New Englanders left their first farms and moved to nearby places where they stood a better chance of getting larger farms. Some migrants, however, moved to distant Long Island or East Jersey or to Connecticut River towns. Massachusetts and Plymouth families competed to seat early Connecticut. One Plymouth observer lamented in 1635 that "the Massachusetts men are coming almost daily some by water, some by land." Cattle raising—a particularly appropriate vocation given the scarcity of labor—required much land. As cattle increased in New England during the mid-1630s, for instance, farmers in Plymouth, Watertown, and Roxbury left for areas that offered more pasture. The search for pastures continued throughout the century, leading to the settlement of the Connecticut River Valley. Although immigrants often persisted for decades, as many as two-thirds of their children moved—at marriage, to set up a farm, or to start a craft business. Most native-born migrants left for nearby villages, and nearly all lived within thirty miles of their birthplaces. By granting large parcels of land to new town proprietors when older towns filled up with farms, New England governments encouraged this movement.[99] (See Map 1.)

Religious radicalism sometimes combined with land hunger to impel New England immigrants to move. Massachusetts authorities thrust Ann Hutchinson and her Antinomian followers into exile (to Rhode Island). Richard Hooker, who disagreed with John Cotton on religious matters, began the Hartford settlement, taking followers, some of whom were land hungry, with him. Others, unwilling to live amidst the saints, moved to Plymouth or Rhode Island, colonies with more open religious polities.[100]

Similar dispersion could be found in the Chesapeake. Ex-servants tramped the countryside searching for farms or for labor paying enough to allow them to rent or purchase land. Often unable to find land near their former master, they moved to nearby frontiers and settled along rivers and creeks before they moved inland. Settlers gradually filled in the land, patenting, for instance, 42,310 acres in Somerset County by 1670 and 210,980 acres by 1690—five times the 1670 total. Able to get land, four-fifths of former servants in Charles County, Maryland, stayed in the county, as did many unable to rise out of the ranks of laborers or tenants. A quarter of surviving immigrants left Charles County, and about two-fifths departed from Lancaster County, Virginia. As opportunities ebbed after 1660, tramping increased, leading county courts to complain that "vagrant persons . . . remove from place to place" to avoid paying taxes. After 1680 migration from the region increased as tobacco prices and opportunities to purchase land declined. Short-distance migration opened new areas for European habitation; the founding of North Carolina and Pennsylvania encouraged poorer men and their families to move greater distances to find land and establish themselves as small producers.[101]

Despite the vestiges of feudalism in colonial charters, private property in land and land speculation became universal, constrained only by customary rights of common use of unenclosed and unimproved lands. The complex English land tenure system thereby broke down. Most land was held in fee simple, a tenure that allowed owners to sell and bequeath their holdings. Free from nearly all feudal dues, land held in fee simple was constrained only by the payment of land taxes.[102]

Feudal vestiges—albeit transformed—remained on freehold land. Most landowners had to pay a quitrent, either to the proprietor (in Maryland and Pennsylvania) or to the king (in other colonies outside New England). A quitrent represented the feudal bond between the owner of in fee simple land and his king; in the colonies it became a small tax on land paid to king or proprietor. The sum collected would have been sizable had everyone paid. But farmers often evaded the tax, and tax collectors or royal officials, underpaid and knowing how unpopular quitrents were, made few efforts to collect them or to bring scofflaws to justice. In Pennsylvania and North Carolina the rates were variable and therefore seemed unjust; in Maryland and Virginia even low rates could impoverish a small landholder in years of low tobacco prices. The quitrent itself, not the level of payment, became the key issue. Colonists knew that the reciprocal obligations that quitrents symbolized in a feudal society had long disappeared, and the payment had become a tax benefiting proprietor or king, not their communities. New England farmers, used to unfettered lands, refused on principle to pay quitrents; those in proprietary colonies objected to paying a middleman, who owed a similar obligation to the crown. Accepting the king as sovereign, farmers in royal colonies evaded rather than attacked the system. In royal South Carolina those who failed to patent land avoided paying quitrents, and partially as a result quitrents were paid on two-fifths or less of the land. This resistance turned land into private property, unconstrained by any feudal obligations.[103]

As we have seen, Indians hemmed in colonists, creating potential land scarcities. Since so little land was available for settler use, one might expect bigger operators to monopolize it and to rent it out to smaller holders. Before 1700, however, large-scale tenancy appeared nowhere. Even the poorest man could farm a piece of land if he was willing to squat illegally on it. Less than a tenth of the household heads in eastern Massachusetts and about a tenth of Connecticut men age forty to sixty rented land; on Long Island one in ten late-seventeenth-century farmers in English towns and one in six in Dutch towns did so. Before the late seventeenth century, New York manor lords had patented at least 2 million Hudson Valley acres. Nonetheless, farmers could easily acquire freehold land, which led manor lord Jeremiah Van Rensselaer to complain in 1671 that it was "no longer possible to get any tenants for the farms"; as late as 1700, the manors had few tenants.

Tenancy did grow in a few places in New England. The Pynchon family owned the best land in Springfield, on the Connecticut River frontier of Massachusetts, and a third of the town's farmers—including some who owned distant, poor, or infertile acres—leased land from them. As Salem developed into a commercial center, fewer families owned land, even on the farming periphery. By 1689 in Salem Village, where three-fifths of the families were farmers, half the heads of household owned land, a quarter lived on family land (but owned none themselves), and a quarter rented. Other Essex County communities probably had similarly high rates of tenancy, as richer men took advantage of the poor and off-season fishermen to develop their land as tenancies.[104]

Tenancy was equally uncommon in the Chesapeake region. One-twelfth of the planters in Charles County, Maryland, leased land in 1660; by 1705, after land prices more than doubled, that number jumped to one in four (and to one in five in the rest of southern Maryland). But a fifth of the Charles County tenants rented land from the Jesuits on good terms. Before 1680 they leased 200- to 500-acre farms for 500 to 1,000 pounds of tobacco (less than half of one worker's crop) or for clearing the land and building houses and barns; after 1680 rents increased but ran for three lives. So much land (more than half) in settled counties remained unpatented in 1704 that nearly all Virginians without a land title could have squatted on unsurveyed land. In 1687 one-third of Surry County householders owned no land, but rather than renting land, some of the landless squatted on unpatented acres on the southern edge of the county.[105]

However many farms they planted in America, the English saw the land as strange, alien, and wild. Indians had named its hills and mountains, rivers and villages, and trees and vegetation. At first adventurers learned Indian names and put them on their maps, as John Smith did in his 1612 map of Virginia. But to symbolize their sovereignty, settlers soon changed them. Sir Walter Raleigh renamed Wingandacola (Roanoke) Virginia (for Elizabeth, the virgin queen) before settlers arrived. In 1602 poet John Beaumont praised the renaming:

> Where Cipo with his silver streames doth goe
> Along the valley of Wingandekoe:
> —Which now a farre more glorious name doth beare.

England, Beaumont continued:

> hath uncontrol'd stretcht her mightie hand
> Over Virginia and the New-found-land,
> And spread the colours of our English Rose.

Later Tsenacomoco, or Tsenacommach, the "densely settled land" of the Powhatans, became Virginia. On the twenty-fourth day after their arrival, Virginia adventurers "set up a cross at the head" of the river the Indians called Powhatan, "naming it King's River, where we proclaimed James, King of England, to have the most right unto it." The river soon became the James. John Smith renamed New England Indian villages and printed them on a map in 1616, four years before the first permanent settlers arrived. Sowocatuck became Ipswich; Accominiticus, Boston; and Passataquack, Hull.[106]

As long as Indians controlled the landscape, the names of places remained contested. A Providence, Rhode Island, deed of 1638—just after Roger Williams and his band arrived—either gave places their Indian names or listed both English and Indian names (Moshausick was Providence; Apaum, Plymouth). The same year, long after intensive English settlement, a Dutch map still dubbed the James River "River Powhatan." Later, Massachusetts praying villages, which retained Indian names, abutted English settlements with English names. Nor did English names (such as Delaware Bay) become fixed for decades, for as long as the Dutch controlled any North American territory, they insisted on replacing Indian names with Dutch.[107]

English place-names, however, did become conventional. By naming colonies, counties, and towns after places they had left behind, immigrants made American land English. More than four-fifths of the counties formed in the colonies before 1700 took the names of British places or British (especially royal) notables; thirteen of twenty-two Virginia and Maryland counties founded by 1660 were named after an English county; six counties honored English monarchs or nobility. At least forty-three of sixty-two Massachusetts Bay and Plymouth towns took the names of English communities, usually from East Anglia, where many immigrants had resided. Towns in newer colonies (such as New Hampshire) were named for New England towns, other British places, or their founder, prominent New Englanders, or English benefactors. In contrast, Indian names—which represented prior occupation by indigenous peoples—were nearly absent. Massachusetts was named after an Indian tribe, but no Massachusetts or Plymouth towns and only two Chesapeake counties took Indian names. Bodies of water usually took on English (or in New Netherland, Dutch) names as well. Nearly a quarter of the ponds and brooks in eastern Massachusetts were named for the owners of the land on which they were found, and as many were named after animals or fish, farm characteristics (cow pond, pantry brook), or the character of the body of water (muddy pond). Naming the hills and rivers made an alien landscape seem English, even when most of the land remained, in settlers' eyes, wild.[108]

The English renaming of Indian places went on everywhere. In New England, as soon as a few English families began to farm in an old Indian village, they petitioned to replace the Indian name (Wannamoisett, for instance) with a more familiar English name (Swansy, in this case). An upper Connecticut River settlement kept the name Pocumuck or Pacomptuck until King Philip's War, but when the destroyed town was resettled after the war, the residents named it

Deerfield. Most topography soon took on English names: two-fifths of the rivers but just a fifth of the ponds, a sixth of the brooks and hills, and a twelfth of the swamps of eastern Massachusetts retained Indian names, even when such English-native names as Indian Brook are included. In New Jersey, Lenape names quickly disappeared, save for six that denoted waterways. Chesapeake planters similarly renamed natural features. The only Indian names that survive in Virginia describe physical features, notably waterways, such as Chesapeake, "mother of waters," or Rappahannock, "quick-rising water." Original owners used Indian words to name only 1 percent of the tracts in seventeenth-century Prince George's County, Maryland.[109]

To make the land theirs, the English used English terms to label the landscape features or, when no term was appropriate, gave English words a new, American meaning. Before 1650 the English in Virginia named features islands, hills, and meadows, all concepts brought from England. They named numerous rivers (a word meaning, as in England, a large stream leading to a sea) and hundreds of creeks (meaning a long, narrow, saltwater stream in England but fresh and saltwater tributaries of rivers in Virginia). The strangeness of the landscape notwithstanding, Virginians rarely borrowed terms from the Indians ("pocosin," meaning swamp, was the most important). New Englanders used similar words, dubbing features meadows, rivers, and gullies; dividing land into commons and wasteland; and labeling species of trees (sometimes inaccurately), on rare occasions transforming Algonquian words such as "pokelogan" or "pecelaygan" (backwater or wetlands) into "logan."[110]

Men of substance named most rivers and counties, but land, even when owned, was but nature until colonists put their imprint on it. Before building houses or plowing fields, settlers named their land. In the Chesapeake planters named their estates after themselves or their wives (Griffin's Chance, Betty's Desire) or harked back to English places or great houses (Epping Forest, Stratford Hall). Marylanders almost never used Indian words (or the word "Indian") in tract names. The land was theirs, not the aboriginal owners'. Owners of nearly half the tracts patented in seventeenth-century Baltimore and Anne Arundel Counties and over a third of the tracts in Prince George's County named the land for themselves, usually attaching their surname to the fact of ownership (lot, purchase, addition), a feature or use of the land (neck, level, hill, pasture), or an economic expectation (chance, desire, hope, folly). Planters used similar words, but not their own names, to identify the rest of their lands and borrowed about a tenth from English place-names. A handful named land after biblical places or (in Prince George's, where numerous Catholics lived) saints. Similar naming patterns could be found in late-seventeenth-century West Jersey, where close to three-fifths of Quaker farmers named their land after themselves, usually combining their surname with terms such as "plantation," "grove," or "meadow." The rest usually named their land after topographical features.[111]

The first English inhabitants envisioned a landscape that they would build by their labor, turning forests into fields and meadows into pasture. Settlers gave grasses, vegetables, animals, and fishes English names. In his 1634 account of New England, William Wood predicted the region's agrarian future. He saw meadows, marshes, wild vegetation, and wild beasts but knew they could be reshaped. The marshes "be rich ground and bring plenty of hay, of which the cattle feed," enough for any planter "though his herd increase into thousands." The ground would bring forth "turnips, parsnips, carrots, radishes, and pumpions [pumpkins]" and "whatsoever grows well in England." By 1634 settlers had changed the land. Mount Wollaston (Quincy) had "a very fertile soil and a place convenient for farmers' houses"; the people of Roxbury "have fair houses, store of cattle, impaled cornfields, and fruitful gardens."[112]

Having conquered and named the land, farmers set themselves and their families to the task of making farms. Before moving to the land, they had to buy provisions and goods—livestock, seeds, axes, hoes, bedding, pots, and dishes—necessary to set up a household and a farm. During their first season they cleared an acre or two of woods, planted crops, built an impermanent house and outbuildings, and put up fences around their crops (a practice mandated by law and by the free-ranging hogs and cattle). Gradually they cleared more land, improved their dwellings, and bought better tools and a greater variety of household goods. To get goods to market, they joined with neighbors to widen Indian paths; to sustain their communities, they banded together to erect churches, meetinghouses, and courthouses. In so doing, they had made a new landscape, at once English and American.[113]

THE SHOCKS TO body and soul immigrants experienced, from the harsh climate to hostile Indians, led them to cling tenaciously to English social norms. To be sure, survival mandated that colonists acclimate themselves to the wilderness, eating new foods, wearing different clothes, and borrowing from the Indians. In making these changes, settlers did not become the new democratic and individualistic people of our folklore; much less did they turn into Indians. Rather, they integrated their new habits into English regional customs, practices, and institutions. Their struggle with nature and Indians in the face of persisting English values lies at the center of their work at farm making and community building.[114]

Patterns of landownership and land use pointed back to England and forward to America. Proprietors tried to control landownership and create an idealized English society on American soil; New England town proprietors distributed land in ways that resembled the customs of their home villages. But high levels of landownership and the dispersion of the population made that impossible. Where English maps showed the feudal tenure of those who held land, colonial maps showed boundaries among landowners. Colonists incorporated the English system of entail (mandating that land be kept in the family) into their laws and duplicated English forms of descent of land in intestate estates, instituting primogeniture or giving the eldest son a double share. But colonists who made wills infrequently practiced primogeniture, and only the land of gentlemen (even in Virginia) was entailed. Settlers used English land-surveying methods designed to make settlement compact but created a thinly settled society of isolated farms, rather than the villages filled with landless laborers found in the mother country.[115]

At first the houses of immigrants bore little resemblance to England's. Their first houses, built after they left their temporary huts or wigwams, contained one room or a hall, a parlor, and a loft and were much smaller than houses in England. They were built in a hurry on posts rather than a foundation, with clapboards and wooden rather than brick chimneys. Only the mid-Atlantic Finns built log houses similar to those in their forest homeland; later immigrants copied the style and took it to the eighteenth-century Appalachian frontier. Whether built of logs or clapboards, these houses were meant to last only twenty or twenty-five years. As late as 1700 many families still lived in such houses. Those of small mid-Atlantic and southern farmers and planters showed little internal specialization and remained small (400 or fewer square feet) and impermanent one- to three-room structures. These impermanent houses served their owners well. Clearing land, herding cattle, and growing market crops took most of the time and capital these farm families had, and by the time the houses disintegrated, they were ready to move to fresh land. After mid-century, however, middling and wealthy farmers built larger permanent homes, with four or more rooms spread over two floors. Prosperous New Englanders lived in wooden homes; substantial southeastern Pennsylvania families moved into multiroom stone houses; rich Chesapeake planters surrounded their main house (still built on post holes rather than foundations) with separate kitchens, dairies, and other structures.[116]

Middling colonists copied the architectural styles of their regions of origin. New Englanders adapted East Anglian and West Country styles; Virginia gentlemen, southern English styles; and Delaware Valley immigrants, North Country or Finnish styles. Builders borrowed from English styles but added colonial innovations, using more wood but preparing it in simpler ways than in England. Middling New England farmhouses resembled those of East Anglia but had root cellars rarely found in England; others used the transverse summer beam construction common in the West Country. By 1700 Chesapeake builders had so radically simplified English construction techniques (borrowing from Africans or Indians) that ordinary houses resembled no English style. To mitigate the heat of summer, soon after arrival planters dug storage pits (like Indians), and after mid-century, middling and rich planters had moved kitchens into separate buildings, perhaps like those in the West Country.[117]

Only laws firmly enforced, immigrant leaders insisted, could prevent the English from sinking into savagery. Colonial charters mandated that "orders, lawes, statutes, and ordinances" be "not contrarie to the lawes . . . of England" (Massachusetts Bay Charter) or "soe farre as conveniently may bee agreeable with the Lawes of . . . *England*" (Pennsylvania Charter). But what English laws should be imposed? English common law—an amalgam of contradictory customs of merchants, courts, towns, and villages—proved unreliable. Nor were parliamentary statutes, themselves at times contradictory, always apt. Colonial assemblies borrowed diverse English customs but innovated when necessary. Fearing disorder, New England towns made their own covenants, signed by male property holders, to regulate social and economic behavior and passed ordinances when necessary. Magistrates kept familiar legal forms and practices and followed English law manuals. Soon colonial laws became so complex and contradictory that assemblies in Virginia (in 1632, 1642, 1652, and 1662), Plymouth (1636), Massachusetts (1641 and 1648), Rhode Island (1647), and later, Pennsylvania and New York brought order to their laws by codifying them.[118]

Magistrates placed English laws into a new context, heightening their original ambiguities, as examples from the Chesapeake colonies illustrate. The harsh penal laws and summary justice of early Virginia, which grew out of disorder in the colony, reflected brutal Elizabethan poor laws, the suppression of vagabonds, and English military justice. But assemblies had to innovate as well. Created by the provincial assemblies, county courts of the Chesapeake colonies gained jurisdiction not only over the petty criminal and civil matters of similar English courts but also over estates, bigamy, and bastardy, matters reserved in England for church courts. They had jurisdiction over nearly all civil matters and debt cases, even those sent to higher courts in England. When the Virginia or Maryland Assembly copied English criminal law, it wrote new provisions to prevent trading with Indians or to punish runaway servants, problems never encountered in England, where no Indians lived and servants worked on annual contracts. Aware of their extraordinary number of orphans, both colonies created orphans' courts (within the county courts). These courts grew out of English practice, but Chesapeake legislatures went beyond English precedent and protected orphans from harm or venal stepparents by specifying in detail how they were to be educated or trained and how estates were to be operated by guardians.[119]

The communities settlers made reminded them of home and helped them adjust to the wilderness. They borrowed communal institutions (county courts, churches, and town meetings) and officials (justices of the peace, sheriffs, selectmen, and ministers) from English practice in their home regions. Just as in England, such institutions mediated between markets where goods were exchanged and households that made wheat, corn, tobacco, or rice. Justices regulated the marketplace, sometimes setting prices and wages, and guaranteed the authority of husbands over wives, fathers over children, and masters over servants and slaves. The form of colonial communal institutions, then, could be traced to England, but the problems they sought to resolve grew out of America, full of forests and Indians.[120]

chapter three

New Lands, New Peoples

Writing in the guise of a simple American farmer of the 1770s, J. Hector St. John Crèvecoeur, a French immigrant who became a gentleman farmer in New York's Hudson Valley, captured the importance of the ownership of land for eighteenth-century farmers. "What should we American farmers be without the distinct possession of the soil? It feeds, it clothes us; from it we draw . . . our best meat, our richest drink, the very honey of our bees comes from this privileged spot." But landownership meant more than prosperity; it created the "only philosophy of the American farmer." A man's labor, not mere purchase, sustained his ownership of land. If he refused to improve his land or if he abandoned it, he deserved to lose it. "This formerly rude soil has been converted by my father into a pleasant farm," Crèvecoeur added, "and in return, it has established all our rights; on it is founded our rank, our freedom, our power as citizens." The American farmer expected to pass his farm on to his son, whom he trained from an early age to appreciate the soil. This abundant land had "enticed so many Europeans who have never been able to say that such portion of land was theirs" to "cross the Atlantic to realize that happiness."[1]

A medieval peasant loved the soil as much as Crèvecoeur's American farmer. He did not own the land, however, but merely enjoyed the right to farm the soil and graze livestock on the commons, constrained by his lord's rights to take rents for the same land. He might bequeath or sell his rights to the land, but he had to pay a fine to his lord. By the late seventeenth century, colonists had abandoned feudal concepts of land tenure. Although commons remained in New England towns, town allocation of land had disappeared. Feudal vestiges such as quitrents or feudal dues paid to manor lords remained, but they had been transformed into land taxes or rents.[2]

Colonists replaced feudal concepts, in Crèvecoeur's words, with ideas "of exclusive right, of independence," and of landownership. "Descended from its great Creator," the land "holds not its precarious tenure either from a supercilious prince or a proud lord." Not even the king had rights to a farmer's land, beyond taxation for public purposes. The consequences, insisted David Ramsay, author of one of the first histories of the American Revolution, were great. "Settled on lands of his own, he was both farmer and landlord—producing all the necessaries of life from his own grounds, he felt himself both free and independent." Land, colonists assumed, was private property, a commodity that could be bought, sold, bequeathed, or leased. Possession of the soil was insufficient; a man had to own it, be the absolute proprietor over it, and be able to prevent others from trespassing on it or using it in any way.[3]

Courts and assemblies, small landholders believed, had to guarantee the secure enjoyment of their landed private property. In conflicts over land titles, they expected officials to support them against Indians and landlords. If officials took their land for mills, roads, bridges, or river improvements that everyone could enjoy, they expected compensation for their improvements or their prior use of the land. Every farmer, twenty-five German settlers in Maryland insisted in the late 1740s, "is secure in the enjoyment of his property, the meanest person is out of reach of oppression from the most powerful nor can anything be taken from him without his receiving satisfaction for it." The 1780 Massachusetts constitution summed up colonial practice: "Whenever the public exigencies require that property of any individual" be taken for "public uses, he shall receive reasonable compensation therefore."[4]

Only landowners could get commercial credit. Real estate—farmhouse, kitchen and barn, arable, pasture, woods, and unimproved acres—were farmers' most valuable possessions. Between 1700 and the 1770s the value of land in New England grew from three-fifths to nearly three-quarters of total property; between 1755 and 1776 land in Prince George's County, Maryland, rose from a third to more than a half the value of property. In 1774 half the value of farmers' property in all the colonies was tied up in land. Land counted most (71 percent) in New England, but even in the slave South, land accounted for almost half (46 percent) of property and was worth more than servants and slaves.[5]

Only men who owned land and directed family labor could claim independence from their superiors. Small farmers expected that virtuous rulers would protect their landed property from the rapaciousness of the greedy and powerful. In contrast, a tenant was required to negotiate with his landlord, an unequal relationship that often made the tenant dependent. Tenants usually could neither vote nor serve on juries. Poor squatters could stay on their land rent-free only until others who were willing to buy or rent the land took it as their own. No wonder, New York Lieutenant Governor Cadwallader Colden wrote the

Lords of Trade in 1761, farmers wanted "to bestow their labour where they think their posterity shall enjoy the benefit of it, rather than on lands, the property of others, however low the rent may be in proportion to the value of the lands."[6]

This chapter examines land acquisition and ownership among eighteenth-century householders. High levels of landownership continued everywhere. In 1774 seven-tenths of colonial household heads (and four-fifths of farmers) owned land, much more than in Britain. One can nonetheless distinguish older, long-settled regions from newer frontiers. Most farm families in older regions still owned land, though the quantities they possessed dropped over time. As isolated coastal settlements became connected colonies, the amount of land available for purchase or inheritance diminished. Since land cost more than most men, especially those just starting out, could afford, poverty grew, tenancy increased, and landlords began to fill their manors with farm families. Many in this growing class of poor had too few assets even to move to a frontier, where greater opportunities beckoned.[7]

But many small and middling farmers, worried about their prospects and seeking cheap land, could and did move to frontiers—the Appalachian foothills and mountain valleys—where high levels of landownership and big farms could be found. "Determined to improve his fortunes by removing to a new district," Crèvecoeur related, such a farmer vowed to buy "as much land as will afford substantial farms to every one of his children." As pioneers cleared land and sent crops to market, they encouraged others to join them. When whites first occupied Indian lands, they either squatted on it or banded together to purchase it and extinguish the Indians' title. The area of white habitation thereby expanded; as migration accelerated, new settlements appeared. After a few years migrants surveyed, patented, or bought the land they farmed; most squatters, however—lacking money—had to abandon their farms.[8]

Hugging the Coast: The Land System of Tidewater Areas

Did eighteenth-century farmers acquire land as easily as their predecessors had? By 1700 densely populated settlements could be found—on the New England coast; along the Connecticut River; on Long Island, Manhattan Island, and the Hudson River as far north as Albany; adjacent to Delaware Bay and to the rivers of the Chesapeake colonies; near Albemarle Sound; and on Carolina's Stono and Cooper Rivers. As dense settlement grew, the proportion of families owning land dropped, many landowners farmed too few acres to feed their families, tenancy rose, and the number of poor, landless men increased. We should not, however, overstate the change. Two-thirds to three-quarters of farmers owned land during the last half of the century in eastern New England, Long Island, New Jersey, and Tidewater Virginia. The rest included children waiting to in-

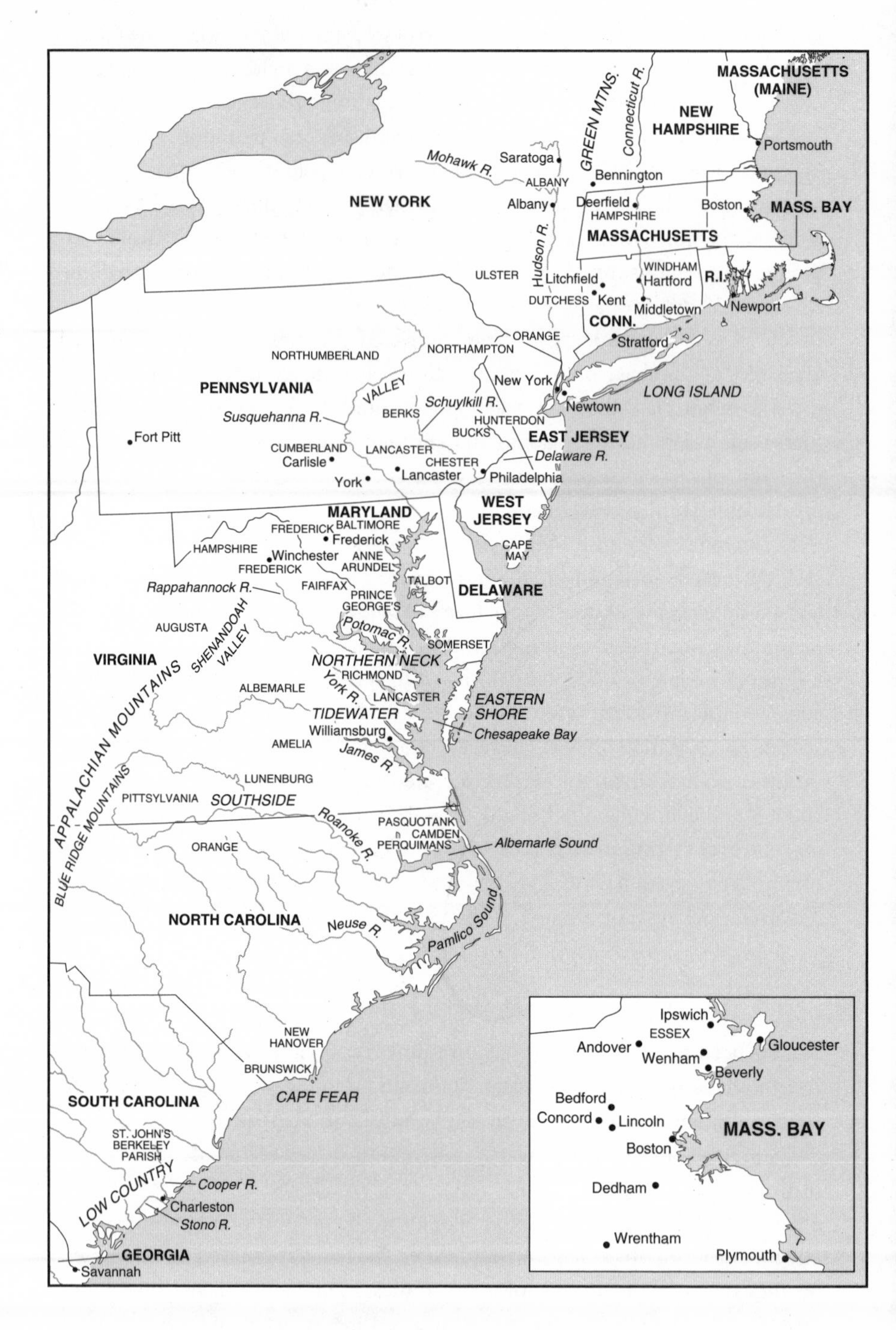

Map 3. British Mainland Colonies, ca. 1770

herit land, recently arrived immigrants, artisans renting their shop and land, or young leaseholders accumulating money to buy land. Many young couples had no land, but just a sixth of farm operators remained tenants their entire lives.[9]

The belief that uncultivated land was widely available sustained early marriage, high fertility, and European immigration; the rapid population growth this regime nurtured, in turn, drove the demand for land ever higher. Between 1700 and 1780 the white population of the mainland British colonies multiplied ten times, doubling every quarter-century, increasing from less than a quarter-million to almost 2.5 million. More than five-sixths of the added people had been born in the colonies, but European immigrants seeking land played an essential role in the backcountry. Colonial governments could not meet such heavy demand for land. By mid-century nearly all the good land in older settlements had been improved; officials could not form counties or towns, make paths, and clear rivers fast enough to provide for every family that wanted its own land. In such circumstances land prices rose in real terms, and the size of farms declined.[10]

By the 1770s population pressure on land, rising land prices, and growing dispossession could be found in every area settled in the seventeenth century from Massachusetts to the Lowcountry Carolinas. The data reported below, however different they appear, show that the proportion of farm families owning land dropped and farm sizes diminished while poverty and tenancy grew. We can point to two differences among coastal regions. As one traveled from southern staple-producing areas to mid-Atlantic and New England general farming regions, farm size dropped. Second, tenancy grew more rapidly in places, such as the Hudson River Valley, where landlords monopolized thousands of acres and forced tenants to sign short-term leases at high and rising rents. Tenancy rose more quickly near cities such as Philadelphia, where urban growth made truck farming profitable. In these places cottagers, who received a house and a tiny plot of land in return for providing labor to their landlord, could be found in large numbers. (See Map 3 for places.)

Population pressure on land became intense in older New England, New Jersey, and Long Island communities. Even in the face of diminishing resources, children wanted to remain close to their parents. As a result, population density rose. Forty-five people—nine families—lived on every square mile of land in rural eastern Massachusetts; by the 1770s population density was almost as high in East New Jersey. After towns had distributed all their land, men could get it only by inheritance or purchase. Sons of small farmers got land with difficulty. Families remained large, and sons usually had to wait until their father died, and even then each son might not get a farm. The father could bequeath all land to one son, cutting the others out; divide it among several sons, none of whom would have enough for a viable farm; or order the estate liquidated and the proceeds split among the children, giving no one a farm. Moreover, rising prices far beyond modest monetary inflation precluded land-poor sons, much less newcomers, from buying land. Land prices in Ipswich, Massachusetts, rose 40 percent between 1720 and 1749 and another 20 percent between 1750 and 1771. Between the 1740s and the 1770s land prices in Andover doubled, while the price of uplands and woodland in Newtown, Long Island, tripled between 1720 and 1770. The price of Connecticut River frontage in Hartford, Connecticut, doubled between 1700 and 1770.[11]

By mid-century landed farmers in New England and East New Jersey often struggled to make a living on inadequate acreage. In 1771 Massachusetts farmers owned, on average, only twenty acres of improved land (perhaps half of their holdings). Two-fifths of them had too little arable, pasture, or hay land to be self-sufficient in food. A fifth held less than eight improved acres, enough land to provide only a small part of the family's diet; they had to seek wage labor to support their families. By the late 1770s farms in six East New Jersey towns averaged less than fifty acres of improved land, and more than a quarter of the landholders owned less than twenty-five acres, too little for subsistence. With such diminished farms, with fields exhausted from a century or more of continuous cultivation, and with commons reduced by distribution to townsmen, farmers in old regions such as Essex County, Massachusetts, neither increased market production of grain and vegetables nor sold more wool, mutton, pork, beef, and butter. So many eastern Massachusetts farmers owned tiny plots of land that many towns had to import grain, vegetables, butter, or meat from outside the region.[12]

Connecticut and New Hampshire, with more unimproved land and internal frontiers, stood in contrast to Massachusetts. In early-eighteenth-century Connecticut, men owned larger farms than their counterparts in Massachusetts and bequeathed more to their children. However, holdings did diminish. Families owned 50 acres, nearly two-fifths less than those of the late seventeenth century. But farmers still had large farms. Farmers over forty owned 120 acres, about half improved, more than men a century earlier. By the 1770s, however, a third of Connecticut's villagers worked mainly as artisans; they owned on average 20 acres, enough to plant a garden and pasture a cow and horse. Land remained plentiful in Exeter, New Hampshire—one of that colony's first villages—until mid-century. The town not only held land in commons until 1740 but distributed land to all sons and newcomers who had arrived by 1725. The least fortunate man received 20 acres; the typical landholder, 70.[13]

Although eastern Massachusetts farm sizes plummeted, most families owned the land they farmed. In 1771 more than four-fifths (84 percent) of Essex County men age thirty-seven to sixty owned land. Four-fifths of Dedham taxpayers owned land in 1735 and 1771, as did seven-tenths of Concord taxpayers in 1749 and four-fifths in 1771. Tenancy was equally uncommon in Bedford and Lincoln, towns settled in the late seventeenth century by migrants from nearby Concord. Just over a quarter of Bedford's householders leased land in 1747, and the proportion dropped to less than a fifth by 1761; a tenth of Lincoln's householders rented in 1760. Nor did tenancy increase. In 1771 seven-tenths of Massachusetts village and small-town householders, including artisans and laborers, owned land, and few eastern Massachusetts towns, except fishing and whaling ports such as Plymouth or Gloucester, had many tenants.[14]

Similarly high levels of landownership could be found in Connecticut, New Hampshire, and East New Jersey. Six-sevenths of Connecticut men held land, and nine-tenths of those between ages forty and seventy farmed their own acreage. New Hampshire colonists faced repeated Indian attacks between the 1670s and the 1720s; but by 1732, after the colony had been at peace for nearly a decade, the population had grown rapidly, and residents of nine older towns (all settled in the 1630s) enjoyed nearly universal landownership: nearly eight-ninths of the householders owned land. But in the ports of Portsmouth and Newcastle only two-thirds of the householders owned land; most held just enough for a house lot and a small garden. During the Revolutionary era two-thirds of taxed men in East New Jersey owned land, but four-fifths of men over age twenty-seven—nearly all the household heads—did.[15]

With opportunities to own land diminishing, young men in northern coastal regions learned a trade, worked as town or farm laborers, or went to sea. In the small port of Beverly, Massachusetts, the proportion of farmers dropped from two-fifths to three-tenths of householders between 1720 and 1800. In the nearby farming village of Wenham, it declined from four-fifths to three-fifths over the same period. By the 1750s youths infrequently became farmers, as the composition of regiments in the Seven Years' War attests. One-seventh of enlistees from the three oldest counties in Massachusetts called themselves farmers; one-quarter, laborers; and one-half, artisans. In the Plymouth region even fewer—one in fourteen—were farmers, but over half were laborers, most on farms. Similar patterns could be found in Connecticut, East New Jersey, Delaware, Long Island, and Pennsylvania. For instance, nearly two-thirds of rural Pennsylvania recruits were immigrants, mainly from Britain. Mostly young ex-servants, just one in twelve of them had worked as farmers; but a third followed a craft, and three-fifths had been laborers.[16]

Plantation sizes in southern Tidewater regions diminished, but farms remained huge by New England standards. Land sales in Prince George's and Baltimore Counties show that mid-eighteenth-century purchasers refused to buy farms under 100 acres, a practice that helped maintain big plantations. Planters in mid-eighteenth-century Richmond and Lancaster Counties, Virginia (Northern Neck counties whites settled in the mid-seventeenth century), farmed an average of about 150 acres. Plantations in Prince George's and Baltimore Counties reflected this decrease. In 1754 landholders in Baltimore County, where some men grew wheat and others tobacco, averaged 190 acres; typical plantations in Prince George's dropped from 260 in 1705 (when the area was on the frontier) to 200 acres in 1733 and 1776. At the end of the century, tobacco plantations were still large, ranging from 150 to 175 acres on Maryland's Western Shore to 250 in Fairfax County, Virginia, an early-eighteenth-century frontier. By 1788 farms in the oldest Virginia Tidewater counties had declined to about 125 acres, but most tobacco planters still owned viable farms. During the 1780s just one in four landowners in sixteen Tidewater counties (ten in Virginia and six in Maryland) had fewer than 100 acres, but most of them had abandoned tobacco for livestock and grain.[17]

Coastal Carolina rice plantations were even larger. Land speculators, thirty-five of whom acquired 100,000 acres (an average of 2,850 acres apiece) by 1738, controlled much of the land along Cape Fear, on North Carolina's southeast coast. Plantation sizes remained large there long after speculators sold their holdings. In rice-growing Brunswick and New Hanover Counties, plantations averaged 350 acres in 1780. Landholdings in St. John's Berkeley Parish, South Carolina, in the 1760s averaged over 500 acres, and average plantation sizes increased rapidly, reaching 1,000 acres by 1793. Mid-century holdings in other Lowcountry parishes ranged from 500 to 1,000 acres.[18]

At first, similarly large plantations could be found in eastern North Carolina. Between 1680 and 1729 proprietary land grants, nearly all located in the Albemarle region, averaged nearly 400 acres. Five influential families, who owned one-sixth of the patented acres, sold their land to smaller producers, but farm sizes remained large. In 1720 typical Perquimans County householders owned 300 acres. As the population grew and plantation sizes dwindled on the colony's northeast coast, farmers replaced tobacco with corn, wheat, and salted meat, products that required less land. In 1735 planters in six counties ringing Albemarle Sound—first settled in the mid-seventeenth century—owned farms that ranged between 175 and 250 acres. By the late 1770s typical farms in four of these counties had only 75 to 175 acres. Farms in Pasquotank and Camden Counties were especially small, compared with most southern plantations: nearly three of ten farms had 50 or fewer acres. Somewhat larger plantations could be found along the Pamlico and Neuse Sounds, areas colonized in the late seventeenth and early eighteenth centuries; during the Revolutionary era, plantations in three counties in those coastal regions averaged between 200 and 225 acres.[19]

The land supply in coastal Maryland, Virginia, and North Carolina fell more rapidly than plantation size. Instead of dividing their plantations into small units, planters bequeathed farms to fewer sons, and big landholders leased surplus acres to landless families. During the eighteenth century, land prices doubled and redoubled in Tidewater Maryland. Between the 1750s and 1770s, a time of no inflation, land prices multiplied four times—from 10 to 40 shillings an acre—in Prince George's and Anne Arundel Counties, Western Shore tobacco-growing areas. Prices rose almost as much in Baltimore (a tobacco and general farming area) and Talbot (a wheat region) Counties. In Baltimore prices rose from 5 to 16 shillings an acre between the 1690s and 1770s; most of the increase came between the 1750s and 1770s, when prices doubled. In Talbot land prices almost doubled between the 1700s and the 1730s (from 5 to 9 shillings) and then nearly tripled between the 1730s and 1760s (from 9 to 24 shillings). Tidewater Virginia land prices rose just as much, reaching perhaps 40 shillings an acre by the 1780s.[20]

The level of landownership inevitably decreased, exacerbated in Virginia by the amount of land (as much as half in older counties) rich men entailed (a legal form that required all land be given to one son, in perpetuity), thus keeping it off land markets. In Prince George's County the proportion of landowners among householders declined from two-thirds in 1705 to just under half in 1776, a number comparable to five other Maryland counties in 1783. Only Somerset County, a poor general farming region where tenants may not have been able to pay rent, deviated from this pattern. Just one-third of the householders rented land there in 1783. A similar pattern could be found in Virginia and North Carolina. In 1787 only two-fifths of householders owned land in Virginia's Northern Neck, a region dominated by the Fairfax proprietorship where huge quantities of land were entailed, but around two-thirds owned land in ten other Tidewater counties. In the early 1780s less than seven-tenths of the householders in eastern North Carolina owned land. Because little unimproved land remained, landless householders had to rent from neighbors or landlords, like the nearly 200 coastal North Carolina and 125 Maryland planters with more than 1,000 acres.[21]

A similar pattern developed in the Delaware Valley, an area increasingly embedded in commercial agriculture. At first southeastern Pennsylvania and West Jersey farms were larger than New England's but smaller than the South's. During the early eighteenth century thousands of families moved to the area. By 1760 more than seven-tenths of the land in the Philadelphia region had been taken up. As the population increased and European demand for the region's wheat grew between the 1730s and 1760s, land prices multiplied from 10 to 40 shillings an acre. As a result, southeastern Pennsylvania farm sizes declined from 250 acres (in 1710) to about 125 acres. Farms close to Philadelphia became even smaller, dropping from 150 acres in 1710 to under 100 acres by the 1760s. The number of households in Marple Township, located at the edge of Philadelphia's market area, multiplied five times between 1715 and 1774. Not surprisingly, farm sizes declined. Farms sold averaged 200 acres before 1750 but only

100 acres by 1775. Farms advertised for sale in New Jersey (mostly in West Jersey) dropped from 300 acres in the early eighteenth century to 200 acres by the 1760s. Newspaper sales exaggerated farm sizes, but during the Revolutionary era, farms in all of New Jersey (including small-farm East New Jersey) averaged about 100 acres.[22]

As the size of Delaware Valley farms dropped, the percentage of men owning land declined, especially during the Seven Years' War, when Indians closed frontiers and landless families stayed put. Holders of 20- to 100-acre farms had too little to give adult children either a separate holding or money to buy or rent land. In 1760 three-quarters of families in Chester County, Pennsylvania, held land (most as owners), but the proportion of landowners declined thereafter, to 69 percent in 1765 and 63 percent in 1775. In Marple Township, in the most developed part of the county, the proportion of householders owning land was smaller, declining from eight-ninths in the early eighteenth century to a little under half in 1760 before increasing again to nearly three-fifths by 1774. Since such a high proportion of the county's land had already been taken up, nearly all the remaining householders leased land, sometimes getting their fields from nonresident landlords who lived in nearby townships.[23]

Most families in older regions eventually owned farms. Tenants rarely constituted half the farming populace. But tenancy did grow, and it became especially important in parts of New York and the Chesapeake colonies dominated by proprietary landlords and in the commercial farming areas near Philadelphia. Until older regions became thickly peopled and cheap freeholds disappeared—in the Tidewater Chesapeake by 1730–60 and in the Hudson River Valley and southeastern Pennsylvania by the 1750s—landlords (a few of whom owned hundreds of thousands of acres) attracted few tenants. Once freeholders had improved most of a region's land, undeveloped acres where squatters could farm disappeared, farm sizes declined so much that fathers could not give sons a farm, land prices rose, and tenancies looked more attractive. By the 1770s half or more of the household heads in the Hudson River counties dominated by big landlords, in parts of southeastern Pennsylvania and southern Maryland, and in Virginia's Northern Neck rented land.[24]

Developmental leases, where tenants paid a nominal rent and in return improved the landlord's holdings, nearly disappeared from coastal regions. Although a few tenants in settled areas of Maryland procured leases for twenty-one to ninety-nine years or for lives at low rents, by the 1760s and 1770s a majority of Pennsylvania, Maryland, and Virginia tenants could find only one- to seven-year leases. They paid a large part of their crops as rent, had to produce what the landlord wanted, and often searched for a new farm each year when the landlord raised rents. Tiring of low returns, the Calverts sold two-fifths of their manorial land between 1766 and 1771, often to the tenants, thereby reducing the number of the province's inexpensive tenancies. After 1760, responding to the declining supply of cheap leased land, Maryland Jesuits bought the rights of some tenants when life tenancies expired and turned them into yearly leases of 2,500 to 3,000 pounds of tobacco, the output of two to three laborers. Land could be rented for less in Anne Arundel County, but even there the annual cost of 100-acre farms rose from 450 to 800 pounds of tobacco over the first half of the century. One can see a similar pattern in Chester County, where tenants leased land for five- to seven-year terms, paying two shillings an acre per year in rent, two-thirds the price of an acre of frontier land.[25]

With so much land along the borderlands, why did so many families work as tenants in overpopulated regions? During wartime, families postponed moving. Even in peacetime, landless families often stayed in older regions. If they expected a landed inheritance, sons wanted to stay near home. Fathers, if they had enough land, settled married sons or sons-in-law on nearby unimproved land they had bought years earlier. A few families expected to save enough as tenants to buy land. Others had neither money for travel nor savings to make up the income they would lose in their first years along the frontier. Still others, especially in commercial agricultural regions, chose to be tenants because they figured they would earn more money by leasing good land near markets than by owning land isolated from trade centers.

As long as land remained available, young would-be farmers stayed near home. Over a lifetime, nearly all families (four-fifths in one New England sample) did move, most for short distances to find work or to marry. In New England from 1700 to 1770, two-fifths to three-quarters of marriages were between people from the same town, and residents of towns within fifteen miles formed nearly all the rest. Migrants born in New England from 1701 to 1740 moved nearly forty miles over their lifetimes. Most moves covered shorter distances, as an analysis of Wrentham, Massachusetts, a farming town twenty-five miles south of Boston, shows. Between the late 1750s and 1779, 430 people (in 234 households) moved into Wrentham; over half were from towns less than ten miles away, and most of the rest came from places less than fifteen miles away. Military enlistees, who could have moved with parents as children or as young adults searching for work, went no further. At least half and as many as nine-tenths of Long Island recruits during the Seven Years' War enlisted in their birth county, and most of the rest hailed from an adjacent county. Between three-fourths and nine-tenths of Seven Years' War enlistees from rural eastern Massachusetts stayed in their birth county; four-fifths of Tidewater Virginia's Revolutionary War recruits stayed in their birth county or moved to adjacent counties.[26]

Not all migrants formed viable households, much less got land; some settled for farm labor. New England swarmed with transients too poor to go to a frontier. Having stayed nowhere for very long, few of them had learned to read, which lowered their chances of hearing of jobs elsewhere. In 1770, when Ebenezer Ware of Wrentham, Massachusetts, hired twenty-one-year-old Amos Turner of nearby Walpole, Turner had gone "from place to place for his living . . . for allmost two years past and hath nothing but his hands to get a living." A growing army of vagabonds like Turner tramped the roads. "Under low circumstances" with "veary litel a staite," they moved from place to place far more often than landed families and traveled in a circuit less than ten miles, seeking farm labor. During the harvest, when farmers needed workers, they were tolerated; in winter, when work was scarce, they were evicted.

Between the 1730s and the 1760s the number of transient single people and families "warned out" of coastal Essex County multiplied 8.5 times, from 200 to 1,700. Those warned were newcomers to town who had no visible property, and the document prevented them from collecting poor relief. Equally divided among men and women and often ex-servants, young Essex County transients sought domestic service, planting and harvest labor, or work in cities. Rhode Island transient heads of household included desperately poor Indians and free African Americans; but four-fifths were white, and half of those had families with children. Widowed, separated, or abandoned women—two-fifths of them with children—headed half the warned families. Whites came from poor families; a third of the women and two-fifths of the men had been indentured servants. Public officials took away many of their children, indenturing them to prosperous families at age seven or eight to serve until they were twenty-one or twenty-two. Even married couples, such as a third of the Essex migrants, could find neither steady work nor a house for themselves and their young children.[27]

Increasing numbers of rural transients wound up in Providence or Boston working on the docks, in ships, or in workshops. Twenty-five transients arrived annually in Boston in the 1720s and 1730s, 65 in the 1740s, 200 in the mid-1750s. By the 1760s and 1770s 300 to 500 strangers—half young and unmarried, the rest couples or families—poured into Boston each year; many were desperately poor. Some had arranged apprenticeships or wage labor; others lodged with family or hoped to get poor relief. Boston attracted transients from a vast region. In 1765 just half came from Massachusetts (many were single women from adjacent areas); most (especially single men) traveled fifty or more miles. Unable to find work, these transients and the urban poor soon left to find farm labor.[28]

Poverty took on a different character in the mid-Atlantic colonies and the South. Mid-eighteenth-century Delaware Valley cottager families signed annual contracts with farmers and agreed to work when needed, getting a cottage and garden in return. They rarely stayed more than a year on any farm; rather, they moved in a small circuit to find employment. The number of Chester County cottagers tripled between 1750 and 1775, from 6 to 19 percent of householders. In the South, slaves performed so much farm labor that sons of small landowners and tenants had fewer chances to find work than those in the North. Single men tramped the roads seeking positions in slaveless families or work as overseers, but families—unlike those in the North—could find no work. Others, including former servants and "abandoned families," Father Joseph Mosely discovered in a trip to Maryland's Eastern Shore, lived "without bread" in "poverty want and misery."[29]

The poor—whether widows or children (the so-called deserving poor) or able-bodied youths—fell into dependence. Some of the deserving poor received pensions, but increasing numbers were placed under a guardian's care. This system worked well through the early eighteenth century because the number of deserving poor remained small. By the 1760s the number of sick and old people requiring lifelong pensions had jumped in older regions. Because local officials wanted to reduce expenses in the face of growing poverty, poor people faced new indignities. Those who boarded with local families not only suffered more often from neglect but were driven to work harder by their master. Many had to live in a poorhouse or workhouse, an affront former householders deemed beneath their dignity. Horrified by the growth of transiency, officials in Massachusetts, on Long Island, and in both Chesapeake colonies built poorhouses before the Revolution; by 1800 Chester County, Pennsylvania, had followed suit. A few Long Island parishes; Middletown, Connecticut; and perhaps Prince George's County, Maryland, auctioned the poor to the lowest bidder. Pennsylvania required those getting aid to wear a badge marked with the letter "P." This public humiliation not only separated the poor from their landed betters but reminded smallholders that only land stood between them and debased poverty.[30]

Landed resources became more scarce in coastal areas. In New England and (to a lesser extent) the mid-Atlantic colonies, farm sizes often fell below what farmers needed for familial subsistence. Most land in the Tidewater Chesapeake had been used until it could no longer produce and then was abandoned; in some places too little land remained to allow families to grow tobacco. As land grew scarce, farm prices multiplied, landownership dropped, and tenancy grew. Men postponed marriage, hoping to save enough money to buy land or to get it from their fathers. Would-be farmers faced difficult choices: use land more intensively and efficiently, become a town artisan, or move to a place with cheap land. The first choice was nearly impossible, given the lack of capital in the colonies. Most men (except New Englanders) found waged labor even in a craft unpalatable. Frontiers beckoned ever greater numbers of families.

Growing scarcity and high land prices near the coast, coupled with plenty of cheap land on the frontier, led to massive migration. Before settlers could begin farming, however, Indians had to be forced off their lands. By 1720 whites had conquered or bought coastal regions and small parts of the piedmont from Massachusetts to Carolina. Indians had abandoned other places, such as Virginia's Shenandoah Valley. More than enough widowed acres remained to satisfy several generations of farmers. Not wanting to face possibly hostile Indians, most migrants moved to these places. But Indians controlled northern New England, central and western New York, northeastern and western Pennsylvania, the Carolina Piedmont and mountains, all of Georgia, and the Ohio country.[31]

As long as fur-bearing animals remained plentiful and colonists had enough land nearby, traders and governments worked to keep land in Indian hands. New York and Pennsylvania officials protected Indian hunting grounds because the beaver and deer skin trades remained a major part of their exports. In the early to mid-eighteenth century, Shawnees and Delawares forced from eastern Pennsylvania and Mahicans expelled from New York and New England settled in the Allegheny and Ohio Valleys, the source of abundant furs. Furs constituted a quarter of the value of the exports of those two colonies to London in the war-torn 1700s; the figure jumped to two-fifths from 1710 to 1749 despite intermittent warfare. Only during the Seven Years' War did the trade dry up, plummeting to just one-sixth of the value of London exports from 1754 to 1758. Rich fur traders took advantage of their knowledge of Indian territory to scout out and buy good farmland (sometimes with trickery) from Indians.[32]

Knowledgeable about European ways from their long contact in the east, Indians worked to protect their lands. Indian translators (some of them women married to English traders or officials or their children, others Christian converts) mediated between European and Indian worlds. Few Indians, except those involved in the fur trade and a handful of interpreters, became fluent in English, though many learned to speak a few words. By mid-century, they dealt solely with traders, missionaries, agents, and officials, the only Europeans who knew their languages and customs. These Indian and English women and men negotiated petty squabbles (hog stealing by Indians, assaults by settlers) and prevented a few violent situations from turning into war. In 1744 Skickllamy, for instance, averted war when he handed over the murderers of three Pennsylvania traders. When settlers, such as Pennsylvanians in the Susquehanna Valley in the 1720s and 1730s or Georgians in the 1740s, began to demand Indian territories, officials abandoned their Indian allies and allocated land to the settlers. When Indian cultural mediators lost credibility with settlers (who wished every Indian dead) and with their own people, vicious warfare resumed, and Indians were forced to leave their lands.[33]

Unlike seventeenth-century farmers, eighteenth-century frontier families rarely traded with Indians, much less invited them into their homes. Rules of hospitality required settlers to feed and house Indians traveling between Indian communities or going to war against Indian enemies; but—fearful of Indians or concerned that they would eat vast quantities of food—many refused, and others acted with ill grace. The few families who lived in the midst of Indians did give drink or tobacco to Indian sojourners. Settlers rarely learned an Indian language. As late as the 1740s, a few older Dutch and English Hudson Valley inhabitants and an occasional Pennsylvania trader's wife could converse with Indians in their own language; but their children did not have these skills, and communication with Indians became increasingly difficult.[34]

Since so few settlers spoke an Indian language, lethal misunderstandings regularly occurred. Any appearance of Indians at a settler's farm caused alarm. In the spring of 1728, soon after a band of eleven Shawnee warriors had appeared in a settlement along the northern Pennsylvania Schuylkill frontier, rumors of violence sent settlers scurrying for safety. When word that another group of Indians had gone to John Roberts's farmhouse reached those who remained, a four-man posse rushed to the scene. When they arrived, they saw a local Delaware family (Tacocolie, his pregnant wife, an older woman, two girls, and a boy) well known to neighboring settlers. Believing (incorrectly) that the Indians had killed two colonists and unable or unwilling to communicate with these unarmed Delawares, the men opened fire, killing the adults and holding the others captive.[35]

Squatters, land speculators, and government officials fought among themselves to see who could most rapidly buy, steal, or conquer tribal territories. As Thomas Pownall (soon to be governor of Massachusetts) lamented in 1755, colonists showed "an insatiable thirst after landed possessions." The British understood why settlers intruded on Indian lands. "Many of Your Majesty's ancient colonies," a 1763 Board of Trade report argued, were "overstocked with inhabitants, occasioned partly from an extremely increasing population in some . . . colonies whose boundaries had become too narrow for their numbers, but chiefly by the monopoly of lands in the hand of land jobbers [speculators]." Even though many settlers might have preferred to stay near home, "the high price of land" forced them "to emigrate to the other side of the mountains, where they were exposed to the irruptions of the Indians."[36]

Indians knew that settlers craved their land but denied that they had a right to it. "We have great reason to believe you intend to drive us away and settle the country" beyond the Ohio, Delaware leader Pisquetomen told an English negotiator in 1759, "or else, why do you come to fight in the land that God has given us?" In 1763 Creek leader Yakatastange insisted that his nation loved "our Lands" for the "Wood is our Fire, and the Grass is our Bed, and our Physic when we are sick." We are "Masters of all the Land," he added, and "own no Masters but the Master of their Breath." But, he lamented, "the White People intend to stop all their Breaths by their settling all around them." Other Creeks dubbed whites "people greedily grasping after the lands of the red man." Knowing that colonial leaders would not stop the flow of settlers, Indians dismissed with contempt claims that the English would protect their land.[37]

The dispossession of Indians started when squatters trespassed on and cleared Indian lands, built cabins, and began to hunt and farm. Squatters occupied Indian territory illegally in Virginia's Piedmont in the 1710s, Pennsylvania's Susquehanna Valley in the 1730s and other Iroquois land in the 1740s, and western Massachusetts, western Pennsylvania, and the Carolina and Georgia upcountry in the 1760s. On occasion squatters overwhelmed small Indian populations. In 1773, 3,000 people lived illegally on Georgia Indian lands; at the same time, 5,000 families farmed without clear title on Indian land in western Pennsylvania, West Virginia, and Kentucky. To keep peace with friendly Indians, local officials on rare occasions forced squatters to leave; but trespassers received little punishment, and many returned as soon as the sheriff and his men left. In the early 1750s a Cumberland County, Pennsylvania, posse led by justice and Indian agent George Croghan evicted sixty families from land near the Blue Mountains. But after giving bond to guarantee their removal, the squatters kept their goods and got money to resettle.

Despite the danger of Indian attack, squatters moved to Indian country in wartime. As soon as the English captured Fort Pitt in 1760, squatters made farms nearby. Frustrated by their behavior, commander Henry Bouquet tried unsuccessfully to evict them, sending troops to disperse them and burn their houses and crops. "For two years past," he complained in 1762, "these Lands have been over run by . . . Vagabonds, who under pretense of hunting were Making Settlements." His efforts notwithstanding, "they still in a less degree, Continue the same practices," angering local Indians. Settlers so often made farms on Indian land that in 1763 the British government ordered trespassers "to remove themselves from such settlements," a command families such as the settlers on southwest Virginia's New River utterly ignored.[38]

Farmers and their leaders sought to legalize this thievery. Despite government prohibitions, land speculators and squatters often bought land from Indians, hoping to extinguish Indian title and gain secure rights to acres they farmed or hoped to sell. For instance, "Every Settler" living on the Philipse Highland (in the northern Hudson River Valley) in 1767 "purchased his own particular Farm from the Indians." In this case, New York officials recognized the title they gave to great landlords for the same lands. Traders and officials who negotiated with Indians scouted out the best land and often pressured their allies to give up vast tracks. Indian trader George Croghan bought 200,000 acres near Fort Pitt from the Iroquois in 1749; in 1773 he added 1.5 million acres in the same region. Settlers and speculators so often bought land from Indians that the Proclamation of 1763 prohibited any "private person . . . to make any purchase from . . . Indians of any lands reserved to them." But colonists—such as the settlers on Watauga Creek, east Tennessee, who bought land from the Cherokees in the early 1770s—ignored this ban. In other cases colonial governments bought land from Indians. In 1724 the Massachusetts Assembly purchased large tracts from the few Mahicans left in western Massachusetts; the first white farmers arrived the next year. Between 1759 and 1762 whites encroached on the lands Mahicans still owned and refused to pay them for it. The court granted much of this land to whites; by the late 1760s Mahicans indebted to whites had deeded over the rest.[39]

At the same time that the first squatters arrived (and at times even earlier), colonial governments brazenly encouraged speculators to buy thousands of acres of Indian land. Such sales were often confirmed by treaties. Between 1715 and 1773 Indians from New England to Georgia signed at least thirty-eight treaties ceding land to colonial governments. They signed sixteen of them between 1763 and 1773, after the British had thrown the French out of North America. Indians in New York, Pennsylvania, Carolina, and Georgia were most often forced to give up their lands. Sometimes whites tricked them into surrendering their lands; sometimes they gave away the lands of other Indian nations. In an infamous example, in 1737 William Penn's sons, heavily in debt, used bogus deeds Indians had supposedly signed decades earlier to justify the "walking purchase" that forced Indians out of northeastern Pennsylvania lands the Penns had already given to speculators. In an equally fraudulent land grab, in 1754 Pennsylvania diplomats signed a deed for the Wyoming Valley in northeastern Pennsylvania and much of western Pennsylvania with Mohawks and Tuscaroras, who had no rights to the land. After the Iroquois gave away part of the Ohio Valley in 1768, Shawnees and Delawares living there protested and eventually went to war. Even fair treaties, such as those negotiated by Indians and the British in 1765–68 to set a border between their lands, left Indians dissatisfied. Whether colonies stole or paid for their land, the Onondaga (Iroquois) council complained in 1756, "they would lose their Lands[,] & the consideration they got, was soon spent, altho' the Lands remained" under settler control.[40]

Violence permeated the western borderlands. Indians going to war against their native enemies passed through frontier settlements, demanding food and committing, as the Virginia Council noted in 1736, "frequent Outrages," including killing several settlers. Seeking to control the land that speculators and settlers had seized, Indians on occasion raided outlying farms in northern New England, western Pennsylvania, the Carolinas, and the Ohio country, taking food and farm goods from poor newcomers and killing a few of them. In turn, traders stole Indian goods; in one instance in the Carolinas they took 330 skins. Amid so much violence, everyone believed rumors of war. Rumors that Indians or settlers were about to go to war to kill one another passed rapidly from trader to Indian or trader to pioneer family, kept relations between Indians and interlopers tense, and on occasion led to a panic, such as one that struck South Carolina in 1751.[41]

Indians and their French allies went to war against the borderlands between 1702 and 1717, 1744 and 1748, and 1754 and 1763. War usually began when colonists intruded too forcibly on Indian lands or murdered Indian families. Indians forced farm families to flee or huddle in the nearest fort, from which they might venture out—with armed guards—to till crops. Few places suffered as much as the Connecticut River town of Deerfield, Massachusetts. Destroyed by Indians during King Philip's War, the village was pillaged again in 1704, when the French and Indians invaded, killing 44 residents, and taking 109 prisoners—two-fifths of the village's people. Both times the town had to be rebuilt. Even after the frontier receded, the town feared invasion. Indians and French attacked an outlying family in 1746; in 1766 Indians assaulted nearby Greenfield. Deerfield was not alone. Any family living near Indian land faced danger. In the 1710s the Tuscaroras, Creeks, and Yamasees attacked outlying Carolina settlements (some on Charleston's outskirts). Forty years later Indians raided frontier farms on the Little Saluda River, in the South Carolina Piedmont, killing at least one family. Nor did attacks on the northern frontier cease. In 1723 Micmacs attacked isolated settlements in Maine, killing farmers in their fields. In 1724 Indians invaded the northwest Connecticut town of Litchfield. In 1745 they raided the northern outpost of Saratoga, New York, killing or capturing a hundred people and sending the remaining farmers scurrying to Albany or New York City. The next year Indians renewed attacks on Maine and New Hampshire.[42]

The years from 1755 to 1758 were especially violent. After Indians and their French allies defeated Edward Braddock in 1755, Indians launched attacks all over the backcountry, from Pennsylvania to Carolina. Seeing how close settlers had come to their hunting grounds, Ohio Valley Shawnees and Delawares sought to push settlement back toward the coast. They devastated the Susquehanna and Delaware frontiers, an area 50–100 miles wide, burning barns and houses; killing or driving away livestock; stealing horses, food, and firearms; killing and scalping settlers and orphaning many children; and taking the remaining women and children prisoners. Indians marauded over the Virginia frontier, where they "captivate[d] and butcher[ed] our out-settlers and have drove great numbers of them into the thicker inhabited parts." Indians burned farms and scalped settlers in long-settled areas such as Winchester town and Frederick County, Virginia, and in the Shenandoah Valley and northern Lancaster, Northampton, Bucks, and Berks County settlements, some only fifty miles north of Philadelphia.[43]

Western Indians disrupted frontier settlement. Between 1754 and 1758 they killed 1,500 settlers and 500 soldiers and captured 1,000 more—3 percent of the region's prewar population. Indian raids (and rumors of raids) led settlers to abandon the war zone, from the Delaware to the Roanoke Rivers, an area 400 miles long and 50–100 miles wide. Half of Frederick County's residents left and sought protection in nearby forts; nearly every settler left Hampshire County, West Virginia, and Cumberland County, Pennsylvania (on the Susquehanna frontier), as did half the settlers in Augusta County, in the southern Shenandoah. Rumors of Indian attack sent "all the People" west of Carlisle to the safety of the town or nearby forts and other frontier residents to camps outside Lancaster and York. "Confusion and Disorder" reigned in Augusta County. "Mothers with [a] train of helpless Children at their heels straggl[ed] through woods & mountain to escape the fury of those merciless savages." Once they reached the nearest forts, they lived in squalid refugee camps. George Washington reported that "Men, Women, and Children who had lately lived in great Affluence and Plenty [were] reduced to the most extreme Poverty and Distress," lacking food and clothing.[44]

Violence continued through the early 1760s. In 1760, after South Carolina went to bloody war against them, the Cherokees retaliated, killing many backwoods families and sending the rest scurrying to Charleston. After the English threw the French out of Canada in 1763, Indians—incensed at intrusions on their lands—went to war twice, attacking forts that regulated trade and protected settlers, occasionally striking small settlements. The Delaware prophet Neolin urged Indians everywhere to forgo European trade goods and alcohol and to seize hunting territories rightfully theirs. Responding to this message, in 1763 Ottawas, Shawnees, Delawares, Iroquois, Hurons, and other nations led by Ottawa chief Pontiac attacked farmers in western Pennsylvania, the upper Susquehanna, Maryland, Virginia, and West Virginia, killing or capturing as many as 2,000, again forcing refugees to huddle in Carlisle. In Greenbrier (on West Virginia's Kanawha River) seventy Shawnees massacred about fifty people who had invited them to dine. The violence continued in the 1770s. In 1774 Shawnees and Mingoes, angered at the cold-blooded murder of peaceable Indians by settlers, attacked surveyors laying claim to their lands and raided farms in the Greenbrier region, southwestern Virginia, and western Pennsylvania, killing or capturing colonists, burning farms, taking cattle and slaves, and sending families fleeing to the nearest forts or east of the mountains. About the same time

Creeks, dissatisfied with the latest treaty with Georgia, attacked outlying farms north of Augusta.[45]

Warfare had little long-term impact on western settlement. With their greater numbers and their southern Indian allies, the English defeated all enemy Indians. In retaliation for raids, settlers in Paxton, Pennsylvania, slaughtered twenty friendly Indians. Facing such massive retribution, Indians sued for peace, each time losing land. Refugees rapidly returned to the farms they had abandoned. Settlers rushed to repopulate western Pennsylvania and Virginia, even as the war continued on other fronts and local Indians still threatened violence. Land speculators and land companies lobbied for hundreds of thousands of acres, even during wartime. Between 1766 and 1775 land theft and sales and treaties (some cemented with alcohol and bribes to Indian signers) opened thousands of square miles to settlers, pushing borderlands south to Florida and west of the Appalachians. Treaties negotiated with the Cherokees and in Georgia document the pattern. Between 1721 and 1768 Cherokees ceded 12,100 square miles, more than two-thirds of it in 1755, but between 1770 and 1775 they were forced to relinquish 48,200 square miles (some of it controlled by other nations), four times as much as in the preceding half-century. In 1733 Indians ceded 1,800 square miles to Georgia, and by the mid-1750s they had added another 500 square miles, more than enough for Georgia's few colonists. In 1766 and 1773 Georgia Indians ceded another 8,600 square miles, multiplying the size of the colony nearly fivefold.[46]

Although reviled by land-hungry speculators and settlers, the Proclamation of 1763 illuminated the continentwide extent of the change. English authorities hoped the line—located at the crest of the Appalachians—would separate colonist from Indian. The proclamation granted colonists the right to farm east of the "heads or sources of any of the rivers which fall into the Atlantic Ocean," a vast area including northern New England, central Pennsylvania, Piedmont Carolina, and half of Georgia. In 1720 few whites lived in these areas; even in 1770, far northern New England, the western Carolinas, and interior Georgia contained few whites. But these lands proved insufficient, and the Proclamation became a dead letter. Hundreds of white farmers poured into the Ohio country and squatted on Indian land. Fearing an Indian war, English authorities denied speculators title to trans-Appalachian land—but at the same time they failed to send an army to expel the squatters. In 1768 colonial authorities negotiated a line farther west that included West Virginia, Kentucky, and the Pittsburgh region—places west of the mountains where a few tiny colonies of white settlers had located. Indians had reason to be concerned. They regularly hunted these areas. New York lands occupied by the Iroquois (and by the thousands of Indian refugees who moved there) and South Carolina and Georgia Cherokee lands lay precariously close to the newly negotiated border.[47]

Once Indians had been forced from their homes, farmers reoccupied the land. Even without hostile Indians, the trip to a new home was hazardous and expensive. Heat, rain, or ice often made roads and streams impassable; with so few inhabitants, settler or Indian, food was difficult to find. Travelers went by foot on narrow Indian paths or rode carts down newly made wagon roads. But horses needed scarce fodder and often turned lame. When swamps and bogs intervened, migrants built canoes to travel swiftly on narrow streams, but such vehicles could carry few of the goods they needed on their new farms.[48]

Only families with assets, moreover, could afford to move long distances. Frontier migration required money for lodging, a wagon, livestock, and farm tools as well as food to feed the family until the first crop came in. Nor could pioneers afford to survey, patent, or buy land until they cleared it and harvested marketable crops, a process that took several years. A family who moved to unsettled Southside Virginia at mid-century spent £24 sterling to travel and buy land, livestock, and food for a year. Frontier landlords or New York or Maryland entrepreneurs sometimes recruited migrants, whose costs were thereby lowered. To reduce expenses, some migrants to western Pennsylvania went "over the mountains in the spring, leaving their families behind to raise a crop of corn, and then return[ed] to bring them out in the fall." Meanwhile, the husband built a cabin and cleared land but—lacking money from the sale of the home place—could only squat on the land. Young couples with small families moved together, but since they had only enough money to get to the frontier, they squatted on land and rarely achieved success. The proportion of families among transients in frontier Hampshire County, Massachusetts, rose from under one-half to over two-thirds between 1739–43 and 1770–74.[49]

When prices were good, propertied families risked the peril of moving to a frontier; during depressions or wars, they stayed put or moved short distances. Boundary disputes (like that between Maryland and Pennsylvania) and the perceived Indian menace, moreover, kept settlers from Maryland and Virginia mountain valleys until mid-century. Just a third of one sample of men born in Massachusetts, Pennsylvania, and Virginia between 1680 and 1719 migrated over long distances, and they traveled only sixty miles on average.

Early-eighteenth-century American farmers nonetheless did populate large frontier areas. They moved from eastern New England to southern New Hampshire and central and western Massachusetts. They migrated north along the Hudson River and moved from Maryland's Eastern Shore or the Delaware Valley toward the Susquehanna River or to southeastern New Jersey. Some repeopled the eastern piedmont Chesapeake, and others went inland from the Carolina coast. Scot and German immigrants landed at Philadelphia but, seek-

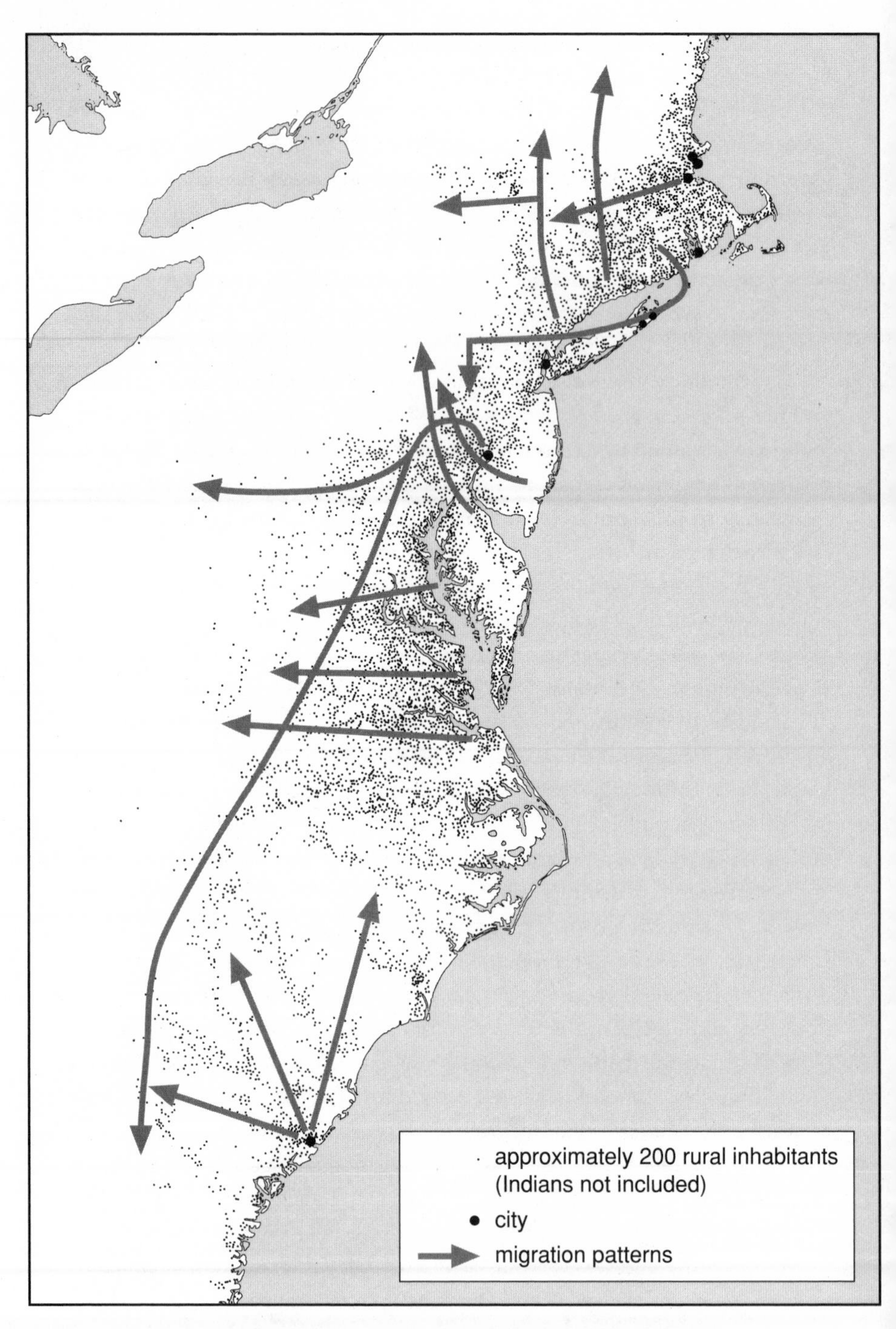

Map 4. European Population in Mainland British North America, ca. 1770

ing cheaper land, moved south to the Shenandoah Valley or northwestern North Carolina. In 1751, when Joshua Fry mapped western Virginia, he discovered that "the Country is not only well peopled as far westward as the Blue Ridge of Mountains, but beyond it even to the Allegheny Ridge." Few people had reached southwest Virginia, and only "about one Hundred Families" lived west of the mountains in present-day West Virginia. The colonization of coastal Georgia had begun, at first by European immigrants but later by Carolina migrants.[50] (See Map 4.)

Mid-eighteenth-century wars reduced migration and kept many immigrants in Pennsylvania for decades. But pent-up demand for land exploded during the 1760s and especially the 1770s, when the size of the settled part of the backcountry nearly tripled, from 17,000 to 49,000 square miles. Half the men born between 1720 and 1739 and two-thirds born between 1740 and 1759 migrated (mostly between 1740 and 1780) in the sample cited above; they usually traveled 100 to 250 miles. Families ventured to the rich agricultural piedmont and valleys between the Blue Ridge and Appalachian Mountains—northern New England, the upper Hudson and Mohawk Valleys, central and western Pennsylvania, the southern piedmont, the Shenandoah Valley, and Georgia.

Migrants followed a north-south axis, splitting into two regional patterns. People born in New York and New England moved north or northeast, up the Connecticut and Hudson Rivers, to Maine, central New Hampshire, southern and central Vermont, western Massachusetts, and northeast New York; they also went south into northern New Jersey and northeast Pennsylvania. Others, like the Ulster Irish who settled in Londonderry, New Hampshire, later moved west to the Mohawk Valley. Moves of over 100 miles were more common in colonies from Pennsylvania south. Although some families moved north into western Pennsylvania, most went southwest, from eastern Pennsylvania and the Susquehanna Valley to western Pennsylvania and from Pennsylvania and the Tidewater Chesapeake to piedmont and mountain districts extending from Maryland to South Carolina. Native-born migrants moved from coastal Virginia or Carolina inland to the piedmont upcountry or (less often) to inland Georgia or the mountain valleys. German, Scot, and Irish immigrants as well as families from coastal areas played a role in this migration. Immigrants went from Philadelphia to the Susquehanna River, then turned southwest on the Great Wagon Road that went through the mountain valleys from Maryland to Augusta; others moved west after landing in Charleston.[51]

By 1775 about 380,000 Europeans and their descendants—70,000 families—had repopulated northern New England, central Pennsylvania, the Maryland and Virginia Piedmont and mountain areas, and the Piedmont Carolinas, more than lived in all the mainland colonies in 1710. The adults among them had come from Britain, Germany, or more settled areas between 1750 and 1775.

Having made farms and gained land titles, they created communities strong enough to attract more settlers and to withstand Indian warfare.

Small enclaves of no more than a hundred white families could be found in East Florida or west of the mountains in Pennsylvania, Virginia, Kentucky, Illinois, Tennessee, and the Mississippi country. In 1775 this vast region contained about 35,000 whites, half of whom lived in western Pennsylvania. The rest lived precariously, threatened by Indians or the environment. The few people who migrated to East Florida, for instance, faced pestilence, high humidity, and heavy mortality. Three hundred settlers in Kentucky could barely feed themselves, much less defend their holdings against hostile Indians.[52]

As lands filled up, frontiers receded westward, population pressures reappeared, and poor families took to the roads. Those with assets left for nearby frontiers or—less often—moved farther west. The few English residents of Lancaster, Pennsylvania, an early-eighteenth-century frontier, for instance, left later in the century for newer frontiers. Their places were taken by German and Scots-Irish immigrants attracted by improved farms. Between the 1760s and the 1780s, after Pennsylvania's upper Susquehanna frontier had been repeopled, local Germans and Scots-Irish moved eastward, across the Delaware River and into northwestern New Jersey, one of the last frontiers near the coast.[53]

A similar pattern could be found in Virginia. The first white (and slave) residents came to Southside (a piedmont area south of the James) in the 1740s from Piedmont counties north of the river. By the 1760s the region had attracted families from depleted Tidewater lands as well as from Piedmont counties settled earlier in the century. Some initial settlers of Virginia's northern Shenandoah Valley—immigrants and planters—moved west from the nearby Northern Neck during the late 1730s, then left for newer frontiers, first in the southern valley, then (by the 1770s) in western Carolina and Kentucky. But most of the first white farmers in the Shenandoah, Virginia's Governor William Gooch reported in 1734, were "Germans and Swissers lately come into Pensilvania, where being disappointed of the quantity of land they expected . . . have chosen to fix their habitation in this unsettled part of Virginia." The first residents of the New River frontier, in southwest Virginia–northwest North Carolina, had arrived in Pennsylvania, gone to the northern Shenandoah, and then moved to New River. During the mid-eighteenth century immigrants and Delaware and Pennsylvania Welsh mingled with Lowcountry planters in the South Carolina backcountry, an area located at first just fifty miles from the coast but later in the Piedmont.[54]

A farmer's decision to migrate was driven by increased land prices in older areas and cheap frontier land. Nathan Birdsey of Stratford, Connecticut, faced a predicament common to New Englanders. He believed it "better to bring my Boys up to Husbandry than to put em out to Trades; but not having Land

Sufficient for Farms for em all," he proposed to "Sell some out-Pieces of Land, & purchase Some . . . in new Towns where Land is good & cheap & ye Title uncontroverted." New England farmers like Birdsey moved to re-create small farms. As the quantity of unfarmed land in Wenham declined, partible inheritance diminished among men who left wills, and authorities often refused to divide land of intestates, preferring to give it to one son but requiring him to pay portions to his siblings. Unfavored sons accumulated little capital but often moved to northern frontiers or nearby towns, hoping to use their cash inheritance to buy land. Similarly, rising demand for tobacco and a growing population drove up land prices in Tidewater counties, impelling poorer but landed planters to move to the Piedmont.[55]

Social constraints against migration among landed families was strong. Neither men expecting inheritances nor women fearing loss of neighborhood sociability went willingly to a wilderness full of hostile Indians, pestilence, and death. Edward Butler, a 1784 Virginia migrant to backcountry Georgia captured this aversion: "New Georgia is a pleasant place," he wrote, "If we could but enjoye it / Indians & Rogues they are so great, / They have almost destroyed it." Those who wanted land "may buy aplenty" but "let your purse be Ere so full, / you may soon have it Emty." After they arrived, pioneers lost crops to weevils and cattle to blood-sucking flies and "musqueators."

Thousands of landed families did go to frontiers. Their decision depended on the father's age, the number of children at home, opportunities at home or on the frontier, and previous moves by family, friends, or neighbors. Adolescents who left home to work on farms or oversee slaves rarely went far. Men moved long distances in their mid-twenties after getting marriage portions or after their fathers died. Recently married, they took wives but few children with them. Long-distance migration slowed once men reached their forties, except for the few who moved with grown children.[56]

Frontier migrants usually moved as members of households, not as part of a kin group or community. Unmarried men sometimes moved alone, and occasionally adult siblings or a widow or widower and adult children traveled together. Frontier families could thus rely on few kindred for support immediately after arrival at a new location. Communal chain migration might have been more common. After early migrants got land, neighbors (often kin of early arrivals) joined them. Quakers who moved from Pennsylvania to Virginia and North Carolina had communal support for migration from their meetings, and on occasion, other settlers moved in groups. Families from Andover, Massachusetts, founded communities in Windham County, Connecticut, in the early eighteenth century and later moved to a cluster of towns in New Hampshire and western Massachusetts. Early residents of Bennington, Vermont, and other Green Mountain villages came from just one or two towns. Similar chain migration can be seen in migration by planters from Tidewater to Piedmont Virginia and migration (often by immigrants) from Pennsylvania to the mountain regions of Virginia and North Carolina. Members of several Pennsylvania Presbyterian churches, for instance, founded daughter churches in the Shenandoah Valley.[57]

Making a Home in the Hills and Valleys: Colonizing the Borderlands

Migrants expected to own frontier land, which seemed free for the taking, as soon as they arrived. Abundant frontier land, Benjamin Franklin argued, allowed an ambitious New England farmer "in a short time" to "save money enough to purchase . . . new land sufficient for a Plantation." Andrew Burnaby, who visited the colonies in 1759–60, echoed Franklin. "Not a tenth of Virginia's land is yet cultivated," he discovered; from this land "every person may with ease procure a small Plantation." A decade later New York governor Henry Moore believed that "every one can have Land to work upon" and lamented that servants "quit their masters" as soon as they completed their terms "and get a small tract of Land," thereby raising the price of labor. North Carolina's Governor William Tryon related in 1761 that the industrious "poorer Settlers . . . sat themselves down in the back Counties where the land is best," but after getting land, they had "not more than a sufficiency to erect a Log House for their families and procure a few Tools to get a little Corn in the ground." During the Seven Years' War an English officer, amazed at the "levelling principle" in Massachusetts, observed that "everybody has property, & everybody knows it." David Ramsay summarized the link between abundant land and universal landownership. "In consequence of the vast extent of vacant country," he wrote in 1789, "every colonist was, or easily might be, a freeholder."[58]

This section describes this process of land acquisition in the vast backcountry that stretched from Maine to Georgia. We will detail differences in land prices, land taxes, and the behavior of land sellers in different places. It is easy to get lost in these details. But the process and especially the result—widespread landownership—was the same almost everywhere. Early settlers were often squatters who could not afford to stay, but those who arrived later with kindred or capital thrived. A substantial majority of pioneers bought land on easy terms from proprietors or land speculators. Once a critical mass of farmers arrived, they began getting land on their own, and the importance of land speculators diminished. The only exception could be found in places—such as New York's Hudson and Mohawk River frontiers—where big landholders insisted on leasing rather than selling land. In these locations, tenancy gradually grew.

Frontier land was hardly free, even if we discount the high costs of travel and starting a farm. Imperial and colonial land policies were profoundly ambiguous. Officials wanted to populate frontiers rapidly, yet they sought tax revenue from newly seated land and indulged in orgies of land speculation that threatened to retard settlement. As early as the 1710s and especially between 1730 and 1745, governors and London policy makers promoted rapid repeopling of the borderlands because they wanted to create a buffer zone, running from Nova Scotia to the Carolinas, to protect settled areas from the French, Spanish, and Indians, who controlled most land west of the Appalachians. To that end, they recruited groups of foreign Protestants, mostly small farmers, gave them large tracts of land, and forgave taxes if they would make farms. But colony and crown expected to gain income from the land and urged governors to assiduously collect quitrents and land taxes. Governors, moreover, worked to build their own fortunes (and those of their class) by granting large estates, thousands of acres in size, to themselves and their favorites.[59]

Proprietors or the crown often granted free land to settlers singly or in groups. The Masonian proprietors of eighteenth-century New Hampshire, who held the southern part of the province, granted town charters to groups contingent on their improving the land. Rather than charge quitrents, the proprietors remained content to enjoy the capital gains they made on a small part of the grant they reserved for themselves. Although settlers in the Granville proprietary (most of North Carolina's northern Piedmont) complained bitterly during the 1750s that land agents charged high survey fees and neglected for years to register their grants, agents did disburse 3 million acres between 1748 and 1763 (half between 1760 and 1762) to 5,000 settlers, in return for a small quitrent. South Carolina's royal government not only continued to grant headrights (50 acres per immigrating household member) but during the 1730s and 1740s gave away 100-acre parcels to individual Protestant immigrants and 20,000-acre tracts to labor recruiters who founded townships. The leaders then redistributed the land to immigrants. During the 1730s and 1740s the trustees of the new Georgia colony gave 50-acre plots to the poor they transported and 50-acre headrights to each free immigrant (or migrant from Carolina) and each servant he brought over, to a maximum of 500 acres; indentured servants got 20 acres. After the crown took over the colony in 1753, the size of headrights increased to 100 acres for the head of household and 50 acres for each dependent and servant. The governor, at his discretion, could sell individuals 1,000 acres for 0.1 shilling per acre. Finally, a settler in West Florida in the late 1760s and early 1770s could claim headrights and land purchases identical to those of South Carolina.[60]

Virginia continued the headright system but also sold the right to seat 50 acres of land for 5 shillings sterling, allowing native-born small planters easy access to land while encouraging speculators to buy rights to large tracts. Seeking to protect the colony's borders and reward his political favorites, Governor Alexander Spotswood gave away many huge tracts (of over 1,000 acres) in the Piedmont. During the 1720s and 1730s Virginia governors gave away 385,000 acres of land in the Shenandoah Valley, nearly all to well placed Ulster-Irish and German immigrants living in Pennsylvania. Unlike the earlier (and later) favored speculators, these men had no intimate ties to prominent Virginia gentry who served as governor or councillor. But they did recruit compatriots and granted them land patents; in turn, they populated the valley with sufficient density to permit the formation of new counties, which could establish militias and protect the region from Indian and French incursions.[61]

Quitrents on land, charged by the crown in royal colonies outside New England and by proprietors elsewhere, rarely retarded settlement or prevented migrants from getting land. To encourage migration to insecure borderlands, Virginia governors, Maryland proprietors, and Georgia trustees canceled quitrents for as long as ten years after initial occupation. Since proprietors (the Lords Baltimore in Maryland or the Penns in Pennsylvania) relied on quitrents for much of their income, they retained only a small part of their holdings, preferring to sell or grant the land. The greater the quantity of frontier or Indian land that came into possession of owners, the higher the proprietors' or crown's potential income. But collection was difficult. In West Jersey proprietors abandoned hope of collecting quitrents and sought merely to sell their lands. In New Hampshire and East Jersey settlers denied the sovereign rights of proprietors and prevented any collection of quitrents. In royal North Carolina and South Carolina, wrangling between governor, assembly, and British authorities prevented collection. In the 1.2-million-acre Selwyn-McCulloh proprietary in backcountry North Carolina, attempts to collect delinquent quitrents led to rioting in 1765 and surrender of the grant in 1767.[62]

Even when proprietor or king gave away land, procuring frontier acres (not to mention making a farm) required money. Colonists had to bear the costs of surveying, patenting, and improving free or cheap land. In Virginia surveying and patenting more than doubled the cost of land, and improvements cost 3 shillings an acre. In the mid-1750s planters who wanted to avoid taxes and added expenses did not patent a million acres of surveyed land east of the mountains. Although European immigrants to South Carolina frontiers got land without surveying or patenting costs, other migrants had to pay £5 in fees per 100 acres. All would-be South Carolina landowners had to travel to Charleston to register their grant, an expensive burden for those who lived in remote places.

Settlers sometimes had to pay for the land itself. Between 1738 and the Revolution, the Lords Baltimore sold vacant land—nearly all west of the mountains in Frederick County or in isolated parts of Baltimore County—for £5 sterling per hundred acres, about half the cost of buying unimproved land from speculators in Piedmont Virginia. The Penns sold land for similarly low prices on the Pennsylvania frontier: 1.4 shillings sterling per acre between 1719 and 1731; 1.9 between 1732 and 1764; and 0.6 from 1765 to the Revolution—years of rapid growth of the frontier population. Even if land had been entirely free, the poor could hardly have afforded to move to the frontier, but the sons of landed farmers and anyone else with access to some capital could meet these costs.[63]

Nearly everywhere migrants went, colonial officials had granted hundreds of thousands of acres to speculators, rich landlords, or big land companies. Although farmers refused to go to such lands if they were distant from settlements and markets, speculators did influence the pace of frontier land acquisition. Speculators who improved land, offered cheap leases or parcels for sale at good prices, or lent money to farmers stimulated rapid settlement. Most frontier tenants negotiated developmental leases. They paid low rents, often a tiny part of their crops, and in return landlords required them to clear brush and forests, build houses and barns, and plant crops and orchards. With low rents they could save money from selling surplus crops to buy land. But when landlords demanded high rents or speculators charged high land prices, they retarded settlement and encouraged those who did come to live as squatters. Eventually borderlands receded westward, and lands controlled by men farmers believed venal stood in the path of migration. Migrants, impoverished by the cost of moving, squatted on speculators' lands and refused either to rent or buy.

During the pre-Revolutionary decades colonies granted millions of acres to land companies organized by rich gentlemen. Located in Indian territory hundreds of miles from settled areas, their lands had little immediate potential. Investors hoped to reap a bonanza after Indians had been forced out and easterners poured in. If the companies made their claims stick, potential settlers would have little choice but to buy land from them. By the mid-1770s the Ohio, Wabash, Indiana, Vandalia, Transylvania, and Illinois Companies had sought grants and had purchased from Indians the southern third of Illinois and Indiana, northern Tennessee, and nearly all of Kentucky and West Virginia. The Proclamation of 1763 not only prevented the companies from selling land, but the few pioneers who made homes in their territories had neither the money nor the inclination to pay. Despite these failures, company activities influenced western settlement by publicizing and sometimes sponsoring exploration. The Ohio Company surveyed some Kentucky land and explored the Ohio country in the 1750s. The Transylvania Company later sponsored Daniel Boone's settlement.[64]

Although neither large manors nor rich speculators dominated the New England frontier, except in parts of Maine and Vermont, communal land distribution and inexpensive grants to proprietors disappeared. Great proprietors enticed hundreds of families (many of them immigrants) to come to the mid-Maine frontier. But because several groups claimed the same land, the newcomers failed to gain control over it. Governors in both New Hampshire and New York granted vast estates in present-day Vermont, mostly to kindred and favorites. Vermont pioneers such as Ethan Allen, who had paid small sums to New Hampshire for town sites, accepted New Hampshire sovereignty and refused to buy land from or pay rents to the speculators New York authorities had given the same acres. New England governments granted new towns to proprietors; but eighteenth-century proprietors, unlike their predecessors, represented only a small majority (seven-tenths in Kent, Connecticut, in 1739) of town families, and absentee speculators sometimes controlled most of a town's lands. Strapped for new sources of income, Connecticut and Massachusetts auctioned vast quantities of frontier land. Absentee proprietors often resold their rights to other speculators or to potential settlers. Resident proprietors divided the land, much like their seventeenth-century predecessors; after land distributions, nonproprietors often bought land from the proprietors or exchanged parcels among themselves.[65]

Proprietors or the crown sold land cheaply or gave it away to favorites so often that speculators acquired millions of acres. Many Virginia farmers bought land from them. Nearly a fifth of the earliest patent holders (1727–45) of Albemarle County, in the central Virginia Piedmont, registered more than 1,000 acres, land which many of them resold. Governor Spotswood himself gained title to 85,000 acres of land in the region. During the first half of the century half the pioneers in Southside Virginia got their first land through the survey and patenting process; a quarter to two-fifths bought 270- to 300-acre tracts from other owners. Men who owned 1,000 or more acres sold much of this land. Virginia authorities gave speculators parcels of land as large as 50,000 to 100,000 acres in the Shenandoah Valley and the Greenbrier region of West Virginia, sometimes in return for bringing families to the region. Speculators who owned more than 1,000 acres in Augusta County, Virginia (southern Shenandoah Valley), rapidly sold their land. Although they increased their holdings from 175,000 to 200,000 acres between 1749 and 1769, at the same time the number of acres taxed in the county grew from 380,000 to 660,000, and the proportion of land they owned dropped from nearly half to less than a third. By 1779 the Beverely and Borden families alone had sold 200,000 acres in 648 tracts they owned in the county.

To sell land quickly before they had to improve it, speculators kept the price low. In western Maryland Daniel Dulany sold thousands of acres to German, Scot, and Irish families; once he even gave 5,000 acres to twenty German immigrants for under a shilling sterling per acre, far below his own cost. He made a killing when the value of the land he kept in reserve multiplied. Similarly, Virginia speculators sold land for about 0.7 shillings sterling an acre in the Shenandoah Valley in the 1730s and 1740s, 2 shillings an acre in Amelia County in the 1730s, 2.1 in Albemarle County in the late 1740s, between 0.6 and 2.3 shillings an acre in Augusta County (in the far southwest) in the early 1750s, and 4.5 shillings—still less than improved land—in Pittsylvania County in the late 1760s.[66]

Speculators sold similarly large quantities of land on other frontiers. Despite serious challenges to their titles, the West Jersey proprietors sold a half-million acres during the middle half of the century to speculators who resold the land to small farmers. George Croghan owned, at various times in the 1760s and 1770s, hundreds of thousands of acres of former Indian land in western Pennsylvania and New York, lands that he resold in smaller parcels to other speculators and settlers. In Orange County, North Carolina, large holders (1,000+ acres) received about two-fifths of the land disbursed by Granville agents, much of which they resold to smaller farmers. Early-eighteenth-century South Carolina speculators acquired hundreds of thousands of acres and sold them at a handsome profit. The richest rice planters claimed 50-acre headrights for all the slaves they imported, and provincial officials and other favorites received grants as large as 6,000 acres, thereby building huge landed estates. They regularly resold these lands to smaller operators. Charleston residents, all nonresident land speculators, placed two-thirds of the advertisements for backcountry land in the *South Carolina Gazette* between 1738 and 1760.[67]

Alone among the colonies, Georgia prevented land speculation—for a while. Georgia's trustees, who founded the colony in 1732 as a refuge for the poor, debtors, and the dispossessed of England and Europe, forbad land speculation and the heavy debts and poor military defense speculation encouraged. They limited landholdings to 500 acres, demanded land be rapidly improved, outlawed land sales, allowed only sons to inherit, and prohibited slavery. For more than a decade neither planters nor speculators came to Georgia. The majority of colonists, whether charity cases whose transportation the trustees paid or private adventurers, owned no more than 50 acres. But the trustees' rules provoked a firestorm of protest from would-be speculators from South Carolina and abroad. In 1741 more than a quarter of the colony's men petitioned the trustees, demanding freehold land tenure (thus allowing land sales), female inheritance (so those without male heirs would not lose their land), and slavery. By the mid-1740s, nine-tenths of new arrivals had financed their own way, and most wanted freehold land. This pressure forced the trustees to abandon the restricted inheritance and enact freehold land tenure. When the crown took over the colony, restraints on landownership disappeared, and feverish land speculation began. Although each settler was entitled to land, he (or she) had to petition the governor and council for rights to particular land. The governor and councillors favored members of their own class, however, and granted the best land to big planters and speculators and small, backcountry tracts to poorer farmers. Speculation became especially rife between 1763 and 1772, when Georgia authorities granted 772,000 acres, more than in the preceding forty years. Much of it went to officers of the Seven Years' War, who got up to 5,000 acres; the rest went to small backcountry farmers.[68]

Astute speculators acted as patrons to their customers, hoping to reap additional benefits by marketing crops, grinding corn, milling lumber, or collecting interest on loans. During the 1720s and early 1730s Robert "King" Carter, who owned 700 slaves and thousands of acres, supplied neighbors in Virginia's Northern Neck with land, servants, and slaves. Daniel Dulany donated land for German churches, provided credit, built mills, and marketed the crops of the German farmers who bought his Maryland land; he developed Frederick town as a marketing center to ensure the area's prosperity. During the pre-Revolutionary era, Masonian proprietors built mills and roads, and the Kennebec Company not only brought immigrants to their Maine holdings but erected mills, constructed roads, and brought missionaries to minister to the community's religious needs. The Leaming and Spicer families, gentlemen who lived in Cape May County, New Jersey—a thinly populated area isolated from the rest of the province by pine barrens—hired local men, leased land on good terms, marketed goods, and lent their laborers to small farmers. Where speculators became developers and devised reciprocal personal relations with residents, they made money. But settlers rejected any hint of debased dependence on a patron. By exacerbating tensions over religion but demanding payment for lands settlers believed they already owned, the Kennebec Company, for instance, faced hostile inhabitants.[69]

Large landholders and speculators dominated New York more than any other colony. Although Long Island remained in the hands of freeholders, landlords and speculators owned nearly all the rest and monopolized frontier lands. In 1700 ten or eleven men owned three-quarters of the province; Rensselaerwyck Manor alone contained 1 million acres and covered most of Albany County. Eighteenth-century governors gave millions of acres to favorites. New York governors helped themselves royally; by 1732, they had given themselves 2.5 million acres. During the 1760s they gave away the vast northeastern frontier, granting speculators some 2 million acres in Vermont. Only lands on the west side of the Hudson River, closer to the Iroquois, remained relatively free of great estates. Smaller speculators, moreover, with thousands of acres controlled much of the land the richest men neglected.[70]

The great landlord families, who owned fourteen manors located mainly on the east side of the Hudson River, refused to sell their holdings, preferring to collect rents and fees. Cadwallader Colden complained in 1732 that such men had prevented the rapid peopling of the colony. Although New York "was settled many years before Pennsylvania, . . . it is not nearly so well cultivated, nor are there near such a number of Inhabitants, . . . in proportion to the quantity of Land." And, he added, "it is chiefly . . . where these large Grants are made where the Country remains uncultivated 'tho they contain some of the best of the Lands, and the most conveniently situated." The big landowners could neither improve their vast holdings nor attract tenants. Wishing to "avoid the dependence on landlords," immigrants and "Young people go from this Province, and Purchase Land in Neighbouring Colonies while" New York's much "better and every way more convenient Lands lie useless." Nor did the manors attract more families over the next two decades. William Smith Jr. lamented in 1756 that "scarce a third part of it is under cultivation," in part because landlords "have rated so exorbitantly high, that very few poor persons could either purchase or lease them."[71]

Colden and Smith ignored the supply of freehold land and the rapid growth of New York's frontiers. Between the 1720s and the 1750s the population of Albany County, the Hudson River frontier, grew 3 percent a year. Although potential settlers avoided the area because it was in the path of French and Indian warfare and close to the Iroquois Confederacy, freehold land could be found there. During the mid-1720s a colony of ninety Palatine German families got cheap hundred-acre farms along the Mohawk River, buying it from the Indians, paying for a patent, and offering a small annual quitrent. The population of Dutchess County, south of Albany, multiplied 8 percent a year between 1723 and 1756, growing particularly rapidly (11–12 percent) in the 1730s and 1740s. Frontier families there farmed existing unclaimed freehold acres around Poughkeepsie and in the county's northwest corner. Others got freehold land in Orange or Ulster Counties, west of the Hudson. These places had fewer large holders, but because their western bounds bordered on Indian territory, they faced repeated Indian raids. As a result, their population grew only 3–4 percent a year between the 1720s and 1750s. Nonetheless, hardy souls could readily buy land there; the Minisink partners—owners of a 170-square-mile tract in western Orange and Ulster—had sold at least a tenth of their holdings by 1763, in parcels averaging 160 acres. Settlers unable to buy land could rent it on easy terms from the big landlords desperate for families to develop their holdings, a particularly appealing prospect to poor immigrants. Landlords not only allowed tenants to live rent-free for as long as a decade and provided livestock and supplies to help them get started; they also lent them money and built mills and other facilities for their use.[72]

During the first half of the eighteenth century, farmers could find sufficient land in New York to meet their needs. But as the Hudson River Valley filled up and the Indian danger receded, thousands of families, some from New England

and abroad, poured into the Hudson and Mohawk River frontiers. Between 1756 and 1771 Albany County's population multiplied 2.5 times, from 17,400 to 42,700. Immigrants seeking freehold land avoided the manors; as late as 1767 there were nearly 500 acres for each tenant on Livingston Manor in southeastern Albany County. Nevertheless, manorial lands sometimes were the most conveniently located, especially for migrants from western Massachusetts. By the 1770s, 6,000 or 7,000 families rented land in New York; the number of tenants at Livingston Manor, for instance, jumped from 50 in 1715 to over 200 in 1750, 285 in 1767, and 460 in 1776. Tenants continued to have the right to sell the improvements they made. But seeing greater opportunities, some landlords raised rents (and insisted on cash payment); charged high fees (as much as one-third the farm's value) when a son took over his father's farm; denied new tenants a rent-free term; and signed new leases only for tenants-at-will, rather than the perpetual leases or leases for lives common earlier in the century. Others, like the Livingstons, kept rental rates the same but accepted payment only in wheat, the price of which skyrocketed as European demand grew in the 1760s and 1770s. Such changes in leasing practice, in part, caused the riots that racked the manors in the 1750s and 1760s.[73]

Elsewhere in the backcountry tenants formed a small part of the population, and landlords knew that they had to give generous leases to get anyone to farm their lands. Virginia gentlemen rented land to many small planters in the Northern Neck (where a third of householders were tenants) and Shenandoah Valley. While most Northern Neck landlords rented land for years, they granted a two-year grace period. Understanding that tenants often could not pay, great planters such as George Washington and Robert Carter of Nomini allowed them to fall years behind. Even so, tenants regularly ran away, and families unwilling to pay often squatted on their lands. Shenandoah Valley landlords, including prominent gentlemen such as Landon Carter and George Fairfax, gave leases for lives or twenty-one years but attracted few families. Landlords made no more profit and attracted no more tenants in backcountry Pennsylvania, New Jersey, and Maryland. To entice tenants, landlords in northeastern Pennsylvania not only had to forgive rents for ten years but give renters seed, livestock, and tools and provide access to mills and markets. In western Pennsylvania in the 1780s—where one in ten farmers rented—the tenants paid rent only on the improvements they made on the property. The Lords Baltimore created only two proprietary manors in the backcountry. In order to get the land developed, they offered leases for three lives at the cheap annual rate of ten shillings sterling per hundred acres; when the leases expired, tenants refused to sign higher leases. Tenants on the two western manors neglected to pay even these small sums; during the 1750s the proprietor earned just £19 sterling, a sixth of what was due him. In northwest New Jersey speculators forced squatters—who had destroyed timber and exhausted the land by constant cultivation—to accept leases with great difficulty, and many of them sold out for low prices as soon as they found buyers.[74]

At the outset of white habitation, most settlers (except in parts of New England) squatted on the land they farmed. Unable to pay the smallest sums for land and eager to evade taxes, many of them neither surveyed nor bought any, even where cheap land with clear titles could be found; a few, like the hunters and bandits of mid-century Piedmont South Carolina, had no intention of gaining legal title. In 1726, James Logan, secretary of the Province of Pennsylvania, complained that Palatine and Scots-Irish immigrants held "near a hundred thousand Acres . . . without any manner of Right or Pretense to it." And, he added, "most of them are so poor, that they have nothing to pay with." Some families, like a group of German immigrants in Frederick County, Maryland, in the early 1740s, contracted to buy land but could not raise money for the purchase. During the 1770s, squatters in the southern Shenandoah Valley frontier refused to buy land from speculator-owners or leave without coercion. So many families wanted the land that chasing off one led others to "settle on the land & make a second & third Ejectment necessary."

Other farmers doubted the title of those who demanded payment or fell victim to disputes between gentlemen over land allocation. East Jersey and New York squatters refused to pay rent to great landlords, insisting that they owned the land. In 1735, ninety-eight squatters who farmed land in thinly settled Hunterdon County in northwest New Jersey ignored the demands of the West Jersey Society to take leases because they questioned the society's title to their farms; they stayed on the land for decades thereafter. Maine pioneers in the 1760s and 1770s refused to buy land from the proprietors, fearful of accepting a title bound to be contested by a competing proprietor. A dispute over the inheritance of the Pennsylvania proprietorship closed the land office between 1718 and 1733, making purchase of frontier land difficult and leading to massive squatting. Similarly, in the 1720s, South Carolina planters engrossed 300,000 acres of land without grants after its land office closed because of a dispute between the former proprietors and the new royal government.[75]

Because of poverty, only a small proportion of pioneer families acquired title to land. As John Dunmore, Virginia's last royal governor, wrote disgustedly in 1774, settlers "did not conceive that Government has any right to forbid their taking possession of a Vast tract of Country, either uninhabited, or which Serves only as a Shelter to a few Scattered Tribes of Indians." Only twelve of the first sixty-five men to arrive in Paxton Township on Pennsylvania's Susquehanna frontier in the late 1720s bought land there. In 1751 no more than three-tenths of the householders in Maurice River Township in frontier Cape May County, New Jersey, owned land; the rest grazed livestock on the nearby woodlands. Just over half the householders in Lunenburg County, Virginia, owned land in 1750, a decade after the first settlers arrived. Similarly, only two-fifths of the householders in Northumberland County, in the upper Susquehanna Valley of the northeastern Pennsylvania frontier, owned land in 1772; 57 percent of the householders in Montgomery County in western New York owned land in 1790.[76]

Too poor to buy land, unwilling to abandon Indian land or to pay high prices to speculators, or unable to distinguish between competing owners, squatters tried their best to protect their farms. When he was a child, Joseph Doddridge (born in 1769) recalled that Pennsylvania pioneers established occupancy by "*tomahawk right*, which was made by deadening a few trees near the head of a spring, and marking the bark of some one or more of them with the initials of the name of the person who made the improvement." Other Pennsylvania pioneers claimed 100 acres of land based on a "corn right," the planting of an acre of grain. Similarly, late-eighteenth-century squatters in Maine established rights to land by building a crude cabin and running a short "possession fence" around the property. Squatters sometimes forced newly arrived settlers to buy these rights. Men who did improve the land believed that such labor gave them rights of ownership.

If squatters owned the land, how could they protect their rights from other, equally worthy farmers who wanted to farm the same land? Squatters on Indian land on the West Branch of the Susquehanna River in the 1770s devised a "Fair Play System" to regulate use of the area's land. They annually elected three extralegal magistrates to record landholdings, determine land boundaries, supervise land sales, and arbitrate controversies between settlers. By the late 1760s, Pennsylvania officials recognized some squatter rights, allowing them first rights to buy land once surveyed or to sell their improvements if they could not afford the land.[77]

Lacking secure title and unable to accumulate enough money to buy their land from speculators who claimed ownership or to build more than a wretched cabin, poor squatter families stayed a few years but inevitably left, desperately seeking better opportunities elsewhere. Governor Dunmore, unsympathetic to their plight, complained that "established Authority" could not "restrain the Americans" who "remove as their . . . restlessness incite them. They acquire no attachment to Place: But wandering about Seems engrafted in their Nature," always imagining that "the Lands further off, are Still better than those upon which they are already Settled." Squatters too poor to buy the land they farmed did rapidly move to more isolated frontiers. During the early 1770s as many as a third of the taxpayers in Paxton Township—most of them tenants, squatters, or laborers—disappeared. Similarly, between 1792 and 1802 four-fifths of householders without land left Turbut Township on the Susquehanna frontier of Northumberland County, Pennsylvania, as did two-thirds of those who held less than 150 acres.[78]

Families who could count on kindred or an ethnic community remained and got land, sometimes from landsmen. Two Virginia examples illuminate the process. From 1745 through 1751 speculators surveyed the vast majority of land along the James and Roanoke Rivers in Augusta County. Hearing of this fine land in Augusta, kindred of the earliest arrivals rushed to secure and survey tracts near those owned by their relations. Four-fifths of the early residents of Augusta with resident kinfolk—but just over two-fifths of those without kin—acquired land. Eleven of the earliest Ulster Irish families in Opequon, Frederick County, Virginia, were members of the same kin group. Bequests, land sales, and exchanges among these Ulster Irish settlers guaranteed that they, their sons, and sons-in-law would get land.[79]

As farmers improved land and the population increased, land prices rose but remained half or less as high as in older regions, making these newly settled regions attractive to middling farmers who had too little land for their families. Since cheap land remained abundant and those who stayed patented land or bought a farm, the proportion of landowners rose. Within a decade of settlement, two-thirds or more of householders owned land in the New York, New Jersey, Pennsylvania, Virginia, and South Carolina backcountry. Close to two-thirds of the families in Lower Township, Cape May County, New Jersey owned land in 1751. In Lunenburg County, Virginia, between three-quarters and four-fifths of household heads owned land in the 1760s, 1770s, and 1780s; similar rates could be found in five other Piedmont counties, and at least nine-tenths of the householders in Albemarle County, Thomas Jefferson's home, owned land. During the 1780s three-quarters of the householders in ten backcountry North Carolina counties and two regions of the South Carolina upcountry owned land. By the early 1790s four-fifths of Northumberland, Pennsylvania, families and more than nine-tenths of those in Balltown, Maine, held land. As the number of householders in Montgomery County, New York, nearly tripled between 1790 and 1795, the proportion of freeholders rose to 71 percent.[80]

Farmers patented large tracks in the wilderness hoping to make farms for themselves and their children. In the 1780s the Virginia backcountry, first seated in the 1750s and 1760s, remained a frontier area, with large quantities of unimproved land. Plantation sizes in six tobacco-growing Piedmont Virginia counties declined from 250–360 acres in the 1760s to 250–300 acres by the 1780s. A similar pattern could be found in the Shenandoah Valley. Augusta landholders in the 1750s and 1760s owned 300–350 acres; by 1787 median holdings had decreased to 240 acres. Land grants in the North Carolina Granville proprietary averaged almost 500 acres in the 1750s. By the early 1780s farmers in ten North Carolina Piedmont counties owned 250 acres, far larger than the farms of eastern Carolina; only a sixth of these farms contained less than 101 acres. Tracts in backcountry South Carolina during the 1770s ranged from 100 to 400 acres, averaging 200 acres.

Maine, Pennsylvania, and Maryland farms were smaller than southern plantations but sizable by settled northern standards. The families of Balltown, Maine—some of whom brought New England traditions of small farms with them—owned nearly 100 acres in 1791, but they had improved less than 10. Holdings in several central Pennsylvania townships averaged between 100 and 200 acres in the pre-Revolutionary decades and about 150 acres in Northumberland County, Pennsylvania, between 1774 and the 1790s; similarly, farmers in Frederick County, Maryland, in 1753 had 160 acres.[81]

Freehold farmers not only acquired land but sought to patent the very best acreage, thereby assuring not only farms for themselves and their children but high capital gains as the frontier filled with farms. Nothing "but a preference to the choice lands," Lord Fairfax insisted in testimony challenging the legality of John Hite's sale of land in the Shenandoah Valley, "would tempt men to become adventurers." Fairfax was right. William Rogers, a farmer who got land in the valley, testified that he "did make search to find" such "Land in order to make a Settlement for himself and family" and "accordingly found a piece he like[d] very well." Hite then told him "he should have it as he had let others have heretofore."

Surveying methods—with the irregular tracts found on southern and Pennsylvania frontiers—compounded the problem of dispersed settlement that the search for the best land began. Only in New England and its colonies (like the contested Susquehanna settlement in Pennsylvania) did rectangular tracts remain the rule; tracts in South Carolina and Georgia townships (granted to various ethnic groups) had more compact shapes, as did the radial division of land found in some frontier Pennsylvania areas. Everywhere else scattered farms rendered defense against Indian attack nearly impossible.[82]

Increasing tenancy and high land prices had long ended the egalitarian days of the seventeenth century, when nearly all free men owned land. Indians, big landlords bent on extracting rent, and land speculators seeking high prices for the unimproved acreage they held initially stood in the way of freehold landownership in frontier areas. But with so much Indian land waiting to be conquered, sons of landowners and European immigrants alike could get small farms. New England farmers easily got land, the increasingly speculative nature of town proprietorships notwithstanding; in other regions, settlers procured cheap frontier acres from governors, proprietors, or speculators, thus remaking a society of smallholders. As Chapter 4 shows, this "best poor man's country" attracted hundreds of thousands of British and German immigrants eager to share in America's bounty. These immigrants, a highly selective group, were essential in settling the backcountry, and like native-born farmers, they came to believe that the ownership of land was their birthright.

chapter four

Deprivation, Desire, and Emigration

As many as 410,000 Britons and Germans came to a mainland colony between 1700 and 1775. David Evans (from Wales about 1704), William Moraley (from London in 1729), and Thomas Fleming (from Ulster in the 1750s) were among them. Evans came to Pennsylvania "to earn money so I could buy plenty of books." But he had to put his book buying on hold. To pay for his voyage, he indentured himself to a farmer and worked "patiently cutting trees and clearing the land." When his term was over, he wandered about, working in Newcastle and Philadelphia before returning to the countryside, where he became a Presbyterian preacher. "Oppress'd by Dame Fortune," unemployed watchmaker Moraley was "very willing to" indenture himself if he "could have some view of bettering my condition." Although he "expected a better fate than to be forc'd to leave my Native Country," he "had rather leave a Place where I have no prospect of advancing." After serving a New Jersey watchmaker, he "roamed about like a Roving Tartar," "went about cleaning Clocks and Watches, and followed the occupation of a Tinker" in and around Philadelphia. He returned to London a failure in 1734. A 1758 letter to Thomas Fleming from his Ulster cousin David Lindsey describes the privation that led Fleming to the colonies, the dread of Indian war that kept his kin home, and the chance to own cheap land that compelled him to overcome his fears. "With wars" Fleming's family had assumed "that you were all dead." But they did hear from Fleming. "The good bargains of your lands . . . doe greatly encourage me," Lindsey wrote, "to pluck up my spirits and make redie for the journey, for we are now oppressed with our lands at 8s. per acre . . . , cutting our land in two acre parts . . . —yea, we cannot stand more." Other family members expected to leave as well. Lindsey's nephew Robert Lindsey wanted to know "if he will redeem himself if he goes over there. He . . . is willing to work for his passage till it's paid."[1]

Immigrants like Fleming, Moraley, and Evans left their homelands to come to America to escape religious persecution, economic depression, population pressure on land, and the increase in landlessness (and with it poverty and wage labor) exacerbated by the enclosures and increased rents of British and German capitalists. They resembled most European emigrants. Living in wretched hovels, barely able to cover their nakedness, hungry much of the time, and oppressed by their superiors, they were poor; many of them—the old, the lame, children in big families—were mired in perpetual destitution. Women and children worked for miserly wages, relied on poor relief or charity, or begged for food. Wherever he went in Ireland or France (in the 1770s and 1780s), English agrarian reformer Arthur Young found the same grinding poverty; reports from Silesia and Bohemia showed the same distress. In famine years, such as the early 1740s, tens of thousands of people died of starvation and disease.[2]

Four-fifths of Europeans lived in the countryside, where they had gradually lost their land. In 1500 more than two-thirds had enough acreage to grow some of their food. By 1800 less than two-fifths still held land; the rest worked for wages, at best farming a tiny plot. The population of Europe grew more rapidly in the eighteenth than the seventeenth century, jumping particularly fast—from 94 to 123 million—in the second half of the eighteenth century. Population growth aggravated landlessness by reducing the size of farms, especially where peasants gave some land to each child. Traditional strategies—postponing marriage, starting cottage industries, migrating short distances, or placing children in farm service—mitigated misery but hardly ended it. Writing at the end of the century, Thomas Malthus linked population growth to the "squalid poverty" and "degradation," the "absolute want of bread" and "stunted . . . growth," and the "epidemic and endemic diseases" he found everywhere. The Irish "lower classes" especially lived "in a most depressed and miserable state," suffering "diseases occasioned by squalid poverty, by damp and wretched cabins, by bad and insufficient clothing, by the filth of their persons, and occasional want." Without options the poor man was "reduced to the grating necessity of forfeiting his independence, and being obliged to the sparing hand of charity for his support."[3]

Britons or Germans could go to America, but only a tiny minority went. Why so few left is baffling, given the opportunities on American frontiers and the ties of kinship, religion, and culture that linked Britons to colonists. "Pity it is that thousands of my country people should stay starving at home when they may live here in peace and plenty," Scot immigrant Dr. Roderick Gordon (of King and Queen County, Virginia) lamented in 1734. "A great many," he concluded, "transported for a punishment have found pleasure, profit and ease and would rather undergo any hardship than be forced back to their country."

This chapter will examine the reasons why so few Europeans came to the eighteenth-century mainland colonies—and the economic and political conditions that motivated those who came. Economic opportunities in Europe, the chapter will show, abounded, belying the gloomy picture Malthus and Young painted: cities grew rapidly; rural industries expanded; wars employed thousands of peasant soldiers; and frontiers in Prussia and eastern Europe beckoned. A few of the desperately poor, unable to take advantage of these opportunities, indentured themselves and emigrated, but most emigrants, as we shall see, took great risks hoping to glean opportunities they thought greater than those at home, in European cities, or on European frontiers.[4]

A Peasant's and Worker's Paradise?

The vast majority of Europeans stayed close to home, moving short distances or taking advantage of growing opportunities in rural industries or flourishing cities. Poor families planted or harvested crops, some traveling hundreds of miles a year. Farm families moved to Europe's frontiers; single men joined the army. Nor were the colonies the only refuge for religious dissenters. Catholics and Protestants could find asylum in a German state or the Netherlands. Those who emigrated to the mainland colonies were more willing to take risks than those who stayed at home or moved within Europe.

Europeans often moved to marry or to find work. Apprentices left home in early adolescence, often to work with an artisan in a nearby town. Adolescent male and female farmhands and female domestic servants left home, moved about the countryside in a circuit of ten miles or less or to the nearest town, and contracted to work. People often married within their village and rarely moved more than ten miles to do so. Once married, poor adults moved more often than those with land. When opportunities became scarce, peasants migrated to a more developed area nearby. Other groups moved to work in country industries or in growing cities or migrated long distances to harvest crops in a distant land or to farm on an internal frontier.[5]

Rural industries—especially textiles—grew mightily in eighteenth-century Europe. Most people in the English Midlands and Yorkshire, the Scottish Lowlands, Ulster, Flanders, southern Netherlands, Languedoc, the Zurich region, the Rhineland, Silesia, and Bohemia toiled in rural industries. Some wove cloth as a sideline to thriving farms; others owned their tools and bought raw materials; most toiled for small wages. By working tiny plots, milking cows and making butter and cheese, engaging in seasonal farm labor, spinning or weaving at home for businessmen, and pooling the family's labor, cottagers could subsist and peasants could increase family income. Landless families relied exclusively on rural industry. In some textile districts—Flanders, northern France, the Zurich region, and Ulster—landless people married at younger ages and had more children than peasants did, knowing that all the family could spin or weave. As industry expanded, textile villages grew into cities, attracting migrants willing to work full time; the population of four English manufacturing towns jumped from 27,000 in 1700 to 70,000 in 1750 and reached 262,000 in 1801.[6]

An English observer lamented in 1758 that "great Multitudes of People, who were born in *Rural Parishes*, are continually acquiring Settlements in *Cities* or *Towns*." And so they did. In 1700, 6.3 million people lived in 555 northern European and British cities with populations over 5,000; by 1800, 11 million resided in 661 cities of similar size. City growth in England was especially strong. The population of London jumped from 575,000 in 1700 to 675,000 in 1750 and 960,000 in 1800. Twenty-two smaller English cities grew at a faster rate, rising from 215,000 in 1700 to 605,000 in 1800. Most city dwellers were migrants. Since most such migrants were young, unmarried men, who moved away or died more often than they married and sired children, cities suffered population losses that only new migrants could replace. The annual number of rural migrants to northern European cities rose from 40,000 before 1750 to more than 70,000 after that date. Between 1700 and 1775 nearly 4 million Europeans moved to cities, ten times the number who reached the mainland colonies.[7]

More than 300,000 late-eighteenth-century Britons and West Europeans, most of them in their twenties and thirties, moved seasonally to find work. In addition, hundreds of thousands of young men, including many foreigners, joined armies, navies, or private sailing ventures for a term of years. Migrants harvested wheat, dug potatoes, cut timber, mined coal, sharpened knives, and peddled goods. Most migrants—and nearly all the women—were farm laborers. Unable to feed their families, poor men living in depressed farming areas left temporarily each year, leaving wives to care for children and tend their tiny farms. To pay high rents, Irishmen harvested East Anglian crops and labored in London while their wives and children "subsist[ed] by begging" and harvested their potatoes. Highland Scots harvested crops in the Scottish Lowlands; the Paris region attracted 60,000 husbands and single adults a year from the central highlands or Alpine regions. Others in mountain France harvested in Spain; the Swiss harvested in Alsace. A hundred thousand migrants harvested in Italy; about 40,000 Galacians (from northwest Spain) harvested in Castile. German peasants (half the people in some villages) left their farms during the slack summer growing season to cut hay in the Netherlands or to work in German cities.[8]

Late-seventeenth- and eighteenth-century immigrants moved to thinly peopled internal frontiers—in Ireland, Prussia, Hungary, and Russia—closer to home than America. More than 900,000 immigrants peopled Europe's borderlands between 1680 and 1800, more than double the number (about 435,000) who reached British North America. As many as 100,000 Britons and an equal number of Scots left for Ireland in the late seventeenth century, and Scot emigration continued into the eighteenth century at a lower rate. Eighteenth-century Swiss and southwest Germans populated eastern frontiers in Prussia, Hungary, Poland, and Russia.[9]

Nearly all migrants moved as individuals or in small groups mainly for economic reasons, but the single largest group fled to escape religious persecution. After the 1685 revocation of the Edict of Nantes that had guaranteed toleration, about 160,000 French Protestants fled into exile in the Netherlands, Switzerland, the Palatinate, Rhineland, Prussia, and England. Mostly city artisans and merchants, they gravitated toward London, Geneva, Amsterdam, or Berlin; half took up farming in their places of refuge. Those unhappy with their fate left their place of refuge and migrated to Ireland, Scandinavia, Russia, the English American colonies (no more than 2,000), or South Africa.[10]

Colonial policies and prejudices kept Catholic immigrants away from the North American colonies. Immigrants expected to own land, but only citizens could get clear titles. Immigrants from Britain qualified for citizenship on the basis of their birthplace, but those from the continent had to be naturalized. A 1740 act of Parliament and local practice allowed foreign Protestants to become citizens after seven years but kept Catholics—who dominated France, Italy, and some German states—from being naturalized. Catholic immigrants faced anti-Catholic sermons, newspaper attacks, and parades. New Englanders were fiercely anti-Catholic; the Seven Years' War triggered anti-Catholic hysteria (aimed supposedly at the French enemy) in Maryland, South Carolina, and even tolerant Pennsylvania. Labor recruiters avoided Catholic areas (except Irish cities); rather than face the intolerance of colonial Protestants, Catholics migrated within Europe or stayed home.[11]

The European emigrants who moved to the mainland colonies in the eighteenth century came from different places than had seventeenth-century emigrants. Over 10,000 Dutch, Swedes, and French Huguenots had migrated in the seventeenth century, but few arrived in the eighteenth. English emigrants (160,000 strong) accounted for almost nine-tenths of the Europeans who reached seventeenth-century mainland British and Dutch North America. The eighteenth-century English stayed home. Only 72,000 (of a mid-century population of 5 million) emigrated to the mainland colonies, comprising a fifth of European immigrants. Few Scots, Irish, or Germans arrived in the seventeenth century. But perhaps 60,000 Scots (of 1.3 million)—almost a sixth of the total—

arrived in the eighteenth century, most after 1760. And 150,000 Irish residents (of 2.5 million) came in the eighteenth century, two-thirds of them Ulster Protestants. Most were descendants of seventeenth-century emigrants from Scotland, and the others, mostly Catholics, came from the south. About 110,000 German speakers arrived in the eighteenth century, mostly between 1730 and 1770.[12]

Economic Development and British Emigration

Varying economic conditions in England, the Scottish Lowlands and Highlands, Ulster, and southern Ireland help explain the differences in emigration from the different parts of the realm. Expanding rural industries, growing cities, and (in some places) still-abundant farmland allowed most people to remain in the kingdom. The English, especially those near London, enjoyed the most vital capitalist economy in Europe. Despite profit-seeking landlords and the decline of reciprocal feudal class relations, many Scots and Irish also enjoyed the fruits of more rapid economic development.

English families had good reasons for staying home. Agricultural productivity rose, leading to more plentiful and cheaper food for landless laborers. Cotton imports and industrial output rose, especially after 1760, and the cost of cloth and other goods declined. As prices dropped, a growing middle class indulged in consumption—of buttons and buckles, china and glass windows, sugar and tea, and fine linens and calicoes—undreamed of by their grandparents. Poor folk gained some benefit from this prosperity. The demand for labor exceeded the supply of youths coming of age because the English population had declined during the late seventeenth century, did not recover until the 1720s, and grew slowly thereafter. Wheat prices dropped in the 1710s, 1730s, and 1740s, making bread cheaper for workers and allowing them to buy more meat and vegetables. City wages rose during the first half of the century and then declined; male rural wages in the south increased through the 1770s, though women's wages had begun to decline by the 1760s. Even after 1750, when wheat prices slowly rose, consumption of white bread and meat grew as fast as the population, and sugar prices declined enough to allow the poor to indulge. Higher wages and lower prices increased the standard of living, as rising life expectancy among all classes indicates.[13]

At mid-century three-tenths of England's land, located in the richest and most densely populated regions, remained in open-field villages or unenclosed fens, marshes, or forests; after two decades of parliamentary enclosure, most of this land was still unenclosed. Poor families in such areas—cottagers with rights to commons and squatters who used wastelands—combined farming with manufacturing. The framework knitting industry developed in unenclosed East Midlands villages where landholders could work part time. In arable districts of Warwickshire, mid-eighteenth-century cottagers, small tenants, and prosperous farmers turned to victualing, shoemaking, carpentry, or tailoring. Cottagers and smallholders enjoyed the rights to graze sheep or cows on the commons and to use the wastelands (for collecting firewood, fishing, hunting, and gathering berries), practices that, in poor times, provided the margin of subsistence. Since they reduced the necessity of families to engage in wage labor, common rights allowed cottager women to add to family income by performing subsistence activities at home, allocate time for leisure, and—insofar as they increased semi-independent subsistence—reduced parish taxes that supported the poor.[14]

Despite growing poverty, there were reasons to stay in Ireland. Vigorous growth of the Ulster linen industry raised per capita income. Linen exports jumped from 6.6 to 20.6 million yards between 1740 and 1770, providing work to a quarter of Ulster's families by 1770. In Ulster's smallholder households the whole family participated, greatly increasing the income from their five- to thirty-acre farms. Husband and wife planted a small flax patch in an old potato field; the men pulled the crop, and the women spread it to dry and beat it to extract the fiber. Then women and children spun yarn, and the husband (and on occasion his wife) wove linen cloth. In Donegal the farmer cultivated "no more Land than is necessary to feed his family," an observer wrote in 1739, depending "on the Industry of his Wife and Daughters to pay by their Spinning, the Rent, and Lay up Riches." Well-off families bought bread and potatoes so they could put more energy into linen making; to increase output, they hired poor men to weave and poor women to spin. Poor families bought flax, spun yarn, and sold it to jobbers to resell to weavers, and some entrepreneurs hired journeyman weavers. In the 1750s the Linen Board, which regulated the trade, started 200 spinning schools for children in the southern provinces; by the early 1760s, linen making had spread rapidly there.[15]

Although most eighteenth-century Catholic peasants and cottiers (cottagers) were desperately poor, much of the populace shared the fruits of growth, consuming more and building new homes. New urban middle classes—five-eighths to four-fifths Protestant—grew. Some Ulster Presbyterians made big profits from weaving and followed linen markets; afraid of losing their middle-class prosperity, they took off for the colonies when markets plunged. Tenants—almost a majority of the rural populace until the third quarter of the century—could get long leases, usually for three lives, or thirty-one years (for Catholics). After 1750 they took advantage of rising crop prices; those with long leases could sell or sublet their land or rent small plots to weavers and pocket the profits. Prosperity brought few benefits to the poor. Women sold surplus eggs and butter and took in spinning; families with a bit more income used more amenities such as tea, the consumption of which nearly tripled between the early 1750s and the early 1770s.[16]

Irish farming improved greatly over the mid-eighteenth century. Although the growing city of Dublin (from 62,000 people in 1706 to 140,000 by 1760) imported much of its food during the first half of the century, tenants living in its hinterland had taken over by the 1760s, supplying the city with flour, potatoes, butter, and meat. By the 1770s they harvested nearly as much wheat as the English and produced more barley per acre. Despite population pressure, many still raised sheep, pigs, and cows. Tipperary grazers raised thousands of sheep and sold wool and mutton; smallholders raised cows; and their wives took butter to market. Farmers in County Mayo (on the west coast) sent large quantities of wheat and malt to Dublin and shipped provisions overseas. The country not only usually fed itself but exported beef and butter to the West Indies.[17]

Except during famines, the mid-eighteenth-century Irish ate a decent diet of potatoes, peas, beans, oatmeal, milk, butter, buttermilk, eggs, bread, a little meat (fowl or pork), and fish (near the coast or lakes). In 1776 Arthur Young found well-fed cottiers northwest of Dublin, with "a cow, and some of them two" and "a belly full of potatoes." "Every cottage," he added "swarms with poultry and most of them have pigs." A family could grow enough potatoes on one acre to feed themselves for a year. Only the most destitute relied on potatoes and milk, and even this diet supplied adequate nutrition. Luckily, potato surpluses fed the growing number of landless laborers. As the Irish diet improved, infant mortality, a key indicator of living conditions, dropped.[18]

Scotland was the mirror image of Ireland. Three-fifths of the country's land—mountain, hill, moor, heath, and waste—was unfit for cultivation, and a cold, wet climate often turned arable to waste. In the seventeenth century, poor harvests, bouts of smallpox and the plague, famine, and the collapse of poor relief forced 200,000 Scots—cottars (cottagers) and smallholding tenants and orphans and widows—to go to Ireland, Poland, and Scandinavia. But eighteenth-century Scotland developed more rapidly than Ireland. In 1691 1.2 million people lived in Scotland. Only 1 million remained in the early eighteenth century after the 1690s famine, but the population grew thereafter. Bolstered by rising marital fertility, diminished crisis mortality, and reduced emigration, the population rose from 1.25 to 1.4 million between 1755 and 1775.[19]

Attracted by opportunities in foreign trade and industry, Scots moved rapidly from countryside to city. The proportion of Scots living in cities rose from one-twentieth in 1700 to one-sixth in 1800. Glasgow's population rose from 13,900 in 1700 to 23,500 (33,500 including suburbs) in 1755 and 46,800 in 1780. Edinburgh had 30,000 people in 1700, 40,000 in 1740, 57,000 (65,000 including suburbs) in 1755, and 70,000 in 1775. More people in 1700 lived in Irish than

Scottish cities; but a higher proportion of Scots were city dwellers by 1750, and in 1800 proportionately more than twice as many Scots lived in towns. As cities grew, nearby farmers turned to provisioning them and their suburbs.[20]

Scots divided their country into two regions—Lowlands and Highlands. Since the Lowlands had longer contact with England, it developed more rapidly. Its tobacco and linen trades were especially important. Annual exports of Chesapeake tobacco to Glasgow jumped from 5.7 million pounds in the 1720s to 18.9 million pounds in the 1750s and to 43.8 million pounds in the early 1770s. Not only did the trade bring large profits to Glasgow, but it provided jobs in central Scotland, at wages higher than elsewhere in the country. The stores that Glasgow tobacco merchants built in the Chesapeake colonies required provisions; to assure an adequate supply of trade goods, the tobacco lords financed eighty-eight industrial concerns and exported between a fifth and a third of the country's linen output to that region.[21]

The Lowlands competed with Ulster in linen production. Linen manufactured for sale more than tripled between the 1730s and the 1770s, from a yearly average of 3.5 to 12.8 million yards, transforming the central Lowlands and the northeast, where weaving was centered. Some spinning families grew their own flax, spun yarn, and wove cloth, but most weavers got yarn from outside spinners. Many women made yarn at home for a linen dealer, but yarn was often in short supply in the 1730s and 1740s; by the 1740s spinning had spread to the Highlands, alleviating the shortage. A few big weaving shops opened between 1740 and 1775, a water-powered thread industry began in the 1760s, and town production rose after 1750. Small farm families, who spun and wove to supplement farmwork, nonetheless made most of Scotland's linen. At midcentury as many as half of Scotland's women (180,000) spun at least part time. Full-time weaving and spinning also grew; many villages along the east coast north of Edinburgh specialized in spinning and weaving, though their workers participated in harvests to augment their wages.[22]

Scottish peasants divided their lands into two fields, a constantly cultivated infield and an intermittently cultivated outfield held as a commons and reserved for pasture. Lowlanders grew barley, oats, kale, potatoes, and a little flax and made linen for local markets; Highlanders grew barley, oats, and after midcentury, potatoes. In pastoral areas tiny plots provided a bare subsistence at best. Beyond the fields lay pastures. Although rising numbers of Lowlanders owned consolidated parcels and lived on single-tenant holdings, many still lived on multiple-tenant farms; held multiple strips of land, each separated from its neighbor by a ridge; and enjoyed common rights to stone, clay, timber, peat, and meadows. They regularly redistributed land, assuring that each family's holdings were roughly equal in quantity and quality. Tenants sublet small parcels of land to cottars. Similar infield cultivation and outfield pasture could be found in the Highlands, but farmers there, unlike Lowlanders, moved stock to summer mountain grazing. Strips everywhere averaged one-fifth to one-quarter acre, thus requiring cooperation in planting and harvesting. As population grew, the size of strips plummeted, in one estate to only one-fortieth of an acre.[23]

Poverty and Poor Law in Britain

Greater opportunities in industry led to earlier marriages and higher rates of population growth throughout Britain. Since newly growing economies could not keep pace with increased population, destitution grew. But the level of destitution was far greater in Scotland (especially the Highlands) and Ireland (particularly the south) than in England, and the English poor law provided relief to the poverty-stricken that their counterparts in Ireland and Scotland did not enjoy. Such circumstances led middling families in Scotland and Ireland to fear that they, too, would fall into destitution.

During the second half of the eighteenth century English poverty increased. Parliamentary act and local agreement led to the enclosure of 1.6 million acres between 1751 and 1775, mostly arable and pasture. But nearly a third had been wasteland used by cottagers and squatters, who had lost their common rights and were forced to rent a cottage without land. Usually they received no compensation, and when they got land, many sold it because it was insufficient to grow food and graze a cow. As the pace of enclosure rose, rural protests proliferated. Between 1748 and 1779 cottagers, laborers, and smallholders mounted at least twenty-five violent protests, fourteen against enclosure, six against draining of fens, and the rest against reduced forest rights. However, protests prevented few enclosures, and displaced cottagers had to support their families by wages; when bread prices or unemployment rose, their standard of living plummeted. Smallholders often lost their land, and the number of nonresident owners increased; in other places, such as the farming and knitting village of Wigston, smallholders, even cottagers, survived by combining farming and industry. As farm productivity rose, demand for farm labor fell; displaced workers sought jobs in local industries or London.[24]

Beleaguered by depressions, an increased labor supply, fewer chances for annual contracts as farm servants, rising bread prices, and dispossession, the poor faced a lowered standard of living. When dearth or high food prices struck, rural laborers and artisans rioted, attacking mills or grain dealers and protesting prices or the export of food, especially in 1740, 1766–67, and 1771–73. Without prospects, these men joined the army; yet unlike their seventeenth-century ancestors, few of them emigrated to the colonies.[25]

If the wages of poor rural spinners (mostly women) and weavers fell, some could rely on the income of family members or their tiny gardens. Even after enclosure, poor women and children continued gleaning wheat fields, a practice that provided food for them and their fowl and cows. When seasonal unemployment or permanent debility struck, relief provided temporary aid. Unable to get annual contracts, more people needed aid as young adults, when they should have been most employable. The proportion of paupers first receiving aid before age thirty-five in Odiham, Hampshire, jumped from one-sixth in the late seventeenth century to over two-fifths by the mid-eighteenth century. Despite the poor-law prohibition on giving aid to any but legal inhabitants, poor migrants often received relief, especially in manufacturing villages. As the number of families without even a cottager plot rose over the century, the proportion of the English population receiving relief (money payments or forced labor in a "house of industry") grew from one in twenty-five to one in ten. Able-bodied men accounted for much of the increase.[26]

In Ireland, unlike England, a growing population during the mid-eighteenth century increased pressure on resources. Because of its open land, Ireland's population was doubled (from 1 to 2 million) by immigration and natural increase in the seventeenth century. In the face of repeated famines, the population declined from 2.2 to 1.9 million between 1732 and 1744, but it recovered rapidly, rising to as much as 2.6 million in 1753 and 4.4 million in 1791. Encouraged by the tenancies available after the famines or by the prospect of finding jobs in the booming Ulster linen industry, Irish women married at age twenty to twenty-two and bore many children. Once the potato was introduced, mortality—especially infant mortality—declined, thereby increasing family sizes. Eventually the division of land among heirs reduced familial subsistence, except in linen districts, where land become an adjunct to cloth production.[27]

Reduced farm sizes and rising rents, exacerbated by famines and population growth, induced some to strike out for the colonies. Famine and high rents drove 3,000 Ulster Protestants to the colonies in 1729. Many long-term leases expired in the 1720s, and landlords wanted to increase their profits from rents, leading tenants—some of them "men of substance and credit"—to flee. Emigrants complained of "rack rents" and of landlords' auctioning off tenancies to the highest bidder; they believed rumors "that if they will but carry over a little money . . . , they may for a small sum purchase considerable tracts of land, and that these will remain by firm tenure as a possession to them and their posterity for ever." When long leases expired in the 1760s and 1770s, rents multiplied two to four times in Ulster while farm size dropped by two-thirds. When leases came due, Ulster landlords refused to allow tenants to relet land to subtenants, thus depriving them of lucrative income. Similarly, rents on the earl of Sherborne's estates in County Kerry, southwest Munster, doubled or tripled in the late 1760s. Landlords raised rents to capture some of the inflated value of land

and crops; tenants, who counted on this income as a critical part of their standard of living and feared being reduced to poverty, complained bitterly of gouging and rack-renting landlords.[28]

Living at the edge of subsistence, Irish cottiers and laborers were far less secure than tenants. Half the populace in the 1680s, the poor increased rapidly to nearly nine in ten by 1800. Rising rural wages did not compensate for rising rents; in the late 1770s Arthur Young found cottiers ill clothed and housed in unfurnished one-room hovels "much worse than an English pigstie." Nor could they pay rent, except for a tiny potato patch, leased by the year. In most places they were "oppressed by many who make them pay too dear for keeping a cow." In southwest Ulster, Young discovered that cottiers spun linen and rented a tiny potato patch; they owned two to four cows, sometimes renting pastures and other times "thieving" grass. In the southwest the poor had better access to land. Forty-three cottiers farmed, on average, 1 acre of potatoes and 2.5 acres in wheat, and four-fifths owned an average of three cows. But in the southeast, Young found that, of twenty-two haymakers, nine held two or fewer acres (five had none); six had no cows and two no hogs; the rest held no more than sixteen acres and had one or two cows. Cottiers often formed partnerships, rented a larger farm, and then divided the land, carefully providing each family with every kind of land and "assisting each other" when they could. In South Munster cottagers who herded cows for the tenant and paid him rent in butter for a cottage and potato patch appeared by the early eighteenth century. By mid-century fewer families had cows and could afford to rent a small potato patch, for rents had risen rapidly but wages had not. Forced to buy milk and food, these families turned to wage labor.[29]

Irish peasants, especially those in the South, lacked the safety net that ensured the survival of the English poor. Although parish rates supported poor relief in parts of Ulster, only deserving orphans and widows got aid, and taxes provided no help in Catholic Ireland. During famines the Irish Anglican Church, municipal corporations, and local gentlemen imported grain, but little reached the malnourished rural poor, who took to the highways or came to town and begged for food. Without the certainty of relief the English poor enjoyed, Irish families had to have a bit of land. During the first half of the century, unemployed landless or land-poor, able-bodied peasants had few choices: leave poorer districts in search of work on large Irish farms, move to Dublin, harvest English crops, or go to America. As linen manufacturing grew, spinning became common, first in Ulster, then less evenly in heavily Catholic regions, providing new opportunities for the poor.[30]

With little land and no poor relief, hundreds of thousands of the Irish died during famines following short crops. These famines occurred with appalling regularity: 1727–29, 1740–41, 1744–46, 1756–58, 1762–64, 1766, and 1770–71.

After poor weather ruined the 1740 wheat and potato harvest throughout Ireland and northern Europe, the worst famine of the century struck in 1741, "the year of the slaughter," with "want and misery in every face" and "the roads spread with dead and dying bodies." Since food shortages spread over Europe, imports could not make up the deficiency. By late 1741 at least 200,000 people had died of hunger, typhus, dysentery, and smallpox. The famine was especially severe in Munster, Ireland's southernmost province, where death claimed three times as many people as in a normal year. Later crop failures killed far fewer people—the worst killed 40,000—and were centered in small areas because enough grain was imported to feed the populace.[31]

Enclosure increased poverty in the Scottish Lowlands. The biggest Lowlands lairds (landlords) controlled one-third to one-half of the land. Most Lowlanders were tenants of a laird or subtenants (cottars) of the laird's tenants. Tenants paid leases in money or in kind; cottars, more numerous than tenants, got a tiny hovel and a kale yard in return for working for the tenant. Some tenants hired servants on six-month or year terms. Rarely given long-term leases, tenants (and their subtenants) could be evicted at the laird's pleasure. Tenants and subtenants stayed on the same farm from two to at most fourteen years. Such low persistence led peasants to devise a collateral kinship system in which households linked themselves in clanlike structures of fictive kin, women kept their family name upon marriage, and families practiced equal partible inheritance.[32]

Lowland lairds first enclosed land around the manor house; encouraged or coerced by the laird, big tenants followed suit. Enclosure then reached the commons, which was divided among owners and large tenants; most tenants and subtenants got nothing. As late as 1754 little land had been enclosed, but during the 1760s and 1770s, lairds all over the Lowlands eliminated strips and enclosed land, ending land redistribution, reducing the number of subtenancies and jointly operated farms, sometimes replacing arable with pasture, and reducing the region's population in the process. Improving tenants—a small minority—created unified farms or rented large farms and thrived. Cottars and subtenants faced destitution and dispossession; even if offered land, they had to sell it, for they lost their essential rights to commons.[33]

As subtenants and cottars lost land, Lowlands wage labor grew. Improving farms increasingly used day labor for harvests, ditching, smithing, or masonry. Many Lowland farmworkers were long-term servants, but unlike English servants, many were married, paid in oats or wheat, and given a tiny plot. Servants signed on for six months to a year, but most left in less than a year. Early in the century, plowing, threshing, processing fertilizer, milking cows, herding, and marketing kept servants busy year round; by the 1770s, more intensive cultivation led tenants to evict subtenants and hire full-time ploughmen instead. In the Borders, tenants hired male shepherds by the year to herd, shear, and castrate lambs and to mow grass and make hay. They hired women for several months to milk ewes and make butter. Big farms needed many temporary workers for the labor-intensive harvest, a need filled by Highland women and cottagers who tramped the Lowlands looking for day labor.[34]

Poverty had long permeated the eighteenth-century Scottish Highlands, but campaigns for agrarian reform increased destitution even more. Highland clan chiefs had vast estates, some covering 200,000 acres. Such big estates required managers—lawyers, chamberlains, and overseers—to encourage market production, collect rents, arrange leases, and evict recalcitrant tenants. Tacksmen—big tenants, often the chief's kin—paid rent and provided labor services (such as grain delivery and road work). They sublet land to tenants who, in turn, rented land to subtenants and cottars. Clansmen believed that they had rights to land under the authority of the chief, but subtenants had few rights, paid much higher rents than the tacksmen, farmed holdings of less than an acre, and faced the tyranny of doubled rents, forced labor services, and ready evictions. Cottars had even less security, working only at harvest and begging for subsistence the rest of the year.[35]

Highlands peasants lived in "hovels which would disgrace any Indian tribe"—thatch-roofed sod houses without chimneys, floors, and furniture. Since there were few towns and little arable land, additional people led to overcrowding. Peasants herded cattle or sheep, caught fish, or grew oats and barley. The economy was marginal in the best times, and the crop failures of 1740–41 ruined the Highlands. An ill-conceived revolt against the crown in 1745 led by clan chiefs (joined by 3,400 kinsmen, tenants, laborers, and servants) impoverished more peasants. As the population recovered, subtenants divided their tiny plots, further reducing the margin of subsistence, with "young ones going about in Rags." Crop shortages again threatened famine and impelled imports of oatmeal, barley, and potatoes in 1763 and 1771–73.[36]

Highlands reformers insisted that only sheep breeding, enclosure, farm consolidation, abolition of subletting, kelp making (seaweed ash fertilizer), fishing, and flax spinning could reduce poverty. Reform began slowly. The 1707 union with England opened English livestock markets to Scots, increasing Highlands cattle exports (traded for English flour), and potato, turnip, cabbage, and kale growing reduced subsistence crises by the 1770s. But neither fishing nor kelp farming employed many men. During the 1730s and 1740s a few landlords enclosed some land. The transportation of defeated soldiers to North America and the confiscation and sale of rebels' land after the 1745 rebellion eroded the power of the clan chiefs, opening land for improvement. Although much land remained unenclosed, Highlands clearances—division of the strips into smallholdings, which sometimes turned arable into pasture—began in the 1760s, throwing peasants off the land and driving them into "beggary, emigration, or near starvation."[37]

Although by 1775 the dispossession of Scot peasants was far from complete, the number of landless families and cottars with tiny plots increased. They found new ways to support themselves, but they earned no more than half the wages of their counterparts in England and had to supplement those wages to support a family. Cottars, who enjoyed the right to graze a cow and cultivate a potato and vegetable patch, made up part of the deficiency, but wives and children of landless laborers had to work. Although women earned just half as much as men, their income from knitting, spinning, and harvesting provided the margin of subsistence, as the increasingly difficult conditions of the Highlands, where fewer women worked, attested.[38]

Despite their growing numbers, poor people consumed adequate calories, protein, and nutrients, except vitamin C and calcium for children and lactating mothers. Three-quarters of their calories came from oats, which were eaten as meal, cakes, porridge, and brose (hot oatmeal boiled with cabbage). In most years cottars and laborers also ate cabbage, kale, potatoes, fish, a bit of meat in broth, a few eggs, skim milk and a little cheese and butter, and ale; Highlanders also ate mutton and more dairy products. When food prices soared in 1708–10, 1739–41, 1755–57, 1762, 1765–66, and 1770–73, wages stagnated, and the higher price of all grains and continued costliness of meat made substitution impossible. The growth of spinning and weaving combined with peasant dispossession to magnify the impact of high food prices; low yarn prices (of the 1730s, 1754–55, and 1772–74) often coincided with high food prices because most Scots had no money to buy cloth. Although the number of days farm laborers worked rose after mid-century, from 100–200 to 270, mitigating some of these costs, a worker able to feed his family one year might not survive the next without aid.[39]

When dearth struck, peasants poured into cities to get charity or public aid. Scottish poor relief, while not as ample as England's, was better than Ireland's. Since Scottish law did not mandate parish rates, aid was voluntary. Financed by Presbyterian kirk courts, church collections, and charity, Lowlands poor relief aided widows, orphans, and some vagrants. When crops failed in 1740–41 and 1756–57, churches and towns took up special collections, landholders accepted voluntary assessments, and kirk sessions bought oatmeal to resell to the poor. Lowlands parishes fed their poor, but aid did not reach all hungry people, as food riots in the spring of 1741 attested. Aid, moreover, provided less than subsistence; the poor had to beg or grow some of their food. Officials often ignored the plight of the able-bodied poor (who had no right to aid) or tried to banish the vagrants, beggars, widows, and orphans—many of whom were Highlanders—who flooded their towns. Unable to get rid of the extra people, towns founded workhouses that gave them a bare subsistence, and sometimes officials imposed an assessment on landowners. Northeast Lowlands and Highlands parishes gave less charity, assessed landowners less often, evicted a higher percentage of the poor, and more often required recipients to wear degrading badges; churches financed aid by selling pews and the effects of poor people.[40]

Decaying Feudalism in German Lands

German-speaking lands stood between a decaying feudalism and a not-yet-born capitalism. During the sixteenth and seventeenth centuries, the Reformation, warfare, market integration and the rise of a peasant land market, and the growth of population, of towns, of an urban middle class, and of a linen industry had thrust Germans into an increasingly modern world. Feudalism, however much eroded by these developments, nonetheless endured into the eighteenth century, providing income for the nobility, protecting some peasant property, and in some places retarding emigration. To understand why Germans migrated to the British colonies, we need to examine briefly the development of the German agrarian and manufacturing economy.

German peasants participated in local markets, or sold grain to cottagers or nearby towns, to pay taxes or fees to lord, state, and church. Villagers aimed at communal subsistence, marshaling crops, livestock, meadow, and woods to that end. Holding too little land to grow all the food their families needed, they engaged in intensive local exchange, combining agriculture and craft, and trading cloth for rye or shoes for harvest labor. Such local exchange allowed them to set up and expand farms despite landlords' practice of evicting peasants who fell into debt in years of poor harvests.[41]

Western German peasants lived in compact villages surrounded by open fields divided into narrow strips. As they reclaimed wastelands or forests or brought pastures into cultivation, they added more, usually shorter, strips, where they grew barley, oats, and rye. Villagers had rights to orchards, pastures, woods, and fields, where their livestock grazed on stubble and fallows. Despite the division of the commons on scattered farms, and enclosure orders in many places during the mid-eighteenth century, peasant families in much of western Germany and parts of eastern Germany continued to farm open fields and enjoy rights to common pasture.[42]

The persistence of open fields and commons led to intense cooperation among the peasantry. Landholding peasants (except women, cottagers, and laborers) chose village officials and cooperated in planting, cultivating, and harvesting. Solidarity went deeper. Spinning bees—winter evening communal gatherings where young peasant women spun and young peasant men socialized—persisted through the eighteenth century. The girls spun thread, making ribbons or yarn, for a "gift called the *Brautrocken*" for a local bride. This sociability was replicated in other daily activities. In 1786 German writer Christian Garve discovered that Silesian peasants "see each other every day, in all their farm labor—in summer in the fields, in winters in the barns and the spinning rooms." With such regular association, peasants were "more adept at intercourse with their equals" than with higher social classes, who had "little more influence" over them. They defended their property with "invective and coarseness" whenever anyone trespassed on their fields or went "through their gardens . . . even when no damage" resulted.[43]

The Thirty Years' War (1618–48), with its orgy of pillaging and death, loosened feudal ties and reduced the German population by a third, from 15–19 million in 1600 to 10–14 million by 1650. Peasants who survived the war did well. With so many abandoned farms, landless and land-poor peasants acquired holdings easily. It took a century for the population to recover: the German population reached 15–18 million in 1700, 17–20 million in 1750, and 26 million in 1800. At first Germans repeopled abandoned lands, but population pressure reappeared, leading many to emigrate east to Prussia and Hapsburg lands. Such opportunities for land, along with the heightened power of the state (Prussia created land banks and distributed free seed to poor peasants) mitigated feudal dues and helped weaken feudal authority. Once the population recovered, however, peasants lost this leverage.[44]

A feudal legal system nonetheless persisted. Peasants lived in a world without private property. They coveted the liberties to stay on the land, use common fields and forests, and bequeath property. When states attempted to make land fully marketable in the mid-eighteenth century, peasants and lords both insisted they owned the land and forests. Eastern lords evicted the peasants, forcing them to become cottagers or wage laborers. Yet very few peasants achieved legal freedom, and lords still collected feudal fees and exacted annual labor services averaging seven weeks. As late as the 1770s, fees, taxes, and labor imposed an onerous burden on the peasantry, ranging from one-twelfth to more than half of the village's farm product.[45]

Most German peasants did keep their farms and pass them on to their children. Peasants in much of southwest Germany practiced partible inheritance, dividing land and other goods equally among all children. As the population surged in the mid-eighteenth century, the number of holdings skyrocketed. Villages became so densely peopled that peasants took up marginal land and farmed ever smaller plots ever more intensively, planting more grains; experimenting with beans, cabbage, potatoes, flax, and clover; or reducing fallow, draining wasteland, or limiting grazing. By the end of the century, most peasants in partible regions held too little land for familial subsistence.[46]

With land scarce and holdings tiny, peasants gradually gathered enough property to marry. But since most sons of landed parents got a portion, no matter how small, adult children saw little reason to work on their parents' farm. Many became servants or farm laborers, boarding in the homes of others while saving for land. Men in partible villages rarely received a portion or inheritance until their mid-twenties, and they married, on average, at twenty-seven, taking women several years younger for wives. Parents provisionally gave plots to sons upon marriage, and a son could combine his tract with personal property his wife received from her parents. Knowing that the land they would inherit might not be enough for a farm, peasants in their thirties and forties traded and bought land, building their estates to support themselves and make a patrimony for their children.[47]

Peasants in northern and eastern Germany gave all their land to one son. The chosen son sometimes had to buy the farm from his father, who used the money to finance his retirement and help his other children. Few peasants sold land in impartible regions, and those who did sold to sons, sons-in-law, daughters, stepsons, or a widow's new husband. As the population increased and land grew scarce, ever fewer children got land but, instead, emigrated or became cottagers. Cottagers made a bare living from their tiny gardens, day labor, spinning, and weaving; they had little work for their children, who left home in early adolescence to work as servants. By mid-century, cottagers had become the biggest class in impartible areas. In Saxony the proportion of landless families rose from one-fifth to two-fifths between 1550 and 1750. In Belm landless cottager families grew from one-third to two-thirds between the mid-seventeenth and early nineteenth centuries. In Prussia, where so many emigrants had found land, the landless class jumped from one-half early in the century to three-quarters by 1770.[48]

With so little access to land, youths stayed at home and postponed marriage until their early thirties. The inheriting son stayed with his parents, helping on the farm and waiting for his father to retire or die. Unable to marry before he received his portion, he sometimes cohabited with his intended and sired an illegitimate child. Lacking opportunities to inherit farms, other children sometimes stayed at home, even after marriage, and worked for their father or brother.[49]

A resurgent textile industry provided a partial solution to underemployment. Centered in marginal farming areas in eastern Switzerland, Westphalia, Rhineland, Swabia, Prussia, and Silesia, textiles employed countless peasants (one-third to two-thirds of some villages) to grow flax or to spin and weave on their farms. A few textile villages grew into small cities that attracted cottagers and smallholders. Although a few peasants made and sold their own yarn, contractors, who bought yarn and had it made into cloth, employed most of them. Textile manufacture served different purposes in impartible and partible inheritance regions. Household industry permitted cottagers and smallholders in impartible areas to split their tiny plots, thus allowing children to marry and begin families and survive by combining vegetable gardens with wage labor. In partible regions, landholders supplemented their crops with sporadic wage labor, especially in the slack winter season. Spinners and weavers—mostly part-time workers who supplemented their farm labor—worked only until they attained subsistence, preferring, in the words of a 1768 observer, to "work less [rather] than earn more by working harder." Despite population growth and reduced landed resources, labor shortages thereby grew, leading officials to employ children in textiles.[50]

As the textile industry grew, families looking for work flocked to textile villages, and farming receded. While the population of farming regions of Zurich canton stagnated between 1700 and 1771, areas of industry and agriculture rose two-thirds. The proportion of landholding peasants in the linen-weaving village of Großschönau, Saxony, declined from one-third of 100 householders in 1587 to only one-twelfth of 400 householders in 1730; cottagers, most of whom spun or wove linen, grew from two-thirds to eleven-twelfths, and lodgers (also weavers) doubled. Entire families spun or wove, using their own wheels and looms, as two Zurich villages in the 1760s show. In Hausen, a herding and cotton-spinning village, three-tenths of the households engaged solely in spinning; a similar number combined farming and spinning, but only a tenth farmed exclusively. Families in Oetwil performed more varied work, spinning and weaving cotton and combing silk. Village women worked almost solely in textiles. Farmwives wove cotton cloth, wives of poor men spun, and wives of textile workers did both. Peasant sons helped out on the farm; farm daughters and children of textile workers and artisans spun or wove with their mothers.[51]

Towns and cities, the hallmark of capitalism, developed more slowly in German-speaking lands than in Britain, and even in 1800 just one in eighteen Germans lived in cities. A few cities grew rapidly. Berlin, the largest German-speaking city, had 90,000 people in 1750 and 150,000 in 1800, making it one-seventh the size of London. Most towns, however, remained small. Krefeld, a village of 400 souls in 1625, became a textile center and reached a population of 1,900 in 1716, 3,900 in 1750, and 6,500 in 1793. The population of Barmen grew from 2,000 to 16,000 over the eighteenth century.[52]

Although crushing poverty, similar to that of Ireland or the Highlands, remained the lot of Germans, they ate enough food to sustain farmwork (two pounds of grain daily, providing 2,800 calories), except when harvests failed. Those unable to grow rye subsisted on oat or barley cakes or porridge, reserving bread for holidays. Peasants tended small fields of legumes, vegetables, and potatoes and raised cattle, pigs, and chickens, adding vegetables, milk, soups, cheese, and a little smoked pork to their diet. Grain prices climbed as the population rose, leaving cottagers and laborers, who had to buy grain, facing malnutrition. When food grew expensive, laborers skimped on housing and clothing. German villages, Thomas Jefferson observed in 1788, "seem to be falling down," their houses made "of mud, . . . all covered with thatch" or "of scantling [timber], filled in wicker and morter, and covered either with thatch or tiles." J. M. von Loen, a friend of Goethe, lamented in 1768 that peasants were "hardly to be distinguished from the cattle they raise"; not only did "children run around half naked and call to every passer-by for alms," but "their parents have scarcely a rag on their backs." Moreover, "their barns are empty and their cottages threaten to collapse at any moment."[53]

As the lot of the poor deteriorated, illegitimacy and infanticide grew. Villages and cities filled with vagrants, prostitutes, journeymen looking for work, petty criminals, and beggars demanding alms. Injured Berlin-born journeyman butcher Johann Kästner (and a female friend) tramped from city to city during the 1770s demanding alms from town fathers and city guilds, paying tiny sums to poor city families for lodging, and stopping to beg food and a night's lodging from peasants en route. To deal with vagrants, cities instituted patrols to expel beggars and set up prisons and workhouses for those who remained. Reformers believed that education might inculcate good habits and reduce poverty, but their efforts were futile. Beggars multiplied, and workhouses employed tiny numbers; the few children who attended schools left at planting and harvest times; and rural people remained overwhelmingly illiterate.[54]

The stagnant economy locked people into perpetual poverty, but the paternal relation between ruler and ruled mitigated disaster. German states avoided mass starvation during the poor weather and crop failures of 1740–41. Prices of rye, the major bread grain, did rise, triggering a few urban food riots; city epidemics of typhus, dysentery, and smallpox; and isolated famines. Yet even though mortality rose, no severe crisis like that in Ireland ensued. Public relief, especially the granaries in Prussia and towns outside Prussia, bread distributed free to the poor, and grain imports, reduced suffering.[55]

Migration and Emigration

Over the eighteenth century the poor improved their living standards in England, the Scottish Lowlands, and Ulster; the Catholic Irish, Scot Highlanders, and Germans sank deeper into penury. Unsurprisingly, Irish, Highlanders, and Germans most often emigrated to North America. Unable to afford the high cost of emigration, the poor—except unmarried indentured servants—stayed at home. Smallholders, fearful of losing their leases or land but owning sufficient property to finance migration, predominated among emigrants.

Youths took the first step in a process that led to emigration to the colonies when they left home to work as servants or moved to towns; but often booming economies allowed them to find jobs without emigrating, and if they did not, they might yet live on family wages, common rights, or poor relief. Like their grandparents, English peasants and laborers moved to escape unemployment or enclosure, or to seek their fortunes. Youths worked as servants in the neighborhood; families tramped the countryside in a small circuit looking for work, or moved to a nearby town, where kin had preceded them, or to London. Many wound up in Lancashire, where the population doubled from the 1660s to 1750, and two-thirds or more worked in manufacturing or mining. Half of the farmers and artisans in Yorkshire (in the far north) in the late eighteenth century moved before siring a child, usually to an adjacent village or town; two-thirds of the laborers, less tied to the land, did so. Three-fifths of the wives of farmers and craftsmen moved, and more of them traveled over ten miles. Despite the prevalence of short moves, long distance migration was common; between 1700 and 1780, over 1 million people moved between counties.[56]

Remarkably few Englishmen and even fewer Englishwomen emigrated to the colonies, just 72,000 between 1700 and 1775, or 12 per 1,000 over the entire period. They hardly represented a cross-section of the populace. Between 1773 and 1776 nearly all English emigrants were unmarried; just 15 percent worked in agriculture, while over half were craftsmen and one-sixth were laborers. Moreover, half the emigrants were convicts, mostly common felons from London and surrounding regions; many were facing execution and received pardons on the condition that they go to the colonies. Overwhelmingly male (four-fifths), young (three-quarters between fifteen and twenty-nine), and unskilled (two-thirds or more), they resented the fourteen-year enslavement they faced and longed only to return to England.[57]

Nearly all the rest (25,000, or just 333 a year) came as indentured servants. So few came, in part, because they heard that servants suffered hard usage and that few men could get land or open a shop. Mostly from London and over nine-tenths men under age twenty-five, servants usually possessed specialized skills needed by large southern planters or city businessmen. Many servants, like William Moraley, had lost their fathers and were left without "friends or relations" in England to help them. They stayed home when London wages were decent and ventured to the colonies when opportunities and wages dropped. Conditions in England in the 1770s especially ignited servant emigration. Coming to London for the reputedly high wages there, craftsmen had great difficulty landing jobs. Hardly London's most debased proletarians, over two-thirds of the 1,895 men who left London as indentured servants in 1773–76 were craftsmen, mostly journeymen who had suffered extended periods of casual employment and unemployment. Even then, the vast majority suffered in London rather than risk colonial service.[58]

Perhaps 9,000 English emigrants (120 a year), mostly farm families and artisans, paid for their own passage. Many, like the Quakers of Cumberland County (in the far north), had kindred, friends, or commercial ties to the colonies. Most years only a few families and individuals emigrated, but between 1773 and 1776, 1,643 free immigrants—three-fifths of them residents of northern counties—left England for North America. Two-thirds of those from Yorkshire came in family groups, more than three times the average. Nearly half of the Yorkshire men were farmers, and only three-tenths were artisans. Threatened by rising rents and fearful of plunging living standards, hundreds of freeholders or substantial tenants sold their property and emigrated to America in 1774. A large majority, recruited by agents or enticed by newspaper articles, went to Nova Scotia to "seek a better livelihood" and to take up cheap land.[59]

Unlike England, Ireland sent growing numbers of people to North America. This was a remarkable change, for "scarcity of people," an English official had lamented in 1686, was Ireland's "greatest want." Ulster had been the internal frontier for Protestant England and Scotland: 180,000 English and Welsh immigrants and 80,000 to 130,000 Scots went to Ireland during the seventeenth century. Although the English and Welsh lived in the towns and on scattered plantations all over the island, the Scots concentrated their settlements in eastern and northern Ulster, thereby dividing the country into two distinct and increasingly antagonistic sections.[60]

Although fewer people emigrated from Ireland to North America in the eighteenth than the nineteenth century, as many as 150,000 people (60 per 1,000 over the first three-quarters of the century) came. Why did descendants of Scot immigrants and smaller numbers of Catholics and Anglo-Irish go to the colonies? Famines and depressions, rising poverty and the lack of poor relief, fear of loss of land and downward mobility, population growth, and rack-renting landlords bent on profits thrust the Irish from their island; opportunities for land attracted them to the colonies. The Ulster Irish more often sought opportunities; the poor, male, Catholic immigrants left to escape difficult conditions.

Two streams of emigrants reached the mainland colonies from Ireland. By the 1760s at least two-thirds of Ireland's population was Catholic. One might expect that Catholic cottiers and townfolk—the most deprived group—would dominate the migrant pool. But three-fifths, mostly Presbyterians, came from Ulster; a tenth may have been Anglo-Irish and Quakers; and the rest were Catholics, mostly from the south. Unlike the prosperous Anglo-Irish landlord, middleman, and urban merchant class, Presbyterians feared losing their property. In the 1770s some Ulster emigrants had little money, but others sold their leases or subleases and took as much as £20 to £40 with them. With this money they paid their family's passage. They anticipated getting land and enjoying religious toleration, welcome relief from the mild persecution they faced in

Ireland, where they were prevented from holding public office and where marriages performed by their clergy were illegal. Poor Catholics, in contrast, had almost no assets, expected few opportunities in the colonies, and confronted prejudice as soon as they landed. Those who could rent a tiny potato patch stayed. But 13,000 poor people, mostly Catholic convicts, transported from Dublin to America between 1718 and 1775 had no choice. Most other Catholic emigrants had lived in a town and learned craft skills before indenturing themselves for colonial service.[61]

Poor harvests, famines, and rising land costs motivated the Irish to leave during the late 1710s, late 1720s, 1740s, late 1760s, and early 1770s, times of the heaviest emigration. A halving of transit costs from the 1730s to the 1770s made emigration more affordable for those paying their own way and may have reduced service time for indentured servants by lessening the contracted debt. About 4,500 Irish emigrated in the late 1710s because of poor harvests, increased rents, and (for Presbyterians) religious persecution. The 5,000–15,000 Ulster emigrants who left during the famine years of the late 1720s gave other reasons—high rents, short leases, tithe collection—as well. The famine of the 1740s led to renewed emigration. Crop failures (1765–67), skyrocketing prices for bread, collapse of the linen industry (1769) that made "weavers turn labourers," and escalating rents impelled 45,000 to leave Ulster and 16,000 to emigrate from the south from 1760 to 1775. One Ulster protester in the 1770s directly linked prohibitive rents with emigration, complaining that tenants faced "the melancholy prospect of being turn'd out of their possessions and obliged to remove their numerous families to America."[62]

Scots migrated to the mainland colonies more often than the English but less often than the Irish, in part because they could find opportunities in towns and the textile industry. To find work, Scots moved from parish to parish, Highlands to Lowlands, and farm to town. Most migrants moved in a ten-mile circuit, but many Highlanders went south each summer, hoping for aid until the harvest. Nearly all adolescent sons and daughters of cottars (but fewer children of tenants) left home to work as servants-in-husbandry, staying a year or two before moving to another parish. Edinburgh and Glasgow attracted thousands of rural folk. Men of humble origins traveled about fifty miles to Edinburgh, and apprentices covered thirty-two miles; but increasing numbers had left the distant Highlands. Over 30,000 young women came to eighteenth-century Edinburgh; many worked as domestic servants while harvest laborers congregated in town. Although most women coming to Edinburgh moved from nearby areas, increasing numbers came from the Highlands, lured by higher wages.[63]

About 60,000 Scots (46 per 1,000 per year) emigrated to British North America between 1700 and 1775. Displaced by improving farmers, angry at rent increases, and expecting more opportunity in North America than on a nearby farm or town, they hoped to assuage their land hunger. Although 1,700 Scots—700 criminals and 1,000 Highland rebels after the 1715 and 1745 revolts—suffered transportation to North America, the rest came seeking land. Just 150 a year arrived between 1700 and 1730, but more—850 a year—came between 1730 and 1760. Emigration quickened in the 1760s and 1770s after smallholders had been evicted or feared dispossession. About 30,000 arrived in the fifteen years before the American Revolution, especially from 1768 to 1775, as many as in the preceding six decades.[64]

Scots who emigrated left usually during times of dearth or depression, such as the early 1740s in the Highlands and northeast, and the early 1770s, which were times of rising rents as well. After the financial crisis of 1772, Lowlands weavers and other artisans were unable to get credit or sell their services; Highlanders flooded the colonies after the crop failures, low cattle prices, and cattle diseases of 1771, which coincided with rent increases and the first clearances. Despite persecution and financial support from the Catholic Church, Catholic Highlanders refused to emigrate to Canada until 1772, and even then few families left.[65]

Lowlanders and Highlanders emigrated at different times. Nine-tenths of the pre-1760 emigrants came from the Lowlands, but 15,000 Highlanders emigrated between 1760 and 1775. Lowlanders spread through the colonies but concentrated in New Jersey, where 2,000 arrived between the 1720s and the 1740s; Highlander settlement centered in New York, North Carolina, and the Canadian maritime provinces. Although Lowlanders organized several land companies in the early 1770s, bought land, and migrated together, the rest came as individuals (including indentured servants to Pennsylvania or the Chesapeake) or in family groups. Highlanders often came as part of larger communal groups under the direction of tacksmen bent on settling big tracts of land they had already purchased.[66]

We know most about the emigrants of 1773–76. Over half migrated in family groups; others followed kindred or neighbors. A few Highlanders "too poor to pay" sold "themselves for their passage, preferring temporary bondage in a strange land to starving for life in their native soil," but six-sevenths paid their passage. Few of the rural poor of the Lowlands emigrated. A fifth of the men (but half the household heads) worked in agriculture; half were craftsmen, mostly in textiles; and a fifth were laborers, most young indentured servants. Textile workers left because they feared poverty; other artisans sought land or higher wages. In contrast, more than two-fifths of the Highlanders were farmers, and more than one-third were laborers, mostly in agriculture. When improving landlords oppressed tenants with increased labor services or raised rents so high that they feared displacement, subtenants and tenants sold their livestock, crops, and household goods to pay for the voyage; tacksmen, fearful that their leases would not be renewed or that subletting would be banned, departed as well.[67]

Dissatisfied German peasants, like those in Britain, could emigrate, but German states, worried about losing population and taxes, put roadblocks in their way. Emigrants had to settle all debts and taxes. Free emigrants had to pay large fees for permission to depart and to take property with them, and serfs—a substantial part of the populace—had to pay manumission fees amounting to 14–25 percent of their property. Unsurprisingly, most legal immigrants were middling peasants who could sell their belongings. A few states forgave taxes for those who moved to neighboring provinces or allowed free emigration. In heavily populated partible inheritance regions, most emigrants had to pay, but authorities sometimes allowed the very poor (who paid few taxes) to leave without paying. Far more peasants tried to leave illegally by smuggling themselves and their property out of their homeland.[68]

Despite these fees, a third of eighteenth-century German and Swiss adults—more than 14 million people—moved within Germany. They sought work as servants or textile workers, went to a nearby town, tramped the countryside looking for day labor or begging, migrated seasonally to harvest crops or mine coal, or joined an army. Rural craftsmen, landless peasants, and those with tiny plots or insecure leases moved more often than those with enough land for subsistence and secure leases. Most went short distances, often within one jurisdiction, thereby avoiding departure fees, but large numbers—more than in Britain—covered long distances, moving eastward or going to a distant city, seeking a place where they could practice their faith. Armies provided opportunities for hundreds of thousands of unmarried, poor young men. Although many of these men were transients or servants, any man without full-time farm or craft labor, even sons of landed peasants, joined these mercenary armies.[69]

Prohibitions against emigration and labor recruiting failed to stem the tide. Seeking to live in German-governed lands, 737,000 Germans (most from the southwest) moved east from 1680 to the late eighteenth century: 300,000 went to Prussia; 350,000, to Hungary; 50,000, to Poland; and 37,000, to Russia. About 335,000 Swiss emigrated over the century; two-fifths joined foreign armies, while others moved to eastern Europe. Governments in Prussia, Hungary, Austria, and Russia offered inducements to attract Germans who shared their religion. Catholics went to Hungary and Poland; Protestants, to Prussia. Frederick the Great used troops to drain fenlands and founded 900 new villages to house peasant and artisan immigrants; in 1770–71 he welcomed 40,000 refugees fleeing crop failures. Russia set up recruiting centers and offered land, livestock, and tax remissions.[70]

Thus Germans had little reason to cross the ocean. About 110,000 Germans—

less than an eighth of the emigrant pool—went to the mainland colonies between 1700 and 1775; most were from the regions of southwest Germany that sent migrants eastward. Similarly, less than one-fourteenth of eighteenth-century Swiss emigrants (25,000) landed in the mainland colonies. Some villages and regions, of course, sent more. The Neckar region of the Palatinate, the Western Palatinate, and Saarland send hundreds of families, but three-quarters to nine-tenths of the emigrants, even from these places, moved eastward.[71]

A few Germans, mostly members of family groups, reached the colonies early in the century, but large numbers did not land until the late 1730s and early 1740s, mostly from southwest German principalities. Swiss emigration peaked between 1734 and 1744, when about half the immigrants arrived; over half (60,000) of German emigrants arrived from 1748 to 1754. German emigration slowed to a trickle in the mid-1750s; during the 1760s, 930 arrived each year, and nearly 1,000 came in the early 1770s. War and peace, along with the lure of eastern destinations, explain the timing of Atlantic emigration. The surge of German emigrants of the late 1740s came after the war of Austrian succession ended, and the flow almost stopped during the Seven Years' War. In contrast, German emigration to Hungary peaked in the mid- to late 1760s, and Germans migrated most heavily to Prussia from 1745 to 1755 and 1763 to 1788.[72]

Why, then, would Germans go to America, rather than Prussia? Emigrants headed for North America left southwest Germany and the Rhineland to escape poor harvests, taxes, and debts. Indebtedness and the lack of prospects (for younger children in impartible inheritance regions and those with little land in partible regions) motivated some; recruiters and the success of previous emigrants enticed others. But recruiters could persuade only those who were dissatisfied with their prospects and unwilling to go east. Turning down Prussian "money for travel and as much land as in America," one south German peasant chose America, with its higher costs. "But, oh, what is a free inhabitant compared to a slave or serf?" And "what pleasure would a man have in" Prussia where "he has to work himself to death for an overlord, and where his sons are at no hour safe from the miserable soldier's life?" Such arguments persuaded few, for most preferred Prussia. Education was crucial. Unlike most peasants, migrants to North America were educated. One-half to four-fifths of adult male German immigrants to Pennsylvania from 1730 to 1775 were literate, able to read about American freedom and eastern serfdom.[73]

The Process of Emigration

How did British and German emigrants find out about the English colonies, decide to emigrate, and finance their voyage? As we have seen, European alternatives to North American immigration abounded, in burgeoning cities, growing industries, or expanding European frontiers. Immigrants to mainland British America were thus a special, self-selected group. Colonial agents scouted villages and towns for Protestant emigrants, exaggerating colonial prospects. Newspaper advertisements and books and pamphlets depicting a colonial paradise (some in German) circulated widely. Leaders of radical sects sought refuge and land in the colonies, promising to settle outlying areas in return for free land. To rid their lands of poor people dependent on charity, officials sent immigrants over. After they arrived, most immigrants moved westward, squatted on Indian land, and fought Indians and land speculators to acquire legal titles. And after they made farms, immigrants wrote home, tempting kin and neighbors to join them.

Colonial governments and private companies competed for foreign Protestants, especially the Germans they believed made reliable farmers. They used a variety of techniques: setting aside land at good terms for immigrants, sending agents to Britain or to German states, circulating favorable pamphlets, and engaging in newspaper campaigns. From the 1680s to the 1720s, William Penn's British and Rotterdam agents scoured the countryside for emigrants, circulating Penn's writings and helping emigrants arrange their trip. In 1727, almost a decade after Penn died, his promotional pamphlets still circulated and attracted immigrants seeking land and religious freedom. Because of his promotional campaign, the Massachusetts governor asserted in 1742, Pennsylvania had "within a few years, most surprizingly increased and flourished beyond all other of his Majesty's Colonies in North-America."[74]

Mid-eighteenth-century South Carolina campaigned as vigorously as Penn to get European Protestants. The colony surveyed nine frontier townships between the 1720s and 1760s and gave them to Ulster Irish, Swiss, and German emigrants. Recruiters got land in return for transporting families; all families received land and most livestock, tools, and food until the first crop came in. The colony relied on recruiters, but raising money proved hard. French-Swiss aristocrat Jean Purry stranded over 300 Swiss emigrants in 1726 when his financing evaporated. Nonetheless, he founded Purrysburg, the first and most successful township, bringing several hundred Swiss farmers in the early 1730s.[75]

Georgia's founders engaged in a similar campaign. During the 1730s and early 1740s they hired the German commissioner of the Society for the Propagation of Christian Knowledge to gather a shipload of Austrian Salzburgers, who had been exiled by their Catholic prince. Georgia founder James Oglethorpe orchestrated a newspaper campaign in London and Edinburgh, published promotional tracts, and sent agents to the Scottish Highlands. Georgia trustees, partly financed by Parliament, paid the passage of emigrants from Salzburg and Scotland and promised to support free emigrants for a year and set up a farm for them complete with cow and calf. Between 1764 and 1767, Georgia (now a royal colony) promised to give a township to any group of forty or more Protestant families. Ulster promoter John Rea brought over 107 people, some of whom settled newly organized Queensborough Township, while others sailed to Charleston and got South Carolina land.[76]

Once a colony's reputation had been secured, private recruiters made repeated efforts to get farmers to emigrate. Irish immigrant Sir William Johnson, the New York Indian trader and land speculator, sent sixteen Ulster Irish families to Glen in 1740 and more in the 1760s. In the 1750s, large Ulster landowner Arthur Dobbs, sometime North Carolina governor, got a huge parcel of land in the colony and brought over several shiploads of emigrants. In 1773, Highlands merchant John Ross mounted an intensive advertising campaign and persuaded 300 emigrants to leave.

A failed recruiting effort uncovers the intensity of competition for immigrants. In 1764, German recruiter Johann Heinrich Christian von Stümpel procured 20,000 acres in Nova Scotia; in return he had to bring 200 settlers within ten years and give them land. While waiting for his grant von Stümpel recruited 300 families, but most left for Russia before the grant was finalized. He then recruited another 400 colonists, who never got to Nova Scotia because von Stümpel ran off with their money. Stranded in London and living in tents, they became objects of charity—and prospects for recruiters for Russia, Maine, Georgia, and South Carolina. South Carolina agents made the best offer; they not only paid the Germans' passage but gave them their own frontier township.[77]

Promotional literature and newspaper accounts extolling new colonies, some wildly unrealistic, circulated throughout Britain and Europe. Jean Purry printed two accounts of his settlement in 1731 praising the soil and fauna and blaming early deaths on the settlers. When he returned to Switzerland in 1733, he published enthusiastic letters from the initial settlers. Pastor John Boltzius, a Salzburger living in Georgia, wrote voluminous and honest letters to his Lutheran superior that were full of practical advice to potential immigrants. Four volumes of his missive were printed between 1735 and 1767. In 1773 an anonymous Scot writer appealed to Highlanders with tales of North Carolina's thick vegetation, rich crops, and abundant land farmed by Highlander emigrants. The next year Alexander Thomson extolled Pennsylvania in a pamphlet aimed at the Scots. In it he lamented worsening conditions in Scotland and related how he had built a large estate for himself and his many sons on fertile land in the central part of the province. With less justification, during the late 1760s Florida promoters printed pamphlets praising the colony's climate and soil, conveniently forgetting that much of the land was an impenetrable swamp. The most dishonest pamphlets inspired some trust, for they were embedded in a rhetoric of American abundance, opportunity, equality, and familial independence that permeated serious European and American works about the colonies. The few people who read these works talked with their neighbors, buttressing promoters' stories and spreading emigration fever.[78]

Successful immigrants returning home on business or for a visit became potent recruiters. "Newlanders" regularly went back to Germany to settle estates and collect bequests for themselves and others—and to praise America. At first they came on their own. But soon recruiters hired them, paying their way. They brought letters from other immigrants that presented a misleadingly glowing account of the colonies, praised colonial freedom from feudal restrictions, and recommended emigration enthusiastically. They told tales of abundant and cheap land, large farms, and low taxes but ignored the steep price of developed land and the Indian threat inland. Despite occasionally cheating trusting emigrants or taking all their money as fees for information about America, Newlanders achieved so much success, especially between 1749 and 1754, that some governments banned them. Irish Quaker immigrant traveling ministers gave similar glowing accounts of asylum and opportunity in Pennsylvania when they went back to Ireland on missionary trips, as did other Quakers returning to marry, conduct business, or visit relatives.[79]

Literate immigrants sent glowing reports home, urging kin and friends to come. Irish Quaker Robert Parke migrated to Chester County, Pennsylvania, with his father and other kin in 1724; in early 1726 he wrote to his sister Mary Valentine in Ireland. He attacked as "utterly false" rumors that they were dissatisfied, writing, "There is not one of the family but what likes the country very well . . . , it being the best country for working folk & tradesmen of any in the world." He reported land prices that grew "dearer every year by Reason of Vast Quantities of People that come here yearly from Several Parts of the world"; their father's purchase of land; the family's bountiful crops of "oats, barley, . . . hemp, Indian Corn & buckwheat"; and the "Extraordinary Plenty" found at the biweekly Philadelphia market "where Country people bring in their commodities." Hoping that his sister might join them, he explained how to get over and what to bring. In spring 1728 Valentine and her husband did emigrate. Similarly, the Georgia Salzburgers, their pastor reported in 1734 soon after their arrival, "show[ed] a great desire to see their compatriots arrive since they hope for much help and a lessening of the burden from them." As the colony prospered, Salzburgers wrote "to their Favourers, Friends, Relations, & Countrymen in Germany & Prussia" relating "what good things & Preferences they enjoy." Still pining for his kin, Ruprecht Steiner wrote his brother-in-law in November 1738, urging him to come and praising the "rich harvest this year" and the "plenty of bread," recounting his property—"seventy head of cattle and 48 acres of good land," enough "so we can eat our own bread."[80]

The most persuasive pamphlets and letters struck a balance between unstinting praise and realistic appraisal, correcting the biased reports of those who attacked emigration while admitting that problems did exist. The climate was mild and the crop yields high, but the mosquitoes were plentiful, forests impenetrable, and winters long and cold. Some letter writers proffered explicit advice about what to bring, what crops to grow, and how to make farms in America. James Whitelaw, agent for a Scot company that bought land in Vermont, told his father how colonists planted corn, girded trees, and tilled land. Disappointed at crop yields, he nonetheless regaled the company with tales of the exotic crops the soil would grow.

Over and over, letter writers called America the "best poor man's country." But servants, some correspondents admitted, faced heavy toil with little compensation. In contrast, those who paid their own way could rent a few acres immediately and soon buy land, unencumbered by heavy taxes or fees. Those who brought savings with them could get land immediately. If land in older areas cost too much, cheap land in the backcountry allowed a man to rent a farm, save money, and acquire hundreds of acres for himself and his sons. Scot factor John Campbell informed his cousin, a medical student, that the high price of land in southern Maryland made it a poor risk unless one had property; but he added that one could work for high wages and soon rent a farm in tobacco country cheaply, raise livestock, get a good crop without great labor, and then buy cheap land on the Virginia frontier. Others insisted that immigrants must be willing to work hard and—once they got frontier land—live isolated from neighbors while they assiduously cleared land, aided only by their sons.[81]

A majority of emigrants could not afford emigration costs—travel to a port, loss of income en route, ship fare. Germans, who had to pay heavy taxes and fees before they could leave, tried to smuggle out property, hoping to sell it and use the proceeds to finance food, lodging, and tolls en route. To pay for the rest, emigrants agreed to become indentured servants or redemptioners. During the early 1770s three-fifths of the southern Irish, one-half to three-fifths of the Germans, one-half of the English, one-quarter of the Scots, and one-tenth to one-fifth of the immigrants from Ulster who came to Pennsylvania were servants.

The redemptioner system resembled indentured servitude. Both were methods whereby immigrants could pay for passage by working as servants for a term of years. Under the indenture system, merchants bartered space and provisions for the right to sell the migrant's labor upon arrival. The shipper risked not recovering his costs, but in turn, the servant could neither choose his master nor negotiate the contract. In contrast, shippers loaned redemptioners the cost of passage. The migrant, who assumed the risk, had two weeks to find kindred or landsmen to pay his debt; if he were unsuccessful, he had to sell himself (and maybe his family) to repay the shipper. This system gave migrants greater choice of masters and terms of service and allowed immigrant families to stay together. As early as the 1740s all German servants—peasants who had enough property to pay manumission fines and finance movement through Europe—but just an eighth of Irish servants came as redemptioners. From 1771 to 1773, this system predominated among British servants who reached Philadelphia: all the Scots, half the English and Ulster servants, and a fifth of the southern Irish signed redemptioner contracts.[82]

Servant recruiters could be found all over Britain and Germany. London—the biggest market—swarmed with them. When William Moraley considered emigrating, he came to the Royal Exchange, read "the printed Advertisements fix'd against the Walls of the Ships bound to America," and was promptly "accosted" by a recruiter who bought him "two Pints of Beer" and "proposed . . . an American Voyage." In the 1770s a few London merchants found servants in register offices and hiring halls, but only the most desperate men frequented them. Most agents scoured public houses or went to Gravesend, London's exit port, and selected servants there from among those who wanted to emigrate. Agents also searched Scotland and Ireland for servants. As early as 1729, "masters and owners of ships" in Ireland would "send agents to markets and fairs and public advertisements through the country." Once he found a prospect, the agent negotiated terms (years served and wages to be paid) and prepared the servant contract. Pennsylvania, Georgia, Massachusetts, and other colonies hired merchant agents in Rotterdam; sent recruiters to the Palatine, Lower Rhineland, and Holland; and advertised in German newspapers. Big Rotterdam merchants hired agents to entice families to emigrate, recruited Rhine boatmen to deliver the emigrants to Rotterdam, petitioned for free passage across Holland to Rotterdam when the emigrants arrived at the border, and outfitted the vessels. Hope and Company even paid servants' expenses—food, shelter, passports, fees, and tolls—on their trip from Germany to Rotterdam. Smaller merchants, unwilling to pay such fees, tried to entice emigrants in Rotterdam with lower prices for the voyage.[83]

Once they agreed to emigrate, servants faced great misery. Cooped together with several dozen (among the Irish) or several hundred (among the Germans) other migrants in the crowded steerage between decks and fed oatmeal, stale bread, a little meat and cheese, and rancid water, some became ill, and a few died en route. Four percent of German immigrants to Pennsylvania died on the voyage, but more than twice as many children (9 percent) died; the annual mortality of 162 per 1,000 adults and 440 per 1,000 children was far higher than in all but the worst famine years. Conditions improved over time, and merchants who wanted to make profits learned to provision ships more adequately. Between 1727 and 1754 the percentage of Germans ill on landing at Philadelphia dropped two-fifths, from one in fifteen to one in twenty-five. Despite this appalling mortality, potential servants judged their risk of death slight and signed on in large numbers.[84]

Arrival brought the new anxiety of sale. Sellers of servants tried to dispose of their charges quickly. Buyers probed for skills and strength, each getting one or two farmers or mechanics; women and unskilled boys languished for weeks. The merchant lumped together those who remained after several months and sold them to a wholesaler at discounted prices. Pennsylvania immigrant Gottlieb Mittelberger, an organist and choirmaster, observed the more humane redemptioner sales. Some fortunate souls found kin or friends to bail them out; "the others, who lack the money to pay, have to remain on board until they are purchased." On sale days English and German buyers "come from Philadelphia" and places as much as "forty hours' journey and go on board the newly arrived vessel." After picking healthy men or families, they negotiated a three- to six-year contract "in order to pay off their passage, the whole amount of which they generally still owe." Families often suffered separation, especially when "the husband or the wife died at sea," leaving the survivor to pay for the deceased's passage. To keep parents and small children together, some families had teenage children take on indentures to pay for others in the family, but most had to serve. During the 1740s two-fifths of German redemptioner families kept children, most under age four, with them without obligation, but three-tenths of families who avoided service paid part of their debt by giving their children up for service. Kin or friends paid the passage of 1 in 15 of the 248 Germans who arrived in Philadelphia on the *Britannia* in 1771; another third settled their debts, and the rest had to put themselves or their children into service.[85]

Once sold, servants marched to their sometimes distant new homes. A third of the German redemptioners who arrived between 1771 and 1773 stayed in Philadelphia; most of the rest went to southeastern Pennsylvania, one-quarter of them to four heavily German counties. Although landsmen bought many servants, German families often suffered separation, as advertisements for lost servant kin in German-language newspapers attest. A third of British servants, mostly Irish, worked in Philadelphia, and another three-tenths were placed in adjacent, heavily British counties; only a sixth went to heavily German counties. Since Chesapeake planters or merchants sought skilled servants, many of them landed in Maryland or Virginia. Servants with the necessary skills—carpenters and ironworkers, for example—sold quickly, but jobbers carried the less skilled from port towns to the Piedmont or the Shenandoah Valley backcountry to find buyers. Those unfortunates who fell into the hands of jobbers endured more sale days of poking and probing after they arrived. Similarly, while most convicts sold in Maryland by Stevenson, Randolph, and Cheston in 1768–75 wound up in Anne Arundel, Baltimore, Queen Anne's, and Kent Counties, close to the sales at Annapolis and Chestertown, the firm sold those without skills in large lots to middlemen who placed them on sale in frontier Frederick County or the Virginia backcountry.[86]

Other immigrants found sponsors willing to pay for their voyage. Some sponsors, eager to have their lands settled, did not demand repayment; others required migrants to sign a servant contract. Merchants, speculators, churches, and governments acted as sponsors, sometimes cooperating to get shiploads of immigrants to settle frontier lands. Germans, Scots, and Irish Quakers often emigrated this way, but Catholics rarely found sponsors. Glasgow and southwest Lowlands merchants enticed many mid-eighteenth-century Scots to New Jersey with liberal leasing policies. Irish Pennsylvania Quaker meetings lent money to poor Quakers to pay their passage. Governments or churches sometimes sponsored large-scale emigration, such as that of the Palatine refugees to New York in 1710 and the English, Scot, and German migrants to Georgia in the late 1730s and early 1740s. At most a fifth of immigrants enjoyed such sponsorship, but it was nonetheless crucial to the settling of colonies. Sponsored immigrants arrived during the early years of a region's repeopling; once they established themselves, they wrote home urging others to come.[87]

The misery of the Palatines began when war reached the Rhineland in 1707. Repelled by this war, heavy taxes, poor weather, and poverty and attracted by promises of cheap land in America tacked on church doors and printed in a widely circulated pamphlet, 12,000 to 13,000 persons—two-fifths Calvinist, one-third Lutheran, and one-third Catholic—fled their southwest German and Swiss homelands in 1709 and poured into London. Nearly all were members of families, and two-fifths were children under fourteen. Arriving destitute, they fell into "a starving condition," overwhelmed government efforts to keep them alive, and inspired a massive charitable campaign to aid them. A more permanent solution had to be found. Eager to rid London of the Catholics, the government sent 2,875 of them to Holland. It also sent 3,000 Protestant refugees (800 families) to Dublin, expecting them to go to rural areas. Some of them refused to leave the city; most of those who did leave returned to Dublin as a result of failure to get free land, mistreatment by landlords and the Irish, and the absence of German-speaking ministers. By 1713 two-thirds of these Protestant refugees had gone back to London in hope of returning to Germany.

By 1710 most of the remaining refugees had reached the colonies. Carolina authorities, who needed men to defend the colony from Indians, eagerly gave land to 650 Swiss emigrants recruited by Christopher von Graffenried. Seeking to reduce further the number of Palatines on the dole, the English sent 2,855 of them to New York but required them to work making tar and pitch to pay off their London subsistence and travel costs, promising each family a forty-acre farm at the end of their term.[88]

Palatine emigrants suffered mightily on North American frontiers but eventually enjoyed modest success. Weakened from destitution in England, more than half of the North Carolina and a quarter of the New York emigrants died en route or soon after they landed. Arriving too late at New Bern (a coastal frontier area) to plant crops and forced to trade their clothing for food, most North Carolina immigrants fell ill. After von Graffenried procured food for them and parceled out land, the settlement began to thrive. The immigrants took the best land, angering the Tuscaroras, who burned the settlement in 1711 and killed or captured seventy colonists. Survivors worked awhile for eastern Carolina planters and then scattered to nearby areas. But a few people, numbering 41 families in 1749, returned to their old farms and squatted there for three decades, until North Carolina authorities finally gave them title to equivalent land.[89]

Palatines in New York suffered as well. Governor Robert Hunter put them on Livingston Manor to make pitch and tar and gave each family a forty-by-fifty-foot lot, smaller than what they had in Germany and hardly the free land they expected. Unwilling to support families when only men and boys could make pitch and tar, Hunter indentured "Orphans and other children whose Parents have numerous family" and cut off widows' pensions. By 1714 Hunter had placed seventy-four children (thirty-one of whom had live parents) with English or Dutch masters. By taking children from their parents "against their Consent," Hunter "depriv'd" emigrants "of the Comfort of their Childrens' Company" and "the assistance and Support they might in a small time have reasonably expected of them." The Palatines protested, refusing to cultivate their lots and demanding the forty acres on the Schoharie, money (£5 per person and £15 per year per family), clothing, and tools they believed owed them, backing down only in the face of the governor's power. When the project ran out of money, Hunter ordered the Germans to "shift for themselves"; without farm tools and facing famine, they survived only by "seeking relief from the Indians." As soon as they could, the Palatines dispersed. Some became Livingston's tenants, 160 other families moved to Schoharie lands they bought illegally from Indians, and the rest went to the Mohawk Valley or to Pennsylvania. New York authorities forced the Schoharie families to quit their farms for refusing to rent or buy them from the speculators who owned the land. After petitioning the Lords of Trade, some finally got land. In 1724 seventy families received a grant near Livingston Manor. By 1731 the Palatines had received 26,000 Mohawk Valley acres and bought another 20,000 acres in the same area.[90]

Like seventeenth-century colonizers, the Georgia trustees (especially leader James Oglethorpe) founded the colony as a utopian refuge for persecuted and poor people. But unlike his predecessors, Oglethorpe aimed to make Georgia exclusively an "Asilum for the Unfortunate"—starving families, unemployed Londoners, debtors recently released from prison, vagabonds forced off the land by enclosure, sailors, and persecuted European Protestants. By prohibiting slavery and limiting land accumulation and inheritance, he hoped to prevent idleness, pursuit of self-interest, and wealth accumulation based on African labor. Oglethorpe saw firsthand in South Carolina the misery slavery brought to Africans, and he was horrified that just a few slaves had led the first Georgia immigrants to become "very mutinous and Impatient of Labour and Discipline," unable to defend the tiny colony against invasion. Without slaves, Georgia could turn "the useless poor in England and distresst Protestants in Europe" into industrious smallholder families. Not only would the emigration of the destitute relieve London's poor rolls, but emigrants would make small farms and send flax, silk, and wine to England. As settler-soldiers, emigrants would defend the southern borders of England's empire.[91]

Although Oglethorpe and the trustees sent a few imprisoned debtors to Georgia between 1732 and 1741, they were most interested in helping the poor. Most of the earliest settlers whose passage and early subsistence in the colony they paid came from the neediest classes. To ensure that only the poorest got help, the trustees rejected "poor persons . . . who were able to earn their bread" and helped debtors settle with creditors. Between 1732 and 1751 they sent 2,122 people to Georgia, half of whom were indigent Britons, mostly artisans or small tradesmen. With the £108,000 Parliament gave to the colony, they financed two-thirds (1,847) of the immigrants who came to Georgia before 1742. In addition, the trustees financed the emigration of many foreign Protestants, but they often required them to repay their passage with a term of service. Aided by the trustees' money and politics, these foreign immigrants, among them small farmers and farmer-soldiers from Salzburg and the Scottish Highlands, were able to set up farms.

In the first years of Georgia settlement, the immigrants faced all the problems—seasoning deaths that carried away as many as a quarter of the early arrivals, heavy debt, labor shortages, wretched weather, and warfare—of new southern colonists, but they made swift progress, clearing and planting 1,038 acres by 1738 (2 acres per surviving man). Happy with these results, the trustees continued to transport 100 or more new immigrants each year.[92]

By the mid-1730s opposition to the trustees by rich adventurers who brought servants to the colony or independent Scot and English settlers had developed. Dubbed "malcontents" by the trustees, they demanded that slavery and large landholdings be permitted. They argued bitterly with the trustees; failing to convince them, they lobbied Parliament to make Georgia appropriations contingent on the introduction of slavery. Influenced by the malcontents, Parliament gave just £16,856 to the trustees between 1745 and 1752. As appropriations lagged, the number of emigrants trustees financed plummeted. Between 1742 and 1752 they sent only 275 people to Georgia, mostly indentured servants who chafed under the trustees' regime; planters and merchants, in contrast, brought in 2,558 persons. Georgia's policies had kept South Carolina planters, who craved Georgia's rich coastal land, at bay. Badly scarred by depression, at least thirty-seven South Carolina planters moved to Georgia in 1747–48, some bringing slaves in defiance of Georgia's ban. As soon as they arrived, they added their voices to the malcontents' challenge of the ban on slavery. Faced with such deep opposition and unable to obtain money from Parliament, the trustees allowed slavery in 1749 and surrendered their charter to the crown in 1752. Immigrants continued to come to Georgia, but most had to work as servants to pay their passage. When freed, they had to compete with a rising tide of South Carolina emigrants, many of whom were large planters who made vast profits off the sweat of their African slaves.[93]

Immigrants landed at Boston, Philadelphia, New York, or Charleston. But most good coastal land had long been cultivated, so a large majority of newcomers lived on frontier farms. Some immigrants, like the unfortunate people von Stümpel cheated, moved immediately after they arrived. Others stayed and then moved to get land. After the naval stores project ended, some New York Palatines walked a few days from the Hudson River to the Schoharie; a decade later, they moved to the Mohawk frontier. In 1723 fifteen of them reached Pennsylvania, walking from the Schoharie to the Susquehanna and then canoeing to their new lands; by 1728 forty-one other families had joined them. Servants sold to backcountry farmers stayed after the completion of their terms; others found their way there after gaining freedom. As pioneers filled in more land, migrants moved farther west. In the early 1770s Alexander Thompson saw "emigrants in crowds" passing his central Pennsylvania farm on their way, he thought, to the Ohio River.

The experience of brothers Michael and Nicholas McDonald shows that even the poorest immigrants might succeed in farming and start a community, if they moved to the backcountry. The brothers emigrated to Philadelphia from Ulster in the late 1750s as servants. During the Seven Years' War, they joined the army and saw the rich lands of Iroquois New York; they returned in 1763 and, with William Johnson's permission, made a farm in Saratoga County. Other Ulster Irish soon joined them, including New York merchant James Gordon, who in 1771 enticed his brother and two sisters to emigrate to his new farm.[94]

Most groups of settlers—unable to get aid to finance their travel—moved by stages over several generations. Emigrants from Kraichtal (in southwest Germany) moved, on average, only once after they arrived in Pennsylvania. At least forty of seventy-one German immigrants who came to frontier Rowan County, North Carolina, in the 1750s and 1760s had landed at Philadelphia. They went first to Lancaster County, Pennsylvania, then moved to Frederick County, Maryland, or the Shenandoah Valley before reaching North Carolina. At first nearly all the German Moravians who moved from Pennsylvania to North Carolina were male immigrants, but by the mid-1760s the Pennsylvania settlements sent mostly the children of immigrants, such as the group that arrived in 1766 with fifteen American-born adolescent girls or young women. A few of these immigrants received an inheritance from the old country that they could use to move from Pennsylvania southward and buy frontier land. Others, like the Moravians, pooled resources, filled a wagon, walked miles on muddy paths each day, and relied on landsmen or fellow believers along the way to provision them.[95]

Immigrants living in the backcountry, even ex-servants, often acquired land. Most, who arrived with no money, had to work years to buy a tract. Some squatted on land for years before patenting it; others got cheap land from governments or immigrant recruiters; many moved near landsmen, who helped them find land. The most successful migrated in family or communal groups, as did most German immigrants before the late 1740s. Nearly all the 1,500 immigrants who came to Pennsylvania in 1737–38 and 1751–54 from Baden-Durlach (a southwest German principality) emigrated as families; almost all single adults came with kin. At least two-thirds had too little money to finance their voyage. Similarly, emigrants from Kraichtal (adjacent to Baden-Durlach) in 1713–75 always traveled in family or village groups, usually numbering seven or more. Once they arrived, most signed servant contracts, worked a few years as laborers, and pooled family earnings. By 1758 all the Baden-Durlach immigrants resident in Lancaster County who had arrived by 1739, but just half of the more recent arrivals, owned land. By 1771 four-fifths of the Lancaster immigrants (two-thirds of the later group) owned land, averaging 100–200 acres. The greater the family's assets in Baden-Durlach, the larger the Pennsylvania farm. Those without land had property in Germany that averaged £13 sterling, barely enough to pay for two trans-Atlantic tickets; those with 100 or more acres had holdings averaging £49.[96]

EIGHTEENTH-CENTURY immigrants were a self-selected group. Despite enclosures, crop failures, and population growth, all but the poorest Europeans could get a subsistence as tenants, farm laborers, or craft workers, and the vast majority stayed or moved within Britain or German-speaking lands. Few American immigrants were desperately poor; they were literate people, more willing to take risks to get land than those who stayed behind. They had already left their villages, moved to a town, or walked though half of Europe to board a ship. Most, forced into temporary bondage to pay their passage, struggled to get the land and make the farms colonial recruiters had promised them. The Ulster Irish and Germans among them usually got land; English and Irish Catholic immigrants succeeded less often.

However much immigrants differed in culture or religion, they all moved to find land to set up farms, a goal their American descendants shared. Surrounded by forests and Indians, immigrants had to abandon some of their old ways and borrow from Indians and earlier immigrants. As Chapter 5 shows, they gradually cleared land; made farms; allocated work to husbands, wives, and children; established borrowing networks with neighbors; and took surpluses to the market. As the population grew, they and their descendants divided farms so often that many of their children and more of their grandchildren sold their belongings and took their inheritance to a new frontier, where they began the process of farm making all over again.

chapter five

Markets and Farm Households in Early America

IN 1775 THE AUTHOR OF *American Husbandry* uncovered a new class of farmers, "the little freeholders who live upon their own property." In New England they made "much the most considerable part of the whole" people. Such a class, unknown in England, had developed "owing to the ease of every man setting up for a farmer himself on the unsettled lands." New Englanders could become freeholders, no matter how humble their origins. "The new settlers," the author learned, "upon fixing themselves in their plantations" became freeholders "at once." At first poor families fell "naturally into a class below them," but wages were so high that they were "able to take up a tract of land whenever they are able to settle it." Small freeholders were significant everywhere. The people of New Jersey "consist[ed] almost entirely of planters," owning mostly "little freeholds, cultivated by the owners"; Pennsylvania was "inhabited by small freeholders" and "by many little ones who have the necessaries of life and nothing more." Those "with no slaves" in the Chesapeake region began by mixing "tobacco planting with common husbandry" and eventually made enough money to buy slaves. South Carolina rice cultivation did require slaves, but because "even the smallest plantation is proportionately as profitable as the largest," a man with two or three slaves could begin farming. "The pleasures of being a land owner are so great," the author concluded, "that it is not to be wondered at that men are so eager to enjoy [them], that they cross the Atlantic Ocean in order to possess them."

Farm making proceeded with ease. In Pennsylvania "new settlers upon the uncultivated parts of the province . . . take up what land they please, paying the fixed fees to the proprietors," or "buy uncultivated spots of other planters, who have more than they want." Then they plant an orchard and garden and "begin their house" with the help of "countrymen" or artisans. They turn next "to work on a field of corn" to get bread for their families. "Doing all the labour of it" but "not yet being able to buy horses," they hire a neighbor to plow the land. When the farmer died, he gave the land—now improved—to his children, dividing it equally among them, to farm much as their ancestors had.

Far from growing produce for the market, the author of *American Husbandry* wrote, northern farmers worked mainly for their own subsistence, and even small southern planters, who did sell tobacco or rice, tried to produce as much food as possible. New England farmers "enjoy[ed] many of the necessaries of life upon their own farms, making food—much of cloathing—most of the articles of building." Though they could not make everything, "they have from the sale of their surplus products" produced "sufficient to buy such foreign luxuries as are necessary to make life pass comfortably." In New Jersey "fish, flesh, fowl, and fruits, every little farmer has at his table in a degree of profusion." Given such self-sufficiency, small farmers bought few manufactured goods. Rather, "the little freeholder, who, probably raising for many years but little for sale, is forced to work up his wool in his family, his leather, and his flax, after which the rest of his consumption is scarce worth mentioning."[1]

This portrait of the self-sufficient, subsistence-land-owning farmer comports in part with what we know about colonial farmers. Most did own land and worked for subsistence before engaging in market production; baking and butchering, spinning and weaving, and hunting and fishing for home consumption did reduce dependence on markets. But freeholders were not as predominant as the author believed, nor were eighteenth-century farmers self-sufficient. The author almost entirely missed the importance of markets and local exchange. Farmers made composite farms, providing for subsistence, trade with neighbors, and sale at market. Yet the author raises important questions about the relationship between land and household formation and between self-sufficiency and markets. This chapter recounts the history of the market, local exchange, farm household formation, and farm labor.[2]

The connections among households and markets stand at the center of agrarian history. Farmers looked backward to subsistence and the perpetuation of their farms and forward to the sale of wheat or flour, milk or beef, and tobacco or rice that made farming possible. Exchange between neighbors mitigated the vagaries of the market, allowing farmers to survive in times of poor prices. Men and women exchanged labor and surpluses and helped one another build houses and raise barns, make quilts and spin thread, and harvest and shuck corn; farmers borrowed money from one another to expand their operations. Eighteenth-century farmers participated more often in markets and bought more consumer goods than their ancestors; but at the same time communal self-sufficiency may have grown, and the density of local (rather than market) exchange certainly did.

Markets required farmers, living on their independent holdings, to allocate labor within the family. By the late-seventeenth century, if not earlier, the subservience of wives was guaranteed, and women and men divided farm tasks in ways compatible with their perceived roles in life. Men cleared land and cultivated wheat, corn, tobacco, or rice; their wives and daughters labored in the gardens, milked cows, baked bread, plucked chickens, cooked meals, and joined the men at harvest time.[3]

To Market, To Market

In the early 1650s Captain Edward Johnson of Woburn, Massachusetts, made clear the importance of markets. Boston attracted farmers from miles around who brought goods not only to feed the city's inhabitants but to sell to merchants for export. Farmers in Dedham were "generally given to husbandry." Their harvests were so rich, "abounding with Garden fruits fit to supply the Markets of the most populous town," that local farmers made "many a long walk," eager to earn the "coyne" and buy the commodities Boston offered. Ipswich farmers had "very good land for Husbandry," and the increase in "Corne and Cattell of late" left them with such big surpluses that they had "many hundred quarters to spare yearly, and feed, at the latter end of Summer, the town of Boston with good Beefe." Sudbury took "in Cattell of other Townes to winter" and drove them to Boston. Andover residents, living on the northern reaches of Massachusetts Bay, complained of "the remoteness of the place from Towns of trade" but nonetheless carried "their corn far to market."[4]

Since farmers had always required a vent for surpluses, agricultural markets were ubiquitous in early America. When farmers produced only for family use and local trade, the surpluses they sold to local shopkeepers often wound up in the West Indies, England, or Europe. Southern tobacco and rice planters, rich owners and poor tenants alike, were so tied to commodity markets that their survival depended on good prices. One must distinguish these markets from "the market" or a market economy. The market implies far more than the existence of marketplaces where people exchange carpentry for corn, wheat for salt, or tobacco for cloth. It conjures up an economy in which supply and demand determine the price farmers get for corn or tobacco and in which international markets (and the desire for profits over a competence) drive decisions farmers make about what to grow. This heavily commercial agriculture, found in southern tobacco and rice regions from the outset, slowly permeated other regions. Far from seeking commercial markets, many colonists paid intermittent attention to market demand beyond their own community, growing as much of their family's food as they could before selling surpluses at local markets. By guaranteeing subsistence first, most colonial farmers participated in markets without being dominated by them.[5]

Markets and farm households formed almost simultaneously. A farm household (what seventeenth-century theorists called a "family") is defined as a unit whose members (husband, wife, children, slaves, servants, and hired hands) pool income. Ideally the father-husband headed the household and allocated tasks to ensure its survival and, beyond that, to attain a competency, settling his family on a secure farm where everyone had enough food, clothing, and shelter. Such a household needed guns, plows, pewter, and innumerable other products made abroad. To acquire these items, farmers had to sell surpluses, however small, at market.[6]

Despite the first colonists' desire to sell surpluses, farm markets took several years to develop in Virginia, New England, and New Netherland. The mostly male immigrants in early Jamestown, who lived in barracks under military discipline or worked on large plantations, had no crops to sell. The first settlers of Plymouth and northern New England engaged in fishing and Indian trade, and those of New Netherland pursued the fur trade. Once they gained a foothold, colonists began a feverish search for a staple crop—one in demand in England, the West Indies, or neighboring colonies—if for no other reason than to permit married couples to start farming. Within a decade of initial settlement, most colonists had abandoned fishing and the fur trade and brought hogs, salted meat, wheat, flour, corn, tobacco, and rice to markets that sustained farm households and permitted farmers to expand operations and buy more land and livestock. Parents then gave or bequeathed such property to their children, allowing a new generation to begin to farm.[7]

The first farm families to arrive in an area bartered with neighbors, but they soon craved regular markets. Knowing that large numbers of settlers would not begin farming until markets were in place, assemblies chartered towns, complete with market squares. By the late 1630s, as immigrants poured in, Boston, Salem, Watertown, and Dorchester, Massachusetts, had initiated marketplaces where farmers could sell produce and cattle and buy imports. Connecticut followed suit in the 1640s. Entrepreneurs set up markets in Philadelphia and Bristol, Pennsylvania, in the 1680s and 1690s, just as the first settlers arrived. As villages such as Boston grew into small cities, farmers sold their grain, butter, fruit, and vegetables directly to townsfolk, more often on the streets than in official markets. Prosperous northern farmers diversified. They reared livestock and harvested hemp, hay, and flax as well as grain, sending small surpluses to market. Seventeenth-century New England women grew hops and made beer for their families, taverns, and businesses. As New Amsterdam (New York City) grew and the fur trade declined, Long Island and Hudson River farmers provided the city with wheat and flour, and by the end of the century, they were sending small surpluses to the West Indies.[8]

In the Chesapeake colonies, where neither towns nor mandated marketplaces materialized, small planters gathered at docks, taverns, courthouses, or the homes of ship captains or planter-merchants (who kept shop) to sell tobacco and buy cloth, rum, and sugar. By the early eighteenth century, villages—where tavern keepers, storekeepers, and artisans set up shop, thus facilitating and increasing local trade—proliferated in all the colonies. By 1800 twenty villages had developed in Kent County, Maryland (a wheat-producing area), alone. Storekeepers moved to every new frontier colonists settled during the eighteenth century, often arriving with the first farmers.[9]

Nearly every farmer bartered farm products with neighbors, frequented local markets, and participated in commercial markets. A complex network of merchants, often related by blood or marriage, linked each colony to London and rural markets to those in Boston, Jamestown, New York City, Philadelphia, and Charleston. Regional differences nonetheless developed. As one traveled from north to south, the proportion of farm families deeply embedded in international markets grew. Most New England farmers, except the Narragansett, Rhode Island, planters and the Connecticut River Valley grain farmers, produced small surpluses and aimed at local self-sufficiency in food. Landowning New York farmers sent grain or flour to market only when they could negotiate good prices. In contrast, Chesapeake planters—no matter how poor—pursued commodity markets. The smallest planter sold a hogshead or two of tobacco to a richer planter, ship captain, or merchant; larger planters sent many hogsheads to England. West Indian emigrants who colonized early South Carolina expected to be enveloped in markets; almost as soon as they arrived, and before they and their African slaves developed rice as a staple, they stocked the range and forests with cattle and hogs and provisioned the sugar islands they had left.[10]

The discovery in the 1620s that Virginia tobacco commanded a high price in England created a vigorous tobacco market that ensnared every Chesapeake colonist. The tobacco trade permitted unmarried men without their own farms—a majority of free colonists—to marry, form households, and get land. As tobacco production rose from 100,000 pounds in the mid-1620s to more than 20 million pounds by the 1680s, tobacco marketing grew ever more complex. At first planters sold tobacco to ship captains representing groups of London merchants and received imported goods off the ship in return. During the 1630s and early 1640s, 300 London merchants bought tobacco, usually in small quantities. The trade thereafter became more concentrated. A handful of London merchants (9 in 1676 and 11 in 1686) purchased more than 250,000 pounds of tobacco; their share of London imports grew from three-tenths to three-fifths between 1676 and 1686. The number of London importers who bought less than 10,000 pounds plunged. The consolidation of the London trade increased opportunities for planters to sell tobacco. The big firms commanded more capital and loaned more money to planters, a necessity when tobacco prices declined and productivity gains failed to keep pace. Rich merchants sent agents to the Chesapeake to look after their affairs or established close relationships with local merchants. After 1650, but especially after 1690, some big planters sent tobacco directly (on consignment) to London tobacco sellers, merchants, or tradesmen and bought manufactured goods in return. Others marketed their neighbors' crops, forming strong patron-client relations that kept poorer families afloat when prices dropped.[11]

Although seventeenth-century New Englanders participated less regularly in international markets, they did sell surpluses. During the 1630s Plymouth farmers provisioned newcomers to Massachusetts Bay. The cattle Plymouth farmers drove to Boston filled Massachusetts pastures, Plymouth corn fed Massachusetts settlers, and Massachusetts farmers planted Plymouth seeds. Settled farmers in Massachusetts soon supplied their neighbors as well. But when immigration dwindled after 1640, grain, livestock, and land prices plunged, and the New England household economy nearly collapsed. Merchants quickly found new markets for corn and livestock in the West Indies, North Atlantic islands, and Iberia, and as Boston's population grew, its landed resources diminished and its need for country provisions grew. Others provisioned the burgeoning fishing and merchant fleets. While many farmers probably began emphasizing subsistence over market production (especially after the wheat blast hit the region in the 1660s), pockets of market agriculture could be found. Narragansett planters sent beef, pork, mutton, and horses to the West Indies; Connecticut River Valley farmers sent furs, grain, and cattle to Boston; middling and rich farmers living near Boston or Salem sold grain, meat, and dairy products to merchants or hawked them on town streets.[12]

The expansion of international markets facilitated farm making in the eighteenth century. Demand for tobacco and rice remained strong, enticing new producers to make those crops. Tobacco production expanded throughout the Chesapeake region, but some areas turned to grain. South Carolina specialized in meat export to the West Indies until that colony's planters improved rice production early in the century. Even the poorest southern planter relied on international markets to guarantee subsistence. Foreign markets grew in other regions as well. As warfare engulfed European bread baskets and English per capita grain production declined during the 1760s and 1770s, new markets opened for American wheat. Pennsylvania farmers took advantage of this market, and they were joined by many others in New York, New Jersey, Virginia's

Shenandoah Valley, and the Carolina Piedmont. At the same time, rice and tobacco prices climbed relative to English manufactured goods, and capital generated in the South grew.

After the late-seventeenth-century depression, local marketplaces, outside the price and production imperatives of a market economy, increased in importance in the North and the backcountry. Despite growing international markets, the per capita value of New England and (to a lesser extent) mid-Atlantic exports declined over the eighteenth century, yet per capita imports rose, suggesting that northern internal markets generated capital through local exchange. New England farmers, except those in the Connecticut River Valley and southwestern Connecticut, drifted into subsistence production and local exchange or shifted from grain production to livestock herding or market gardening, leaving substantial grain deficits in the coastal towns that had to be filled by mid-Atlantic farmers. Hudson River farmers exported some wheat and flour, but most prospered on intensive local exchange, a pattern that extended (along with cattle and hog driving) to the entire backcountry, stretching from Pennsylvania to Georgia.[13]

As farm income rose, the colonies saw an explosion of importation and consumption of manufactured goods. During the seventeenth century, colonists had imported mostly capital goods needed to build farms. Between 1699 and 1774 (when the population multiplied 8.6 times) merchants imported and sold rapidly increasing quantities of woolens (7 times the value, despite price declines), window glass (28 times the weight), and earthenware (8 times the number of pieces). Growing international and local markets stood behind these increases in imports, as a discussion of each colonial region, from New England to the backcountry, shows.[14]

New England's agricultural surpluses declined during the century after 1650. In 1771 many towns in eastern and central Massachusetts made too little grain or pastured too few cows and pigs to feed their people, much less send surpluses to Boston. Since cheap grain was available from the mid-Atlantic and Chesapeake colonies, farmers grew less; as forests disappeared, the firewood and lumber industries atrophied. In 1771 two-fifths of Massachusetts farmers needed to trade to feed their families. Families farmed some land in Maine fishing villages, but not enough to feed themselves; the "country people," the Reverend Thomas Smith noted in April 1775, "were flocking in" to Falmouth "to buy corn and other provisions." As farmers in the older parts of New England divided their land among many sons, the number of farms with less than sixteen acres of improved land grew. Since these small farms rarely fed a family, household members had to labor for more fortunate farmers or find alternative work to make up the deficits.[15]

These deficits made farm diversification essential. Eighteenth-century New

England farmers replaced some wheat with corn, and bread with other foods. Eastern Massachusetts farmers put close to nine-tenths of the land they farmed in pasture rather than crops, to make more butter and meat for their tables or wool for their homespun clothing. Middlesex County farmers experimented with new crops, growing increasing quantities of potatoes, beans, turnips, onions, and carrots for the family; they expanded their apple orchards and made more apple cider, apple butter, and vinegar. Holdings of pigs also grew, allowing farm families to make more salted pork. Farmers who made small surpluses of vegetables, cider, butter, or pork traded these goods with neighbors or exchanged them for wheat or flour.[16]

Despite grain deficits, the number of farmers selling big surpluses grew in the eighteenth century. None was exported, save for dried meat and livestock sent to the West Indies, and one-tenth of the grain and pork consumed in the region was imported from other colonies. Local markets grew rapidly. Boston's population rose, as did that of coastal ports, reducing space for pasturing cows, raising chickens, or planting gardens. The 25,000 to 30,000 people in mid-eighteenth-century Boston, Salem, Providence, and other towns relied more and more on local farmers for meat, vegetables, fruit, butter, and flour.[17] Poorer fishermen and farmers in coastal villages also had to buy food. To take advantage of these markets, middling and rich farmers increased their surpluses. Farmers living near Boston and in the Narragansett region intensified market gardening, selling milk, cheese, butter, vegetables, and fruit at city markets. Connecticut Valley and some Middlesex County farmers sent grain or flour to Boston. Cattle wintered in western and central Massachusetts (on that region's surplus grain) before they were driven closer to Boston for fattening.[18]

New England's marketing opportunities grew rapidly during the eighteenth century. As farmers had dispersed during the seventeenth century, village centers had atrophied. Although New Englanders continued to live on isolated farms, villages reappeared in the eighteenth century. Boston remained the largest market, but its population stagnated at about 15,000 after mid-century because it lacked a staple-producing hinterland. Other towns developed their own farming hinterlands. In Connecticut appeared five towns of more than 500 residents, each with a dozen or more stores, taverns, and artisan shops, and eight village centers (population 100–500), with competing stores and taverns. Innumerable crossroads, each with a tavern, a mill, a church, and perhaps a store, emerged between the larger places. Similar marketing centers were scattered throughout New England. During the 1760s and 1770s farmers took advantage of this competition for their surpluses, traveling fifteen or twenty miles (a day's trip) to sell their goods, some hiring or driving wagons to Boston or Salem.[19]

The explosive growth of demand for bread in the West Indies, England, and southern Europe raised wheat and flour prices, thrusting mid-Atlantic farmers—unlike New Englanders—into international markets and making them part of a complex commodity chain that stretched from farm to miller, miller to exporter, and exporter to European consumer. Farmers in a vast region—the Hudson River Valley, East Jersey, the Delaware Valley and upper Chesapeake Bay, and western Maryland—switched to wheat. Philadelphia had the largest hinterland—the Delaware Valley, New Jersey, and part of Maryland—and exported nearly three-quarters of the region's flour and bread and two-thirds of its wheat in 1768–72. New York attracted wheat and flour from the Hudson Valley, and Baltimore drew on northern and western Maryland and western Pennsylvania. Philadelphia flour exports grew as fast as the population in the 1730s and 1740s and a third faster than the population in the 1750s and 1760s.[20]

Soon after the first settlers had arrived, Philadelphia's merchants had sent wheat to the West Indies, and that trade slowly increased. But when European demand for bread grew after mid-century, Pennsylvania's market centers expanded. Philadelphia wholesalers—central players in this market and town system—sent a vast array of imported English goods, including cloth and buttons, pewter and china, and tools and nails, to storekeepers in villages and towns in the city's hinterland. Farmers living near Philadelphia drove wagons full of trade goods to the city; others traded at uncounted crossroads, a hundred villages of 100 to 300 people, and ten towns of over 1,000. Some farmers set up stalls in town markets to sell produce, but most dealt with local storekeepers and bargained with competing merchants for the best deal, selling their wheat, lumber, butter, or homespun cloth and buying imported cloth and other goods. Most farmers had their wheat ground at a local mill. Some sold flour to storekeepers; others to a grain broker or miller. Farmers who traded with millers used the cash they got from the miller to buy manufactured goods. Miller and storekeeper then sold flour to Philadelphia merchants, sending it by water to the city. This huge increase in the number of marketplaces and consumer goods, along with the growth of international demand for wheat products, inevitably led farmers into production of staples for the market.[21]

Few Delaware Valley farmers, however, specialized in grain production. Rapidly growing city and town produce markets led nearby farmers to diversify. As the labor needed to handle and ship flour grew, Philadelphia's population rose from 13,000 in 1750 to 29,000 in 1770. The population of places with over 1,000 people grew from 15,000 to 40,000—nearly a fifth of the region's total—during the same time. By the early 1770s urban folk annually consumed 6,000 tons of wheat, flour, corn meal, and bran; 1 million gallons of milk; 350,000 pounds of butter; 5 million pounds of meat; and 5,000 tons of potatoes and turnips. Nearly all these products had to be brought in from the countryside; few city residents made any food, though some kept hogs or chickens and a few had a cow. To meet this growing demand for food, farmers living within a day or two of the city sold small surpluses of vegetables, fruit, butter, and milk or fattened cattle on hay they made. The growth of urban market gardening and dairying reduced the risks inherent in staple production and enticed farmers to take advantage of grain markets.[22]

New York's farming hinterland—Long Island, the Hudson River Valley, East New Jersey, and part of Connecticut—was less deeply involved in market agriculture than the Delaware Valley. Dutch and English farmers from as far away as Albany had long sent grain surpluses to the city for town use and export, but they did not increase production substantially when European demand grew. New York City marketed proportionately only half as much grain, flour, and salted meat as Philadelphia. Wheat output grew around Albany, but the city's merchants continued to trade furs. Great landlords sent thousands of bushels of wheat to New York, but this output—rents paid in kind or surpluses bought from tenants—represented a tiny portion of the production of tenant farmers. Although a few landowning farmers in Ulster County (near the Hudson River) sold surpluses large enough to fill Benjamin Snyder's sloop with £1,200 of wheat in 1775, most people who traded with him sold him no grain. Farmers near the city—in Harlem, Queens, and adjacent New Jersey—took part in urban markets more often, sending grain and flour, fruit and vegetables, milk and butter, meat and leather, and hemp and potash to the city for domestic use and export to the West Indies and Boston.[23]

These differences defy easy explanation. The soils and climate of the New York and Philadelphia hinterlands resembled one another, and both cities had a large export-oriented merchant class eager to buy rural produce. New York's population grew rapidly over the century and reached 22,000 by the 1770s, second only to Philadelphia's, providing many opportunities for market gardening. The activities of New York merchants and great landlords resolve the puzzle. Merchants and townsmen sought to keep farm prices so low that farmers refused to grow surpluses; intermittent attempts to inspect flour (rejecting unmerchantable flour) and prohibitions of the export of flour in times of poor crops further dampened farmer desire. Farmers withheld grain and flour until merchants agreed to higher prices. Until 1763, when Canada entered the empire, the fur trade contributed a major portion of New York's exports, and fur traders and their merchant allies helped the Iroquois control central New York's rich farmland. The great landlords, who owned the best land west of the Hudson, attracted few tenants until late in the colonial era, for farmers could inherit, buy, or squat on land nearby. In sharp contrast to Philadelphia merchants, who invested growing amounts of capital in shipping and trade, rich New York merchants and gentlemen invested in land. Tenants had little incentive to produce surpluses beyond what they needed for rent, taxes, and subsistence; moreover, with little in-migration, there were few youthful laborers (abundant in Pennsylvania) to help harvest larger wheat crops.[24]

As Chesapeake tobacco prices plunged below production costs in the late seventeenth century, producers in marginal areas with poor soil for tobacco (the lower Eastern Shore and the Norfolk region) replaced it with larger corn crops. Even where tobacco output remained high, corn crops increased, and wheat cultivation began. Oronoko producers aimed to make large tobacco crops; producers of the sweet-scented tobacco of the James and York River basins reduced production to improve their crops. At the same time, small planters (especially in marginal areas) increased home production, spinning thread, making tools, or shoeing horses, and others took up craft work full time. When tobacco prices rose in the second third of the century, marginal producers returned to tobacco production, but diversification continued apace. Local exchange networks thereby intensified: more small planters bartered grain or traded craft services. Planters living near the small (but growing) towns of Williamsburg and Annapolis fattened cattle and hogs for quick sale and supplied city dwellers in each place with 300,000 pounds of meat a year. But the small city populations needed too little grain, fruit, and vegetables to stimulate market gardening. When new opportunities for selling wheat to southern Europe emerged during the mid-eighteenth century, more tobacco planters switched to wheat or corn. Tobacco production on the upper Eastern Shore declined after the early 1740s, as Philadelphia merchants bought more of the area's wheat for use in urban markets and export abroad. Planters throughout the region, even those who grew large tobacco crops, sold small quantities of grain to supplement their tobacco earnings.[25]

Notwithstanding the growth of grain markets, most Chesapeake planters sold tobacco. Tidewater planters remained confident of a good market for their high-quality tobacco. Responding to rising prices, thousands of new producers, pioneer planters in Virginia's Piedmont, entered the tobacco market during the middle half of the century, sending exports soaring from under 20 million pounds in 1700 to over 100 million pounds by 1770. Although rich planters and planter-merchants still consigned tobacco to London merchants, highly capitalized London, Glasgow, Liverpool, and Bristol mercantile houses (importing 1 million pounds or more) increasingly dominated the trade. By the 1760s the top five or six Glasgow firms imported a quarter of the Chesapeake crop. Firms in Glasgow and, to a lesser degree, Liverpool and Whitehaven established chains of stores in tobacco country where planters could sell their crops. These firms imported nearly three-quarters of the tobacco grown in the rapidly expanding Upper James region. Other planters traded with local merchants, who exported tobacco themselves or sold it to Scot wholesale merchants in Virginia towns and got cargoes of manufactured goods for their stores in return. The economies of scale the biggest Scot merchants achieved reduced the cost of imports they sent to their stores and permitted storekeepers to pay good prices for tobacco, extend credit to poor planters, and carry insolvent planters on their books when prices plunged.[26]

New marketplaces—crossroads settlements, hamlets, and towns—proliferated in tobacco country during the eighteenth century. After a few hours' ride, nearly every planter in the Tidewater and parts of the newer Piedmont could reach a crossroads, complete with tavern, store, blacksmith, church, and perhaps county court. By requiring that all tobacco be brought to central warehouses to be examined, the tobacco inspection acts (of 1730 in Virginia and of 1747 in Maryland) centralized marketing and paved the way for hamlet and village growth. Merchants set up stores adjacent to warehouses, and tavern keepers and artisans joined them to take advantage of trade. Some crossroads, especially Tidewater county seats where legislatures put inspection warehouses, grew into villages of several hundred people. Located at the eastern edge of the Piedmont, Baltimore, Alexandria, Fredericksburg, Richmond, and Petersburg grew larger, attracting tobacco from nearby areas and wheat from a vast Piedmont and mountain hinterland. Wheat, which required more processing than tobacco, contributed to the increased size of these towns. Small planters living near villages haggled over the terms of credit and the price of their crops; those residing near the larger towns, where competition between local merchants and Scot storekeepers was intense, had even greater opportunities to bargain for the best deal.[27]

By the 1750s crossroads, where stores, gristmills, and inns clustered, began to appear at county seats and next to ferries in the tobacco-growing Piedmont. But no towns developed until after the Revolution. Since the Virginia Assembly refused to place inspection warehouses there, tobacco had to be shipped to towns at the heads of navigation of rivers. Tavern keepers, local traders, and Scot storekeepers representing the big Glasgow tobacco firms opened shops throughout the Piedmont, sometimes at crossroads but as often in the open countryside on roads or at isolated plantations. By the late 1760s Scot merchants were operating at least twenty-nine large stores in Southside, the region of the Piedmont south of the James River, and dominated its tobacco trade. Another nineteen could be found in the Richmond–Petersburg area, the closest places with tobacco inspection warehouses. Storekeepers bought tobacco from frontier planters (mostly producers of a single hogshead) and sold them manufactured goods in return. After the sale, the Scots took all the risk, hiring wagons to take the crop to Petersburg for inspection and export.[28]

Planters who left Barbados for South Carolina in the late seventeenth century expected to produce for export. Unable to sell a tropical commodity like sugar, they turned to growing markets for deerskins in England and salted pork and beef in the West Indies, where sugar cultivation took up nearly all the land. By 1682, a decade after initial white occupation, Carolinians had "many thousand Head" of hogs and cattle, and the richest planters owned several hundred cattle and smaller numbers of hogs. Let loose by their owners, hogs and cows foraged in the meadows, marshes, and forests all year long. Slaves and white servants drove some back to cowpens at night, but most stayed in the woods, necessitating a semiannual roundup. Once captured, the cows were driven to Charleston, where they were butchered and sent to islands. After rice replaced livestock as South Carolina's most valuable export, cattle owners moved to upcountry South Carolina and to Georgia. In the 1760s and early 1770s slaves owned by Georgia cattlemen herded 40 to 200 cows, enough to make 1,000 barrels of salted meat.[29]

By the early eighteenth century planters in coastal Carolina and Georgia specialized in rice production. Rice exports grew from 3 million pounds in 1715 to 70 million pounds in the early 1770s. Rice accounted for over three-quarters of the value of the region's farm exports in the 1750s; even after some big planters began exporting indigo (for dyeing cloth) in the early 1770s, rice comprised two-thirds. Slaves, who made up two-thirds of the population as early as 1730, produced nearly all the rice. The two-thirds of the planters who owned ten or fewer slaves grew some rice but could not compete with the few hundred large planters who took full advantage of rising labor productivity in rice cultivation and the economies of scale that ownership of fifty or more slaves entailed. A few small planters in Charleston's hinterland diversified their farms and turned to market gardening; butchering livestock; growing grain for the Charleston, Savannah, or West Indian markets; or craft production. These markets were limited, however, since only 20,000 people, half of them low-consuming slaves, lived in the two towns, and colonies farther north now dominated the West Indian trade. Charlestonians, unable to rely on surrounding or even backcountry planters, looked to the mid-Atlantic colonies to fill their larders with increasing quantities of bread, flour (9,000 barrels by the 1760s), onions, potatoes, and apple cider.[30]

Pioneers who moved to the vast backcountry, which stretched from Pennsylvania's Susquehanna Valley through the Carolina and Georgia Piedmont and mountains, found markets with greater difficulty than other colonists. During the 1730s and 1740s pioneers in western Maryland and the Shenandoah Valley lived far from tobacco warehouses and feared rolling that fragile crop over long distances. German immigrants in western Maryland raised livestock but increasingly specialized in wheat production. Seeking familial subsistence, Shenandoah Valley farmers sold provisions to new arrivals and transients and herded livestock to coastal markets for export to the West Indies. Understanding how transitory such markets were, local justices had roads built and maintained to ease grain marketing. By 1760 farmers ground their wheat at one of the more than sixty gristmills in the region and sold the flour to a storekeeper located at one of the seven small valley towns, who then had the flour transported to Fredericksburg or Alexandria. When European demand for wheat skyrocketed in the 1760s, valley farmers exported more than 250,000 bushels of wheat and 2,900 tons of flour, over ten times more than in the early 1740s. Nearly three-fifths of the wheat and three-tenths of the flour went to new markets in southern Europe, and much of the rest went to New England. Farmers living in the western Shenandoah Valley, furthest from markets, grew wheat and corn and raised pigs but mostly relied on cattle. Most farmers there, who owned fewer than twenty cows, sold one or two a year to a local merchant, who then drove the large herds he acquired to market.[31]

The Carolina and Georgia backcountry developed less rapidly than the Shenandoah Valley. Even in the 1770s farmers had yet to find an easily transportable commodity in demand in Charleston or overseas. Backcountry North Carolina farmers drove cattle to Philadelphia or Charleston, but the great distances reduced the volume of the trade. During the 1750s and 1760s Moravian immigrants in central North Carolina experimented with several crops, including wheat, but supported themselves through the fur trade, exchanging craft labor and store goods for deerskins and for neighbors' surplus wheat and corn. South Carolina and Georgia backcountry farmers herded cattle but also grew tobacco and indigo and sent wheat or flour to Charleston or Augusta, the only sizable backcountry town. Only ten small marketplaces such as Camden, South Carolina, or Hillsborough, North Carolina, which had a gristmill, a lumber mill, several stores, and taverns, developed in the backcountry. Richer farmers, who owned wagons, probably bought up their neighbors' surplus crops and transported them with their own to Charleston. Faced with difficult transportation and little competition for their produce, these backcountry farmers had little incentive to increase market production.[32]

Both a Borrower and a Lender Be?

Matthew Patten, born in Ulster in 1719, moved to New Hampshire in 1728 and to Bedford (on the Merrimack River) in 1737. From the 1750s through the 1770s this prosperous farmer and surveyor ran a diversified farm with his sons, trapping animals, making and repairing tools, catching shad and salmon, and planting rye, corn, wheat, peas, potatoes, and flax. Trade took him all over New Hampshire and to Boston, but he mostly relied on neighbors. What he could not do himself—milling rye or shoeing horses—he hired others to do. He marketed little grain but lent large sums of money, traded farmwork with neighbors, and got corn from men or butter from women. He demanded commercial notes when he lent money but sought no cash when he swapped labor and goods with neighbors. This amalgam of commercial and noncommercial exchange belies any notion of either full market imbeddedness or self-sufficiency, but it does point to the crucial significance of credit and debt.[33]

Borrowing and lending, buying and selling, and barter and exchange among farm folk and between farmers and merchants were ubiquitous, as Patten's daily activities attest. The ordinary furnishings of a farm—land, livestock, beds, linens, and pots and pans—all cost money that immigrants could accumulate only by marketing farm produce, working for a farmer who did, or trading with neighbors. Favored native-born whites inherited enough property, real and personal, to outfit a farm; the rest had to work for it in ways similar to immigrants'. In addition everyone had to pay taxes and quitrents, buy pewter or earthenware, and procure salt and sugar. Only farmers with crops or home manufactures to sell or labor to exchange could get cash, command credit, or trade with neighbors.

Seventeenth-century farmers, who usually lacked cash, required loans or credit to pay taxes, buy land and livestock, and purchase imported nails or cloth. Chesapeake planters used mercantile credit to finance farm operations until their tobacco was ready; New England farmers relied on credit to buy town sites from Indians. But credit was scarce, as it continued to be on new frontiers. Rich planters, big farmers, and local merchants, with their access to English capital, made loans to poorer men. On occasion, a single patron—such as the Pynchons of seventeenth-century Springfield, Massachusetts; the Hudson Valley Livingstons; or the Dulanys in eighteenth-century western Maryland—was the only source of local credit. The Pynchons and the Livingstons controlled nearly all the good land in their areas and thereby increased the dependence of local farmers on them; men who argued with them got little or no credit and soon had to depart. To conserve their cash, farmers swapped goods, recording their bargains as "book credit" (debts listed in an account book but not secured by a mortgage or bond) or making oral contracts that had to be recorded at court to assure repayment.[34]

During recessions, when English merchants (in whose hands credit originated) called in loans to local merchants, local merchants or rich farmers had to reduce the amount of new credit they granted neighbors and collect long-standing debts. In such times farmers could lose their land; even if they held on, they could not borrow to expand or give farms to their sons. Farm laborers, moreover, could not start their own farms. Credit contracted in the late seventeenth century, affecting the entire country. From 1675 to 1700 the rate of dispersion of the European population (measured in square miles settled) declined by four-fifths from its 1650–75 level, despite the settlement of Pennsylvania and the Carolinas. Tobacco prices plunged, forcing marginal producers out of production. New Englanders, forced from outlying areas by King Philip's War, could not rebuild, even after the Indian threat receded. When times improved at the turn of the century, farmers moved to more promising frontiers, attracting loans from merchants and gentlemen.[35]

Eighteenth-century southern farmers had more sources of credit. Chesapeake planters and planter-merchants loaned money or granted book credit to friends and neighbors. Many had ties to British merchants, but others generated capital from their own farm and mercantile activities. Merchants followed suit: increasing numbers were willing to grant twelve months of credit. Small Chesapeake tobacco planters were especially fortunate, for Scot representatives of the great Glasgow tobacco houses carried them from one crop year to the next. Rather than "depend for payment on the real ability of the people," Glasgow merchant Archibald Henderson wrote in 1766, storekeepers "trust to the labor and industry of many" men of little property, granting them credit to buy "household furniture and working tools." Because merchants competed for customers, small planters, getting "goods upon credit from different storekeepers," accrued large debts. In Southside Virginia during the 1760s and 1770s, Scot storekeepers universally extended credit (a tenth of the customers paid cash) and received payment in the planter's tobacco crop or from others indebted to the planter. Lowcountry Carolina small planters had less access to credit. Great planters commanded large sums from British merchants, their Charleston representatives, and Charleston merchants to buy slaves and luxuries, but these men loaned money or provided book credit to lesser planters irregularly. Small operators had to rely on the patronage of great planters for credit. Credit was even less plentiful in the backcountry, where farmers got it when buying imports from tavern keepers and storekeepers. North Carolina Moravians, for instance, set up a store and granted book credit to outsiders—even though it was against their principles—because there was no other way to continue the trade in furs necessary to their survival.[36]

As commercial agriculture in the northern colonies grew, local sources of capital increased. Rich men, who had built fortunes in land speculation, mercantile trade, or commercial agriculture, loaned money on long terms to commercial farmers. Entrepreneur Sir Peter Warren of New York City occasionally lent money to prosperous farmers; smaller landowners in Dutchess County, New York, took out 519 mortgages (some for £10 or less) with city or local men. Merchants (or farmer-merchants, tanners, or millers) such as Elijah Williams of Deerfield, Massachusetts, advanced farmers money to buy manufactures and pay taxes; months later they took wheat, corn, or butter as payment and, in turn, were linked to creditors in Boston, Philadelphia, New York, or England. One in twenty householders in Massachusetts in 1771 kept shops, providing customers with many places to borrow money or get book credit. Farmers increasingly borrowed money from one another. Between 1710–19 and 1760–75 the proportion of Dutchess County decedents who showed debts owed them jumped from one-third to three-fifths; during the 1760s nearly three-fifths of Burlington County, New Jersey, decedents held negotiable bonds or notes. In York County, Pennsylvania, on the Susquehanna frontier, Quaker farmers lent money to others in their meeting, as did Presbyterians and Moravians to a lesser extent.[37]

Using their land as collateral, northern farmers could borrow small sums (at 5 percent interest) from land banks run by the colonial governments. In 1714 the Massachusetts land bank lent to 110 men in Middlesex County; two-thirds were farmers, and the rest were craftsmen. In 1717 it lent a new emission to 78 additional Middlesex householders, three-fifths of whom were farmers and one-third, artisans. Pennsylvania's land bank, which operated from 1726 to 1755, was especially successful, making 3,421 loans (137 per year); seven-tenths were small sums ranging from £30 to £80 lent to farmers who owned 50–250 acres.[38]

By the late seventeenth-century, mortgages had become a major source of credit for farmers, who used the money to buy more land or livestock, build new structures, or put up fences. Between 1650 and 1750, Middlesex County residents took out 619 mortgages; 77 percent went to rural landowners. Nearly two-thirds of rural borrowers were farmers, and most of the rest were craftsmen; they borrowed from clergymen, lawyers, and rich farmers and artisans. Despite the oft-stated short terms of the instruments, farmers regularly got them extended an average of seven years. The mortgage market was more vigorous in rural New York, where merchants accumulated more money to lend. Landholders in two Dutchess County precincts alone took out 329 mortgages between 1754 and 1770, averaging about £170 sterling. Nearly three-quarters of the mortgagors were dubbed yeomen or farmers, but nearly half the lenders were merchants, most from growing New York City.

In the Chesapeake region, where commercial credit from merchants and storekeepers was plentiful, planters rarely mortgaged land to expand their operations. Instead, during debt crises (such as that in 1772), creditors forced them to take out mortgages to pay for goods they had bought. Between July 1767 and June 1771, on average, just 36.5 Southside Virginia planters mortgaged land each year. As the economy weakened, that number jumped 2.5 times (to 94) in 1771–72 and then—after the crisis hit—nearly tripled (to 184).[39]

Everywhere in the colonies, a system of local exchange or barter developed. Based on neighborly reciprocity, this "borrowing system" grew out of commercial exchange. At first local exchange was very limited. Early Chesapeake planters emphasized tobacco so much and engaged in so little craft activity or cloth manufacture that they had little to trade but their labor. Farmers in mid-seventeenth-century Massachusetts were so short of labor that they limited the borrowing system to farm produce. Farmers did help one another in time of need, lending money or tools. As a Virginian put it in 1650, "no man will ever be denyed the loane of Corne for his house-spending and seed till the harvest." The borrowing system grew as a substitute for commercial exchange where markets were lacking or declining. When tobacco prices plummeted in the Chesapeake in the late seventeenth century and when wheat rust and poor soil limited New England wheat and flour exports at mid-century, the borrowing system expanded.[40]

By the mid-eighteenth century the borrowing system had become a strongly ingrained custom, a necessary part of economic and social life. More important in guaranteeing household sufficiency than commercial markets, the borrowing system sustained local trade. Rather than selling their produce or labor power, farmers bartered them, sometimes working for one another thirty or forty days a year. A cash-poor farmer received manufactures from a storekeeper or wheat from a neighbor without making immediate payment, but he eventually paid the storekeeper corn or butter and the neighbor help at harvest time. Or a farmer who owed a neighbor money gave corn to another neighbor to pay for help that neighbor had given the first one months earlier. Even when farmers brought tobacco or wheat to market, the exchange sometimes involved neither cash nor commercial notes, and the creditor—a neighbor, a gentleman, or a merchant of long acquaintance—willingly granted book credit and took the market risks himself. These exchanges reduced the farmer's need for cash; diminished the possibility that farmers would become heavily indebted; mitigated the impact of recessions, poor weather, and bad crops; and allowed communities to become more self-sufficient.[41]

Daily rituals of reciprocity on farms and in stores stood at the center of the borrowing system. Frolics, barn or house raisings, spinning bees, and harvests were occasions for work and neighborliness, making loans, and repaying debts (each seen as a "gift"). Farmers helped one another at harvest or weeding time and borrowed and loaned food and plows while their wives bartered butter for spinning or cheese for eggs and made country cloth together, as several examples suggest. During the early 1730s, when "he was a little boy [12 or 13 years old] just able to pick up wheat," John Evans, who lived near the Patuxent River in Prince George's County, Maryland, helped at his uncle Samuel White's harvest. "There was an abundance of people present," including kin and neighbors, helping to get in the crop and waiting to hear White relate which of his children would receive which of his slaves. In April 1797 Erkuries Beatty, a prosperous farmer living near Princeton, New Jersey, engaged carpenters to build a barn. He sent his hired hand to find neighbors to help, and the next day "we got it [the barn] up very well without any accident and the cow shed too. . . . —we had 32 people at the raising and gave them all supper, at least twice as many people as we wanted, but I was happy I had so many friends to help me." These friends probably included women (Beatty was unmarried) who cooked the meals.[42]

Farm women played a key role in these rituals, trading among themselves. They borrowed pork, loaned a skein of yarn, tended a birth or a barn raising, and traded (on occasion in their own name) with storekeepers. Martha Ballard's diary suggests the complexity of such female rituals. On September 3, 1788, Ballard, a Maine farm woman and midwife, reported that "Mrs Savage... has spun 40 double skeins for me since April 15"; in return Ballard's daughter Dolly "wove her 7 yds of Diaper," and Ballard "let her have 1 skein of lining warp." As this example suggests, making homespun cloth, a ubiquitous female activity, required cooperation with neighbors, for households rarely owned the sheep, flax, cotton, wool cards, spinning wheels, and looms needed to make cloth. To finish a linen and wool sheet might entail trade among four women: the sheep owner, the flax producer, the spinner, and the weaver. But women traded with men as well, exchanging the butter they made for grain, for example. Matthew Patten regularly got butter from neighboring women, and late-eighteenth-century Delaware Valley women became so proficient at making butter that some of them marketed it in Philadelphia for cash.[43]

By increasing services and goods available to farmers, rural artisans—carpenters, shoemakers, blacksmiths, tailors, and weavers—deepened rituals of mutual obligation. Farmers rarely saw these artisans, who lived next to them in the open country, as a separate class. Middling and rich farmers bought everyday shoes or furniture from local artisans and had their plows sharpened, their horses shod, and their barrels and hogsheads made by a local blacksmith or cooper. Since the vegetables, grain, and butter artisans made rarely fed their families, trade of smithing or carpentry for neighbors' corn or butter benefited craft families, too.

Frontier areas supported few artisans; as the population increased, artisans grew more numerous. Only a tenth of the early settlers of Rowan County, in the North Carolina backcountry, had craft skills, but that proportion had doubled by 1759, a decade after the first colonists arrived. Local merchants or big farmers set up artisans in business and paid them a wage or received rent, a share of the profits, or the right to market their products in exchange. Artisans often worked alone (or with family help) but occasionally hired an apprentice or journeyman. They owned or rented small farms or worked as slave overseers or farm laborers during planting and harvesting season, plying their trade (like Kent County, Maryland, blacksmiths or Connecticut furniture makers) in fall and winter. Nonetheless, artisans (especially in capital-intensive trades such as tanning, smithing, or weaving) often passed on their tools and skills to their sons (or apprenticed them to a neighbor practicing their trade), thereby creating artisan or artisan-farmer dynasties.[44]

Farmers, moreover, dabbled in carpentry or tanning, and their wives spun thread and wove cloth. Half of mid-eighteenth-century Essex County farmers owned craft tools, and nearly three-tenths of householders in Burlington County, New Jersey, and Chester County, Pennsylvania, and three-quarters in Kent County, Maryland (on the Eastern Shore), owned carpentry tools. Two-thirds of planters in the tobacco-growing Chesapeake and one-half in the Low-country owned carpentry tools, and the largest planters employed slave artisans, mostly woodworkers.[45]

A few farmers, such as the Lanes, a family of tanners in mid-eighteenth-century New Hampshire, or the handyman-carpenters of western Connecticut, combined farming with craft, holding each of equal importance. This "union of manufactures and farming," as Tench Coxe, George Washington's assistant secretary of treasury, put it in 1794, was common on Delaware Valley farms, "where parts of almost every day, and a great part of every year, can be spared from the business of the farm," but it could be found as well in New England. Two-fifths of the farmers who died in Middlesex County between 1730 and 1780 had also been employed as craftsmen. The services these farmer-artisans provided were essential in rural areas. As Coxe noted, they "make domestic and farming carriages, implements and utensils, build houses and barns, tan leather, manufacture hats, shoes, hosiery, cabinet-work, and other articles of clothing and furniture, to the great convenience and advantage of the neighbourhood."[46]

Since kinfolk and ethnic groups often clustered together, participants in the borrowing system often shared ties of ethnicity, religion, or kinship. Maryland and Pennsylvania Germans and Long Island Dutch borrowed mostly from one another, even though English colonists lived nearby. Pennsylvania Friends had thick networks of credit and debt that increased in intensity in the 1750s and 1760s. As new generations came of age and seated themselves on land their fathers had acquired, and as children from neighboring households or nearby settlements married one another, dense kinship networks of landed families developed. Cousin marriage, which was common in the Chesapeake colonies and the backcountry, increased the density of these networks. These neighboring kinfolk—parents and adult children, brothers and sisters, aunts and uncles, and cousins and in-laws—strengthened the borrowing system through reciprocal gift giving, lending, and borrowing. An all-encompassing neighborhood borrowing system emerged, its participants linked by neighborliness, religion, ethnicity, and kinship.[47]

Shopkeepers, found in every rural community, played an essential role in the borrowing system. In 1771 there was one storekeeper for every five families in small and middling Massachusetts towns; but half of them carried stock of less than £20, and another third had less than £100. The larger storekeepers, such as the Scots of Southside Virginia, had far more imported merchandise, including numerous kinds of cloth, ribbons, pins, and needles; spices (chocolate, sugar, and pepper) and liquor; and a wide range of tools. Rather than just redistributing local goods, these men improved the standard of living of their customers.

To attract customers, storekeepers gave cash-poor farmers book credit, advancing them money to pay taxes, acquire livestock, or buy cloth or sugar. To retain valued customers, merchants reduced prices, lengthened terms of repayment, or kept their accounts open for years without requiring full repayment. Most merchants (62 percent in court suits for debts in Dutchess County), moreover, charged no explicit interest. Such a system placed a premium on knowledge of the creditworthiness of customers. Merchants participated in the borrowing system by trading the goods they received from one neighbor to pay another neighbor for other goods or services, thus keeping butter, flour, or grain in the community. During the 1730s a Dutchess County storekeeper swapped goods for work (half the time) and farm products (a third of the time) but rarely received cash (just 5 percent of the time). Cash increased in importance during the 1760s, as indicated in the records of a Dutchess storekeeper who received cash from two-fifths of his customers but still accepted farm goods, work, and wood in payment. Storekeepers took local meat, dairy products, and vegetables (such as the shad, turnips, and onions that one Deerfield, Massachusetts, store bought in the mid-eighteenth century) and swapped them for exotic imports (such as Mediterranean raisins and figs and West Indian sugar). Four-fifths of the goods Benjamin Snyder sold in his Ulster County, New York, store from 1774 to 1776 were locally produced.[48]

In staple economies nearly everyone borrowed from storekeepers or merchants to procure manufactured goods until the crops came in. During the early 1770s more than half the householders in Lunenburg County, deep in the Virginia Southside, owed money to Scot storekeepers. Although less than half (46 percent) of those without slaves (three-fifths of the householders) ran up debts at the stores, three-quarters of the small slaveholders (those who owned 1–4 slaves over age sixteen) and four-fifths of the larger slaveholders (who owned five or more slaves) patronized the stores. A third of the slaveholders, moreover, owed money to more than one storekeeper.[49]

The borrowing system helped define the bounds of the community. Neighbors created their own customs of exchange and decided what barn raising or husking bee they should attend, when butter or sugar should be lent, when and how obligations should be settled, and whom to include in or exclude from the rounds of reciprocity. The system thus cemented friendships and incited conflict as neighbors fought with one another over slights real or imagined. For instance, competition over notching logs at a Chester County, Pennsylvania, house raising in 1743 led neighbors David Allen and William Armstrong into mutual insults and fisticuffs. At Erkuries Beatty's barn raising, the relationship between Beatty and his neighbors was reciprocal and equal. Those who showed up knew they had a neighborly responsibility to lend a hand, and Beatty, in turn, understood his obligation to feed them. As members of farming families, they knew they could call on each other again. Every farmer, farmwife, son, daughter, storekeeper, carpenter, miller, or tavern keeper could participate as long as he or she showed signs (by owning land, signing a long-term lease, or living on parents' land) of staying in the community.

Many could not take part in the borrowing system. Gentlemen, great planters, or wealthy merchants might participate, but only on the basis of equality, not as a patron or creditor who demanded farmers display subservience. Rural laborers, who depended on wages for subsistence, had no surpluses to trade. Transients, whether peddlers who trod the countryside with imports they bought in the city and for which they had to receive prompt payment or youths or families tramping a circuit in search of work or lodging, did not stay in a place long enough to reciprocate. Widows, old or sick men, the lame, and orphaned children might receive relief from parish, town, or county; when hired by farmers, justices, or overseers of the poor, they worked for their keep as subordinates, not equals.[50]

Slaves traded produce from the garden plots their masters' granted them, eggs or poultry from the fowl in the quarter yards, fish or shellfish they caught, mats or bowls they had made, and even corn, rice, hogs, and beef with the master, other slaves, servants, white neighbors, or townsfolk. Although slaves (especially in the Lowcountry) devised a borrowing system among themselves, and an occasional slave might have a book credit account with a local storekeeper, nearly all transactions with whites were exchanges for cash or goods, not reciprocal gifts. Slave men in the Carolina Lowcountry most often traded with masters, but several hundred slave women went to Charleston (and Savannah by the 1780s) and sold bread, eggs, fruit, vegetables, berries, and baskets for cash. Similar, but less extensive, exchange could be found in small Chesapeake cities such as Norfolk and Baltimore. Whites who traded with slaves often treated them with contempt, a sentiment the slaves sometimes returned; masters often construed the trade of slaves among themselves and with poor whites as theft of their property.[51]

The borrowing system, with its neighborhood interdependence, nourished the small property holders' ideal of farm independence. No farmer was autonomous, but his participation (and that of his wife) in neighborhood exchanges allowed him to choose when and how he would buy and sell goods at the market. If his family did not make shoes, maybe his neighbor did; if his wife ran low on apples, she might borrow some from a neighbor and return an apple pie or cornbread later. A man who headed such an independent household had the right, he believed, to vote, to have his voice taken seriously, and to have his property and his position as patriarchal head of his family protected.[52]

Nonetheless, local storekeepers linked farmers who rarely sold goods for export indirectly to commodity markets. Tied as they were to colonial wholesalers or English merchants, storekeepers had to pay their debts promptly in bills of exchange or farm goods. To compensate for the months between sale and payment and the inevitable "doubtful" and "desperate" debts, most marked up their prices. Storekeepers inevitably sued farmers for collection of debt, and assemblies established ways to prove the validity of book debt. As the sources of local and commercial credit increased during the first half of the eighteenth century, court suits for debt skyrocketed in places such as Dutchess County; Hartford County, Connecticut; and Richmond County, Virginia. Cases involved neighbors as often as strangers. But creditors faced strong limits on their ability to collect debt. A storekeeper who wished to build business with cash-poor farmers gave customers as much time as they needed to settle debts; in some places, such as Piedmont Virginia, debtors abandoned a merchant who demanded repayment for a nearby competitor, eager for the business. Most debtors settled with creditors informally, underwent mediation, or failed to contest a debt case brought by the creditor (permitting the debt to enter the public record). Court action usually stopped there. If a creditor seized a debtor's property, imprisoned him, or sold him into temporary servitude to pay the debt—all allowed by colonial laws—he ruined his relationship with farmers, destroying his customer base. When pressed by suppliers, creditors did execute writs to seize property, but they rarely had farmers thrown into debtor's prison.[53]

When credit collapsed, wholesalers called in debts owed by mercantile and farm debtors, leading to a flurry of court suits wherever their agents operated. Between 1750 and 1754, when grain prices plummeted and credit contracted in Philadelphia, local merchants called in their debts, and rural land foreclosures skyrocketed, from 18 in 1748–49 (when credit was plentiful) to 119 in 1753–54; a new crisis in 1765 led foreclosures to leap to 250. The international crisis of 1772 had a similar impact in the Chesapeake colonies. Debt cases more than doubled in frontier Pittsylvania County, Virginia, between 1768 and 1774, and the proportion of writs of execution among debt cases rose from just over half to more than three-quarters. At the same time, both the number of debt cases in Prince George's and Somerset Counties (one was a tobacco county; the other, less dependent on the staple) and the number of debtors in Maryland prisons jumped.

When many local or British merchants failed, as they did when international credit broke down, less money entered the economy and fewer goods changed hands. As commercial credit disappeared, the amount of money that merchants and big planters could lend dwindled, and the borrowing system approached collapse. The bankruptcy of a single merchant could wreak similar, if less widespread, havoc. If a storekeeper failed to collect most of his debts and defaulted on Boston or London loans, he could not buy new manufactured goods and would lose his line of credit. Even if he remained in business, he would no longer have money to loan on long terms.[54]

During the mid-eighteenth century, just as in earlier times, farmers sold goods at market, but a minority—located in the staple-producing South and in pockets scattered throughout the North—were fully integrated into a market economy. Small farmers participated in a borrowing system, separate from yet dependent upon commodity markets. Eighteenth-century farmers fell into three groups: staple-producers, who directed their labor toward meeting market demands for tobacco, rice, or grain and who used informal networks as an adjunct; surplus producers, who sent small surpluses to market but depended heavily on nonmarket exchange; and craft farmers, who grew some food but got the rest by trading food for home-manufactured goods or working for other farmers. How a particular farm household related to the market depended not only on the farm's location and the market for its crops, but on the land the farmer owned and the household labor he commanded.[55]

Household Formation

Colonists regarded any form of social organization other than a household, whether communal agriculture (like that found in Plymouth Colony in the early 1620s) or an all-male barracks society (typical in early Virginia), as unnatural. Only households could practice the forms of patriarchy, family discipline, and wifely subordination God mandated. Plymouth Colony abandoned common property and common cultivation in 1623. William Bradford, the colony's leader, explained that the communal system had bred "much confusion and discontent and retard[ed] much employment." Communal ownership and work had led "young men, that were most able and fit for labour and service," to "repine that they should spend their time and strength to work for other men's wives and children without any recompense." Their wives, for their part, deemed being "commanded to do service for other men, as dressing their meat, washing their clothes, etc.," as "a kind of slavery," and "neither could many husbands brook it." In contrast, the division of land "made all hands very industrious."[56]

More than a century and a half later, New Englander Mayo Greenleaf Patch (1766–1818) expected to head such a farm household. But his family was poor and tramped from town to town with little prospect of getting land. He gained respectability only by seducing and marrying a seventeen-year-old girl in 1788.

His father-in-law made him a householder. He gave the couple a small plot of land to use, built them a house, and provided Patch with a shoemaker's shop. Patch repaid his wife's family by gaining control over their land and misusing it. Most husbands were less venal than Patch, but their households structurally resembled his. Free families, who lived, produced, and consumed in households like Patch's, were units of production. Household members—husbands, wives, sons, daughters, hired hands, servants, and slaves—pooled the income they made from farm work, craft labor, and local exchange. Like Patch, husbands controlled all family property, including land and chattels wives inherited. As legal heads of household, husbands made every economic decision, from determining the farm's mix of subsistence and market crops to delegating tasks to wives, children, and servants. Fathers pursued family subsistence and aspired to give their children the resources to make farm households. While affection among family members was common, open or suppressed conflict among husbands, wives, and children (like those of Patch's family) may have been as common.[57]

How did nearly all white men come to expect to be landowning farm householders? Colonial household government grew out of contested English traditions and colonial demographic and economic circumstances. In England patriarchal landed households had diminished in importance; most immigrants had lived in somewhat egalitarian cottager and urban laborer families, where husbands and wives worked for wages. Other immigrants, especially in New England, had lived in more bourgeois homes, where husbands ruled with patriarchal authority but women demanded more autonomy. Colonists got land more readily, married younger, and bore more children than the English, circumstances compatible with patriarchal *and* egalitarian households. Communal norms about family life slowly developed, the result of thousands of compromises between husbands and wives. By the early eighteenth century, a new kind of household had emerged, characterized by subsistence and market production, male field labor and female domestic production, male authority and female subservience, and shared authority over child rearing.[58]

Colonists, as much as the English, connected household formation to marriage and presumed that only men able to support a family could wed. Unmarried men were supposed to be dependents in the homes of others. Before contemplating household formation and marriage, single immigrants who paid their own passage had to find land and build a house; servants had to wait until their terms ended. Little stood in the way of New England immigrants. During the 1630s most women wed in their teens; most men, by twenty-six. But after they completed their terms, Chesapeake servants had to postpone marriage and work as laborers for several years to accumulate money to buy land, a more difficult task when tobacco prices plunged after 1680. Even then many men postponed marriage (until age thirty) because so few women immigrated. Since so many men sought wives, native-born women were forced to marry as young as age sixteen. Unable to find wives, many men set up households as bachelors, but such homes lacked the vegetable gardens and household furnishings a wife could bring to the operation.[59]

When native-born generations came of age, chances for early marriage multiplied. Benjamin Franklin viewed colonial chances to marry as superior to England's and implied that a new kind of productive household developed as a result. He linked abundant land to early marriage and large families. Indians, "having large Tracts," he insisted in the mid-1750s, "were easily prevail'd on to part with Portions of Territory." "Land being thus plenty in America, and so cheap as that a labouring Man, that understands Husbandry, can in a short time save money enough to purchase a Piece of new land sufficient for a Plantation, whereon he may subsist a Family." Since such a landholder was "not afraid to marry," "marriages in America are more general, and more generally early, than in Europe." Marrying so early, American women bore eight children, twice as many as women in Europe. Since land was cheap and plentiful, their adult children readily procured some. But more than "the greater Fecundity of Nature" was involved; men with land provided their children "Examples of Industry . . . and industrious education; by which the Children are better able to provide for themselves, and their marrying early, is encouraged from the Prospect of good Subsistence."[60]

Franklin accurately compared marriage in England and its colonies. Englishmen postponed marriage until they were nearly thirty, after a term of service-in-husbandry and years of wage labor. They usually married after they had acquired a cottage and rights to use a tiny piece of land or secured city employment. Women married at a younger age (twenty-five or twenty-six), but they, too, had often been servants-in-husbandry or had worked as spinners or laborers before marriage. At a time when few people practiced birth control, a woman's first marriage at age twenty-six rather than twenty reduced the number of children in a family by three. Cottagers and laborers must have welcomed such small families, for children could perform little household labor and were therefore more of an economic drain than an asset.[61]

The descendants of English immigrants, who had far more land at their disposal, jumped quickly into marriage. During the seventeenth century native-born New England women married at age twenty and men at twenty-five, much younger than their counterparts in England. By the end of the century similar patterns had appeared in the Chesapeake region and in North Carolina. Pennsylvania's Quaker pioneers were almost as young (the men were twenty-six and a half; the women, twenty-two). As land became more expensive in older areas during the eighteenth century, the age at first marriage climbed to twenty-three for women and twenty-six for men. To care for aged parents, sons of long-lived New Englanders postponed marriage; to look after younger siblings, orphaned sons in the South did the same. But in frontier areas such as backcountry South Carolina, with its abundance of unimproved land, men married at twenty-two and women at nineteen. Despite the pressure to postpone marriage in older areas, enough families moved (or expected to move) to frontiers to keep marriage age low. Early marriage echoed throughout the economy, affecting fertility and family labor, inheritance, and relations between spouses—and these social patterns in turn influenced the marriage pattern.[62]

Colonists soon expected to marry at young ages and refused to let the high land prices in older areas stand in their way. The children of poorer landed householders living in older regions increasingly conceived children before they married. At first officials punished nonmarital sexual activity, hoping to eliminate the practice. The seventeenth-century New England couples (numbering at least 163) who engaged in intercourse (or a suspicion of sexual play) as a prelude to marriage were hauled before the court for fornication or lewd behavior. Chesapeake authorities whipped servants who bore bastards or, more commonly, added years to their terms. Such communal control disappeared in the eighteenth century. Although only about one in fifteen free seventeenth-century women were pregnant on their wedding day, Chesapeake servant girls often established sexual relationships before their terms expired. (At least a fifth of the female immigrant servants who came to Charles County, Maryland, between 1658 and 1705 bore an illegitimate child.) By the pre-Revolutionary era the proportion of women pregnant on their wedding day had jumped to more than one of every four. During the eighteenth century, premarital intercourse was usually a prelude to marriage and perhaps a way to gain parental approval for the match. Not a single Quaker woman bore an illegitimate child; unmarried women gave birth to only 1 percent of the children born in Middlesex County, Massachusetts, in 1764.[63]

Early marriage and household formation came at the end of a complex but hardly inevitable process. The millions of acres settlers conquered allowed men to marry young. But the high ratios of land to labor that ready access to land created could have led colonial rulers to impose serfdom on the rest of the populace. No matter their class, however, colonists demanded personal freedom. Serfdom and feudalism had so withered away in England that no one turned servants into serfs, and every attempt to impose feudalism in the colonies failed. Colonial household formation was predicated upon plentiful land and this expectation of freedom. As long as land remained abundant, marriage and household formation continued on a generational cycle. The conjugal couple bore many children, used their labor to build capital in land, and repaid them for their efforts with a dowry (of linen or livestock) or inheritance that (along with work on nearby farms) allowed them to marry, find land on a new frontier, and begin the cycle again.[64]

Married women bore children every other year throughout their twenties and thirties, giving birth to seven or eight children if they remained married through age forty-five. Marital fertility was low in the seventeenth-century Chesapeake region and—after an increase toward the end of the century—probably declined again. It stayed high nearly everywhere else. Few families—except possibly Quakers and scattered families in New England in the late seventeenth century (when the region was plagued by depression) and the late eighteenth century (after land supplies had dropped)—practiced birth control. These high levels of fertility meant that most children lived with many siblings and most middle-aged men headed large households. Late eighteenth-century households had three or four children; young couples had one or two children. Even though some children died in infancy, by the time a couple reached their early forties, five or six children lived at home; thereafter, the number declined as children set out on their own.[65]

Although household heads often bought an indentured servant, employed wage labor, or owned a slave, husband and wife and their children were the core of the household. With most of their kinfolk in England, New England immigrant households rarely included adult siblings or a widowed grandparent. In the Chesapeake region in the seventeenth century, where adults married late and died young, widows and widowers usually remarried, creating families that included stepchildren and half-siblings as well as their own offspring. By the early eighteenth century nuclear family households, with no kin other than the married couple and their children, predominated everywhere, but extended households (with other kin) began to appear. Seven of ten Guilford, Connecticut, households in 1732 and six of ten in Prince George's County, Maryland, in 1776 (four-fifths when units with stepchildren, boarders, or unrelated orphans are included) were nuclear in structure. Extended households were most common in New England. In 1764 more than three of ten Massachusetts farm families shared houses, and most of them were probably related. At particular times in their life cycles, individuals often lived with kin. Newlyweds might board with parents long enough to begin a family, thereby creating a three-generation household; a child (and her family) might live with parents as they aged; or an aged parent might move into a child's house. As many as four of five people over sixty-five in Prince George's County in 1776 lived with one of their children.[66]

Farmers viewed household members not as individuals but as parts of an organic whole, each with assigned tasks. Productive relations inside the household combined with the legal subservience of wives and children, long marriages, and the obedience children owed fathers and servants and slaves owed masters to create a system of patriarchal family government. In return for this willing subordination, patriarchal theory insisted, the husband-father-master had to lead his family in prayers, feed and clothe them, educate his children and train them in farming, and provide marital portions when they married. If husband, wife, and children maintained this familial hierarchy, the state allowed the husband-father-master to direct and correct household members without interference. Patriarchal theory granted joint authority to father and mother over children but insisted that fathers make all important decisions, especially those concerning sons.[67]

The timing of the transition to domestic patriarchalism depended on demographic realities. In New England the abundant acres early town proprietors owned combined with high adult life expectancy, long marriages, and large families to increase paternal authority and instill the practice of patriarchal ideals from the outset of settlement. Law and state action sustained New England patriarchs. Town fathers placed orphans with families whose heads could act as patriarchs; law mandated that children receive paternal permission before marriage and decreed (but never enforced) the death penalty for children who assaulted a parent; magistrates punished wives (and husbands) when patriarchal order broke down, even ordering fathers to whip recalcitrant children or fining drunkards who failed to support their families.[68]

A similar process was completed in the Delaware Valley by the early eighteenth century, but it took much longer in the Chesapeake region for patriarchal theory, much less practice, to emerge. High death rates of seventeenth-century immigrants reduced the security couples needed to maintain the reciprocal obligations of female obedience and male support essential for patriarchy. When immigration declined and adult life expectancy improved early in the eighteenth century, the length of marriages rose to twenty years or more, and family sizes grew. Freeholders could then form patriarchal families. The spread of slaveholding, with its mandated absolute dependence, further strengthened patriarchal ideology in the South. By 1774 patriarchal households could be formed in every colony. In that year dependents of male household heads—wives, children, servants, and slaves—constituted an astonishing four-fifths of the population and more than half the adults.[69]

Men gained independence when they formed households and married; women exchanged subservience to fathers with dependence on husbands when they wed. Soon after settlement, all colonies adopted laws that subsumed the legal identity of a wife into that of her husband. Husbands governed their wives, had the right to demand access to their bodies, and by custom could discipline them. Married women lost the right to make contracts or sue or be sued unless they brought suit with their husbands. (Until 1665 New England wives could and did go to court themselves in slander cases.) Unless protected by a prenuptial trust (marriage settlement), a wife turned over control of all the personal goods, slaves, and land she brought into marriage to her husband during his lifetime. Marriage settlements, usually made by a rich widow before remarriage, safeguarded her control of her property and her children's inheritance during marriage. Wives without a marriage settlement retained only the right to veto sale of land they owned or would inherit from their husbands as their dower right, and husbands often disregarded that right. In return for legal submission, wives expected husbands to support them and had the dower right to one-third of a husband's property upon his death. This common law of coverture enshrined the dominance of husbands over their wives. Only femme soles—women who never married and widows—had property rights.[70]

These points should not be misunderstood. Female subservience was socially mandated, but the relationship between spouses was reciprocal, if unequal. Middling and rich farmwives increasingly expected companionate marriages based on love and respect as much as economic necessity. As New Englander Benjamin Wadsworth put it in 1712, husbands and wives should not only "cohabit and dwell together" but "have a great and tender love and affection toward one another" and "be patient one toward another." Husbands owed wives love, fidelity, and economic support and should refrain from disciplining them excessively. As Wadsworth argued, "Wives submit yourselves to your own Husbands, be in subjection to them," but husbands should discipline their wives "lovingly and kindly" and not "curse them, reproach, defame and belie them, beat and strike them."[71]

A farmwife, moreover, functioned as a "deputy husband"; she gained autonomy in kitchen and garden, shared authority over children, and sometimes represented him in economic matters. She had to learn enough about farming to take over if her husband deserted her or died before a son reached adulthood. Despite legal prohibitions, wives occasionally hired day laborers, bought servants, made bargains, or paid debts without the guidance of husbands, and (as we have seen) farmwives fully participated, on their own, in local borrowing systems. Widows and some deserted wives regained the legal rights of a femme sole, and widows frequently were sole executors of their husbands' estates; older widows rarely remarried and often ran successful farms or other enterprises.[72]

The practice of kin naming suggests that the identities of children were subsumed under those of their parents—especially their fathers. Parents in some New England villages—Hingham, Massachusetts, and Windsor, Connecticut, for instance—named first children after themselves more often than others in the lineage and often named a child after a deceased sibling, practices that placed children strongly within conjugal families. Gradually parents dropped the custom of kin naming, thus accentuating individuality over kin identification. But in Plymouth colony, New York, and the Chesapeake, parents generally named children after grandparents, aunts, or uncles as often as after themselves, suggesting identity with lineage as well as the conjugal family.[73]

Children worked on the farm until they married, and by custom and law they could wed only with parental permission; servants required the permission of their masters. The Society of Friends went further and insisted that both the family *and* the meeting grant permission. The sequence of marriages among a couple's children in Hingham, Massachusetts, shows the ubiquity of paternal control there. If children had a free choice, one would expect to find no relationship between birth order and marriage age or between marriage age and the father's longevity. But Hingham daughters rarely married out of birth order, and sons whose fathers lived into old age married one or two years later than sons whose fathers died young. The preservation of family property, moreover, led children to marry cousins, other blood kindred, or in-laws, a common practice in the Chesapeake colonies, among the backcountry Scots-Irish, and perhaps in New England.[74]

The political link between authoritarian father and absolute monarch that sustained patriarchal families in Europe never developed in the colonies. In return for obedience to state and monarch, European states upheld the absolute power of the father to run his family and discipline its members. Few English Lords emigrated; colonial patriarchs were landowning men heading autonomous households; they participated in government equally (in a legal sense) with their betters. After 1640 the English monarch, constrained by Parliament, provided neither theoretical nor practical legitimacy for patriarchs. Attempts by early New England magistrates to enforce the link between the state and patriarchal order (for instance, comparing adultery or murder of a husband to treason) invariably failed. Lacking this theoretical link to an all-powerful state, patriarchal theory (not to mention practice) constantly faced disintegration. The central contradiction of colonial farm families—that the husband's independence was sustained by the labor of wives and children at the same time that family members had to show due subordination to him—not only fed familial struggles but threatened to destroy patriarchy when wives or children rebelled against male authority.[75]

We will never know how often farm husbands and wives fought, for few such disputes were recorded. When a wife withheld obedience or (worse) assaulted her husband, he had the right to beat her. Some men would abide no challenge to this right. In the mid-1660s, when a local magistrate in York County, Maine, berated a man for "sleighting & abusing" his wife, the man replied indignantly, "What hath any man to do with it, have I not power to Correct my owne wife?" In seventeenth-century New England, where Puritan magistrates valued family harmony, authorities prosecuted 128 husbands for abusing their wives for such sins as refusing to feed a pig. Eighteenth-century New England women seeking a divorce often cited cruelty as a legal grounds. A wife's show of independence or a husband's tyrannical behavior must have triggered most marital arguments. Hannah Heaton, an eighteenth-century Connecticut farmwife, joined a revivalistic congregation and spent a day away from home in church, much to her husband's disgust; he fought repeatedly with her and once met her at the door by saying, "I hoped you was dead and if it was so ide yoke up my team and sled you to hell." Jane and Jeremiah Pattison, a middling planter family in early eighteenth-century Calvert County, Maryland, bickered over the use of the property of Jane Pattison's first husband. But arguments soon escalated into violence, with Pattison "beating her with Tongs" and turning "her out of Doors . . . destitute of Cloaths and almost naked." On occasion such wife beating led to the wife's death, as in one 1733 Pennsylvania case when an enraged husband killed his wife with a stale loaf of bread.[76]

A physically abused farmwife had few options. She might return abuse for abuse (as did fifty-seven wives in seventeenth-century New England) or refuse to work or wait on her husband, but she risked continued abuse or bodily harm. After much mistreatment, she might show her bruises to her servants, neighbors, or pastor, but few wished to involve themselves in domestic disputes. She might go to court to stop the abuse, but courts often sided with husbands and masters. The Accomac County, Virginia, court did when confronted in 1668–69 with charges against rich gentleman Henry Smith, whose wife, Joanna, accused him of violence and whose female servants accused him of violence and rape. An abused wife might try to conform to her husband's demands, but she did so only by giving up those shreds of independence she had earned through her rebellion. Elizabeth Perkins of Willingburg, New Jersey, put up with "the most notorious scenes and disorder" and "often met with treatment which would have shocked a savage of the Ohio." She might run away and turn to her family or neighbors for support, depriving her husband of her labor, as Perkins did in 1767. But as unhappy eighteenth-century Nantucket wife Abigail Gardiner Drew knew, if she ran away, she would lose use of "that property I have endured so many hardships to obtain" and be forced "once more in to the wide World bereft of Interest and friends." Since her labor was crucial to the farm, running away or refusing to work might, at best, lead to a domestic truce. Nearly all unhappy couples did stay together, struggling daily to support themselves and their children.[77]

Legally tied to their husbands and with little recourse to law in the face of violence, unhappy wives separated from husbands with difficulty. Desertion was the most common way for women to end a farm marriage. Informal severing of the marriage had to suffice, for courts and legislatures rarely granted divorce (in New England) or allowed legal separation (elsewhere). Even in Connecticut, the most liberal colony, where 839 couples gained divorces between 1750 and 1797, the divorce rate remained low through the 1770s, at 0.062 to 0.081 per 1,000 per year. Most divorcing couples were members of the mercantile, trade, gentry, or artisan classes. A handful of New England farmwives or husbands sued for divorce each year. At most three-tenths of the 229 divorce petitions in eighteenth-century Massachusetts came from farm families, less than half their representation in the colony's population. Knowing that poverty awaited them and their children if they succeeded, Massachusetts women waited at least four years after marital trouble began to file for divorce (men, in contrast, waited a year). Nevertheless, farmwives used defenses in separation and divorce suits that often resulted in favorable decisions. They complained that husbands committed adultery, deserted them, refused to support them, or beat them without mercy—grounds singly or together for divorce or separation in one or more colonies. They couched their petitions in a servile language of subordination that belied their quest for independence. Jane Pattison insisted that she had been "a very dutiful good wife"; Sarah Whelsher told the New Haven County court in 1723 a woeful tale of her husband's reckless spending, adultery, and desertion that left her and "his poor Children" to "Ignomie, reproach, shame, and poverty."[78]

Farm husbands and wives almost always remained civil enough to maintain good household order. Every year each farmer had to decide what crops to grow, what craft activities to pursue, where and how to trade, and how to divide household labor. The range of possibilities was wide, from striving for self-sufficiency to producing almost exclusively for markets. Seventeenth-century farmers searched for a staple crop, neglecting to a degree subsistence production. By the eighteenth century, small farmers, even tobacco or rice planters, mixed subsistence and market production, making food and clothing for the family while selling surpluses to neighbors or merchants. These strategies aimed at the perpetuation of the farm business into the next generation. Husbands and wives shared this goal, even when they struggled over how to implement it.[79]

Small farm households, no matter the region or crop mix, pursued investment strategies that changed little over the colonial era. Before starting a farm, the couple needed bedding, cooking and eating utensils, linens, a table and chairs—goods often part of a dowry or an inheritance. Although the couple gradually acquired more household goods, investments in land, bound labor, and especially livestock took precedence over comfort.[80] Ex-servants and laborers in the seventeenth-century Chesapeake began to buy livestock even before they formed households. As soon as a family set up a farm, they got a cow or two, a few swine, and some dunghill fowl. Cattle were scarce in the early seventeenth century; but by mid-century they abounded. Herds of cows grazed the open range and hardy swine lived off the forests. Middle-colony and southern farmers each acquired a horse; New Englanders bought oxen. In the seventeenth-century Chesapeake, fathers, godparents, and other relatives gave children livestock as gifts; in all colonies daughters of middling and sometimes poor men got livestock as part of their dowries. Ownership of these animals did not assure subsistence, but at least a family could plow their fields and put eggs, milk, butter, and meat on the table. Next, the farm family cleared land and planted subsistence and market crops. With crops in the ground, they could expect to pay off the debts they had accrued in buying livestock, and they looked forward to gathering food for the following winter and having a small surplus to trade with neighbors for the many foodstuffs they could not grow.[81]

Having acquired the means of subsistence, farm families sought the security that only landownership could bring. Lucky newlyweds received land as an inheritance, gift, or dowry, but most had to build capital or wait for an inheritance. Opportunities to get land remained good in the seventeenth century. Men in Windsor, Connecticut, got land (most by inheritance or gift) on average before they reached thirty. Land acquisition was a high priority for poor immigrants. In seventeenth-century Maryland most former male servants worth less than £50—which usually meant they had some furnishings, kitchen goods, or livestock—owned land. As land prices rose in the eighteenth century, opportunities for landownership dropped, and in 1774 less than one-third of householders worth less than £100 held land. Once bought, land was the costliest item in most farmers' portfolios. A third of the wealth of all householders in seventeenth-century Maryland was held in land; the value of landed property in southern New England rose from about one-third of farm assets to nearly two-thirds between the 1630s and 1760s. In 1774 more than half (56 percent) of the wealth of middling colonial landowners (those worth £100 to £399) was held in land.[82]

Thus although small southern planters (but many fewer northern farmers) bought servants or slaves, a big majority of the capital-producing assets most farmers owned was held in land and livestock. Farmers added to their holdings through their fifties, buying or inheriting more livestock and land. But when their children reached adulthood, they began reducing their wealth by disbursing their assets as dowries or gifts. In 1774 the wealth of men in New England and the mid-Atlantic colonies grew from about £90 for men in their twenties to £650 for men in their late fifties, before slowly declining. The nonlanded wealth of young seventeenth-century Maryland fathers was £161; men with grown children had property worth £238; men over sixty lost wealth. In late eighteenth-century Hingham, Massachusetts, men in their twenties owned real estate worth £188, but that average jumped to £1,228 for men in their seventies.[83]

Since children of rich men had more opportunities to get property, inequality among free men appeared in farming communities during the seventeenth century and increased thereafter. The richest tenth of seventeenth-century New England and Long Island farmers owned one-fifth to one-third of the wealth of their communities, but the richest tenth of southern planters owned one-half. In the eighteenth-century northern colonies, the richest tenth of rural householders—merchants and gentleman farmers—owned one-third to one-half of the community's wealth; in the slave South the richest tenth of great planters owned two-thirds or more. The economic growth of the mid-seventeenth and mid-eighteenth centuries combined with a more gradual increase in inequality allowed middling landowning farmers to live more comfortably than their seventeenth century ancestors. Although small farmers and planters could not hope to acquire riches, they gained enough wealth over their lives to eventually attain the wealth of their fathers.[84]

Nothing reduced the chances that a father could pass on his farm to his children more than his early death, common throughout the colonial era but epidemic in the seventeenth-century South. In building a farm the family may have accumulated more debts than assets, and the land and livestock left to the family was often too little to support the family, much less to provide an adequate inheritance for each child. A younger widow therefore often remarried, potentially allowing her new husband to waste the estate or abuse her children. But colonial orphans' courts regulated guardians closely; venal stepparents rarely prevailed, and surviving children usually collected the small sums due them. Surviving children of poor deceased fathers whose mothers remained widows were often bound out to learn a trade or housewifery; orphans lived with relatives or, more often, were bound out until adulthood.[85]

Since farmwives usually married older men, they lost their spouses more often than farm husbands. As a femme sole, a widow regained the property rights she had lost at marriage. If her husband died without writing a will, her dower rights guaranteed her a life interest in one-third of her husband's land and ownership of a third of his movable property. Most landed men made wills, and they gradually reduced the assets they left wives. In the Chesapeake colonies and South Carolina, most men gave wives more than their thirds, often granting all their personal property and land for life. In older Chesapeake counties the proportion of men bequeathing more than dower fell from two-thirds in the late seventeenth century to one-half by the 1770s, but three-quarters of those who lived in counties founded in the eighteenth century continued to do so until the Revolution, albeit limiting their interest in property to the term of their widowhood. Northern farmers allowed widows to take their thirds or gave them less; when they gave more, they usually restricted it to widowhood or the minority of the children. In two Connecticut towns the proportion of men granting wives more than their dower share of land dropped from 15 percent in the 1750s to 9 percent in the 1770s; during the 1760s and 1770s in rural Massachusetts, a quarter of rich men (over £750) with grown sons bequeathed wives less than dower, but only a tenth of the poorest farmers (less than £150) followed suit. In Bucks County, Pennsylvania, the proportion allowing widows more than dower dropped from two-thirds in 1685–1730 to two-fifths in 1731–56.[86]

Widows found running farms difficult. Their husbands' creditors often took them to court and won assets from the estate. At the same time, a widow could go to court to sue for money owed her husband. After she paid her husband's debts, collected what was owed him, and distributed bequests to adult children, the value of her property declined by a quarter (and by two-thirds in New England); only among the New York Dutch, who withheld inheritances from adult children until the widow's death or remarriage, did women control a greater portion of the family holdings. Many seventeenth-century widows fell into destitution; eighteenth-century widows controlled at most one-ninth and usually one-twentieth of colonial wealth. Moreover, courts (especially in early New England) sometimes tried to supervise the widow's finances. Only rich widows succeeded in hiring workers and managing the family's farm, opening a tavern, or operating a mill; the rest remarried or depended on their children. In eighteenth-century rural Philadelphia County, widows were 2 percent of the taxed population, and less than a sixth of them stayed on the tax rolls for eight years. Widows with young children commonly remarried. Two-thirds of the women widowed in King Philip's War, most of whom were young, remarried, as did two-fifths of the women widowed under age fifty in seventeenth-century Wethersfield, Connecticut. At least three-quarters of seventeenth- and early-eighteenth-century Middlesex County, Virginia, widows under age fifty and half of those under forty in eighteenth-century Hingham, Massachusetts, also remarried.[87]

Widows with adult sons neither remarried nor ran farms; instead they lived in their children's homes or depended on their sons' labor for their food. A growing portion of Pennsylvania and New England farmers bequeathed widows the use of a room or house and obligated their sons to provide her with provisions for subsistence and sale. In Chester County, Pennsylvania, the proportion of widows who got a house and provisions rose from a quarter in the 1750s to a third in the 1760s and 1770s; in 1765–71 a fifth of Massachusetts widows with adult children obtained similar inheritances. This was less common in the Chesapeake region, but two-thirds of widows with only adult children (versus two-fifths who had only minor children) were given less than their dower right in Lowcountry South Carolina, a prescription for dependence. A son required to provision his mother might run the farm cooperatively, but the arrangement could lead to conflict. Sarah Brooke of Calvert County, Maryland, complained in the 1750s that her son Roger cared for her reluctantly, and that she had "no Body to do anything for me not so much as to bring me a Coal of Fire to light my Pipe."[88]

Although young men could easily get cheap land, gifts and bequests of land from fathers to sons and (less often) daughters often made the difference between success and failure. An implicit contract regulated such transfers. Fathers used their children's unpaid labor to improve the farms and kept ownership of most land until they died, when they gave it to widows and adult children. Most allowed married children to use their land, but only rich farmers gave them land at marriage. (Some, like increasing numbers of eighteenth-century Essex County, Massachusetts, farmers gave them land a few years after marriage.) This strategy preserved farmers' independence into old age but perpetuated their sons' dependence, even into middle age. Paternal attempts to control adult sons, especially in New England, led to intergenerational strife and propelled sons to migrate to frontier lands to establish their independence as adults.[89]

Nonetheless, the family goal of maintaining the farm over generations and of helping each child get a farm remained paramount. Despite the common-law rule of primogeniture among intestates, which prevailed in the southern colonies and New York, and laws in New England, Pennsylvania, and New Jersey, which mandated that the eldest son get a double share, fathers who wrote wills usually subdivided their land. Seventy percent of sons in Bucks County, Pennsylvania, got land, as did four-fifths or more in two Connecticut towns and three-fifths in four eighteenth-century Maryland counties. Three-quarters of the German settlers in Augusta County, Virginia, divided property (including land) equally among sons. How fathers divided land among their sons depended on their wealth. Small landholders, who had just enough land to make one farm, tended to give it to one son or liquidate it and divide the proceeds. Middling and rich landholders divided their land among most or all of their sons, sometimes giving the eldest a larger share. As land became more expensive and farms became smaller in Maryland, primogeniture rose from less than a tenth in the seventeenth and early eighteenth centuries to one-sixth in the mid-eighteenth; it probably increased in coastal Virginia as well.

Larger landholders or those with few or no sons bequeathed land to daughters. In four late-seventeenth-century Chesapeake counties, with their high mortality and small families, between one-third and one-half of daughters got land; once family sizes grew, the proportion dropped to only one-tenth to one-fifth. Similarly, one-fifth of daughters in Bucks County acquired land from fathers. Two eighteenth-century Connecticut towns were exceptions: three-fifths of daughters there received land, though most of them got far less than their brothers.[90]

Northern intestacy laws gave eldest sons a double share of personal property (livestock, slaves, and household goods) and other children a single share; southern law mandated equal division among children. Will-writing fathers typically divided other property among all children, giving equal shares to each child or bequeathing eldest sons a double share. Daughters often received a different kind of inheritance. In eighteenth-century Bucks County, two-fifths of daughters were bequeathed cash exclusively, three times the proportion of sons. More than two-thirds of sons but only one-half of daughters inherited cash and tangible property. Massachusetts fathers gave household items to daughters, tools to sons, and livestock to all children. Chesapeake daughters increasingly received slaves rather than land. In Charles County, Maryland, daughters got more slaves than sons did; they usually received slave women while their brothers received slave men.[91]

This bilateral inheritance system accentuated the importance of marriage. A man could not easily build a farm on his own. Husband and wife stocked a farm together; each brought significant property into the marriage or could expect to receive it when their fathers died. Men added land, livestock, and slaves to their wives' household goods, livestock, money, and slaves. By giving daughters property, fathers permitted much of it to leave the male line, adding to the holdings of sons-in-law and building farms for other families. Over time, however, this practice helped create bilateral kinship networks.

The rapid growth of the colonial population, the result of early marriage and high fertility, filled the land with farms but reduced the number of acres available to each. With this population pressure on land, land prices rose so high that fathers bought too little of it to give every son a working farm. To overcome this problem, farmers either gave all or most of their land to a single son or sold their holdings and moved to a frontier, where they could buy cheap, abundant land for each son. Others stayed put but turned to provisioning growing seaports and wartime armies or placed their sons in prosperous artisan trades. Farm intensification, job diversification, inheritance, and migration thus perpetuated a small farm economy and led to the rapid settlement of piedmont and mountain regions.[92]

Rising land prices, combined with the gradual reduction in the price of imported manufactured goods, led middling farmers to improve their houses and buy a wider variety of consumer goods. Like their seventeenth-century ancestors, most farm families lived in one- or two-room houses, each room serving many purposes—cooking, dining, sleeping. Some middling farm families built larger houses with three or more rooms spread over two floors. These houses, with a hall, a parlor, a kitchen, and perhaps a chamber (bedroom) contained more varied furniture and allowed greater room specialization. Five examples suggest the extent of change. In the richest grain-surplus regions of Massachusetts, half or more of the farmers lived in two-story houses by the end of the century; even in poorer Worcester County, farmers lived in houses ranging from 600 to 1,200 square feet (with two to five rooms). In 1798 the log homes of Germans in Lancaster County, Pennsylvania, averaged over 1,000 square feet. That same year the wealthiest fifth of families in frontier Mifflin County, Pennsylvania, owned two-story houses of over 1,000 square feet; the homes of the next fifth averaged about 600 square feet. By the late 1760s a third of the tenants living on proprietary manors in tobacco-growing parts of Maryland cooked in separate kitchens. Mid-eighteenth-century Virginia planters still lived in tiny wooden, log, or frame houses, with at most four rooms, of no more than 600 square feet, but most (like their late-seventeenth-century ancestors) continued to build kitchens, tobacco houses, and perhaps several other dependencies.[93]

Increased consumption of imported consumer goods before the Revolution, prompted by the falling prices of cloth and consumer durables, made the house more central in the lives of farm families, even if it remained a tiny, impermanent structure. Middling seventeenth-century farmers ate simple meals with wooden bowls and spoons and pewter plates. Sitting on chairs at a table, their mid-eighteenth-century descendants ate fancy meals and drank tea or coffee with family and friends, using knives and forks and ceramic plates, bowls, and tea and coffee services. These implements appeared in wealthy homes by mid-century and in a big minority of the houses of middling farmers by the 1770s. As the variety of cloth increased and its quality improved, the sartorial standards of farm folk grew. Ownership of religious books had always been common in New England but spread rapidly around mid-century among middling Chesapeake planters. A few farmers and planters, especially but not exclusively the rich, indulged their tastes for luxury goods such as watches, clocks, wigs, and silver.[94]

Rising consumption helped divide rural society into sharply delineated classes. Rich seventeenth-century families owned more land, livestock, and slaves than their poorer neighbors, but they lived in similar housing and bought similar consumer goods. The richest eighteenth-century colonists built ornate mansions, bought fine china, and wore elegant clothing, making themselves into gentlemen and ladies; they entertained in genteel parlors and provided hospitality to refined travelers. The household amenities middling farmers and farmwives bought made their houses more comfortable and enhanced female-dominated ceremonies surrounding meals, allowing farm women to become arbiters of diet and custom. However, the modest purchases of middling farm families hardly presaged the start of a consumer society, much less a consumer revolution. Mid-eighteenth-century farmers spent proportionately *less* of their income on imported manufactures and exotic foods than their grandparents had. Like their ancestors, they sought a competency (albeit extended to include tea and mirrors) and used capital they accumulated to buy land, livestock, and labor.[95]

Eighteenth-century poor rural laborers, tenants, and land-poor farmers could afford few consumer goods and lived much like their ancestors, nor did every middling farmer participate in this rising consumption. Only half the householders in Newtown and Woodbury, Connecticut, where numerous furniture makers lived, owned a chest with drawers in the 1760s and 1770s, and at most a quarter had desks or clocks. During the 1760s in Burlington County three-fifths of prosperous farmers but just over one-quarter of middling farmers and one-fifth of poor farmers owned fine ceramics. Small slaveholders (in places such as St. Mary's County or the Shenandoah Valley) often drank tea but rarely owned tea equipment, and most had no fine ceramics but used pewter instead. Frontier families in Bedford County, Virginia, often lacked chairs and tables, much less owned fine ceramics. They owned knives and forks, drank tea, and consumed sugar, but they drank out of wooden or pewter vessels, not ceramic teacups.[96]

Small farmer households looked quite different on the eve of the Revolution than they had a century earlier. The family consumed a greater variety of food; owned more manufactured ceramics, knives and forks, and other goods; and lived in larger houses than their great-grandparents. Just as important, farmers had invented and sustained new traditions of lineage and inheritance, landownership, market participation, subsistence production, and division of labor between husbands and wives. Middling farm families prospered if they followed the social rules these new traditions mandated. Perhaps the most important, and most contested, social rules concerned the ways husbands, wives, children, and hired hands were supposed to work together for the greater good of the family.

Family Labor and Hired Hands

Labor, not land, was the most precious commodity on early American farms. Holding much unimproved land but lacking all but primitive hand tools, farmers could not subsist without help. Labor outside the family remained scarce in the North and expensive in the South. Northern farmers hired casual labor, got help from neighbors, or occasionally leased an acre or two to a cottager who worked in return. Some seventeenth-century Chesapeake planters bought servants, and eighteenth-century small southern planters often owned a slave or two. Families whose children were too young to work sometimes employed a hired hand by the year. But most farmers relied exclusively on their wives and children for daily work. Although each household member—husband, wife, son, daughter, hired hand, or servant—had a prescribed role, the precise work each performed was the subject of negotiation and even conflict.[97]

Although some tasks, such as milking cows or preparing meals, required daily labor, most farmers paid closer attention to the seasonal rhythms of the crops they grew when allocating labor. The number of growing days dropped from 200 to 160 as one moved north and west, but all farmers and planters (save those in the Lowcountry) confronted a similar annual cycle of work and leisure. The year began for farmers in early spring, when they plowed fields, sowed seeds, planted grain and flax, and started vegetable gardens. Wheat required no more labor until harvest, but most crops needed careful nurturing. In April or May farmers transplanted corn and tobacco, and they spent the summer keeping the fields free of weeds. The harvest began in July (for winter wheat), continued through late summer, and was over by November, after hogs had been butchered, hay mowed, winter wheat sowed, corn husked, vegetables stored, and tobacco hung to dry. Seventeenth-century Maryland tobacco planters spent three-quarters of their working days from April through November making tobacco and corn; by the eighteenth century these crops required nearly nine-tenths of workdays. A festive and relaxing season (November in Massachusetts and New Jersey, and December or January in Maryland and Virginia) followed the harvest. Farm families spent their time repairing farm buildings, cutting firewood, caring for livestock, and attending holiday festivities and weddings.[98]

The division of labor on seventeenth-century farms reflected English gender conventions and colonial conditions. With so much land to be cleared and so many farmhouses to be built, families struggled for survival and subsistence. The niceties of bourgeois life and home manufacture—cooking elaborate meals, washing large wardrobes, spinning or weaving, and making candles—fell by the wayside. Hard labor—felling trees and making corn, wheat, or tobacco—took precedence. At first, families paid little attention to livestock, setting hogs on the woods and using cattle for meat more than for milk or butter, but soon women turned to dairying, much like their English sisters. Masters put servants and slaves to work making crops, no matter their sex. Wives everywhere helped husbands hoe and harvest corn, reducing time they could spend caring for children or cultivating gardens. Wives of small Chesapeake planters joined husbands making tobacco, their burdens sometimes lightened by indentured servants.[99]

Eighteenth-century farm men spent most of their working hours in the daily rhythms of sawing wood and herding livestock and in the seasonal cycles of slaughtering livestock, preparing fields for crops, mowing grass for hay, and sowing, harrowing, planting, transplanting, weeding, and harvesting corn, wheat, rice, and tobacco. They cleared forest land for crops, an annual activity everywhere and a pressing task in frontier areas. Because they were always short of labor but often had much unimproved land, farmers universally used land without fertilizer until it no longer produced decent crops; then they moved to another piece of land and let the old field lie fallow for several decades. Farm husbands themselves labored in the fields, and they supervised the work of sons, servants, slaves, and seasonal laborers. They used their spare hours in craft employments, visiting friends, or selling and buying goods.[100]

No eighteenth-century farm could succeed without the labor of farmwives. A 1782 poem by Ruth Belnap of Dover, New Hampshire, shows the intensity of the labor farmwives performed every day. As soon as she awoke, she must "haste to milk my lowing cow. / But, Oh!," she adds, "it makes my heart to ake, / I have no bread till I can bake." After milking and baking, she turns to churning butter, swilling hogs ("for hogs must eat or men will starve"), spinning yarn, and cooking dinner. Through the summer, "I toil & sweat, / Blister my hands, and scold & fret," probably in the vegetable garden; at harvest time, "Corn must be husk'd, and Pork be kill'd."[101]

Since eighteenth-century farm men did field labor, their wives spent more time "raising small stock, dairying, marketing, combing, carding, spinning, knitting, sewing, pickling, preserving," as well as tending fowl, gathering fruit, and cultivating vegetable gardens. New England gardens became more extensive as rising numbers of wives grew beans, potatoes, onions, and carrots, boosting the quantity and variety of food on the family's table while creating small surpluses. Twice-daily milking during the forty weeks a year cows gave milk, picking fruit, and annual planting, cultivating, and harvesting of vegetables were as tiring and time consuming as field labor. New England farmwives churned most of their butter in the summer but made most cheese in November, after the harvest. When demand for livestock and dairy products grew during the 1760s and 1770s, wives milked more cows and made more butter. At the same time, Delaware Valley wives established intricate local trade networks with other women, selling eggs they gathered and butter and cheese they churned.[102]

Middling farm women, like their English counterparts, were not supposed to plow fields, cultivate field crops, or bring in the harvest. Early-eighteenth-century tobacco colonies prohibited free white women from tending tobacco, but to ensure their family's survival, wives and widows of poorer planters ignored the law. The spread of slaves throughout the population, agricultural diversification, and higher tobacco prices reduced the number of tobacco workers by mid-century. The adoption of oxen in late-seventeenth-century New England eliminated most women's field labor there. In the early eighteenth century, middling northern farmwives joined husbands at wheat harvest time, but as scythes and cradles came into common use later in the century, women no longer mowed grain or grass. They did help reap and bind grain, however. Daughters and wives of cottagers and poorer, younger, and frontier farmers worked in the fields, but field labor by females, including unmarried women, in older regions, declined by the end of the century. Only a quarter of seventy day laborers hired on one Delaware Valley farm in the first decade of the nineteenth century were women. But the "wives and daughters of German women" in Pennsylvania, Benjamin Rush discovered in 1789, still "frequently forsake, for a while, their dairy and spinning and join their husbands and brothers" in the labor of "cutting down, collecting, and bringing home the fruits of their fields and orchards."[103]

Farm women, even slaveholders' wives, universally prepared food. Their daily labor began at the family's well or at a nearby stream, where they hauled heavy buckets of water for cooking. Then, in the barnyard, garden, orchard, cellar, or kitchen, they got fruit, vegetables, and meat; milked cows; collected eggs; baked bread; churned butter; and made cheese. Occasionally or annually they preserved fruits and vegetables, brewed cider or beer, and gathered honey from beehives. They fixed the day's meals. Seventeenth-century wives made simple stews, soups, porridge, or corn mush in a large pot over an open hearth. Such cooking remained common in frontier areas and among poor farmers. But eighteenth-century middling farmwives served varied meals, with more vegetables and a steady supply of salted pork and beef. Standing over the fireplace in northern houses or southern kitchens, they cooked separate dishes meant to be eaten with a knife and fork (rather than a spoon).[104]

The home industries of women—candle and soap making, sewing, spinning, and weaving—played a key role on small farms. Although women made only a small portion of the cloth their families needed, spinning and weaving were essential farm manufacturing tasks. Sewing imported cloth into dresses or pants or knitting stockings and mittens from woolen yarn, farmwives had always made some of the family's clothes. Seventeenth-century New England authorities urged towns to maintain herds of sheep, and by the 1650s, one-third to two-fifths of eastern Massachusetts householders owned sheep, assuring enough wool to make winter clothing. With the diffusion of spinning wheels and looms in the eighteenth century, farm women increasingly spun thread and yarn, carded wool or linen, or wove cloth. Though found everywhere, even on large plantations where more and more slave women spun and wove, increased home production of yarn and cloth was most pronounced in New England, where at least two-thirds and often four-fifths of the households had wheels and one-quarter of the households owned looms.

As discussed earlier in this chapter, making homespun cloth was a complex cooperative activity. Because few eighteenth-century families owned all the materials necessary to make cloth, they had to buy most of the cloth they needed, either from abroad or from local weavers—women in New England, immigrant men in southeastern Pennsylvania. As families bought more imported and locally made cloth, the variety of their clothing grew, increasing women's hard work washing clothes.[105]

The experience of Chester County, Pennsylvania, farmers illustrates the intricate relationship of homemade cloth, the weaving trade, and imports. As country cloth made by weavers (some of high quality) replaced some imports, families had more cash for other expenditures. But country producers supplied only a tiny proportion of the forty-five yards of cloth (for clothing, blankets, farm linens) a typical family of six consumed in a year. At best, half the county's householders had sheep, and the wool they produced, if distributed to each family, would have made just half a blanket or a pair of mittens and stockings for each person. Thus, county residents had to import woolen cloth. At the same time, farmers and weavers grew flaxseed for export and imported fine Irish linen. In all, at least four-fifths of the cloth county storekeepers stocked had been imported.[106]

Amidst these time-consuming farm chores in field and cabin, colonial women spent much time bearing and rearing children. They bore children every twenty-four to thirty months from marriage until around age forty. A farmwife was pregnant, nursing, or caring for a child under five for two-thirds of her married life. Since there were so many children, custodial care took precedence over nurturing and moral training. Farm mothers nursed their infants for twelve to eighteenth months, sending them to wet nurses only if illness dried up their milk or prevented suckling. They took care of small children while they were pregnant or nursing. Wealthy matron Pamela Sedgwick was contented if she could "still a squalling infant and settle a matter of great contention among a company of unruly boys." Farm women had little help with child care from older daughters until they reached their late thirties, and even then, as they trained daughters in housewifery, they had to supervise child care. Although child care consumed much of the farmwife's day, she did not consider child nurture a special vocation. Rather, bearing and raising children were crucial parts of the *productive* roles of farmwives because older children provided essential farm labor.[107]

Parents and public authorities expected children to work; as the Massachusetts General Court put it in 1641, "Masters of families should see that their children and servants should be industriously implied." Parents trained children to master skills appropriate to their sex and age. Children as young as five to eight (in places such as Essex County) helped their parents with farm labor; Massachusetts farmer-clergyman Ebenezer Parkman employed his eight sons beginning at five or six. By early adolescence they worked full days. The median age of beginning Boston apprentices was nine. Seventeenth-century Essex

County boys plowed at age eleven, and brothers worked together on the farm, directed by the eldest son. Virginia and Maryland laws allowed boys of ten or twelve to make tobacco. Eighteenth-century New Hampshire farmer Matthew Patten set his sons to farmwork between eleven and thirteen.

Throughout the colonial era both boys and girls helped clear land, hoe corn, and harvest crops. With increasing diversification and home consumption, children's (especially girls') tasks multiplied. Eighteenth-century boys hunted and fished and worked with fathers, tending corn or tobacco, making barrels, or shoeing horses. Girls looked after younger siblings and helped mothers prepare meals, spin yarn, knit socks or gloves, sew and repair clothes, weave cloth, milk cows, and make soap. Children of small slaveholders performed the same tasks; children of large slaveholders worked less intensively, but even they helped oversee the slaves' labor. Children often continued to work on the parental farm until their early twenties, when some entered the labor force and began to acquire property.[108]

Although fathers often taught children to read and write, schooling took little time from farm chores. To master religious precepts, parents expected children to read the Bible; to run farm businesses boys (but often not girls) learned to write and compute. Schooling extended over years but ended before adolescence, and the total time in class rarely amounted to more than a year. As data on adult literacy show, schooling was more common among New Englanders and mid-Atlantic Friends, where an increasingly dense free populace and religious imperatives allowed schools to thrive, than in the slave South, with its scattered free population. Half of farm men and a third of farm women in seventeenth-century New England could read and write; by the mid-eighteenth century nearly all adults were literate. Despite initially declining literacy, Friends showed similar patterns. Literacy in the South was less extensive. Half of late-seventeenth-century Virginia white men could read and write, and that percentage grew to two-thirds by the mid-eighteenth-century; North Carolina's white men reached similar levels, but only a quarter of the women were literate.[109]

A new style of child nurture, which accentuated the mother's role, developed among rich and some middling farm families in the early to mid-eighteenth century. This new role grew out of the concern mothers shared about the proper feeding of infants. While present throughout the colonies, such concerns were especially common among the child-centered families of Pennsylvania Friends. Quaker wives may have participated less regularly in market exchange than other women, and they took on new roles as spiritual mentors to their children. By example and precept, a Quaker mother "ruled well her own house, having her children and household in subjection with gravity." But neither ideas of a separate female sphere, with its high valuation of domestic work, nor the practice of intensive child nurture reached many farm families. Rather, the large number of children farmwives bore and the work parents required children to do sustained patriarchal households.[110]

The number and kind of workers the head of the household directed depended on time, region, and the father's age. Everywhere parents and children provided most of the labor on small farms. The number of family laborers grew from husband and wife to include several children and then declined as children left home. Nearly every small farmer shared the experience of John Abbot of late-seventeenth-century Andover, Massachusetts. Married in 1673, he sired six sons and two daughters. By the early 1690s his three eldest sons were working with him; as his oldest children married, others replaced them. When Abbot died in 1721, only one son and one daughter remained at home.

Northern farmers counted less on hired or bound labor than southern planters. Only one-half of households in eighteenth-century southeastern Pennsylvania owned a servant or a slave, rented land to a cottager, or hired seasonal labor. Small southern planters supplemented family labor with bondsmen and -women. Mid-seventeenth-century Chesapeake planters owned a servant or two. When the supply of servants dwindled late in the century, they turned to slaves. Eighteenth-century small planters owned one or two adult slaves; hired hands and servants worked on few plantations. Enticing servants proved difficult for South Carolina and, later, Georgia small planters, but as early as the 1720s they owned several slaves.[111]

Far from having moral reservations about using unfree labor, small farmers bought servants or slaves whenever they could afford them. New England immigrants often brought indentured servants with them, but once immigration stopped in the 1640s, the supply of servants dried up, and only the richest farmers employed any. In 1689 about a third of the families in Bristol, Rhode Island, owned one or two servants. Even fewer worked in other areas; in Massachusetts in 1687 just 2 percent of householders in Topsfield and 5 percent in Dedham had any. Families began to employ young orphans. In 1681, 22 of the 112 households in Dedham had taken in out-of-town orphans. Eighteenth-century New England towns expanded this system of child labor. Selectmen often took local children from poor parents and indentured them to more prosperous families. Towns, moreover, sometimes held hiring fairs where local adolescents could be hired as annual servants-in-husbandry by families that needed their labor. Farmers and rural craftsmen avidly hired Boston apprentices and orphans. Between 1734 and 1806, the Boston Overseers of the Poor sent out 838 orphan children to farming towns and small seaports. Most of the orphans worked for younger farmers or craftsmen who had been married but a few years and whose children could not yet provide much labor.[112]

By the mid-eighteenth century, indentured servitude had nearly disappeared from New England, and slavery was just as uncommon. A mere 1 to 3 percent of the people of Massachusetts and Connecticut lived in slavery, and many were clustered in port towns. In Rhode Island a tenth of the rural population was enslaved in 1774, and only one in ten families held slaves, with two-thirds owning just one or two. Similarly, in Fairfield, Connecticut, where a twentieth of the population was enslaved in 1790 (the highest in Connecticut), a seventh of the households owned a slave, almost always one or two.[113]

Unfree labor was more plentiful on eighteenth-century mid-Atlantic farms, but even rich men rarely owned more than one servant or slave. Slaves provided most unfree labor in rural New York and accounted for one-fifth to one-twentieth of the population of rural New York counties in 1756 and 1771. Slaves outnumbered servants among unfree laborers as early as 1700 in Newtown, Queens County; in 1755 about two-fifths of Newtown householders owned a slave or two. Pennsylvania's unfree work force was more diverse. Farmers hired Philadelphia apprentices and orphans, bought indentured servants and German redemptioners, or held slaves. Servants were plentiful in the late seventeenth century but declined thereafter. Nearly two-fifths of the householders in Chester County owned servants in the 1690s, but only one-fifth did so by the 1710s, and a tenth by the 1760s. Fewer farmers owned servants elsewhere in the region; most of the 35,000 redemptioners and servants who arrived in Philadelphia in the mid-eighteenth century stayed in the city. Slaveholding was more common in New Jersey (where 7 percent of the population was enslaved) than in Pennsylvania (2 percent). Prosperous farmers nonetheless used slaves; nearly two-fifths of the residents of eastern Chester County owned slaves in the 1690s, and as late as the 1760s one-tenth held slaves.[114]

While only rich northern farmers owned unfree laborers, small southern planters often owned indentured servants in the seventeenth century and slaves in the eighteenth. With few sons to help with tobacco, early- and mid-seventeenth-century Chesapeake planters often bought English indentured servants. Between 1658 and 1670, three-fifths of householders in southern Maryland owned bound laborers, nearly all of them servants. In 1678 two-fifths of the households in Surry County, Virginia, owned servants; twenty years later only three-fifths as many still held servants. Planters continued to hire servants, especially those with craft skills, but their numbers had dwindled by the mid-eighteenth century. In 1755 servants and convicts constituted a quarter of Maryland's adult unfree labor force; during the mid-eighteenth-century, servants comprised three-tenths of North Carolina's unfree labor force.

At first few planters could afford slaves, but during the first half of the eighteenth century, two-fifths of householders in the Tidewater Chesapeake and one-quarter to one-third of those on the Piedmont frontier owned slaves, usually employing two or three adults. By the 1770s one-half to three-quarters of planters owned slaves, using four or five slaves of all ages. Although fewer planters in backcountry Carolina had slaves, by the 1770s, upcountry South Carolina farmers (except the poorest third) held one or two. Slaves predominated in the population of the Lowcountry even before planters began exporting rice. Ownership of blacks was universal in South Carolina and Georgia rice country, where slaves made up the vast majority of the population. As early as 1726 more than three-quarters of the planters in St. George's Parish, South Carolina, owned slaves; nearly half of them held five or fewer Africans. In South Carolina as a whole the proportion of slaveowners who held ten or fewer slaves fell somewhat, from two-thirds in 1721–24 to one-half in 1729–31. By the 1760s virtually every plantation operator had slaves. Big planters with surplus slaves and slaveholding widows unwilling to run their plantations themselves often hired out slaves to their neighbors, some of whom were small planters who used the slaves in their fields, gardens, and workshops.[115]

Small planters—heads of household, sons, and their several slaves—still produced much of the Chesapeake region's tobacco far into the eighteenth century. At the end of the seventeenth century, units with two to five workers (a household head, a son or two, and a servant or two) made two-fifths to one-half of the tobacco. In 1725 three-quarters of the workers in frontier Stafford County, Virginia, were small white planters and their sons. Even in long-settled Norfolk County, only about half the workers were slave women and men, while a quarter were heads of household, a ninth were sons and brothers, and a twelfth were servants. After 1750, when the number of big planters increased, plantations with one to five laborers (a household head, a son or two, and a slave or two) made two-fifths to three-fourths of the tobacco; the greatest predominance of small planter production was in frontier regions. Similarly in the 1750s, southern Maryland planters who made less than 5,000 pounds of the leaf (with roughly one to five laborers) marketed four-fifths of the tobacco.[116]

Slaves performed virtually every household and farm task. Whenever they made new plantations, southern planters set slaves to clearing land. Along seventeenth- and early-eighteenth-century Chesapeake and Carolina Lowcountry frontiers, slaves herded livestock, often dominating that activity. Their eighteenth-century descendants continued to care for livestock. We have already seen how slaves made tobacco and rice, tasks many of them (especially the women) learned in Africa, where they had grown both crops. Slave adults and children (starting at age six or seven) worked in grain, rice, and tobacco fields. But the contribution of slaves to small planter households went beyond cultivation of staples. Slave women made cloth, raised fowls, and milked cows. Northern slave men cultivated corn, grew vegetables, and tended livestock for local consumption and trade to the West Indies; in slack times they chopped wood, made fences, and cleared land. Northern slave women increasingly became domestic laborers and helped farmwives with their chores. Carolina slaves hand-milled rice (a time-consuming task), built ditches and fences, grew indigo, collected tar and pitch, and made their own shoes and clothing. Chesapeake slaves made corn as well as tobacco. Eastern Shore and northern slaves worked with their masters growing wheat and oats, cultivating flax, and performing farm chores. Slaves (especially those owned by larger planters) in both the Lowcountry and the Chesapeake grew some of their own provisions in garden plots and hunted small game to supplement their diets.[117]

The slave workers smaller planters and northern farmers owned were integrated into the household economy. Most seventeenth-century slaves worked on small units and lived in the master's house (sleeping in a loft) or in an outbuilding, rather than in a slave cabin or barracks. They worked in the grain and tobacco fields with the servants and the master's sons, under the master's direct supervision. Most eighteenth-century slaves lived on large units in separate quarters or slave villages, isolated from the master and often supervised by a white overseer or slave driver. Those few slaves small Chesapeake planters owned still worked alongside their masters and lived either in or close to his house. Slaves in Monmouth County harvested wheat alongside white day laborers, indentured servants, and free blacks.[118]

Northern and Chesapeake families incorporated the servants they owned into the farm economy and put them to work digging potatoes, milking cows, making fences, hauling produce, carrying water, or gardening. A seventeenth-century ballad describes the never ending labor of female servants in the Chesapeake. Kidnapped in England and sold in Virginia, the servant girl became "weary, O" from cultivating tobacco with the "Axe and the Hoe." With "Sorrow, Grief, and Woe," she "played" her "part / Both at Plow and Cart." When not in the fields, she collected "Billats from the Wood" and got "water from the spring / Upon my head." In return, she claimed, her master gave her a miserly diet and let her clothes run threadbare. Similarly, in 1780, a farmer who lived "about half a day's journey from Philadelphia" advertised for a servant girl, seeking an "affable . . . single Woman of unsullied Reputation" to "manage the female Concerns of a country business, as raising small stock, dairying, marketing, combing, carding, spinning, knitting, sewing, pickling, preserving."[119]

Patriarchal theory deemed servants and slaves, like the master's children, subordinate members of the household, and thus under the father-master's discipline. But servants and slaves, lacking the incentives to good behavior that an inheritance made, constantly disrupted good household order. Because they saw themselves as part of a dispossessed class with little control over their own labor, they refused to work or to work speedily enough, feigned illness, ran away, stole household goods or livestock, or disobeyed orders from master or mistress. Sometimes servants and slaves united to protest harsh masters or ran away together. Heads of household whipped disobedient unfree laborers (especially slaves), sometimes without mercy. Chesapeake servants brought charges of cruel treatment (vicious assaults and rape) to the courts; servants and slaves at times showed their bruises to neighbors. Neither courts nor neighbors intervened, except in cases of severe beating or murder. Servants waited out their terms, but slaves, lacking any other options, rebelled more forcibly; they ran away to towns, conspired together for their freedom, or even killed an overseer.[120]

Although northern farmers on occasion bought a servant, they hired day laborers mainly at harvest. The supply of wage laborers was small, and farmers usually owned enough land to keep their adolescent children at work. Most youths worked a few years, but they soon got land. Once freed, New England's few indentured servants either refused to work for wages or demanded, a disgusted John Winthrop wrote, "unreasonable terms." Nor did neighbors, who were busy with their own farms, provide much help. Men did work for wages on Salem's farming periphery, but elsewhere in Essex County families searching for workers found few, and those demanded "extortionate" wages.

The supply of wage laborers probably increased in the eighteenth-century North. Busy farmwives in places such as the northern Chesapeake increasingly hired women to care for children; to cook, clean, or garden; to spin, sew, or weave (mostly for slaves); or to market produce. But northern farmers hired men and boys far more often than they employed women. They received a customary wage based on their age and gender (women and boys got less than men), the kind of work performed, and the season. Some farmers employed adolescent boys or men in their early twenties on longer contracts to supplement family labor. During the 1730s and 1740s Ebenezer Parkman hired adolescents and young men on three- to six-month contracts during the growing and harvesting seasons; from 1751 to 1764, after his sons were old enough to do field labor, he ceased using contract workers.[121]

The development of wheat farming increased the demand for harvest laborers, as the example of Maryland's northern Eastern Shore suggests. Unlike corn, wheat required little work after planting but had to be harvested as soon as it ripened. Wheat farmers needed workers for only the two- to three-week harvest. Late-seventeenth-century (1675–1715) Eastern Shore tobacco planters hired small numbers of former male servants as day laborers, mostly (seven-tenths of the time) to cultivate or pack tobacco or to husk corn. Early in the eighteenth century some planters replaced tobacco with wheat, and those who still grew tobacco used slaves instead of white workers. Demand for harvest labor skyrocketed at the same time that the number of native-born adolescents looking for work grew. Most male day laborers performed tasks necessary for wheat production. Between 1716 and 1735 they spent half their days plowing, reaping, and threshing wheat. Day laborers continued to harvest wheat between

1736 and 1783 (three-tenths of their time), but their tasks became more varied. Many helped build and repair the gristmills essential in a wheat economy (a sixth of their time), while others cut trees or worked at the sawmill (a fifth of their time).[122]

On occasion seventeenth-century farm families, in both the Chesapeake region and New England, hired married couples (sometimes servant couples) to work for them. These couples apparently lived with or near their employers. In 1666 an unmarried Talbot County, Maryland, planter hired Alexander Davis and his wife. Alexander was to plant crops and make fences; his wife, to "doe all theire houshould imployment, and to make and mend, theire Lining." Similarly, in 1640 a Massachusetts family employed a couple (probably as servants-in-husbandry); the man apparently was to work in the fields, while his wife was to "tend the cattle" and feed day laborers at haying time, among other tasks.[123]

Perhaps building on these occasional instances of the hire of married couples, a new group of wage laborers—cottagers—emerged in the Delaware Valley in the mid-eighteenth century. Each year they rented a house and a small garden and collected firewood and grazed a cow on the owner's lands. In return the cottager and his wife had to work for the landowner, especially at planting and harvest, when the cottager (and his wife) might work nearly full time, as did John Rock and his wife on the Brinton farm in the early 1790s. Artisan-cottagers provided employers with skilled labor; the wife spun yarn, and the husband made cloth or shod horses. Cottagers appeared in the mid-eighteenth century when Delaware Valley land grew expensive and commercial wheat cultivation, with its heavy requirements for harvest labor, expanded to meet foreign demand. In 1750, before the surge in wheat production, cottagers headed one-twentieth of Chester County, Pennsylvania, families. By 1783 (after the surge), they headed more than one-third of the county's families.[124]

Renting a cottage might prove just as beneficial for a man just starting out as for the landowner, for much of his time would be his own. In 1774 Scot immigrant and Presbyterian divinity student Hugh Simm urged his brother Simon, a weaver, to emigrate. Once he arrived, his brother recommended, Simon should leave his family in the city "and endeavour to find a house" on Long Island "with 1 or 2 acres of land within 10 or 15 miles from N-york." Once Simon had found land and "set in order your house and planted your land," Hugh suggested, he should use the farm as a base to find employment, by publishing "in the News the different kind of work you are Skilled in." If he did so, Hugh had no doubt that "in a Short time you will have a Sufficientcy of employment."[125]

How, then, should we describe the division of labor on late colonial small farms? Family labor predominated, but more fortunate farmers bought a servant or a slave, hired harvest workers, or hired neighbors' children for months at a time. Law and custom mandated that husbands be patriarchs over their wives and children, farm laborers, and hired hands. But their patriarchal authority became diluted by the necessity of day-to-day cooperation in farm labor and child rearing, by wives' involvement in farm business (im preparation for widowhood), and in the allocation of essential tasks to each member of the household. Within this patriarchal system, then, wives enjoyed a bit of authority over children, and southern wives of small planters gained enhanced authority over the servants and slaves their husbands owned.

ABUNDANT LAND, markets, and the borrowing system structured the households small farmers made. Farmers aimed at landownership, communal sufficiency in food, a comfortable subsistence, and the perpetuation of the farm across generations. Able to get unimproved land easily, youths married young and had many children. Expansion of farm output—needed to build estates for children—required labor, but since land was plentiful, laborers for hire were hard to find. Family labor, along with neighborly exchanges of labor, partially offset this scarcity. To finance planting, farm expansion, and neighborly barter, farmers sold surpluses to merchants and received credit in return. When land grew more scarce in older regions, farm families moved to newer frontiers and began the process again.

A growing sense of well-being pervaded pre-Revolutionary America, especially among landholding farmers. Able to make much of their own food, they willingly traded for their other needs; well-fed, they grew taller and stayed healthier than any others of European descent. Adults in New England farm families, for instance, annually consumed an average of 3.75 gallons of imported rum, 2.5 pounds of imported tea, 5–10 pounds of sugar, 5–32 gallons of domestic cider, 150–78 pounds of locally butchered meat, 46 gallons of milk, and 15 pounds of butter. In addition, each family consumed 365 pounds of cheese a year. With crops or services to sell and growing markets for wheat, tobacco, and rice, they bought more consumer goods. Willing to work hard clearing land, they reaped the benefits of capital gains, borrowed, and made loans. As foreign and domestic markets grew, they accumulated more wealth. Not surprisingly, economic growth—measured by levels of per capita wealth—grew throughout the colonies during pre-Revolutionary times.[126]

epilogue

The Farmers' War and Its Aftermath

Rebellion against Great Britain turned Mary Fish's life inside out. A well-off, forty-year-old widow with three children, in 1775 she married rich widower Gold Selleck Silliman. After they married, Mary moved to Gold's Fairfield County, Connecticut, farm. While he managed his wife's inherited estates, she learned about his farm, knowing that the coming war could turn her into a farm manager. Little did she realize how traumatic the war would become. When soldiers spread smallpox and dysentery throughout coastal Connecticut in 1776, Mary took her family to safer New Haven, but she nonetheless fell ill with dysentery. While Gold led the state militia, Mary ran the farm, entertaining militia officers, housing refugees from war violence, directing the labor of several slaves and her adult stepson, drawing accounts, and collecting rents on her first husband's farms.

In 1777 when the British landed on the coast near her home, Mary Silliman retreated inland. While Gold Silliman led troops that year, their first child was born. Gold was kidnapped by Loyalists in 1779 and held prisoner on a Long Island farm. Six months pregnant when Gold was captured, Mary endured the birth of their second child and the care of their two sons alone. "I had a large family," she remembered, "& the care & weight lay on me"; day by day she "lived in constant alarm. The dreadful fright I had the night of his capture made me feel like the timorous roe, & I started at every noise, fearing the enemy." Countless farm women shared Mary Silliman's loneliness, hard labor, and suffering, and many who lacked the advantages enjoyed by the wife of a rich lawyer-farmer endured far more. Mary Silliman commanded enough men to run her farm and live comfortably, but when sons or husbands in poor families went to war, women and children struggled with the farm alone, facing destitution when they could neither grow enough food nor borrow it from neighbors.[1]

This chapter tells the story of farm families during and after the war years. As the war ground on for seven years, the farm economy nearly disintegrated, foreign trade disappeared, local exchange atrophied, labor became ever more scarce, and the destruction and terror of war, perpetrated by both sides, threatened tens of thousands of farm families. Unable to exchange with equally poor neighbors, much less trade surpluses at market (as they had done before the war), farmers could barely feed their families, and for them the differences between Loyalist and Whig partisans, each out to take their puny supplies of food, faded into the background. After the war, farm families rebuilt the farm economy, and their lives gradually returned to normal.

Disruption of the Agrarian Economy

Warfare tore apart the farm economy of early America. By middle age most farmers owned land, and landless would-be farmers had moved to frontiers newly wrested from Indians. The prospect of owning frontier land and thereby achieving a measure of economic independence attracted not only colonists' children but hordes of immigrants. Farmers had land, but labor (except that of family members, poor youths, and a few immigrant servants) was scarce. With so little help, farmers used land extensively, girding trees, planting under stumps, abandoning exhausted land, and letting pigs run wild. Farmers strove to grow the family's food; nonetheless, few families attained familial sufficiency. Instead they traded with neighbors and sold surpluses at the market or made tobacco or rice for foreign markets.

Embargoes, privateering, and blockades cut off foreign trade and immigration, while rebel and British armies and irregular troops and their Indian allies destroyed thousands of farms, killing or stealing tens of thousands of cows, horses, sheep, and swine and sending farm families fleeing to safety. At the same time, the youths, poor men, and recent immigrants farmers usually hired to plant and harvest crops enlisted. Farm sons and some farmers served as well, forcing women to cultivate corn, wheat, and tobacco. By eliminating both the foreign markets for crops and the labor needed to maintain production, the war thrust farmers into subsistence farming and home production of manufactures.

By early 1776 both the Continental Congress and the British Parliament had banned trade between their countries. British tobacco imports, which had averaged 100 million pounds in the early 1770s, dropped to 9.3 million pounds between 1777 and 1782, a tiny 1.6 million pounds a year. Between 1777 and 1780 little, if any, rice reached Britain; in 1781, after the British occupied Charleston and its hinterland, Britain imported rice and other goods worth a fifth of prewar levels. Long dependent on the mainland colonies for nearly all the food slaves and whites consumed, British West India planters bought American meat and flour in neutral French and Dutch West Indian islands. Such imports (and British supplies) could not, however, replace American imports, and as long as the war continued, island slaves went hungry.[2]

Rebels attempting to open new markets faced a British blockade and privateers. A 3,000-mile coastline punctuated by countless bays, inlets, and rivers prevented the British navy from closing down rebel shipping. But even a few ships could wreak havoc with trade. The British navy shut down trade in the Delaware and Chesapeake Bays in the winter of 1776–77; by 1779 it had so completely cut off Philadelphia trade that tonnage plunged to a seventeenth of prewar levels. By intermittently blockading Savannah and Charleston, the British cut off most South Carolina and Georgia trade. Privateers—private vessels licensed to capture and sell enemy shipments—combined with the blockade to reduce farm exports to a trickle. The British navy and Loyalist privateers seized enough ships to eliminate almost all tobacco and flour exports, force some merchants out of business, and send freight rates and insurance sky-high (20–30 percent of the value of the goods). American privateers, which seized 2,000 British vessels worth £18 million sterling, similarly reduced British and Loyalist wartime commerce.[3]

Despite the blockade and privateers, Americans tried to market surpluses in Europe and the French West Indies. But these attempts failed miserably. More than four-fifths of the tobacco Britain imported had been re-exported to the continent. Most of it went to France and the Netherlands, each of which received 20 million to 30 million pounds; between 1778 and 1780, however, Chesapeake merchants sent only 2.2 million pounds a year to France. Lacking markets, planters stopped growing tobacco; as late as 1782–83 Virginia authorities inspected only 18.6 million pounds, three-tenths of the average exports (60 million pounds) between 1768 and 1773.[4]

Mid-Atlantic grain farmers, who had sent 12,500 tons of bread and 167,000 bushels of wheat annually to southern Europe and 16,000 tons of bread to the British West Indies, had no better luck in opening new markets. Grain, meat, and flour exports to the French West Indies rose after the 1778 treaty between the United States and France. But Congress prohibited grain and flour exports the same year, reducing the trade to smuggling. After Congress lifted the embargo (in 1780 to Spanish islands, fully in 1781), flour reached Cuba and Hispaniola, but the British capture of neutral Dutch port St. Eustatius in 1781, combined with the renewed blockade of the Chesapeake Bay, eliminated much of the West India trade.[5]

Farmers, then, had to rely on domestic markets to sell surpluses. Contending armies, which consumed vast quantities of wheat, flour, pork, and beef provided a large potential market. Between 1777 and 1781 at most 84,000 people served in military units. If mid-Atlantic farmers had provided wheat to make a pound of bread per day per soldier, they could have sold about 45 percent of prewar exports. Military demand for meat, fresh vegetables, milk, fodder, horses, and wagons exceeded colonial surpluses. The colonies exported 7 million pounds of meat a year; 80,000 soldiers required from 16.6 million to 29.1 million pounds (4 to 7 pounds per man per week) of meat a year. Forage for the thousands of horses and other livestock the military used or consumed created a large market for hay.[6]

The military food market, however, was neither reliable nor lucrative. The British stationed no more than 20,000 men in North America. Loyalists added 5,000 to 10,000. The Continental army, which during peak months had 30,000 men, grew and shrank randomly, losing half of its strength only to build up again. Between 1779 and 1781 the average size of patriot forces plummeted from 24,000 to under 10,000. The French army never numbered more than 7,800 soldiers (plus 4,200 sailors).[7] Farmers could not even count on this market. Military authorities refused to pay market prices or offered farmers worthless paper money. The British sometimes gave hard money, but despairing of buying local supplies, they hired men to grow vegetables, cultivated their own gardens, and imported forage from Canada and food from Britain. At first the French army, which paid gold for vegetables, flour, and horses, had better luck and quickly secured all the provisions it wanted, but it, too, soon ran out of hard money.[8]

Farmers resisted trading provisions to the Continental army for paper money, deeming it "no Better than oak leaves & fit for nothing But Bum Fodder." To keep up with inflation they demanded ever higher prices or cash on delivery, tore up contracts when prices rose, sued army officers for payment, and went from market to market to sell to whoever offered the best price. After 1775 few farmers willingly accepted Continental money. When soldiers forcibly took wheat or a cow, they often gave certificates for future payment rather than paper money. Commissaries and their agents paid farmers at least $95 million ($3.7 million in specie) in certificates, nearly two-fifths the nominal value of the money Congress printed. Before 1778, certificates—most of which came from the mid-Atlantic states—were almost worthless, handwritten documents lacking details of the transaction; the printed forms used later listed quantities and prices of goods requisitioned. Nonetheless, farmers took certificates no more willingly than Continental currency.[9]

Some farmers did take advantage of military demand. Those living near the British lines in Rhode Island, New Jersey, or New York avidly traded with the military for hard currency, exchanging provisions and livestock for manufactured goods. Delaware Valley farmers sold to the British army (or the army seized) vast quantities of goods in the summer and fall of 1777. Other farmers who lived near harbors where British battleships lay boldly went aboard to truck. In the winter of 1778, farmers who refused to sell to Americans took food to British-held Philadelphia, risking confrontations with partisans and confiscation if caught by patriot army patrols. But the Continental army's best efforts barely stemmed the avalanche of goods farmers sent to Philadelphia.[10]

Even more farmers supplied the patriots. New Englanders, located far from the center of fighting, slaughtered cattle and sent hay and provisions to feed the Continental army. Hudson Valley landlords sold large quantities of grain to the army, as did Ulster County farmers, who lived close to army camps. New Jersey farmers sold provisions for the army during the winter of 1778–79. In the winter of 1778 Maryland farmers sold such large numbers of cattle that state agents temporarily ceased purchasing livestock. Between 1781 and 1783 Shenandoah Valley farmers, protected from war by two mountain chains, exchanged at least 232,000 pounds of flour and 566,000 pounds of beef for state certificates.[11]

Most farmers could ill afford any requisition of grain or livestock; but when they kept surpluses, both armies accused them of hoarding to make big profits and took their grain, flour, and forage, leaving them with just enough for subsistence. Ignoring evidence of dearth and seeing grain where none existed, generals, assemblies, and soldiers protested farmers' greed. General Washington complained to New Jersey governor William Livingston in January 1778 about "the avarice of your farmers, who like their neighbours are endeavouring to take every advantage of the necessities of the Army." Connecticut private Joseph Plumb Martin agreed. In May 1780, fed up with a lack of provisions in New Jersey, he joined a mutiny against country and officers, "venting our spleen . . . at ourselves for our imbecility in staying here and starving . . . for an ungrateful people who did not care what became of us, so they could enjoy themselves while we were keeping a cruel enemy from them."[12]

No farm family, however, could subsist on its own labor or achieve self-sufficiency. Farmers had to trade the surpluses, which the armies confiscated, with neighbors to acquire food, fiber, clothing, or labor. By disrupting the borrowing system, requisitions impoverished farmers who were already struck hard by embargoes and plunder. Repeated army incursions stripped farm country of the means of subsistence, much less surpluses. Even as he railed against "an ungrateful people," Private Martin noted that "there was not the least thing to be obtained from the inhabitants" of New Jersey, "they being so near the enemy, and many of them seemed to be as poor as ourselves."[13]

Inflation of Continental and state paper money, however, had little impact on most farmers. Congress printed $241 million between 1775 and 1779; but by December 1780 the money traded at 100 to 1, an inflation rate of 1,000 percent, and millions of dollars of state currencies depreciated as rapidly. Unlike laborers, who had to buy what they consumed, farmers benefited from inflation. They kept their books (and traded with neighbors and storekeepers) in a money of account, the purchasing power of which declined much less. Between 1775 and 1779, farm prices of Massachusetts corn, hay, and dairy products rose by one-half; from 1775 to 1778, Maryland crop and livestock prices grew two-fifths, and Virginia farm prices jumped 300 percent, still far less than Continental currency.[14]

Some farmers forced local merchants or rich farmers to accept paper money, knowing that using inflated money would cheaply clear up their debts. From 1776 to 1790 farmers paid cash for three-fifths of the value of goods purchased from Ulster County, New York, merchant William Pick; during 1777 farm tenants of Loyalist Philadelphia lawyer James Allen often paid off "six or seven years" of rents due in sterling "in continental money at the old exchange." "Yet," he added, "I dare not object, though I am as much robbed of my property as if it were taken out of my drawer." Other farmers accepted paper money for crops but discovered—like the 185 Mecklenburg County, Virginia, planters who complained to the state assembly in May 1777—that merchants deemed paper money "of little or no value and absolutely refuse to receive" it even for "debts due British merchants."[15]

Inflation did reduce the appeal of taking vegetables, fruit, and flour to city markets. These markets had grown mightily during the pre-Revolutionary decades, especially around Boston, New York, Philadelphia, and Charleston. City governments, their citizens awash in paper money, tried—without success—to regulate food prices. When Philadelphia authorities fixed food prices in 1779, Delaware Valley farmers and millers, who refused to accept paper money at set prices, kept produce out of Philadelphia, preferring to sell at higher prices to French forces, Maryland buyers, or the Continental army. New York City, occupied by the British for nearly the entire war, proved an exception. Between 1778 and 1783 the British pumped £6.4 million sterling into New York's economy, some of it for army provisions; paper money maintained a steady exchange with sterling. At the same time the real prices of flour and pork jumped by 70 to 200 percent. Such high profits led farmers in the city's hinterland—upper Manhattan Island, Staten Island, Long Island, the lower Hudson Valley, western Connecticut, and eastern New Jersey—to increase market production of hay, wood, wheat, vegetables, fruit, butter, beef, and pork.[16]

As markets disappeared, small farmers pursued subsistence production. Chesapeake tobacco planters turned to corn and wheat, grew tobacco and hoped the war would end, or cut back market production of tobacco, corn, and wheat and tried to supply their families with food, salt, cider, and cloth. Hard pressed by the war, New Jersey farmers diversified by growing corn and wheat or raising fowl, pigs, cows, and sheep. Northern backcountry grain farmers grew more vegetables and fruit and made more butter and meat. Farmers near Bennington, Vermont, grew corn, wheat, rye, potatoes, pumpkins, turnips, flax, and hay; raised cattle, sheep, and hogs; made butter, milk, and cheese; and gathered berries. Lacking forage over the winter and seeking to protect cattle from danger, neighbors may have mowed hay from several fields and stacked it in small fields they all used, and they probably stored potatoes communally as well.[17]

The Revolutionary War, then, turned the lives of farmers—even those far from the battlefield—upside down. Having lost their foreign and coastal markets, they tried to make more of their own food. Without the money that market exchange brought, local exchange of goods and labor became more difficult but more essential. Barely able to make ends meet by their own labor and neighborly barter, they faced repeated demands from both armies for provisions, clothing, and shoes. A serious labor shortage, as we shall see, compounded these problems.

Who Plowed the Fields and Made the Corn?

Since labor available to farm families declined steeply, farmers could not meet even the reduced demand for flour, meat, vegetables, and fruit. Not only did the Continental army, militias, and privateers take sons and husbands from the farm, but the military attracted poor, footloose youths whom farmers had hired as yearly servants or harvest laborers. And wherever British troops marched, slaves ran away, hoping for freedom. Most men stayed in the labor market, but farmers, unable to predict the size of the labor pool, reduced their output. Such decisions had a devastating impact on corn and wheat production. Corn required regular pruning and weeding during the growing season; after planting, wheat could be left in the ground untended, but it had to be harvested quickly when ripe.

Before the Revolution, immigrant servants had labored on some mid-Atlantic wheat and truck-garden farms or as craftsmen on Chesapeake plantations. The war ended immigration, but during the first years of the war, many servants still had time to serve on their contracts. To escape harsh masters, some servants joined British or American armies, much against their masters' wishes. Desperate for men, the Continental army enlisted servants, offering them freedom and land in return for service, but urged states to pay masters for the labor they had lost. Five states allowed servants to enlist without the permission of masters but mandated compensation; New York and (later) Maryland insisted servants gain masters' permission to enlist but paid no compensation. More than a seventh of Maryland's convict servants joined General William Smallwood's regiment, but only in frontier Pennsylvania, where farmers occasionally mobbed recruiters, did servant enlistment cut farm labor drastically. As servants earned their freedom, the servant supply plunged, and rather than work on the farm, freed servants often joined the Continental army.[18]

The Revolutionary War disrupted the use of slave labor even more. On the eve of the war at least half of Chesapeake and backcountry Carolina farm operators owned slaves. An even higher proportion of families held slaves in the Lowcountry, where nine-tenths of the population was enslaved. The labor of northern slaves was almost as essential. During the 1770s nearly two-fifths of eastern Chester County households owned a slave or two. More male slaves than free white laborers or indentured servants worked on New Jersey farms, and slave women worked in both the house and the fields. The outbreak of war halted the Atlantic slave trade and the movement of slaves within the South. While rapid natural increase supplied the Chesapeake region, backcountry Carolina planters clamored for more. But Congress prohibited the trade; the British, who had supplied all Africans to the colonies, sent no more during the war, and no other European power took up the slack.[19]

Slaves took advantage of war to seek their freedom. Encouraged by British commanders, slaves of rebellious masters ran away wherever the British army invaded—New York, New Jersey, the Delaware Valley, Virginia, and the Carolinas. In late 1775 John Murray, the earl of Dunmore and Virginia's last royal governor, promised freedom to any slave man who joined his army; British Generals William Howe and Henry Clinton repeated the offer later in the war, promising freedom to all runaways. These declarations spread throughout slave communities, leading ever more slaves to head for the British lines. During the war at least 26,000 slaves fled bondage, rendering the labor supply of the South and parts of the mid-Atlantic states precarious. During the invasion of Virginia 5,000 slaves (2 percent of the colony's slaves) reached British lines. As many as 19,000 South Carolina and Georgia slaves (nearly 17 percent) escaped to the British lines, the Carolina backcountry, or Florida, and more than 2,000 others left North Carolina, Maryland, and the North. Many fleeing slaves died from hunger, disease, or battle wounds. Although masters recaptured a few, they had little hope during the war of recovering those who reached enemy lines. To protect runaways from the fury of former masters, the British took over 3,000 survivors with them when they left New York at the end of the war.[20]

While servants and slaves ran off, armies and militias cut deeply into family and hired labor, often leaving farms in crucial planting or harvesting seasons to women, children, and older men. About 200,000 men enlisted in the Continental or state forces, serving 396,000 terms and repeatedly disrupting labor supplies for a season or a year. As many as 50,000 more Americans joined the British army or Loyalist partisan bands. In total, in the new states about a third of the men (and half of the men of military age), white and black, fought during the war.[21]

Youths, poor men, and immigrants in their teens and early twenties, the backbone of hired farm labor, predominated among the recruits; older men with family responsibilities usually refused to enlist (except early in the war). More than two-thirds of the men mobilized in 1775 in thirteen New Hampshire militia companies called themselves husbandmen, farmers, or yeomen. Half were under age twenty-three; nearly all were unmarried sons who worked at home or did day labor on nearby farms. While farm sons enrolled in militias for short terms throughout the war, the Continental army attracted mostly poor, landless men: laborers, servants, recent immigrants, slaves, and free blacks. Rural laborers made up two-fifths of New York's third regiment in 1775 and two-thirds of Delaware soldiers in the southern army in 1782. The families of three-fifths of the men in the New Jersey line held no land and fell into the poorest third of taxpayers. The army vigorously recruited immigrants for long-term service, sending recruiters to immigrant neighborhoods and circulating recruiting broadsides in German and English. A quarter of Continental soldiers had been born in Ireland, and another eighth were from Germany. By the end of the war, close to half of the men in some units had been born abroad.[22]

Before the war, these youths and poor men had made the difference between bare subsistence and a decent competency on small farms. Continental army recruits often served terms of several years or more; militia recruits stayed a few months at a time, serving perhaps a year in total. Soldiers thus deprived farmers or their families of their labor for two or more growing seasons, and even the shorter terms of militia recruits interfered with the seasonal rhythms of farm life. Farmers never knew if a son or a hired hand might be forced to leave at planting, weeding, or harvesting times or if they could hire day labor at harvest. The call for Massachusetts men to fight at White Plains, New York, came in 1776 during the corn harvest in Concord; fighting around Augusta, Georgia, in late October 1781 threatened to ruin the harvest unless the militia returned home. The Continental army had a particularly difficult time maintaining a militia during the 1776 harvest. Many New Jersey, New York, Vermont, and Connecticut soldier-farmers and farm laborers demanded furloughs to harvest wheat in July and hay in September.[23]

Young slave men rushed to join contending armies, believing that military service would bring them the freedom they craved. Throughout the war 5,000 slaves enlisted in the Continental army; some joined on their own, while others came as substitutes for their masters. Thousands of slave runaways joined British armies or were impressed by them, further reducing the farm labor supply. A thousand slaves (including some women and children) reached Dunmore's British camp in 1775, and at least 300 men fought the rebels, took part in foraging, or participated in plunder. Despite the death from disease of a large number of the men, slaves served the British in later southern campaigns. In 1779, 750 slaves helped defend Savannah from recapture by a French fleet; by July 1780, British regiments included 1,500 black soldiers.[24]

Poor women—otherwise available to harvest crops, churn butter, and care for children—filled military camps. Women constituted an eighth of the British army, and their children nearly a tenth. Many had followed their husbands across the ocean, but British and Loyalist units also recruited American women and succored wives of Loyalist refugees. All these women worked as nurses, cooks, washerwomen, and seamstresses and sold liquor. The wives among them took care of ill husbands, and on occasion women joined men in plundering. Twenty thousand women, including wives of officers and privates and a few female hawkers, followed the Continental army. Wives stayed in camp, despite filth and poor rations, because the war had turned them into refugees or made them too poor to succor their families alone. Army women provided essential support services. They labored as hospital nurses, cooks, washerwomen, seamstresses, and tailors. Some carried ammunition for canons and water to swab them down, and a few were officers' servants. They lived under army discipline and got a stipend and rations. The army hired every nurse it could find but faced shortages, for few women wanted to expose themselves to disease and filthy camps.[25]

By necessity women took over the farm when their husbands were away, adding the burden of hiring laborers and working in fields to their rounds of gardening, milking, churning, spinning, sewing, and caring for children. Women had always helped with harvests and taken over when their husbands died or went away for business. But the war multiplied the number of women thrust into unaccustomed male roles, forcing them to plant, cultivate, and harvest alone while their other responsibilities increased. The disappearance of imported cloth and inflation mandated frugality and an increase in home industry. They had to buy needed goods and collect debts owed them, but they often had little detailed knowledge of the value of the farm or their husbands' assets.[26]

Knowing that women could not manage alone (any more than men could), farm families struggled mightily to find workers. They pursued a range of strategies. Farm men evaded military service or shaped it to family needs by deserting, hiring substitutes, or refusing to enlist. In unsafe areas (such as the Iroquois and Ohio countries) they used armed militiamen to protect and cultivate farms. They hired prisoners of war or expanded the cottager system. But none of these strategies prevented the endemic scarcity and even hunger that stalked city and countryside alike.

Once war weariness set in, farm men regularly deserted, especially during planting and harvesting seasons, hoping to salvage the family's crops and prevent hunger. Some New Jersey Loyalist militiamen deserted in mid-August 1777, just before corn and vegetable harvests. One in five Continental soldiers deserted during the war, often within weeks of enlistment; a third left in 1778, and a quarter disappeared in 1781. A tenth of Virginia soldiers on two muster rolls deserted—men with land more often than laborers. Soldiers from Worcester, Massachusetts, rarely deserted, but those who did left in July (for the haying season) or during the fall harvest. Harbored by their families and protected by their communities, deserters rarely faced punishment for their behavior.[27]

Families often kept one son at home to help aged fathers or widowed mothers, or they hired substitutes (sometimes their sons or younger brothers) to serve in place of their men, thus conserving family labor at planting and harvesting seasons. Nearly half (46 percent) of soldiers in Lancaster and Northampton Counties, Pennsylvania, enlisted as substitutes, as did three-tenths of 1778 enlistees in Charles County, Maryland, and at least a fifth of New Jersey's and a sixth of Virginia's enlistees. Over a third of Virginia substitutes replaced fathers, brothers, or other kin, and a similar pattern could be found in frontier Pennsylvania (one in fifteen) and in Massachusetts. When planters in Charles County could not persuade (or did not have) kinfolk to serve for them, they hired poor, landless wage laborers; several Massachusetts towns similarly sent British deserters in place of their men.[28]

The men who remained at home disobeyed military orders. In the summer of 1776, New York, New Jersey, and Connecticut militiamen refused to enlist in the army, and men already mobilized often deserted. Militiamen in Essex County, New Jersey, told their commander in late June 1776 that "if they are take in at this Season . . . they may as Well knock their famalys on the head for that they will be Ruined." In August 1777 some Baltimore County men—refusing to take up arms because they could not "leave our plantations untilled, nor our wives and children to mourn in our absence"—ran to the woods to avoid recruiters. Once they had served terms totaling about eighteen months, Virginians refused to sign up again. In March 1781, 132 Orange County planters—most of whom owned few or no slaves—complained about an eighteen-month army draft and instead sought three-month militia terms. To serve so long not only meant "a seperation for that length of time from our dear families" but also "losing our Care of two Crops [and] must in all human probability come to misery and ruin."[29]

While men in settled regions evaded military service, frontier families begged for soldiers to protect them as they worked. As they fought Indians, rangers (militiamen) not only guarded farmworkers but occasionally herded livestock and cultivated crops. During the summer of 1778 farmers in Bedford County,

Pennsylvania, held "their weapons with one hand and labor[ed] with the other." While serving in the Pennsylvania militia in the late summer of 1781, James Huston's company stayed two weeks at "Standing Stone on Juniata River" and went out "to assist the country people in putting in their grain and hay." When frontier women thought it safe, they worked—sometimes in groups with the remaining men—in fields clustered around a fort. Even when attack imperiled borderland communities, women helped with the work, as did Vermont women in 1779 who stayed in towns in the path of the British, despite an order from the general assembly to abandon their farms.[30]

Finally, farmers searched for new supplies of labor, using women, cottagers, soldiers in camp or on furlough, exiled Loyalists, or war prisoners. Southeastern Pennsylvania farmers hired more cottagers. Farm families, used to exchanges of labor, relied on neighbors to help with the crop. During the 1776 harvest in Norwich, Connecticut, "the Men being universally gone" and "the Burden being so heavy on Those who are left," fifty young women "assembled together" and voluntarily husked about 700 bushels of corn.[31]

Farmers eagerly hired prisoners. Mid-Atlantic and Maine farmers employed captured Hessians. German families in isolated areas increased the family labor supply by allowing daughters to marry runaway Hessian prisoners. Others, desperate for labor, hired British prisoners of war attempting to escape. American officials rarely allowed "Convention Prisoners" (those who surrendered at Saratoga) to work outside prison camp, but in 1781 and 1782 numerous farmers in the Lancaster, Pennsylvania, area employed them. Piedmont Carolina planters also hired British prisoners for the harvest, prompting General Nathanael Greene to forbid prisoners from seeking employment outside camp.[32]

Farmers bid up the pay of the men they did hire, and they themselves worked only for market wages. Officials repeatedly complained that competition from farmers prevented them from hiring teamsters, hospital workers, and artisans. Pennsylvania farmers refused to send wagons to cart goods to Valley Forge because the state offered 30 shillings in Continentals, while they could hire out wagon, driver, and team for £3–£4 specie. In May 1779 Moore Furman failed to hire teamsters because they got "higher Wages at home." Even after the army hired teams, they could find few workers because "farmers and country give more wages than we do."[33]

Labor shortages combined with military requisitions and occasional poor weather and insect infestations to reduce production. Lessened production led to shortages of bread, meat, and salt (essential for curing meat), most pressing in the towns but appearing on occasion in the countryside. By the summer of 1778 the service of farmers in the military had reduced the quantity of land cultivated in the mid-Atlantic states and with it the supply of grain and fodder. As military demands for beef and wool jumped, the number of livestock as well as the supply of butter and meat plummeted. By the middle of the war, shortages of grain had become common. Dearth reached Dutchess County, New York, in late 1776; Ulster and Dutchess Counties the next year; the Philadelphia region in 1777–78; New Jersey in early 1779; Livingston Manor (in the mid–Hudson River Valley), where "A great Cry for bread" rang out, in late 1779; and northern New Jersey and parts of New York in the spring of 1780. Throughout the war, hunger stalked Maine lumbering and fishing villages, which were used to importing their food. Scarcity hit the southern states, too. Salt riots occurred on Maryland's Eastern Shore in 1776 and in the central North Carolina Piedmont during the summer of 1777. Indian warfare left Kentucky pioneers hungry in 1777, and dearth reached southern Virginia in early 1779 and Delaware by the summer of 1780. Intensive militia foraging in the early 1780s led to shortages of fodder and provisions in the Carolinas.[34]

In times of dearth, poor men with no sons at home and especially wives or widows of soldiers beseeched military, local, and state officials to pay their subsistence and turned to neighbors for help when officials refused. Although many New Hampshire towns, for instance, gave firewood, corn, beans, potatoes, and salt to wives of Continental soldiers, they supported only legal residents they deemed worthy. And gentlemen, sometimes relishing their role as patrons and protectors, tried to forestall hunger, even keeping their provisions from the army to feed their neighbors.[35]

The Continental Congress tried to forestall dearth and save supplies for the army through legislation. Fearing widespread privation, Congress prohibited the export of grain and provisions from June 1778 to November 1779. To conserve flour for bread, in October 1778 it mandated that no wheat be used as forage. Knowing that the embargo hurt poor New Englanders, in 1778 and 1779 Congress asked states to allow grain export to Massachusetts and Rhode Island and tried, with some success, to get rice from South Carolina to New England through the blockade.[36]

"The want of men to do the work," Richard Durfee of Tiverton, Rhode Island, remembered, "who had mostly gone into public service, was not the only or main difficulty in the way of tillage of the earth." The town was a constant battleground. If a farmer did "plant his land, he was never sure of receiving his crop" but "was in constant jeopardy of losing it through the hostile attacks of the enemy" and "by the numerous wants and necessities of our own army, which lay encamped" in town. Unable to make a living, townsmen "entered the regular army" in great numbers. Without men to till the soil and with both sides taking what they did produce, dearth haunted the town. "Some people" had to grind "flaxseed and cobs together to make bread" or make "potato bread," stew "sweet apples," and grind "cornstalks to obtain the juice to boil down as a substitute for molasses." As Tiverton's example suggests, the ravages of war aggravated shortages, leaving ever greater numbers of women and children hungry.[37]

The Ravages of War

Westchester County, New York, had been, Continental army surgeon James Thatcher wrote, a "rich and fertile" country, but in 1780 it stood "in ruins." Many families had "abandoned their farms," the Tories escaping to New York City and the patriots going "into the interior." The few who stayed "find it impossible to harvest the produce. The meadows and pastures are covered with gras of a summers growth, and thousands of bushels of apples and other fruit are rotting in the orchards." Timothy Dwight, American chaplain in 1777, remembered that farmers' houses there "were in a great measure scenes of desolation. Their furniture was extensively plundered or broken to pieces. The walls, floors, and windows were injured both by violence and decay, and were not repaired . . . because they were exposed to" repeated "injuries." Their farms stood deserted; "their cattle were gone; their enclosures were burnt" for fuel; "their fields were covered with a rank growth of weeds and wild grass." Afraid of "new injuries and sufferings," those who stayed never ventured out except for "a rare and lonely excursion to the house of a neighbor."[38]

From 1775 through 1782 Westchester County became a no-man's-land, and its 4,000 families suffered constant insecurity and plunder. Neither side dominated the county, but both wanted to control it because of its strategic location on the Hudson River north of New York. Armies took provisions, leaving families too little to last through winter. Contending armies, militias, and partisan bands raided friend and foe alike, pilfering personal property, stealing livestock, burning barns and houses, and cutting trees and fences for firewood. Bandits looted what armies and militias left behind. In this violent atmosphere both sides tried to gain farmers' allegiance. Far from winning the hearts of the populace, indiscriminate looting left farmers numb, frightened, disaffected, and neutral.[39]

By 1778 the contending forces had turned many families into refugees, especially in the no-man's-land between the armies. If refugees stayed in the country, they suffered new violence. In the fall of 1778, patriots appeared at a Tory refugee camp near Morrissania and "plundered & burnt the huts of the Refugees," leaving 490 "poor men," most with "large families" without lodging or food. A Welsh officer reported that the British army gave these people tents, food, and soldiers' pay as long as "they continue[d] under the present circumstances"; at the same time the army in New York collected money for their support. In 1780 more Loyalists turned refugees, for patriot authorities made them move from the battle zone to the interior, allowing them to take just a six-month supply of food.[40]

Although the destructiveness of warfare in Westchester County was unique, combatants destroyed the property of farmers nearly everywhere. Nothing was safe. Footloose and hungry young men marching to war stole food; British and patriot army supply officers took grain, horses, cattle, and slaves; patriot, French, and British troops (and their Indian and German allies) plundered friend and foe alike. Contending governments confiscated farms; armies burned grain, houses, and barns, or turned their horses onto growing crops. Loyalists and Indians burned down outlying patriot settlements; patriot irregulars and armies destroyed Loyalist and Indian communities. Sometimes, as in Valley Forge and the Carolinas in the early 1780s, several types of destruction occurred simultaneously: armies raided for provisions or fodder, and men on raids plundered farms against orders.[41]

Victims of military action equated requisitions with plunder and raids because soldiers in all cases took their food and livestock or burned their homes, leaving them destitute or homeless. Military authorities, in contrast, distinguished plunder by soldiers (which they prohibited) from requisitions (which General Greene compared to a tax that "the people must agree to pay for the common benefit resulting from a military force") and raids (which they called acts of war). In August 1778 Scot Major Patrick Ferguson urged raids that would leave the country devoid of supplies; he expected that the troops would live off the country while on the march. General Washington, who forbade plunder, nonetheless allowed soldiers on "scouting parties" to take the enemy's public stores, provisions from farmers trading in Philadelphia, or cattle from Tories as compensation for their "extraordinary fatigues, hardship and danger." He required that they hand over the booty to officers to be sold at auction, and the proceeds were then to be divided among the men.[42]

To preserve horses and livestock for their own use and deny livestock and fodder to their enemies, the Americans and the British alike not only requisitioned them (often without recompense) but moved them at the first rumor that the enemy was nearby. Patriots regularly ordered cattle removed to safe havens, giving such orders in 1775 in Chelsea, Massachusetts, and Rhode Island. In the summer of 1776 General Washington ordered that "all the Stock & Horses, except what is absolutely necessary for the Support of the Inhabitants be removed from Long Island, Staten Island, & the adjoining Coasts," as well as the New England coastal islands. The Continental army continued to order livestock and provisions moved away from the British throughout the war: from March 1777 through 1780 from the Jersey coast; from the Wilmington area, Darby Creek (Pennsylvania), and Kent County, Maryland, during the Valley

Forge winter; in May 1780 from parts of Maryland's Eastern Shore; and from the North Carolina coast in late 1780.[43]

Removal of livestock and horses caused great hardship. Small farmers relied on cows for milk, cattle for meat, oxen for plowing, and horses for transportation. Just before their livestock was to be moved, families in Chelsea, Massachusetts, urged that they be allowed to keep "one pair of working oxen to a famaly and a few Milch cows." The army did not object, but farmers "would have to take the utmost care that they do not fall into the enemy's hands." The army was stingier in New York, where it proposed that inhabitants keep no more than three milk cows for a large family, two for a middle-sized family, or one for a small family. Local notable Benjamin Kissam, who directed the removal of thousands of cattle and sheep from one Queens County neighborhood, reported that "some among the poor sort . . . , must be left to starve," for they had counted on slaughtering a cow they had been fattening to see them through the winter.[44]

The British army often foraged for sheep, cattle, and fodder. In August 1775 they took 2,000 sheep and 150 cattle off Fishers' Island (near New London, Connecticut) and 823 sheep and 62 cattle from Gardner's Island (off Long Island). In early 1777 New Jersey Regulars went on twenty-two foraging campaigns; foraging on eastern Long Island in early 1779 quickly stripped the area of fodder. After the British took over Long Island, they not only took cattle the Continentals had left behind but insisted farmers sell goods to them at a set price, on pain of confiscatory taxation. Island Loyalists (including refugee farmers they put on farms) suffered repeated confiscation without payment. The British came out of Philadelphia throughout the winter of 1778–79, on occasion burning houses while taking provisions, livestock, fodder, and straw from farmers in the city's suburbs, Bucks County, and New Jersey. The British pursued similar strategies in the South. With forage in short supply after the conquest of Charleston, they took horses and cattle from outlying islands and destroyed crops there in the fall of 1781; marching through the Carolinas, General Charles Cornwallis tried, with meager success, to obtain provisions from farmers.[45]

Farmers viewed Continental army requisitions as theft of food, fodder, cloth, wagons, or teams the family needed for subsistence or local exchange. Such seizures, General Washington knew, led to the "most pernicious consequences." They not only "excited the greatest alarm and uneasiness, even among our best and warmest friends" and spread "disaffection, jealousy and fear in the people," but "never fail . . . to raise in the Soldiery a disposition to licentiousness, plunder, and Robbery, difficult to suppress afterwards." Provisioning expeditions in the Delaware Valley by both armies fulfilled Washington's prophecy, and bandits masquerading as soldiers stole what the armies and soldiers left behind. The Continentals, desperate for clothing, requisitioned all the cloth, clothing, and blankets inhabitants of Bucks, Chester, Philadelphia, and Northampton Counties could spare. Officers gave certificates for future payment for all goods taken, but gleanings were slim. The "disaffected," General Washington reported, "hid their goods," hoping for higher prices, and "our friends had, before, parted with all they could spare." Even force turned up little more.[46]

The Continental army's efforts to feed and clothe itself led to civil war with farmers. Through the winter of 1778, army commissaries and hungry soldiers rambled through the country, seizing provisions, straw, fodder, and livestock "under the Shadow of the Bayonet & the appellation Tory." Local officials impressed teams of horses and drivers to cart provisions to Valley Forge, paying them too little even to maintain their carts; foragers burned what fodder they could not take. Bands of soldiers "insulted and abused" farmers while they took forage and left them without "a reasonable share for the subsistence of their families"; they came back and took more, even when shown a certificate stating that the farmers had "furnished their Quotos." Officers, who knew that the Delaware Valley had become barren, continued to depend on it for the army's sustenance. Fearing starvation, farmers hid their grain and cattle in neighbors' barns, the woods, or swamps. They refused to thresh their wheat, sell cattle, or lend wagons to the army. They sold grain to speculators or took it (along with livestock and horses) to the British in Philadelphia, where they got trade goods and hard money. Although farmers saved some cattle, horses, and flour for use or the Philadelphia trade, armies swept the country clean, whipping civilians who traded with the enemy, taking everything they could find, and threshing wheat that farmers tried to keep off the market.[47]

The Continental army managed relations with civilians more amicably during the frigid 1779–80 winter at Morris Town, New Jersey. The winter started badly. From late December through mid-January, hungry men plundered food and stole poultry and stock. By early January heavy snow blocked the roads and made getting supplies almost impossible. With neither bread nor meat for days, the freezing, ill-housed, and ill-clad men faced starvation. Generals Washington and Greene, however, wanted no repeat of the Valley Forge experience. Greene asked inhabitants and the militia to clear the roads; Washington sent officers to urge magistrates to go with the army to inventory cattle and provisions and persuade farmers to part with them. Farmers brought provisions and cattle to be weighed to a central location, where they received certificates for the market price of their goods. Only if farmers refused to part with their surpluses would the army impress the goods, and even then it could take no milk cows.

New Jersey farmers willingly helped provision the army because the army treated them with respect, and because they knew that if they helped clear roads, Pennsylvania provisions would reduce their burden. Moreover, the army asked for goods farmers could spare and spread the requisitions over the state, giving higher quotas to more productive counties. When farmers lacked cattle, the army accepted buckwheat, rye, and corn instead. "Very Hospitable" farmers, General William Irvine reported in early January, gave what little they could spare, but they gave him "a thousand apologys, particularly scarcity for their families." By January 24 General Washington reported that the camp was "tolerably well supplied with provision," because farmers "more than complied with the requisitions." Farmers continued to help supply the camp through winter and early spring, but without pay for their efforts, they quickly became restive.[48]

The reasons for the army's difficulties at Morris Town are not difficult to discern. With the great inflation of Continental money, the system of military requisitions collapsed. In 1780 Congress set quotas for essential war goods—grain or flour, cloth, beef, and bacon—that each state had to meet. The requisitions failed to meet the army's needs, however. General Washington calculated deficits of 100,000 barrels of flour, 11,600 tons of hay, and 756,300 bushels of grain fodder. But states failed to come up with even these quotas. Since local collectors would not leave farmers, who had tiny surpluses, in a state of starvation, they refused to take what Congress demanded. Because the army never knew what supplies states would provide, it had to "subsist the Horses of the Army by force." Such force gave "rise to Civil disputes" with citizens and magistrates and even worse conflict in the Carolinas in 1780 and 1781, where the Continental army and its British counterpart denuded the countryside of provisions and livestock.[49]

From the beginning of the war to its end, soldiers plundered farmers' crops, fodder, livestock, and personal property, despite repeated prohibitions by both American and British commanders, because line officers undercut the orders of their superiors. British and Hessian officers let most offenders go unpunished, and many supported campaigns of terror as the best way to bring the colonies to submission. Sympathizing with the hardships of their men and viewing all Americans as rebels, they looked the other way when their men stole food or even personal property. American officers similarly justified plunder, arguing that they took only from Tories, who richly deserved the punishment. American line officers often stood aside while their men plundered; even worse, they bought goods soldiers had plundered or—in a few cases—led their men in plundering expeditions.[50]

Soldiers not only took food but looted the property of farmers. Everywhere they went, they demanded food and lodging or stole food farmers needed for their families, leading them, as General Washington told his men in July 1777, to "dread our halting among them, even for a night." Recruits usually had no food for their trip, often 100 miles or more, from home to army camp. With a few dollars of worthless Continentals in their pockets, they stayed hungry or took from farmers. They often suffered hunger even after they arrived. Troops—loyalist, English, patriot, and French—regularly stole food from nearby farmers to supplement their meager rations and burned fences to cook or keep warm. Hungry Continentals, for instance, roamed the countryside around Valley Forge (in the winter of 1778) and Morris Town (in the winter of 1780), robbing cornfields, stealing provisions farmers preserved for winter, and extorting food and drink. Wherever the army encamped—Valley Forge, New Jersey, New York, or the Carolinas—civilians complained that soldiers looted their property. The men stole from poor and rich alike but often targeted rich Loyalists. Angry, cold, and wearing little but rags, at Valley Forge the troops lashed out at comfortable farmers they believed had abandoned the cause, stealing gold and silk clothing, breaking furniture, and wrecking houses near the camp.[51]

British, Hessian, and Loyalist soldiers plundered as much as the patriots. Lurid stories notwithstanding, Hessians looted no more than other royal soldiers and for the same reasons: hunger, cold, need for forage, and hatred of rebels. Contemptuous of the patriots' rebellion against authority and angry at hostility directed at them, they turned on their tormentors; unwilling or unable to distinguish between friend and foe, they plundered patriot and Loyalist alike. Envious of productive farms and the riches they saw around them, they targeted the biggest farms (even those of noted Loyalists in Rhode Island and New Jersey), where they destroyed luxuries and carried off booty.[52]

All early modern armies, with their battles on farms and in villages, ruined civilian property. Razing of farms by armies as they fought, gathered forage, or searched for provisions was extraordinarily widespread. Armies and militias especially targeted men of substance who led opposing movements and supplied their neighbors, but small farmers suffered as well. Indian and Loyalist warfare on the borderlands reduced New York and Pennsylvania frontier settlements to rubble; retaliatory marches by patriot militias and the army ruined Loyalist settlements and Indian villages. British marines and army troops raided and burned villages and settlements from New England to Georgia. Contending armies and partisan bands roamed New Jersey from 1776 to 1778, Virginia in 1778 and 1781, and the Carolinas from 1780 to 1782, spreading destruction wherever they marched.[53]

The most severe British raids took place in Connecticut, where the British regularly gathered provisions, took livestock, or destroyed military stores before returning to their ships. Nearly all the raiders landed by boat from Long Island or New York and marched as far as twenty-nine miles inland. Usually getting some advance warning, most residents fled, leaving empty villages. Between 1777 and 1781 thirteen Connecticut coastal towns suffered British raids, several more than once. During the April 1777 Danbury raid, for instance, soldiers destroyed military stores (including 3,000 to 4,000 barrels of meat and 1,000 barrels of flour) and burned down 22 dwellings (of 400 in town) and 22 barns.[54]

American troops indulged in similar, if less extensive, raiding. To retaliate for raids on Connecticut's coast, Connecticut and Westchester County patriots commissioned privateers to invade Long Island and seize any goods plundered from Connecticut; while there, they looted homes of local Tories and Whigs. In 1777 they raided Sag Harbor, led by a refugee who had fled the town for Connecticut. Privateers stole clothing, silverware, and luxury goods during fall 1778 raids and later sold them in Connecticut. Another group of raiders destroyed 300 tons of hay at Coram during a foray in 1780, and a fourth party of privateers attacked East Chester in 1781.[55]

Partisan warfare, with its pillage and plunder, hit New York, New Jersey, and the lower South most severely. Between 1776 and 1780 armies and partisans created havoc in New Jersey. In 1776 and 1777 the British and Hessians burned mills the enemy might use and destroyed houses that would protect them from British fire, but at the same time they slaughtered livestock, burned hay and grain, and looted property. During early 1777 the two sides sent large armies through central New Jersey in search of forage and cattle. The raids devastated farming. The "track of the British army," the Continental Congress reported, "is marked with desolation and a wanton destruction of property, particularly through Westchester County" and "Newark, Elizabethtown, Woodbridge, Brunswick, Kingston, Princeton, and Trenton"—all the major farm towns on the road from New York to Philadelphia.[56]

The backcountry Carolinas and Georgia suffered more than any other place except Westchester County from partisan warfare. Five-sided warfare, involving British regulars, Loyalist and local patriot militias, the Continental army, and Cherokees increased the carnage, as partisans on both sides plundered livestock, stole slaves, burned farms, and massacred civilians. Early in the war Loyalist Rangers from East Florida repeatedly raided Georgia for cattle to supply the British at St. Augustine, but the British conquest of Charleston opened an extraordinarily violent civil war. Loyalists savored the British victory, clamoring "for retributive Justice." The British kidnapped Lowcountry civilians and took their slaves and invaded the Georgetown District three times, burning fifty plantations. Loyalists started a fierce civil war in the upcountry in late 1780, burning houses and plundering. In response patriot forces went on a rampage. They plundered British stores previously taken from unwilling farmers, burned houses, killed cattle and sheep, stole horses, whipped slaves, insulted civilians, and murdered overseers in Loyalist strongholds, turning the area, a Loyalist merchant complained, into a place of "confusion, robbery, and murder."[57]

As the war moved north in late 1780, the two armies spread havoc throughout North Carolina and Virginia. During the fall of 1780 the British spent five weeks around Charlotte, North Carolina, foraging, burning, and plundering as they moved. When Cornwallis's army moved northwest of Charlotte in January 1781, it ravaged the countryside, burning houses and provisions. Even worse, Tory partisan bands, formed in the wake of the invasion, pillaged the backcountry. What the Loyalist militias failed to plunder, the occupiers destroyed or took in raids and foraging expeditions. Patriots responded with equal savagery. As a result, neither side could find enough food to support armies, nor did enough food remain to feed civilians. Indiscriminate plunder and wanton murder reduced many citizens "to beggary," one resident reported to General Greene in July 1781. Residents trying to feed their families refused to give supplies to either side. With mass starvation, especially of poor families, a possibility, the general ordered General Andrew Pickens to impress "necessities" from those with "plenty" and give them to those in need.[58]

The Chesapeake region suffered much after the British invaded in 1779, 1780, and 1781. When the British landed in 1779, they burned several ports on the James River and plundered plantations, taking as many as 1,000 slaves. Throughout the first half of 1781 the British drove off and killed cattle along the York River. They attacked farms all over the James River basin and the Northern Neck, destroying munitions and fences, stealing cattle and horses, burning at least 6,000 hogsheads of tobacco and much grain, ruining plantations of gentlemen, and setting slaves free. Everywhere they targeted the rich, sometimes leaving the farms of the poor untouched. The Continental and French armies—thinking everyone a Tory—behaved little better.[59]

The Balance of Terror

Wartime violence terrorized farm families, and although women suffered the most, victors tortured and terrorized men as well. After the battle of Bennington, victorious patriot soldiers tied Loyalists (probably soldiers) to horses driven by slaves. New England Tories, too, reveled in terrorizing their captives. In July 1777 a party of Tories and Indians invaded Samuel Churchill's farm in western Vermont. Although they heeded the pleas of women to save the house, they humiliated Churchill, tying him to a tree for hours and yelling at him over and over, "Tell us where your flour is, you old rebel!" Torture became a fine art in the North Carolina backcountry. In October 1780 patriot militiamen beat a Wilkes County Moravian named Schemel and his wife and then plundered their farm.[60]

Invasions forced farm women left alone to defend their homes and children from men bent on plunder, murder, and rape. Backcountry women took their children and hid in the orchards or woods at the first hint of attack, sometimes watching as the men burned their homes. More than plunder, farm women feared rape at the hands of enemy soldiers. Whereas American troops occasionally raped enemy women (especially Indians), British troops did so more often. During the 1776–77 campaigns an epidemic of rape by British troops spread from Staten Island through New Jersey. In one particularly horrifying case, British soldiers repeatedly raped thirteen-year-old farmgirl Abigail Palmer, despite her grandfather's pleas and her own screams. The men returned and raped Palmer's aunt and a friend and finally dragged Abigail and a young friend to their camp, where more men raped them. Two years later similar sexual assaults broke out in Fairfield and New Haven, Connecticut, when the British raided those towns. In 1781 and 1782 British or Loyalist troops assaulted a few girls near Hillsborough, North Carolina.[61]

Military victory by hostile forces perpetuated terror and resistance to it. In the spring of 1776, as soon as American forces occupied western New York, Captain Joseph Bloomfield reported, they interrogated "Scotch, Dutch, & Irish Tories." "It is very surprizeing," he observed, "to see what a consternation & fright the Tories . . . are thrown into. Those miserable Wretches are afraid to be seen by any of the Soldiery" and "behave with the greatest servility imaginable . . . & make all the Promises for their future good behavior that can be desired, upon which they are usually dismissed." Most "signed the Association" or "gave bonds for their future good Behavior." Still, many stood unbowed. The Americans imprisoned 52 of the 195 men ("the most dangerous and incorrigible") interviewed. Even the meek resisted. Fear, Bloomfield thought, prevented the women from selling "fresh Provisions, Butter, Eggs" to the occupiers; more likely they hid from the soldiers in silent protest.[62]

Indian warfare struck far greater terror than armies and militia alone. These battles grew out of earlier wars, in which settlers and Indians had pillaged and massacred one another while they fought over rich farmlands. The struggle between Tory and Whig intensified an earlier war without boundaries between Indians and settlers, in which every person became a combatant and in which revenge answered revenge in a perpetual cycle. Tories, British regulars, and Indians stood against the Continental army, militias, settlers, and their few Indian allies, each vying to outdo the other in destructiveness. The Americans (but rarely the British) burned Indian villages, destroying their corn, beans, and squash. This scorched-earth policy led to retaliation against American soldiers and frontier communities. Hundreds of civilians—Indians and settlers—died, many of them victims of brutal massacres. Racial slaughter fed racial hatred, leading to yet more violence.[63]

Warfare in the Iroquois country illuminates this barbarity. The British, Tories, and Indians repeatedly struck the Mohawk River frontier, rich and well-peopled lands close to Indian territory. In July 1778 they raided Springfield, burning all the houses and barns. From September to November 1778 John Butler's Rangers, 800 strong, attacked the Mohawk Valley, burning corn and farms. Retaliating for Sullivan's raid the previous fall, John Johnson led Loyalists and Mohawks in May 1780 in "Burning and laying Waste everything before them." They killed cattle, burned or took corn and flour, destroyed 120 houses and barns (burning 60–70 sheep living in them), and stole 70 horses. In September and October they returned to burn 200 houses and destroy 600,000 bushels of grain. The next October another expedition plundered the area, destroying 100 houses and 80,000 bushels of grain.[64]

American retaliation left the Iroquois country desolate. Many Iroquois had adopted European farming and housing styles and thus had much to lose. Ordered by Washington in the summer of 1779 to lay waste all the Iroquois settlements, General John Sullivan and his men systematically destroyed Indian communities. The troops stole skins, plates, and kitchen utensils; burned forty villages, 160,000 bushels of corn, bean and potato fields, along with fruit, vegetables, and fowl; chopped down orchards; and set cattle and horses loose. With their crops burned, their men subsisting on green corn and defeated in battle, Indians (and the Tories who lived near them) fled villages before the onslaught.[65]

Murder of noncombatants, most often done to avenge earlier deaths, accompanied raids. Cherokee raids in 1776 on the North Carolina frontier killed 37 people; the next year the Creeks killed 20. Both the British and the Americans paid for enemy scalps, thus encouraging the killing of civilians. At Cherry Valley, Pennsylvania, in 1778 the Senecas murdered 31 women and children, retaliating for patriot massacres of their people. They took 30 to 40 surviving women and children with them as hostages. Indians and Tories at Wyoming killed 210 Yankee defenders and settlers. The Americans tortured, killed, and scalped as many noncombatants as the Indians. In 1777 American militia murdered 4 Shawnee neutrals, held as hostages, as revenge for a white man the Indians had killed. A year later the Westmoreland County, Pennsylvania, militia carried out a "Squaw Campaign," deliberately murdering a man, 4 women, and a boy. In 1781 Colonel Daniel Brodhead's men executed 16 warriors they captured; in April 1782, after hostile Indians had repeatedly murdered settlers, enraged frontier militia massacred and scalped 93 friendly Delawares (90 of them were Moravians—29 men, 27 women, and 34 children), as they sang hymns and prayed.[66]

Refugees and Émigrés

Wartime violence turned many families into refugees desperate to find food and shelter. Knowing what destruction lay ahead, farmers abandoned their homes at the first rumors of invasion. Brave souls who defended their homesteads watched as soldiers plundered their possessions, burned their houses, and even slaughtered their families and neighbors. Unable to survive on their ruined farms, they, too, took to the roads in search of asylum. Refugee families not only abandoned their farms, reducing labor and farm production, but burdened their hosts, who often had to find them food and clothing.[67]

Loyalists rushed in as soon as the British captured Boston, New York, Philadelphia, Newport, Charleston, and Savannah; other refugees reached Canada, leaving thousands of farms uncultivated. The largest number of refugees, some 25,000, crowded into occupied New York City. They came from upstate New York, New Jersey, Virginia (masters and slaves), and other states, taking over homes patriots had abandoned. The city's population jumped from 5,000 in the summer of 1776 to 11,000 in February 1777. When the British abandoned Philadelphia, perhaps 3,000 refugees—most already thrust from their homes—poured into New York, raising the city's population to 25,000 by 1781. In 1783, after the British evacuated Charleston, the population of the city jumped to 33,000 (plus 10,000 soldiers). Lacking any means of support, 758 to 868 refugees, mostly women and children, required British aid to survive. During 1783, with peace at hand, 29,200 refugees sailed away into exile, leaving the city to the returning, victorious Americans.[68]

Patriots dispossessed as many southern Loyalists. Fearing Indians less than Whigs, western North Carolina Loyalists went to Kentucky. When patriots recaptured South Carolina's frontier Ninety Six District in 1781, Loyalist troops—some with families—marched to Charleston; after retaking Augusta that September, patriots made Loyalist women and children join husbands and fathers in occupied Savannah. Thousands of others fled coastal plantations for Florida, the Natchez region, Savannah, or Charleston. When the British abandoned Charleston in 1783, 9,127 people (5,333 of them were slaves) left for New York or the West Indies. Florida became a haven for thousands. Between 1775 and 1783 West Florida's population doubled and East Florida's multiplied five times. The Spanish takeover of East Florida at the end of the war forced 2,948 residents (with 6,240 slaves) to flee again, most to the West Indies.[69]

Patriot families fled the British in equally large numbers. A 1775 raid on Falmouth, Maine, drove inhabitants from town, and patriots abandoned the Penobscot area in 1779 after the British burned their homes and stole their cattle. Many patriots left Boston for Cambridge, Worcester, or Providence during 1775; 12,000 more fled British rule and the American siege in 1776. As contending armies fought over Charlestown, Massachusetts, in summer 1775, residents left; the British then burned the town, driving townsfolk "into the wilderness for security from British violence." As the British moved toward New York, patriot families fled before them. After the British conquered Long Island in 1776, 1,000 families (a fifth of the island's people) fled to Connecticut. In August 1775 and February, March, and April 1776, rumors of imminent invasion emptied New York City. The situation became so precarious in August 1776 that General Washington ordered the evacuation of infirm men, women, and children. Only 5,000 of the 22,000 residents of New York remained after the British finally occupied the city. Many crowded into adjacent New Jersey or Orange County, New York, where nearly every farmer took in one or two refugees.[70]

British actions turned just as many southerners into refugees. In 1775, after the British took over Norfolk, patriot families fled to nearby Suffolk County; the British fired on the town before leaving it in December 1775, creating more refugees. After the British occupied Charleston and demanded loyalty from the planters, many fled into North Carolina and Virginia, taking their slaves with them, and General Cornwallis sent others to St. Augustine. In 1779, when the British invaded backcountry Carolina, women fled to their neighbors or escaped to the mountains. Some patriot families in the Charlotte area, a local planter wrote, "abandoned their habitations," some fleeing "with such of their property as they could carry," while others fought Cornwallis.[71]

Invasion by Indians, Tories, and the British emptied much of the northern and trans-Appalachian frontiers. The defeat of the Quebec expedition in 1776 led Vermont farmers to flee, as did rumors of a British invasion in the spring of 1777; in July, after the surrender of Ticonderoga, other Vermont families scurried toward Albany. In March 1779 an Indian attack in Westmoreland County, Pennsylvania, "struck the inhabitants . . . with such a panick that a great part of them were moving away"; a similar panic in November 1780, after Tories and Regulars invaded western New York and Vermont, depopulated the region. When the British and Indians plundered the Mohawk Valley in October 1781, the expedition's leader reported, "The Inhabitants fled precipitately in the Night." The population of Tryon County, at the epicenter of this conflict, dropped from about 10,000 to 3,500.[72]

The southern frontier, with its widely scattered settlements, suffered severe depopulation. The 1776 Cherokee invasion of the Cumberland Valley sent women and children scurrying several hundred miles to southwest Virginia; they returned later that year but fell prey to new attacks and again fled to Virginia, never to return. Only the hardiest frontier families, numbering several hundred, ventured into Kentucky or Tennessee in 1777–80, hoping the Indians would not strike again. When Indians went to war, as they did in 1777–78 and 1782, even the most hardy folk left. As competing militias and armies trod about the backcountry plundering and (on occasion) murdering the residents, farmers took to the roads. Everywhere Cornwallis marched in the Carolinas, farm families fled, carrying or carting whatever worldly goods they could. Similarly, after the Kentucky militia foolishly attacked a larger invading British force,

Colonel William Christian reported that "numbers of People are now on the Road, moving out."[73]

Both sides confiscated land and chattels of enemy families on territory they controlled. After looting farms abandoned by patriot evacuees, British armies turned them over to Loyalist refugees. The British gave forty Long Island farms to 468 refugees, who grew provisions for the occupying army. They seized many estates of patriot refugees, Continental soldiers, and rebels in South Carolina and Georgia, saving a fourth of the estates' annual income for wives and children but selling crops and livestock that remained. Patriots refused to let families of Loyalist soldiers stay on their farms. At best officials took the land and returned a dower right to the woman or tried to support the family out of the estate; at worst, presuming the woman as guilty as her husband, they thrust her into exile. By 1780 New York officials gave wives of refugee Tories just twenty days to sell their estates and leave for Canada or face charges of treason; even worse, they had to leave sons over age twelve at home to serve in the Continental army.[74]

Every state confiscated and sold the land of British subjects or Loyalists who had fled or had fought with the British. Massachusetts banished and seized the lands of 309 people, 144 from Boston; New York divided confiscated estates into more than 300 farms and rented them to tenants. Rich Loyalists, who lost hundreds of thousands of acres, suffered the most, but small farmers lost land as well. Seven of the 35 Loyalists who lost land in Worcester County, Massachusetts, in 1779 were yeomen, while 23 were gentlemen.[75]

Their property seized, many Loyalists of modest means had little reason to stay in the country. Exiles moved to Canada, the West Indies, or Great Britain. Between 35,000 and 49,000 Loyalists migrated to Canada. At least 14,000 people, including many slaves, moved to the West Indies; another 7,000 to 8,000 reached Great Britain. Permanent Loyalist exiles numbered at least 60,000, comprising 2.5 percent of the American population, proportionately five times greater than the number of émigrés thrust into exile during the French Revolution. These émigrés represented a cross-section of the rural populace. Half the people (1,368 of 2,782) who claimed compensation for their losses from the British government were farmers, mostly men of small means. Nine-tenths of the New Brunswick emigrants were born in the colonies, and the vast majority were modest farmers or craftsmen. Seven-tenths of Ontario refugees came from the northern New York frontier; nearly half had been born abroad (predominantly Highlander Scots), and many of the rest were children of immigrants.[76]

Rebuilding Farm Markets

The war left much of the country—the Philadelphia area, Long Island, Rhode Island, the South—in shambles. Farmers spent the 1780s and 1790s recovering from war damage. They rebuilt burned-out houses and barns, expanded livestock herds, replaced torn out fences, replanted orchards, cleared new fields, paid debts ignored during the war, remade the borrowing system, and sought new foreign markets. But export markets reopened slowly, recovery of herds took years, profits dropped as farm prices plummeted after 1785, and specie remained scarce. Despite a brief postwar boom fueled by pent-up demand for imports, per capita income plunged from about $1,400 (in 1994 dollars) in 1774 to $800 in 1790; even after the prosperous 1790s, per capita income in 1805 reached only $1,200.[77]

Commercial farming began to recover as soon as the war ended, but southern planters, their farms ravaged by war, suffered greatly. Rice planters did not fully recover their markets. During the 1790s exports averaged three-quarters of the prewar total. Although backcountry Carolina and Georgia planters expanded tobacco production, growing English demand for cotton, along with the cotton gin, led eager Piedmont Carolina and Georgia planters to switch to that crop. As a result, cotton production multiplied during the 1790s, from 1.5 to 35 million pounds. To increase tobacco and cotton production, frontier southern planters craved more slaves. The heavy demand for those commodities gave them sufficient credit to buy almost 120,000 Africans before January 1, 1808, when Congress prohibited the practice. Most slaves worked in the Georgia backcountry; the rest, in Piedmont South Carolina, Tennessee, Mississippi, and Louisiana.[78]

The Chesapeake region, with its diversified economy, recovered more rapidly. Flour exports from western Maryland and the Susquehanna Valley boomed after the war, averaging 174,000 barrels during 1792–99, nearly four times the prewar amount. Throughout the 1780s productivity per worker in tobacco regions returned to prewar norms. Virginia exports jumped from 18,592 hogsheads in 1782–83 to almost 60,000 per year from 1786 to 1792, nearly the prewar average. Maryland planters exported an average of 27,000 hogsheads by 1791–92, compared to 30,000 before the war. But the French Revolution and the ensuing wars between France and Britain disrupted the tobacco trade. Between 1793 and 1798 Virginia warehouses inspected an average of only 35,700 hogsheads.[79]

Farmers in the Delaware, Hudson, and Mohawk Valleys—the granaries of colonial North America—resumed production as markets reopened in southern Europe and in the British West Indies. In 1773 Philadelphia merchants exported 266,000 barrels of Delaware Valley and upper Eastern Shore flour. In 1784 and 1785 the port sent 218,000 barrels to market (four-fifths of the prewar total); flour exports rose to 359,000 barrels in the early 1790s, a third more than before the war. New York, with a less populous hinterland, had always exported less than Philadelphia, but by 1788 New York farmers were exporting 322,000 bushels of wheat, 183,000 bushels of corn, and 62,000 barrels of flour annually.[80]

Despite mercantilist restrictions, the British and French West Indian markets regained their importance for mid-Atlantic and Chesapeake grain farmers and New England beef producers. New England and Pennsylvania farmwives exported 2.5 million pounds of butter and 1.8 million pounds of cheese in 1796, at least three-quarters of it to the West Indies, far more than before the war. The French islands provided the best markets. In 1790 they took three-quarters of the beef, more than two-thirds of the fish, and one-quarter of the flour exports; the British islands took an equal quantity of flour but little fish and meat. Nonetheless, in 1785–87 the British islands did import an average of 176,000 bushels of corn, 93,000 barrels of flour, and 17,000 barrels of bread.[81]

During the war the borrowing system—so central to small farmers—nearly collapsed. If farmers never expected an immediate return on the "gifts" they gave, they nonetheless expected an eventual return. The war destroyed those possibilities. Farmers living in war zones had little to trade with neighbors, and with so little available labor, they swapped work with neighbors less often. Outside war zones, the export trade (which brought the money and manufactured goods essential for borrowing) collapsed, leaving each farmer with less to barter. The end of the war brought the rapid remaking of the borrowing system. Hired hands returned, making welcome additions to farm labor and allowing farmers to diversify production and swap work at planting and harvest seasons. A surge of cloth imports let women return to their dairies and vegetable gardens, adding to the stock of goods they could give, lend, or exchange.[82]

The experience of wheat farmers in the mid–Hudson Valley during the postwar decades suggests the intensity of such exchanges. These farmers regularly borrowed from one another, lending or giving provisions and boots, oxen, and homemade cloth. Farmers helped clear land and harvest timber so a neighbor could build a new farmhouse. At the same time, farm women spun and wove for each other, and their daughters attended spinning bees, extending their families' resources without spending scarce hard money needed for imported goods. Money rarely changed hands in these exchanges, and farmers who kept account books sometimes neglected to record the value of the exchange.[83]

With the economic constraints that Britain had forced on the colonies gone, merchants began to build a national commercial system with new capitalist institutions such as banks. Rich northeastern farmers, gradually joined by their middling neighbors, took part in the orgy of development, expanding farms, founding agricultural societies, buying stocks and notes, and hiring wage labor at market rates. They lent money to men from other towns, reducing reliance on neighborly exchange; they searched for the best market for their surpluses and bought much of what they had made before at the growing number of local shops.[84]

At the end of the war, a fever for American frontier land burned hot throughout Europe and every new state. Families with too little land or none at all dreamed of making farms on fertile frontier lands they believed were empty. Pioneers rushed to northeastern and western frontiers, searching for land, many squatting at first. State or federal governments held sovereignty over this land and sold much of it to land companies and speculators, but Indians controlled nearly all of it. Before anyone could gain control over the frontier, Indians had to be removed.

Since the Americans had not defeated them in war, Indians insisted on negotiating treaties as independent nations. Nonetheless, most of the Iroquois, who had fought the patriots, lost their lands in 1784 treaties; other Indians signed treaties giving away millions of acres. At the same time, squatters rushed onto Indian lands in New York, Pennsylvania, Kentucky, and Ohio, making long-term Indian occupation impossible. The United States, pressed by settlers and their representatives in state governments who denied that Indians had any right to the soil, wanted land cessions or sales. But most Indians repudiated treaties already signed, and from 1782 through the early 1790s Georgia, Kentucky, Tennessee, and Ohio Indians went on the warpath, angered by treaties, encroachment on their lands, and attacks by settlers. By 1800, after vicious warfare, Indians had lost their lands in Pennsylvania, New York, and much of Ohio.[85]

Emigrants from both Europe and the East rushed in to take Indian lands. As soon as peace returned, immigration resumed and quickly surpassed prewar totals. As many as 200,000 immigrants came during the 1780s and 1790s, half from Ireland and most others from England, Scotland, and Germany. Battered by poor harvests, the Irish clamored to come to America, where promotional literature told them they could earn high wages and easily get a farm on the boundless frontiers. Since the British could no longer dump convicts and since more families accumulated money needed for passage, the proportion of servants among immigrants plummeted from one-half to less than one in thirteen. At first immigrants stayed in their port of arrival, but without the necessity of serving a term, many could seek work, look for a farm to rent, or buy land soon after they landed. Two-fifths of the 3,129 German redemptioners who arrived in Philadelphia between 1787 and 1804 fanned out across the country, providing essential farm labor to small and middling farmers. Small numbers of Welsh, Scot, German, and French immigrants reached the New York frontier during the late 1790s.[86]

Land hunger, intensified by publicity, drove landless or land-poor families to move. Half of the householders in St. Mary's County, Maryland, in 1790 were tenants, and landowners held an average of just 100 acres, barely enough to sustain a family in tobacco country. Many northern New Jersey farmers did not get enough land to feed their families until they were in their late forties and fifties. Such men listened avidly to the glowing accounts of the bountiful land in Kentucky or Ohio that awaited exploitation by industrious farmers. New Jersey men saw ads for land north of Albany for sale "in small lots of 200 acres and upwards, very suitable for poor farmers." Small planters in Maryland read ads hawking 480,000 acres of land by lottery or on easy terms, most in Kentucky, between 1789 and 1799.[87]

Groups of kin sometimes moved together; other migrants traveled in communal groups or followed predecessors to the area, sometimes founding towns in the process. Andover, Massachusetts, residents founded Andover and Denward, Maine; settlers from Martha's Vineyard began Vineyard, Maine; Dighton, Massachusetts, men planted Richmond, New York. Advance parties found fertile New York lands, returned home, and enticed neighbors to come with them. Yankee pioneers wrote home extolling their newfound land, and big New York land companies reached out to communities and kin groups. Southern plain folk, like northern farmers, often followed recommendations from pioneering kindred and friends and migrated as families or joined groups of kin or neighbors on the trip westward. In 1779, for example, a colony of neighbors, mostly from Berkeley County, Virginia, moved to Strode's Station; in the 1780s and 1790s many poor Catholic tenants from St. Mary's County, Maryland, moved to the fertile Bluegrass region; and a Baptist congregation in Spotsylvania County, Virginia, numbering 500 to 600 souls, sold their property and moved as a group in 1781 to central Kentucky.[88]

The Revolutionary War had ended frontier migration and turned frontier families into refugees. After the war, refugees rapidly returned, joined by families from older regions, to repeople western New York and Pennsylvania, the Ohio country, and the old Southwest. Usually they went in short steps, moving from one frontier to another. During the 1780s and 1790s, New Englanders moved northeast to Maine and Vermont, west to New York and Ohio, and southwest to the Wyoming country and western Pennsylvania. New Yorkers moved across the state and into western Pennsylvania and Ohio. If New Englanders moved northeast, Ohio attracted eastern families, who traveled north- or southwest to reach the staging area in Pittsburgh. Chesapeake planters moved farther than any other postwar migrants, populating backcountry Georgia, Kentucky, Tennessee, and southern Ohio. The direction of long-distance migration did change markedly. Before the war migrating Chesapeake whites went southward down the Great Wagon Road to backcountry Carolina and

Georgia; after the war they moved in a more westwardly direction, toward Kentucky, Ohio, and Tennessee, first stopping in West Virginia or Pennsylvania.[89]

The Revolutionary War had kept most farmers confined to older areas, but migration of white families after the war changed the face of the nation. In 1780 only one-fourteenth of its whites lived on frontiers. Between 1774 and 1784 the number of acres farmers cultivated, political economist Samuel Blodget estimated in 1806, grew only 12 percent, from 4.9 to 5.5 million. But peace with Britain unleashed an orgy of farm development. Amidst raging Indian wars, families swarmed onto new land; by 1790 the number of acres of improved land leaped 45 percent (to 8 million). As this land grab continued through the 1790s, improved acres rose 29 percent (to 10.3 million). The increasing availability of land led to rapid population growth. By 1800 the white population of northern New England, western New York and Pennsylvania, Ohio, Kentucky, Tennessee, and backcountry Georgia had reached 921,000, more than one-fifth of the country's whites.[90]

Nearly all this land lay in frontier areas. State governments confiscated hundreds of thousands of acres from Loyalists, most of them in older areas, but men of small means got very little. Even if officials had split these vast tracts of confiscated land into 200-acre tracts, few families—only 900 in Virginia and Georgia—would have gained land in this way. Rich men, moreover, took the bulk of Loyalist land state authorities sold. Twelve men acquired a third of the 128,300 acres Georgia seized from 166 Tories; four men got 9,310 of the 14,519 acres South Carolina confiscated from John Bailey; and just eighty-seven people bought the 51,000 acres Virginia took from Loyalists.[91]

New York and Maryland confiscated more Tory land. Maryland confiscated and sold 245,000 acres from proprietary manors, half located in places that had been settled by the early eighteenth century. Tenants had enjoyed secure and renewable leases over three lives. Although the state honored leases, sale of the land eventually evicted every tenant unable to buy his farm. The state sold farms on eight manors totaling 41,625 acres to 199 different purchasers. Small farmers predominated on all but Queen Anne's manor, where one buyer acquired 2,375 acres, or half the land. More than three-quarters of the buyers on seven manors had been tenants. Only on Monancy Manor in Frederick County did new men, many of them rich speculators, buy up the land. On other manors, located in productive general farming country, richer tenants did outbid poorer men.[92]

Only in New York, where officials seized 2.5 million acres from Loyalist landlords, could land have been redistributed. Speculators, however, wound up with most of it. To gain a right of first bidding, a tenant had to submit twelve testimonials proving his patriotism (or pay a third of the price with gold or silver money), and before receiving title he had to repay all rents owed his landlord. Many tenants could neither find enough men willing to vouch for their patriotism nor come up with the cash to pay off debts. The state sold 160,500 acres in rapidly growing Albany and Tryon Counties to 323 people, an average of nearly 500 acres each. Tenants bought more land in the lower Hudson River area. In Dutchess County 401 people bought 455 lots, almost all under 500 acres. On Philipsburgh Manor, a 50,000-acre estate with 270 tenants in war-torn Westchester County, small farmers, mostly manorial tenants, bought almost four-fifths of the land. Nonetheless, 90 former tenants, a third of the estate's farmers, failed to buy. Some had Tory leanings and thus did not qualify; others were too poor to command the credit needed to make a down payment.[93]

State and national governments understood land hunger and enacted land bounties to get men to enlist. In 1776 the Continental Congress promised 100 acres to each private who enlisted in the Continental army for at least three years. But since the states owned the land, Congress either had to buy it (an impossibility, given its empty treasury) or cajole states to donate some. The states refused, and the United States redeemed almost no land certificates until 1797. Eight states also granted land bounties, allocating on average 300 acres to privates. Most bounty land wound up in the hands of speculators, who bought federal and state certificates for a few cents per dollar. As a result, few soldiers redeemed their certificates for frontier land. Georgia proved exceptional, however. Eager to have white families people its frontiers, it gave bounties totaling 3 million acres to 9,000 veterans (only 2,000 of them natives of the state). Many men took advantage of the offer; 500 settled on land there with state certificates in 1783 alone.[94]

States also gave away frontier land as headrights, sold it cheaply to squatters, or allocated it to families they deemed deserving. Georgia offered headrights to any emigrant; Connecticut gave away a half-million acres in its northern Ohio reserve to victims of British coastal raids. More often, states permitted squatters to buy the land they occupied. In 1777–78 Virginia allowed squatters to preempt 400 acres of Kentucky land for 2.5 cents per acre, thus encouraging massive speculation, as rich men unloaded worthless paper money for the rights to land. Kentucky sold land at 20 cents an acre and granted liberal credit. Between 1777 and 1781 North Carolina gave settlers 640 acres per household; in addition, wives and children received 100 acres each (at 40s. per hundred acres).[95]

Land speculators and land companies held ownership rights to most unimproved western lands. Rich men (and their companies) fought over who among them would control the vast patrimony beyond the mountains. Pretending that neither Indians nor squatters lived on the land they coveted, they consolidated their holdings in Maine, New York, Pennsylvania, and Kentucky; they bought millions of acres from Indians and cash-strapped state legislatures; they drew maps and formed states almost no one recognized; and they bribed state legislators into making larger and larger grants.[96]

Such speculation sometimes raised the price farmers had to pay or forced families to buy land they already considered their own. Some land companies and individual speculators held onto land, waiting for its value to increase. Other companies—especially those whose lands lay distant from transit routes—built towns, roads, schools, and churches to attract families, thereby playing a crucial role in repopulating frontiers. Study of Appalachia, New York, and Ohio reveals the various strategies speculators and companies pursued.

The most intensive land speculation in the country took place in southern Appalachia. As late as 1800, speculators, large planters and merchants, land companies, and northeastern investment capitalists owned three-quarters of the region's land. North Carolina granted eleven men some 320,000 Tennessee acres in 1788–90 alone; Virginia allocated a half-million West Virginia acres each to five big speculators. Uninterested in development, these men held on to land until they could reap huge profits. In contrast, local planters and merchants, who owned three-fifths of absentee land in Kentucky and who were more connected to their regions, sold more of their land at cheaper rates than landholders in Tennessee and West Virginia.[97]

Starved for funds, New York sold 22,000 square miles, almost half of the state's land, to developers. Two land companies—Phelps and Gorham and the Holland Land Company—held 6.2 million acres, much of the western half of the state. In 1791 Alexander Macomb bought 3.6 million acres in the state's northwest corner. Smaller developers gobbled up most of the 2 million acres in the state's military tract reserved for veterans. These fertile lands lay far from the Hudson River, the only route to Albany or New York markets. Knowing they could not sell isolated land, developers platted towns; built roads; cleared streams; recruited ministers and teachers; constructed churches, schools, docks, taverns, and mills; and publicized land sales in newspapers, handbills, and maps. Understanding that farmers could not afford large tracts, they sold small farms for a small down payment. But business remained slow until the early nineteenth century, as pioneers filled in lands further east with good access to markets.[98]

By the 1787 Northwest Ordinance, the United States acquired control over all land north of the Ohio River. The federal government expected to fund its budget by selling land to farmers, but settlers avoided buying federal land. Such slow occupation can be readily explained. Much fertile, unimproved acreage in Vermont, Maine, New York, Pennsylvania, Kentucky, and Georgia lay closer to settled areas, and this land could be acquired cheaply from speculators or governments. Indian wars kept settlers out until 1794. And the Federalists who controlled the national government wanted to encourage commercial development in compact townships. They kept land prices high by requiring purchasers to buy 640 acres, far more than families needed. They also refused to allow squatters to preempt lands they farmed, and they failed to open a land office in Ohio.[99]

Speculators and land companies directed Ohio's early settlement. To replenish its treasury, the confederation government sold 1.1 million acres to the Massachusetts-based Ohio Company and to New Jersey land speculator John Symmes. When they gave up their western land claims to the confederation, Connecticut and Virginia reserved 7.2 million acres for their veterans (in reality, those who bought bounty certificates at a heavy discount). In 1795, before settlers arrived in the reserve's isolated Lake Erie location, Connecticut sold their land to the Connecticut Land Company. The two companies, Symmes and Virginia's government, controlled nearly a third of Ohio's land. The Ohio Company and the Virginia military reserve, both dominated by speculators, distributed land more efficiently and cheaply than the U.S. government, and by 1800 numerous farmers had patented land in both places. Since both groups owned lands near the Ohio River, they avoided the developmental costs that New York land companies bore.[100]

Like their pioneer ancestors, the first families to reach trans-Appalachian frontiers acquired title to land with difficulty. Many squatted illegally on Indian land before land developers, states, or the federal government acquired it. Having used what little cash they had in moving their families and belongings over the mountains and unable to pay for patents on free military bounty land, much less buy any, they fell into tenancy or (more often) squatted on land, hunting, herding cattle, and making small crops until the owners chased them away. Those lucky enough to stay made surpluses, built savings, and bought land. More prosperous farmers, unable to buy a farm large enough for their growing families in the East, moved to the area, often paying the squatter for improvements his family had made. Landownership thereby rose to levels (two-thirds of heads of households) found in older regions.[101]

Despite early capitalist development in the East and land speculation, farm families did form households, acquire land, and make new communities predicated on producing surpluses and trading with neighbors, thereby replicating colonial small farm society. As long as unimproved land could be stolen from the Indians, the cycle of land development and land scarcity in older areas, Indian removal from their farms and hunting grounds, migration to new frontiers, pioneer squatting, followed by purchase and development of land, could be repeated endlessly.

afterword

Toward a Farmers' Nation

Farmers came into the Revolution with a long tradition of economic independence. A substantial majority owned the land they worked, farm families produced much of the food they ate, and farm women manufactured some cloth and sewed all the family's clothing. Farmers knew that their labor, not location or fertility of the soil, gave land its value. Yet farmers' independence, built inside families and communities, was no more than a well-honed myth. No farmer could do without a wife, who cultivated the garden, helped with the harvest, milked the cows, swapped eggs and butter with neighbors, and cared for the children. Farm men, too, traded goods and labor with neighbors, thus distributing surpluses across the community. They sold surpluses at market to buy what they could not make and to get money to pay their taxes. They obeyed the dictates of the rule of rich gentlemen and followed communal norms.[1]

Even as they depended on their wives to sustain their independence, husbands demanded the right to rule within their families. At best a husband resembled a benevolent parent, awarding rights to an obedient wife, recognizing her contribution to the family economy, and allowing her to keep the fruits of her own labor. At worst a husband became a petty tyrant, lording over wife and children, harshly disciplining them for their misbehavior (real or imagined), taking their income, and leaving the family to work while he went to town or tavern. Most husbands, knowing that the success of the farm depended on cooperative labor, resembled a benevolent parent more than a petty tyrant. But however well or badly he behaved, a husband knew that the common law gave him this power over his wife.

Their independence, farmers knew, gave them rights as voters and citizens. Colonial farmers willingly deferred to gentlemen on political issues, but gentlemen, in turn, had to respect farmers' status as husbands and citizens and act as patrons. If gentlemen refused to grant them freehold land, forced them into permanent tenancy, passed high taxes, or failed to protect them from bandits or Indians, small farmers insisted that their labor assured their rights of ownership of the soil. If gentlemen refused to listen, farmers turned to violence. They rose up against their betters in New York and New Jersey in the 1740s and 1750s, protesting tenancy, over-powerful landlords, and insecure land title. Rural violence especially spread in the 1760s and early 1770s, just as Whig gentlemen began their rebellion against Britain. Regulators in North Carolina challenged corruption and high taxes; South Carolina Regulators fought bandits while demanding the establishment of local government; anti-rent rioters on New York manors demanded freehold land; the western Pennsylvania Paxton Boys insisted on protection from hostile Indians but showed their contempt for all Indians by slaughtering innocent, peaceable tribesmen.[2]

As the Paxton uprising shows, frontier farmers built their independence by throwing Indian farmers and hunters off the land. When white farmers encroached on Indian land, Indians chased off their livestock and on occasion murdered isolated families. Vicious retribution and revenge resulted, leaving at times swatches of desolation, destruction, and death in its wake. But the end result was always the same. Indians gave up more and more of their coastal lands and retired across the mountains; white families moved onto the rich lands the Indians abandoned and started to make farms, gradually filling the land and leading to repeated rounds of conflict farther west.

When the Revolution began, gentlemen worked hard to mobilize farmers to support their cause. Whig leaders espoused an ideology that emphasized popular sovereignty, the right of representation, and the opposition to the willful attempt of the British to reduce independent men to abject dependence and slavery. Farmers read this set of ideas as giving them authority to make policy that guaranteed their control over land and family. They joined the Revolutionary movement enthusiastically where gentlemen had treated them well before the war; they stayed loyal to Britain when Tory gentlemen they respected adhered to the motherland; they engaged in bitter, violent civil war in every place—the Hudson River Valley, New Jersey, and the Carolinas—where rural uprisings had taken place.

Wherever they lived, farmers endured destruction, violence, and bloodshed during the war. Armies and partisan bands stole crops and livestock, burned farms, insulted women and children, and forced men to swear loyalty to causes they abhorred. In a flash, farmers lost everything they had taken years to build. Tens of thousands had to abandon their farms and tramp the roads searching desperately for food and a roof over their heads. Bitter men joined partisan bands to avenge the wrongs done them, spreading the vicious cycle of violence and destruction. Large numbers of families, however, struggled to stay neutral, hiding in the woods or abandoning their farms at the first sound of tramping feet.

The ideology of the American Revolution and the violence of the war together created a democratic class of small property holders. After sacrificing all for their country, yeomen stopped deferring to gentlemen and insisted on democratic decision making. They justified the democratic political dominance of their class by using the same labor theory of value that had sustained their property rights. Viewing themselves as the most independent and virtuous class, small landowning property holders believed they had a right to rule. In the ferment of the 1770s and 1780s, they sometimes succeeded in electing farmers to legislatures, who on occasion helped to pass laws issuing paper money whose inflationary impact allowed farmers to pay their debts in depreciated currency. More often, gentlemen and capitalists dominated legislatures, and rural folk turned to the venerable tactics of the people-out-of-doors, closing courts, chasing away tax collectors, and denying the legitimacy of faraway authorities. When new uprisings broke out in western Massachusetts, Maine, and western Pennsylvania in the 1780s and 1790s, state and federal governments violently suppressed them.[3]

However much Federalists and capitalists wanted to build a manufacturing economy where big farmers chased smaller ones off the land and into wage labor, the United States remained overwhelmingly rural, and a majority of free families worked on small farms. Just 3.8 percent of the people lived in cities of 8,000 or more in 1770, and despite rapid growth in New York, Baltimore, Philadelphia, and Boston, that percentage dropped during the war, recovered by 1800, and grew slowly thereafter, reaching 8.5 in 1840. In 1800 three-quarters of all Americans worked on farms and plantations; many of the rest provided the craft work farmers needed. The proportion of farmers dropped slowly; as late as 1840, two-thirds of Americans still labored in agriculture, most on small farms. While the number of farmers plummeted in New England, where cities, rural industry, and capitalist farmers began to predominate, at least a half of all workers labored in agriculture everywhere else, a number that rose to three-quarters to four-fifths in the South (save Maryland) and the trans-Appalachian states. Such continued predominance of agriculture suggests that opportunities for small-scale farming persisted for decades after the Revolution.[4]

In Pennsylvania, where an 1800 census lists occupations, half of employed men called themselves farmers, but that number excludes farm laborers (and probably cottagers) as well as artisans who farmed. Farmers constituted more than three-fifths (63 percent) of the employed men in newly seated Lycoming County, in the north-central part of the state; another 9 percent were farm laborers. But the proportion of farmers in long-settled Bucks County dropped from two-thirds before the war to two-fifths (44 percent) by 1800, while another 11 percent worked as laborers, many on farms. Others worked in the growing iron industry but probably still worked their farms as well.[5]

The defeat of yeoman popular democracy was only the first stage in a century-long struggle of farmers to maintain their communal and patriarchal social order. When President Jefferson's envoys purchased Louisiana in 1803, they reset the cycle of Indian warfare and removal, frontier migration, and farm making this book has described. Yeomen thus reinvented their small-farmer class, remaking a world of patriarchal families, food-producing farms, local exchange, and self-sufficiency. They built economies where small farmers dominated numerically and where—as voters—they helped shape public policy on contested issues such as Indian removal, the distribution of land, and banking.

Jefferson himself understood both the symbolic and the material importance of that vast acquisition. The purchase of Louisiana, he told Congress in October 1803, would "promise in due season . . . an ample provision for our posterity, and a wide-spread field for the blessings of freedom and equal laws." He elaborated on his vision in an August letter: "The best we can make of the country for some time," he wrote, "will be to give establishments in it to the Indians" living east of the Mississippi "in exchange for their present country, and open land offices in the last, & thus make this acquisition the means of filling up the Eastern side." Only then would people and new states follow west of the river. This empire of freeholders, spreading endlessly into the west, made an old land forever new, turned potential wage laborers into independent farmers, and sustained an agrarian way of life—based on energetic labor by the entire family, subsistence production, neighborly exchange, sale of surpluses, and movement to new lands—for more than a century.[6]

notes

INTRODUCTION

1. Merrill Peterson, *Jefferson*, 290.

2. Theories swirling around the transition to capitalism are beyond the scope of this volume. For a small sampling, see Marx, *Capital*, vol. 1, chaps. 27–28; Hilton, *Transition from Feudalism to Capitalism*; Aston and Philpin, *Brenner Debate*; Fox-Genovese and Genovese, *Fruits of Merchant Capital*; Wallerstein, *Historical Capitalism*. For alternative views, see Gilje, *Wages of Independence*; Rothenberg, *From Market-Places to a Market Economy* (neoclassical economics); Innes, *Creating the Commonwealth* (Weberian definitions of capitalism). I present my own view in *Agrarian Origins*, chaps. 1–2. E. P. Thompson, *Making of the English Working Class*, is the essential classic on that topic.

3. The classic work remains Bailyn, *Peopling of British North America*, but see also his comments in "Challenge of Modern Historiography," 14–17, where he makes migration history central to the history of the modern world.

4. Galenson, *White Servitude*; Bailyn, *Voyagers to the West*; Wokeck, *Trade in Strangers*.

5. Chapters 2 and 3, below, provide data. Kain and Baigent, *Cadastral Map*, 269–76, 285–89; Bond, *Quit-Rent System*; and Kim, *Landlord and Tenant*, provide examples of the three-sided conflict among farmers, speculators, and the state.

6. Fox-Genovese, *Within the Plantation Household*, chap. 1, and Kulikoff, *Agrarian Origins*, chap. 2, and "Households and Markets"; Vickers, "Competency and Competition"; Bushman, "Markets and Composite Farms," esp. 364–67.

7. Problems of exchange and markets have been central to recent historical debate. See, for instance, Rothenberg, *From Market-Places to a Market Economy*, chap. 1, and "Market and Massachusetts Farmers"; Vickers, "Competency and Competition"; Merrill, "Cash Is Good to Eat," and "Putting 'Capitalism' in Its Place"; Bushman, "Markets and Composite Farms." For my own views, see Kulikoff, *Agrarian Origins*, chaps. 1–2.

8. Lawrence Stone, "Revival of the Narrative"; Maza, "Stories in History." For the examples cited in the text, see David Hackett Fischer, *Paul Revere's Ride* (esp. xiv–xvi); Dayton, "Taking the Trade"; Demos, *Unredeemed Captive*.

9. Marx, *Eighteenth Brumaire*, 97.

10. Bender, "Wholes and Parts"; Painter, "Bias and Synthesis"; Rosenszweig, "What *Is* the Matter with History?"; Ross, "Grand Narrative."

PROLOGUE

1. Sir Thomas Smith, *Discourse of the Commonweal*, ix (first published 1581), 11, 17, 19, 22, 37, 39–41, 49, 57–58 (first quotes on 11, 37); John E. Martin, *Feudalism to Capitalism*, 163 (second quote); Laslett, *World We Have Lost*, 36–37.

2. Tawney, *Agrarian Problem*, 245–46.

3. Marx, *Capital*, vol. 1, chaps. 27–28; Tawney, *Agrarian Problem*; and Alice Clark, *Working Life of Women*, are classic accounts of social disorder in early modern England. Levine, *Reproducing Families*, chap. 2, is a recent synthesis.

4. Snooks, "Great Waves," 57, 70–71, 77–78 (economic growth); Overton, *Agricultural Revolution*, chap. 3 (little growth before eighteenth century); Clay, *Economic Expansion*, vol. 1, chaps. 3, 6, and vol. 2, chap. 8; Kerridge, *Agricultural Revolution*; Walter, "Social Economy of Dearth," 79–128; F. J. Fisher, *London and the English Economy*, chap. 4; Hoskins, *Provincial England*, chap. 7.

5. On regional differences, see Thirsk, "Farming Regions." On cultural diffusion, contrast David Hackett Fischer, *Albion's Seed*; Merchant, *Ecological Revolutions*, chaps. 1–4; and Kulikoff, *Agrarian Origins*, chap. 7.

6. Peter Jeremy Piers Goldberg, *Women in England*, 169–70.

7. Brief statements of the medieval social order may be found in Hilton, *Class Conflict*, 1–6; Hilton, *English Peasantry*, 13; Hilton, "Crisis of Feudalism"; Robert Brenner, "Agrarian Class Structure," 30–37, 46–53; and Postan, "Feudalism and Its Decline," 78–87. Richard M. Smith, "'Modernization' and the Corporate Medieval Village Community," 140–79, esp. 140–46, 161–79, and Hanawalt, *Ties That Bound*, chap. 1, point to continuities with later times, while attacking MacFarlane, *Origins of English Individualism*, for calling medieval England "modern."

8. Levine, *Reproducing Families*, 14–30; Mark Bailey, "Peasant Welfare"; Fryde, *Peasants and Landlords*, chap. 2; Dyer, *Standards of Living*, 109–40, esp. 119–23, and *Lords and Peasants*, chap. 4; Britnell, *Commercialisation*, chaps. 4–5; Gregory Clark, "Economics of Exhaustion," 75–76. Examples are from Razi, *Life, Marriage, and Death*, chap. 2, esp. 94–99, and "Abbots of Halesowen," 151–64; Mate, "Agrarian Economy"; Poos, Razi, and Smith, "Population History," 304–6; H. S. A. Fox, "Exploitation of the Landless," 539–59.

9. Hatcher, *Plague, Population, and the English Economy*, and "England in the Aftermath," 3, 6–19; Keen, *English Society*, chap. 2; Edward Miller, "Introduction," 1–8; Britnell, *Commercialisation*, 156–58; Gregory Clark, "Economics of Exhaustion," 75–76. On conversion of arable land to pasture, see Bruce M. S. Campbell, "Fair Field Once Full of Folk," 63–64; Campbell, Bartley, and Power, "Demesne-Farming Systems," esp. 135–37, 142, 150, 154, 157–58, 163–64, 171, 177–78; and David L. Farmer, "*Famuli*," 218–19.

10. Hilton, *Decline of Serfdom*; David L. Farmer, "*Famuli*"; Dyer, *Standards of Living*, 146–50; Fryde, *Peasants and Landlords*, 11, 35–36, 39–40, 116–19, 123–34, 140, 211–18, 268–70, 282; Hatcher, "England in the Aftermath," 20–25, 28–30; Razi, "Myth of the Immutable English Family," 39–41; Britnell, *Commercialisation*, 192–94; Sherri Olson, *Chronicle*, 163; Steinfeld, *Invention of Free Labor*, 22–24, 28–38, 60–64, 204–5 (exaggerates coercion).

11. Hatcher, "England in the Aftermath," 17.

12. Scholars are so interested in the complexities of feudal land tenure (Harvey, conclusion, 328–38; Tawney, *Agrarian Problem*, 281–310) that they sometimes overlook its difference from private property, but see Cohen and Weitzman, "Enclosures and Depopulation," 168; North and Thomas, *Rise of the Western World*, 63–64; Robert C. Palmer, "Origins of Property," 382–88, 395–96; Hampsher-Monk, "Political Theory," 407–9; Libecap, "Property Rights," 231–35; and Robert C. Allen, *Enclosure and the Yeoman*, 58–62. On freehold land, see Zell, *Industry in the Countryside*, 11–12.

13. We cannot discuss regional differences in land tenure, much less the legal complexities of feudal tenure. Case studies include Richard M. Smith, "Some Issues Concerning Families," 1–20, and "Coping with Uncertainty," 43–57, esp. 46–49; Hanawalt, *Ties That Bound*, 120–21, 135–36, 297, 151; Dyer, *Lords and Peasants*, chap. 14, and *Standards of*

Living, 140–43; Howell, *Land, Family, and Inheritance*, chap. 10; Bruce M. S. Campbell, "Population Pressure"; Razi, "Erosion of the Family-Land Bond," and "Myth of the Immutable English Family," 24–34; Faith, "Berkshire," 121–58; Lomas, "South-East Durham," 290–316; Andrew Jones, "Land and People"; Mate, "East Sussex Land Market," 46–60; Whittle, "Individualism and the Family-Land Bond"; Elaine Clark, "Charitable Bequests."

14. Hilton, *Class Conflict*, chaps. 4, 6 (esp. 74–77), and *English Peasantry*, chaps. 1, 2, 9; Dyer, *Lords and Peasants*, chaps. 10, 14; Razi, *Life, Marriage, and Death*, 99–109, and "Intrafamilial Ties," 378–81, 388–89; Blanchard, "Industrial Employment," 241–57, 269–75.

15. Hanawalt, *Ties That Bound*, chaps. 2–10, presents the most cohesive picture of post-plague medieval families, but see also Le Patourel, "Rural Building," 843–64, and Dyer, *Everyday Life*, 136–61. On peasant communities, see Ault, *Open-Field Husbandry*; Bonfield, "What Did English Villagers Mean?," 105, 115.

16. Judith M. Bennett, *Women in the Medieval English Countryside*, chaps. 1, 7, and *Ale, Beer, and Brewsters*, chap. 3; Hanawalt, *Ties That Bound*, chaps. 2, 8, 9; Hilton, *Class Conflict*, chap. 12; Helena Graham, " 'A Women's Work' "; Peter Jeremy Piers Goldberg, *Women, Work, and Life Cycle*, 137–49, and *Women in England*, 167–80; Leyser, *Medieval Women*, 142–58; Ault, *Open-Field Husbandry*, 13–14; Simon A. C. Penn, "Female Wage-Earners," 10–13; Michael Roberts, "Sickles and Scythes," esp. 4–6, 14–15; Christopher Middleton, "Sexual Division of Labour," 152–55, and "Peasants"; Poos, Razi, and Smith, "Population History," 336–40.

17. Dyer, *Everyday Life*, chap. 9; Simon A. C. Penn, "Female Wage-Earners"; Hilton, *English Peasantry*, 22–24, 51–53, and *Bond Men Made Free*, 153–54, 170–74; Lachmann, *Manor to Market*, 16–17; Peter Jeremy Piers Goldberg, "Marriage, Migration, Servanthood," 149–55, and *Women, Work, and Life Cycle*, chap. 4; Richard M. Smith, "Marriage in Late Medieval Europe," 35–36, 41–42.

18. Judith M. Bennett, *Women in the Medieval English Countryside*, the best study, stops before the plague. For the post-plague era, see Judith M. Bennett, "Medieval Peasant Marriage," 193–215; Hanawalt, *Ties That Bound*, 199–203; Finch, "*Repulsa Uxore Sua*," 11–14, 21–28; Leyser, *Medieval Women*, 106–22, 180–86; Peter Jeremy Piers Goldberg, "Marriage, Migration, Servanthood," 156–60, and "Female Labour"; Simon A. C. Penn "Female Wage-Earners," 9–10; Casey, "Cheshire Cat," 228–32; and Richard M. Smith, "Coping with Uncertainty," 54–61. On merchet, see Poos, Razi, and Smith, "Population History," 316–19.

19. Homans, *English Villagers*, chaps. 5–6; Tawney, *Agrarian Problem*, chap. 3; Tate, *English Community*, chaps. 2, 5; Thirsk, *Rural Economy*, chap. 4, esp. 35–37, 43–49; Titow, "Medieval England and the Open-Field System"; Hoskins and Stamp, *Common Lands*, chap. 4; Ault, *Open-Field Husbandry*, 11–54; Birrell, "Common Right"; Everitt, "Farm Labourers," 400–412. McCloskey, "Persistence of English Common Fields," rejects communal interpretations of open fields, insisting that open fields were devised to reduce risk.

20. J. L. Bolton, *Medieval English Economy*, chaps. 4–5; Hilton, *English Peasantry*, 42–51; Biddick, "Missing Links," 282–98; Elaine Clark, "Debt Litigation," 250–71; Dyer, *Everyday Life*, chap. 13, esp. 269–81; Britnell, *Commercialisation*, 157–62; Masschaele, "Multiplicity of Medieval Markets," 256–69; Bruce M. S. Campbell, "Fair Field Once Full of Folk," 67–69; David L. Farmer, "Marketing the Produce," 363–66, 394–95, 398– 99, 417–24; Leyser, *Medieval Women*, 146. Agnew, "Threshold of Exchange," distinguishes between markets and "the market."

21. Hatcher, "England in the Aftermath," 19 (quote); Dyer, *Everyday Life*, 31–45, 175–

77; Fryde, *Peasants and Landlords*, 125–27; DeWindt, "Redefining the Peasant Community"; Razi, *Life, Marriage, and Death*, 117–24, and "Myth of the Immutable English Family," 34–39; Richard M. Smith, "Demographic Developments," 48–49; Beresford, "Review," 8–12; Raftis, *Tenure and Mobility*, chaps. 6–8, and *Warboys*, 130–52, 264–65; Peter Jeremy Piers Goldberg, *Women, Work, and Life Cycle*, chap. 6; S. Sherri Olson, *Chronicle*, 42–43, 166, 174.

22. Britnell, *Commercialisation*, 229–30 (between 0.5 and 0.6 percent per annum, 1300–1470); Snooks, "Great Waves," 56–57, 77–78 (0.29 percent per annum, 1066–1688; −0.7 percent, 1302–71; 0.1 percent, 1372–1491). Snooks stands alone in his absolute insistence that income (and, by implication, the standard of living) failed to rise in the fifteenth century.

23. Hanawalt, *Ties That Bound*, 55–59; Birrell, "Peasant Deer Poachers"; Dyer, "English Diet," 197–214; Dyer, *Standards of Living*, 151–60; Dyer, *Everyday Life*, chaps. 5, 7; Hatcher, "England in the Aftermath," 16–17, 25–27; Campbell, Bartley, and Power, "Demesne-Farming Systems," 133–34; David L. Farmer, "Prices and Wages," 434–525, esp. 490–94, 508–24; Britnell, *Commercialisation*, 168–71, 196.

24. For general surveys, see Keen, *English Society*, chap. 3, and Goheen, "Social Ideals." On the debate over the crisis of feudalism, see John E. Martin, *Feudalism to Capitalism*, chaps. 3–4; Hilton, "Crisis of Feudalism"; Bois, "Against the Neo-Malthusian Orthodoxy"; Robert Brenner, "Agrarian Roots," 232–36, 264–73. McIntosh, "Local Change," argues for great change in the 1465–1500 era. On agricultural change, see Thirsk, *Rural Economy*, 35–40.

25. Roger Burrow Manning, *Village Revolts*, 230–31 (first quote); Thirsk and Cooper, *Seventeenth-Century Economic Documents*, 32–33 (second quote). For protests, see Roger Burrow Manning, *Village Revolts*; McRae, *God Speed the Plough*, 31, 36–37, 41–43, 54–55, 64–67, 74–75, 86–91.

26. For the transition to capitalism see Marx, *Capital*, vol. 1, chaps. 27–30; Dobb, *Studies in the Development of Capitalism*, chaps. 2–3; Hilton, *Transition from Feudalism to Capitalism*; Aston and Philpin, *Brenner Debate*; Levine, *Reproducing Families*, chaps. 2–3; Charles H. George, "Making of the English Bourgeoisie"; and Mark Gould, *Revolution in the Development of Capitalism*, 135–79. MacFarlane, *Origins of English Individualism*, insists that a transition to capitalism—if it happened—occurred before the Norman conquest.

27. Macpherson, *Democratic Theory*, 123–31; Seed, *Ceremonies of Possession*, 20 (Lawson quote); Cohen and Wietzman, "Enclosures and Depopulation," 162–63, 170–76; McRae, *God Speed the Plough*, 30–31, 36–37, 40–43, 89–90, 124–25, 178, 182, 192–95 (quotes on 31, 124); Dahlman, *Open Field System*, 153–68; Hoyle, "Tenure and the Land Market," esp. 8–11. On law surrounding property, see Aylmer, "Meaning and Definition of 'Property.'" On peasant-initiated enclosures, see Tawney, *Agrarian Problem*, 57–97, 147–73, and Steve Hindle, "Persuasion and Protest," 44.

28. Wrigley and Schofield, *Population History*, 207–15; Wrigley et al., *English Population History*, 519, 532; Robert C. Allen, *Enclosure and the Yeoman*, 62–64; Zell, *Industry in the Countryside*, 21; Steve Hindle, "Persuasion and Protest," 37, 41, 44–47, 66; McIntosh, *Community Transformed*, 9–11.

29. The debate over enclosure began with Tawney's discussion of sixteenth-century peasant land consolidation, growth of large farms, enclosure, and dominance of improving landlords (*Agrarian Problem*, 57–97, 147–230). Recent works such as Wordie, "Chronology of English Enclosure"; Yelling, *Common Field and Enclosure*, chaps. 4–6; and Overton, *Agricultural Revolution*, 148–56, place most enclosures in the seventeenth (and eighteenth) century. John E. Martin, *Feudalism to Capitalism*, 132–40, and "Sheep

and Enclosure" (Midlands); Lazonick, "Marx and Enclosures"; Robert C. Allen, *Enclosure and the Yeoman*, 30–32, 41–44; and Fryde, *Peasants and Landlords*, 187–208, reassert the importance of Tudor enclosures. See David Grayson Allen, *In English Ways*, 21–22, 30–33, 45–54, 63–64, on perpetuation of open fields.

30. Wordie, "Deflationary Factors" (most sophisticated analysis); Brian Manning, "Peasantry and the English Revolution," 133–39; Hoyle, "Tenure and the Land Market," 8–11; Mate, "East Sussex Land Market," 60–63; Clay, *Economic Expansion*, 1:85–91; Tawney, *Agrarian Problem*, 301–10; Peter Bowden, "Agricultural Prices," 687–94; Outhwaite, *Inflation*, esp. 9–14; Zell, *Industry in the Countryside*, 44–51.

31. Humphries, "Enclosures, Common Rights, and Women"; Woodward, "Straw, Bracken, and the Wicklow Whale"; Everitt, "Farm Labourers," 400–412; Steve Hindle, "Persuasion and Protest," 48–50, 66; A. Hassell Smith, "Labourers in Late Sixteenth-Century England," 367–72, 382–88; McIntosh, *Community Transformed*, 122–25, 128–29.

32. On size of holdings, see Tawney, *Agrarian Problem*, 32–33, 64–65 (12 percent of a sample of Tudor and early Stuart manors were cottagers, 27 percent owned less than 5 acres, and 27 percent owned 5–19 acres); Margaret Spufford, *Contrasting Communities*, chaps. 2–5, esp. 73, 100, 149 (33 percent in three Cambridgeshire communities, 1560–1607, were cottagers; 8 percent owned less than 2 acres; 32 percent owned 2–13 acres; and 22 percent owned 13–25 acres); Hunt, *Puritan Moment*, 16–20 (46 percent in Essex County owned 5 or less acres, and 25 percent owned 6–20 acres); Zell, *Industry in the Countryside*, 97 (30 percent in Kent weald held less than 10 acres; 38 percent, 10–29 acres).

33. Alice Clark, *Working Life of Women*, chaps. 3–4, esp. 65–73, 112–18; Shammas, "World Women Knew"; A. Hassell Smith, "Labourers in Late Sixteenth-Century England," 11–52; Woodward, *Men at Work*, 108–13, 131–37, 236–41. On cottagers, see Everitt, "Farm Labourers," esp. 396–403, 412–17, 425–29; Walter and Wrightson, "Dearth and the Social Order"; Clay, *Economic Expansion*, 2:28–36; and Barley, "Rural Housing," 761–66. On wages, see Peter Bowden, "Agricultural Prices," 598–601, 605–9, and Woodward, *Men at Work*, chap. 6.

34. On wages, see n. 33, above, and Wordie, "Deflationary Factors," 37–39, 44–46 (links wages to standard of living). On diet, see Walter and Wrightson, "Dearth and the Social Order"; Shammas, "Food Expenditures"; Shammas, "Food Budget"; Shammas, *Preindustrial Consumer*, 77, 121–31, 137–50; Outhwaite, *Dearth, Public Policy, and Social Disturbance*, 9–34; Jütte, *Poverty and Deviance*, 72–74, 77; Woodward, *Men at Work*, 3, 13, 147, 209–22, 226–41, 282–85; Mendelson and Crawford, *Women in Early Modern England*, 270–71; McRae, *God Speed the Plough*, 65; Tusser, *Five Hundred Points*, 18–19, 23–24, 65, 178; and Chappell and Webworth, *Roxburghe Ballads*, 6:525.

35. Sreenivasan, "Land-Family Bond," challenged MacFarlane, *Origins of English Individualism*, on this issue; in turn, Hoyle, "Land-Family Bond," 158–73, critiqued Sreenivasan, but Sreenivasan, "Reply," is persuasive. McIntosh, *Community Transformed*, 100–104, finds most land sold outside the family, but even here the proportion that stayed in the family nearly doubled from 15 to 28 percent between 1488 and 1617.

36. Brian Manning, "Peasantry and the English Revolution," 139–45, summarizes conflicts over the wastelands; Tawney, *Agrarian Problem*, 282–301, documents landlord evictions. See also Overton, "Re-establishing the English Agricultural Revolution," 7–10, 14–15, 17–18, and Campbell and Overton, "New Perspectives," 76–87 (productivity change); Dahlman, *Open Field System*, chap. 2 (link between specialized agriculture and enclosures); Zell, *Industry in the Countryside*, 37–41 (leasing); McIntosh, *Community Transformed*, 109–22 (case study).

37. On Gloucestershire, see Tawney and Tawney, "Occupational Census," 36, 47–50

(excluding servants and townspeople). Gregory King's 1688 distribution of the English population (Laslett, *World We Have Lost*, 36–37) counts "freeholders of the better sort" as 3 percent of the household heads and all freeholders as 14 percent. Lindert and Williamson, "Revising England's Social Tables," 387–91, reduces these proportions to 2 percent and 9 percent. On agrarian villages with more yeomen, see Levine and Wrightson, *Making of an Industrial Society*, 154–58; Wrightson and Levine, *Poverty and Piety*, 32–37; Friedeburg, "Reformation of Manners," 354–56; and Hey, *English Rural Community*, 52–57.

38. Mildred Campbell, *English Yeoman* (classic account); Wrightson, *English Society*, 134–36; Glennie, "In Search of Agrarian Capitalism," esp. 23–29 (yeoman land accumulation); Erickson, *Women and Property*, 40, 42–43. On borrowing, see Margaret Spufford, "Limitations of the Probate Inventory," 165–72; McIntosh, "Money Lending," 564–70; Holderness, "Credit," 102–5; Tittler, "Money-Lending," 256–57, 260; and Muldrew, "Interpreting the Market," 174–80.

39. Harrison, *Description of England*, 132–34 (quote on 133); Sir Thomas Wilson, *State of England* (1600), 18–20 (quote on 19). For yeoman income, wealth, and savings, see Laslett, *World We Have Lost*, 36–37; Lindert and Williamson, "Revising England's Social Tables," 392–94; Lindert, "Probate Sampling," 662–64; Hey, *English Rural Community*, 54–55; Hoskins, *Provincial England*, 150–55; Margaret Spufford, *Contrasting Communities*, 36–40; Margaret Spufford, "Limitations of the Probate Inventory," 165–72; Margaret Spufford, *Great Reclothing*, 116–18; and Shammas, "Determinants of Personal Wealth," 682–88.

40. Thirsk, *Economic Policy*, chap. 5; Margaret Spufford, *Great Reclothing*, chaps. 4–7; Clay, *Economic Expansion*, 2:22–36; Berger, "Development of Retail Trade"; Hoskins, *Provincial England*, chap. 7; Barley, "Rural Housing," 734–60; Machin, "Great Rebuilding."

41. Erickson, *Women and Property*, 40, 42–43, 65–70, 74–76; Roger Burrow Manning, *Hunters and Poachers*, 114–15, 118–19, 122; Outhwaite, "Progress and Backwardness," 8–12; Margaret Spufford, "Peasant Inheritance Customs" and *Contrasting Communities*, 85–90, 99–101, 104–11, 159–64; Bonfield, "Normative Rules"; Hoyle, "Land-Family Bond," 169, and "Tenure and the Land Market," 16–17; Zell, *Industry in the Countryside*, 17–19, 29, 96–108, 151; Everitt, "Farm Labourers," 412–42; Woodward, "Wage Rates"; McIntosh, *Community Transformed*, 51, 103–6.

42. Everitt, "Marketing of Agricultural Produce," 467–80, 490–502, 507–16, 531–32, 543–57; Margaret Spufford, "Importance of Religion," 38–39; McIntosh, *Community Transformed*, 93, 131, 144–55.

43. Tax records from 1522–24 used by Cornwall, *Wealth and Society*, esp. 16–17, 28–33, 44–45, 48, 83–85, 107, 113, 172, 175–79, 190–91, 198–215, and Hoskins, *Age of Plunder*, 32, 44–46, count one-third to three-fifths as poor, but they cannot distinguish independent laborers from youthful servants or find those who held land by manorial tenures (much less those who leased for short terms), making the meaning of the assessments far less clear than they claim. After reviewing the evidence, Cornwall posits a laboring class of 23.9 percent, more plausible but probably still too high.

44. Marx, *Capital*, vol. 1, chaps. 27–30 (classic account); Lazonick, "Marx and Enclosures," 7–17; Levine, *Reproducing Families*, chaps. 2–3; Wrightson, *English Society*, 23–38, 133–39 (alternative view); Gregson, "Tawney Revisited," 18–42 (new class system came much later in the north of England). On population, see Wrigley and Schofield, *Population History*, 207–12, 532, 571. Estimates found in Lachmann, *Manor to Market*, 16–17; Everitt, "Farm Labourers," 419–21; and Levine, *Reproducing Families*, 40–41. The 1560 figure mediates between Lachmann's data and the higher Cornwall and Hoskins

estimates by halving the latter two (see n. 43, above). For information on laborers, husbandmen, and yeomen, 1560–1620, from wills, see Takahashi, "Number of Wills," 209–11.

45. For the 1650s, see Timmins, *Made in Lancashire*, 67–69, 73–74; for the 1688 estimate, see Lindert and Williamson, "Revising England's Social Tables," 386–94 (a reworking of King's 1688 estimates). I allocated family heads proportionately to cities (all those over 5,000 in De Vries, *European Urbanization*, 30, 270–71) and market towns (Chartres, "Marketing," 409–12, taking away the 28 towns De Vries lists, assuming that half the rest existed only on paper, and allocating the remainder to places of 2,5000, 1,000, and 500, yielding some 70,000 households in places of 1,000+). I further assumed that no farm families lived in London or other towns and that those King called "laboring people and outservants" and "vagrants" were wage laborers, and those called "cottagers and paupers" retained some rights to land; see Everitt, "Farm Labourers," 396–442, 454–65.

46. Spufford and Takahashi, "Families, Will Witnesses, and Economic Structure," claims a classless community prevailed, despite strong economic differences among households. The authors' evidence of social interaction, but not their interpretation, is persuasive. See also Muldrew, "Interpreting the Market," 178–80; Cornwall, *Wealth and Society*, 209–10.

47. Clark and Souden, *Migration and Society*, 28–35; Laslett, *Family Life and Illicit Love*, 65–75, 98–101; Peter A. Clark, "Migrant in Kentish Towns," 117–63, esp. 124–46; Souden, "Migrants and the Population Structure," 134–67; Zell, *Industry in the Countryside*, 80–85; Woodward, *Men at Work*, 100; Margaret Spufford, "Importance of Religion," 61–63; Peter Spufford, "Comparative Mobility"; Roger Thompson, *Mobility and Migration*, esp. chap. 7 (reports in great detail the individual and generational persistence [within a ten-mile radius] of English Puritans); McIntosh, *Community Transformed*, 25–34, 107–8; Ben-Amos, *Adolescence and Youth*, 92, 95.

48. Clark and Souden, *Migration and Society*, 11–48, esp. 28–35; A. Hassell Smith, "Labourers in Late Sixteenth-Century England," 16–21; Kussmaul, *Servants in Husbandry*, chap. 4; John Patten, "Patterns of Migration," 111–29; Millward, "Emergence of Wage Labor," 21–39; Woodward, *Men at Work*, 54–55, 68–71, 97–103, 162–64; Ben-Amos, *Adolescence and Youth*, 1–2, 69, 77–80, 84, 86, 88, 130, 135–41, 151–53, 158–61, 166–68, 216–21, 227–29, 266, 298–99; Kitch, "Population Movement," 70–73; McIntosh, *Community Transformed*, 61.

49. Kussmaul, *Servants in Husbandry*, is the best analysis, but see also Ben-Amos, *Adolescence and Youth*, 2, 69–77, 163; A. Hassell Smith, "Labourers in Late Sixteenth-Century England," 14–18, 33–36; McIntosh, "Servants and the Household Unit," 3–23, and *Community Transformed*, 39, 53–65; Laslett, *World We Have Lost*, 66–73; Zell, *Industry in the Countryside*, 79–80; Mendelson and Crawford, *Women in Early Modern England*, 264–68.

50. Beier, "Poverty and Progress," 201–39, and *Masterless Men*, esp. chaps. 1–6, pp. 207–26; Slack, "Vagrants and Vagrancy," 361–79; Mikalachki, "Women's Networks," 52–69; Garthine Walker, "Women, Theft," 81–91; Mendelson and Crawford, *Women in Early Modern England*, 293–98; Cornwall, *Wealth and Society*, 216–30; Jütte, *Poverty and Deviance*, 147–50, 169–70; Kent, "Population Mobility," 35–51. On the beggar boy, see Chappell and Webworth, *Roxburghe Ballads*, 3:323–28 (quotes on 323, 325, 327).

51. Mendelson, "'To Shift for a Cloak,'" 7–8, 10–12 (quotes on 10–11).

52. Pound, *Poverty and Vagrancy*, chaps. 1–4, 7; Slack, *Poverty and Policy*, chaps. 2, 6–7; Beier, *Masterless Men*, chap. 9; Bush, "'Take This Job,'" 1391–95; Outhwaite, *Dearth, Public Policy, and Social Disturbance*, 35–36, 39–42; Alice Clark, *Working Life of Women*,

74–92; Mendelson and Crawford, *Women in Early Modern England*, 289–92; Hunt, *Puritan Moment*, chap. 3; Jütte, *Poverty and Deviance*, 120–23; Everitt, "Marketing of Agricultural Produce," 573–86; Sharp, "Popular Protest," 277–84; Solar, "Poor Relief."

53. Levine, *Reproducing Families*, chaps. 2–3, and Wrigley, *People, Cities, and Wealth*, 153–74, give alternative models of this change.

54. Nef, *Conquest of the Material World*, chaps. 2–3; Cantor, *Changing English Landscape*, chap. 6; Jack, *Trade and Industry*, chap. 2; Clay, *Economic Expansion*, vol. 2, chap. 8; Thirsk, *Rural Economy*, chap. 13; Clarkson, *Pre-Industrial Economy*, 86–92; Roger Burrow Manning, *Hunters and Poachers*, 114–15, 133; Tawney and Tawney, "Occupational Census," 35–58; Wrigley, "Changing Occupational Structure," 9–15; Ripley, "Village and Town," 171–77; Zell, *Industry in the Countryside*, 116–21, 135–39, 145, 155, 164–79.

55. Sharp, *In Contempt of All Authority*, chap. 6; D. C. Coleman, *Industry in Tudor and Stuart England*, 25–32; Kussmaul, *General View*, chap. 6; C. B. Phillips, "Town and Country," 110–17; Alice Clark, *Working Life of Women*, chap. 4; Julia de Lacy Mann, *Cloth Industry*, chap. 4; Zell, *Industry in the Countryside*, 92–95, 164–86; Wadsworth and Mann, *Cotton Trade*, chaps. 2, 4; Heaton, *Yorkshire Woollen and Worsted Industries*, chaps. 3, 6; Hunt, *Puritan Moment*, 238–47; Supple, *Commercial Crisis*, chaps. 3, 5.

56. Nef, *Rise of British Coal Industry*, 1:7–24, 411–48, 2:135–97, 380–89, and Levine and Wrightson, *Making of an Industrial Society*, chaps. 1–3, are indispensable, but see also Levine, *Reproducing Families*, 97–100; Clay, *Economic Expansion*, 2:46–52; Dietz, "North-East Coal Trade," 280–94; and Chaytor, "Household and Kinship," 30–33.

57. Keen, *English Society*, chap. 4; Phythian-Adams, "Urban Decay," 159–85, esp. 165–80; John Patten, *English Towns*, chap. 3; Overton, *Agricultural Revolution*, 138.

58. Wrigley, *People, Cities, and Wealth*, 133–74 (esp. 134–37, 142–49); Souden, " 'Rogues, Whores and Vagabonds' "; Horn, "Servant Emigration," 51–95; Wareing, "Migration to London," 356–78; John Patten, "Patterns of Migration," 111–29; Peter Earle, "Female Labour Market," 331–45; Elliott, "Single Women," 90–92; Finlay, *Population and Metropolis*, chap. 1; Finlay and Shearer, "Population Growth," 37–53; Ben-Amos, *Adolescence and Youth*, 95–97.

59. Levine, *Reproducing Families*, 80–82, 143–45; Peter A. Clark, "Migrants in the City."

60. Hill, *Change and Continuity*, 219–38 (evokes the plight of wage laborers); Sreenivasan, "Reply," 176–77.

61. Mary Abbott, *Family Ties*, 71, 87–88.

62. Tusser, *Five Hundred Points*, xi–xx, 9–16 (quote on 13), 31–33 (detailed tool lists suggest substantial farmers), 202–11, 314–28; McRae, *God Speed the Plough*, 5, 146–51; Erickson, *Women and Property*, 54; Chappell and Webworth, *Roxburghe Ballads*, 2:117–19. Overton, *Agricultural Revolution*, chap. 1, describes sixteenth-century farming in a way Tusser would have recognized.

63. Tusser, *Five Hundred Points*, 99 (quote), 163–64, 167–77; Cressy, *Birth, Marriage, and Death*, 260–62, 287–89, 297 (definition of husbandman; yeoman on 287); J. A. Sharpe, "Plebeian Marriage," 77; Mendelson and Crawford, *Women in Early Modern England*, 205–6, 269, 303–4, 307; Wiltenburg, *Disorderly Women*, 89; Chappell and Webworth, *Roxburghe Ballads*, 7:185–87.

64. Tusser, *Five Hundred Points*, 21–22, 51, 95, 122, 124, 137, 160, 164–65; Kussmaul, *Servants in Husbandry*, chaps. 3–5; McIntosh, *Community Transformed*, 39, 59.

65. Tusser, *Five Hundred Points*, 27–127.

66. Alice Clark, *Working Life of Women*, 43–64; Michael Roberts, " 'Words They Are Women,' " 143–44, 153–55; Amussen, *Ordered Society*, chaps. 4–5; Mendelson and Crawford, *Women in Early Modern England*, 37–42, 48, 257, 269–76.

67. Ben-Amos, *Adolescence and Youth*, 40–47, 59, 75, 149; Pamela Sharpe, "Poor Children as Apprentices," 253–63; Mendelson and Crawford, *Women in Early Modern England*, 86–87, 92, 96; McIntosh, *Community Transformed*, 34–39, 49.

68. Tusser, *Five Hundred Points*, 84, 98–99, 134–39 (quotes on 135, 137); McRae, *God Speed the Plough*, 149–51, 206–7, 210–13, 218–21.

69. Houlbrooke, *English Family*, 105–9; Chaytor, "Household and Kinship," 34–51; Christopher Middleton, "Women's Labor," 181–200; Alice Clark, *Working Life of Women*, 42–64; Thirsk, foreword, 9–14; Kussmaul, *Servants in Husbandry*, chaps. 3, 5; Richard Wall, "Age at Leaving Home," 180–202; Ben-Amos, *Adolescence and Youth*, 62, 64; MacFarlane, *Family Life of Ralph Josselin*, 205–10.

70. Tusser, *Five Hundred Points*, 19 (quotes), 126, 173; Chappell and Webworth, *Roxburghe Ballads*, 3:601–3 (cow).

71. Chappell and Webworth, *Roxburghe Ballads*, 1:122–27 (quotes on 122–24).

72. Mendelson and Crawford, *Women in Early Modern England*, 273–74; McRae, *God Speed the Plough*, 152–56 (quote on 155); Kitch, "Population Movement," 70.

73. Alice Clark, *Working Life of Women*, 65–149; Cahn, *Industry of Devotion*, 40–60; Michael Roberts, "Sickles and Scythes," 1–28, and " 'Words They Are Women,' " 135–38, 144–47; Shammas, "World Women Knew"; Peter Earle, "Female Labour Market," 336–44.

74. Levine, *Reproducing Families*, 80–86; Levine and Wrightson, *Making of an Industrial Society*, 172–205; Richard Wall, "Women Alone in English Society," 304–5, 310–14; Pamela Sharpe, "Literally Spinsters," 46–65; Underdown, "Taming of the Scold," 135–36; Barbara J. Todd, "Remarrying Widow," 54–92; Ben-Amos, *Adolescence and Youth*, 142–43.

75. The vast literature on early modern English family and women focuses on a later period or emphasizes the ruling class. But see Levine, *Reproducing Families*, chaps. 2–3; Schellekens, "Courtship," 437–41 (illegitimacy rises until 1600, then probably stays the same); Amussen, *Ordered Society*, chaps. 4–5; Underdown, "Taming of the Scold," 116–36; Ingram, " 'Scolding Women,' " 50–72; Quaife, *Wanton Wenches and Wayward Wives*, chaps. 3–5; Michael McDonald, *Mystical Bedlam*, chap. 3, appendix C; Lawrence Stone, *Road to Divorce*, 5, 141–45; McIntosh, *Community Transformed*, 67–69.

76. Quaife, *Wanton Wenches and Wayward Wives*, chap. 3; Laslett, *Family Life and Illicit Love*, chap. 3; Levine and Wrightson, "Social Context of Illegitimacy," 158–75; Mendelson and Crawford, *Women in Early Modern England*, 109–11; Cressy, *Birth, Marriage, and Death*, 321–22; Ben-Amos, *Adolescence and Youth*, 200–206. On child abandonment and infanticide, see Hoffer and Hull, *Murdering Mothers*, chap. 1; Gowing, "Secret Births," 87–115; Fildes, "Maternal Feelings," 142–43, 146–50; Dolan, *Dangerous Familiars*, 123–25, 129–30, 144–45.

77. Houlbrooke, *English Family*, 67–69, 72–82, and "Making of Marriage," 343–48; Greaves, *Society and Religion*, 155–77; Carlson, *Marriage and the English Reformation*, 107–10, 116–21, 138–39; Gouge, *Of Domesticall Duties*, 449–52, 563–65; O'Hara, " 'Ruled by My Friends,' " 14–22, 28–33, 41; J. A. Sharpe, "Plebeian Marriage," 73–75; Cressy, *Birth, Marriage, and Death*, 243–45, 254–58, 272–75, 308–9; Mendelson and Crawford, *Women in Early Modern England*, 113–21; Pamela Sharpe, "Literally Spinsters," 55–58; Elliott, "Single Women," 84–97; Peter Earle, "Female Labour Market," 338–46; Lawrence Stone, *Road to Divorce*, 52–54, 57–58, 99.

78. Peter A. Clark, *English Alehouse*, chaps. 3, 6–7; Wrightson, "Alehouses, Order, and Reformation," 1–13; Carlson, *Marriage and the English Reformation*, 106, 110; Ben-Amos, *Adolescence and Youth*, 192–95; Erickson, *Women and Property*, 94.

79. Laslett, *Family Life and Illicit Love*, 31–46, and Richard Wall, "Household," 494–

501, deny the prevalence of these households (especially those headed by women), but they fail to take into account that transients rarely appear in parish registers or census listings. On behavior and ideals of plebeian families, see Lawrence Stone, *Family, Sex, and Marriage*, 91–93, 142–49, 191–93, 468–74, and *Road to Divorce*, 5, 141–45; E. P. Thompson, "Happy Families," 44–50 (slashing critique of Stone); Levine, *Reproducing Families*, 72–75, 106–23; Mendelson and Crawford, *Women in Early Modern England*, 133–37, 146 (quote); Wrightson, "Two Concepts of Order," 21–46; Hunt, *Puritan Moment*, 51–54, 74–75; Margaret Spufford, *Small Books and Pleasant Histories*, chap. 7; J. A. Sharpe, "Plebeian Marriage," 69–90; Underdown, "Taming of the Scold," 119–21, 126–32; Mack, *Visionary Women*, 71–73; Cunningham, "Employment and Unemployment of Children," 118–21, 146–48; Willen, "Women in the Public Sphere," 562–65.

80. Deal, "Widows and Reputation," 382–92; Foyster, "Laughing Matter," 5–21 (first quote on 6); Wiltenburg, *Disorderly Women*, 4–5, 29–34, 73, 82–87, 91, 94, 97, 100–104, 107, 153–56, 159–60, 165–73, 213–21, 252–55; Dolan, *Dangerous Familiars*, 7–9, and chap. 1 (house traitor proverb on 21); Anthony Fletcher, *Gender, Sex, and Subordination*, xix–xx, 4–8, 16–21, 27, 70–71, 77, 192–200, 230–31, 243; Mendelson and Crawford, *Women in Early Modern England*, 60–63, 128, 211, 213–15, 256; Stretton, *Women Waging Law*, 196–201; J. A. Sharpe, "Plebeian Marriage," 78–83, 85. Sources include Tilley, *Dictionary of Proverbs*, 722–25, 741–49 (quotes on 725, 745, 723), and Chappell and Webworth, *Roxburghe Ballads*, 2:183–88, 7:192–93, 8:696–98.

81. Wiltenburg, *Disorderly Women*, 108–10, 126–31, 187–90, 198–202, 221–23 (ballad quotes on 109–10, 127, 129); Tilley, *Dictionary of Proverbs* (quote on 742); Dolan, *Dangerous Familiars*, chap. 1, esp. 21–22, and 88–89, 99–101, 104; Mendelson and Crawford, *Women in Early Modern England*, 136 (yoak fellow), 140–45, 240; Amussen, *Ordered Society*, 127–29, 167–68; Foyster, "Laughing Matter," 12–15; J. A. Sharpe, "Plebeian Marriage," 88–89. Wiltenburg treats depictions of female violence as farce, ignoring their clear darker side.

82. Hill, *Society and Puritanism*, 472 (quote); Erickson, *Women and Property*, 9; Foyster, "Laughing Matter," 15–16; Gouge, *Of Domesticall Duties*, 159–63. Issues of social control of the poor are beyond the purview of this volume, but see McIntosh, *Community Transformed*, 240–58, and Friedeburg, "Reformation of Manners," 347–85, for examples.

83. Erickson, *Women and Property*, 19, 83–89, 100, 104–7, 115–19, 123, 126, 130–33, 144–49, 162, 178–82, 193–95; Peters, "Single Women," 325–45; Mack, *Visionary Women*, 56–57, 87–124; Whittle, "Inheritance, Marriage, Widowhood," 33–72; Stretton, *Women Waging Law*, esp. 38–42, 73–79, 84–86, 92–102, 108–19, 134–54.

84. Margaret George, *Women in the First Capitalist Society*, chaps. 7, 11; Mendelson and Crawford, *Women in Early Modern England*, 309–10; Anthony Fletcher, *Gender, Sex, and Subordination*, 182–83; McIntosh, *Community Transformed*, 44, 289–91; J. A. Sharpe, "Plebeian Marriage," 75–77; Gouge, *Of Domesticall Duties*, 16, 21–22, 78–79, 114, 224–25, 228–29, 255–59, 268–73, 281, 286–87, 290–93, 298–99, 350–53, 357–59, 377–79, 402, 405–7, 484–87, 514–17, 529–31, 536–37, 552–53; Chappell and Webworth, *Roxburghe Ballads*, 3:301–6, 7:188–89.

85. Gouge, *Of Domesticall Duties*, 2, in Epistle Dedicatory (first quote), 270 (second quote), 286 (third quote), 389–93 (wife beating). For Gouge, see George and George, *Protestant Mind*, chap. 7; Margaret George, *Women in the First Capitalist Society*, 3, 33–34, 144, 146, 251; and Dolan, *Dangerous Familiars*, 33–34.

86. Louis B. Wright, *Middle-Class Culture*, chaps. 7, 13; Henderson and McManus, *Half Humankind*, esp. 3–32, 47–98; and Woodbridge, *Women and the English Renaissance*, esp. chaps. 3–5, 9, 11, describe the controversy. Margaret George, *Women in the*

First Capitalist Society, esp. intro., chap. 7, places the controversy into the context of capitalist development and bourgeois class formation. See also Mack, *Visionary Women*, 52–53 (quote on 52), and Anthony Fletcher, *Gender, Sex, and Subordination*, 12, 60–61.

87. My position owes much to Fox-Genovese, "Property and Patriarchy." See also Schochet, *Patriarchalism*, chaps. 3–8; Hill, *Society and Puritanism*, chap. 13; Ezell, *Patriarch's Wife*, 1–35, 129–44; Lawrence Stone, *Family, Sex, and Marriage*, chap. 5; Amussen, *Ordered Society*, chap. 2; Jonathan Goldberg, "Fatherly Authority," 3–32, esp. 3–9; and Anthony Fletcher, *Gender, Sex, and Subordination*, chap. 11.

88. Amussen, *Ordered Society*, 38–48; Ezell, *Patriarch's Wife*, chaps. 2, 4; Cahn, *Industry of Devotion*, chap. 4.

89. The vast literature on these works begins with analyses of "Puritan" or Protestant ideals (Hill, *Society and Puritanism*, chap. 13; George and George, *Protestant Mind*, chap. 7) and then shows the ubiquity of similar ideals among Anglicans or Catholics (Greaves, *Society and Religion*, chaps. 6–7; Amussen, *Ordered Society*, chap. 2; Ezell, *Patriarch's Wife*, chaps. 2, 4; Davies, "Sacred Condition of Equality," 563–80; Margo Todd, "Humanists, Puritans, and the Spiritualized Household," 18–34).

90. Seaver, *Wallington's World*, 33–34, 41, 74–93, shows one family that approximated the ideal; A. Hassell Smith, "Labourers in Late Sixteenth-Century England," 19–32, shows the impact of wage labor; Margaret George, *Women in the First Capitalist Society*, chaps. 8–9, illuminates dysfunctional families.

91. Kussmaul, *Servants in Husbandry*, chaps. 3–5; Mendelson and Crawford, *Women in Early Modern England*, 92–108; Houlbrooke, "Making of Marriage," 343; Rushton, "Property, Power, and Family Networks," 212–14; Amussen, *Ordered Society*, 38–41, 157–61; Schochet, *Patriarchalism*, 66–73; Chappell and Webworth, *Roxburghe Ballads*, 3:318–22.

92. Durston, *Family in the English Revolution*, chaps. 2–3, 5; Shanley, "Marriage Contract and Social Contract," 78–85; Roderick Phillips, *Putting Asunder*, chap. 3.

93. Schochet, *Patriarchalism*, chaps. 9–11; Fox-Genovese, "Property and Patriarchy"; Joyce Oldham Appleby, *Economic Thought and Ideology*, chaps. 2, 6–7.

94. Joyce Oldham Appleby, *Economic Thought and Ideology*, chap. 6; Hatcher, "Labour, Leisure, and Economic Thought," 73–84, 92–95; Amussen, *Ordered Society*, chap. 5; Wrightson, "Two Concepts of Order," 21–47, 299–307, esp. 24–25, 34–35, 300–303; E. P. Thompson, *Customs in Common*, chap. 6 (development of industrial worker discipline, mostly after 1700).

95. Chappell and Webworth, *Roxburghe Ballads*, 1:111–15 (quotes on 112).

CHAPTER ONE

1. Mitchell, *Winthrop Papers*, 1:295–310, 2:107–49 (quotes on 119, 121, 138–39, 142–43); Dunn, "Odd Couple," 6, 9–10, and "Experiments Holy and Unholy," 272–74. On the English background of Winthrop's decision, see Foster, *Long Argument*, chap. 3, esp. 108–14; Edmund S. Morgan, *Puritan Dilemma*, chaps. 2–5; Rutman, *Winthrop's Decision*; Bremer, "Heritage of John Winthrop," 540–47. Bridenbaugh, *Vexed and Troubled Englishmen*, chaps. 11–12, is a through, if dated, analysis of emigration from England.

2. Howard Mumford Jones, "Colonial Impulse," 131–61, esp. 146–52, remains the best analysis of promotional pamphlets, but see Quinn, *Explorers and Colonies*, 105–17; Edmund S. Morgan, *American Slavery*, 30–33, 65–70. Kenneth R. Andrews, *Trade, Plunder, and Settlement*, 32–34, insists that such arguments were often "cant," and Mildred Campbell, " 'Of People Either Too Few or Too Many,' " argues that social commentators were divided on the need for colonies to relieve excess population. For examples, see

Horn, *Adapting to a New World*, 64; Sir George Peckham, *True Reporte* (1583), 49–50; Hakluyt, *Discourse*, 71, 82–85 (quotes); John Smith, *Works* (*Description of New England*, 1616), 1:343–48; Donne, *Sermon Preached* (1622), 272 (quote); John Hammond, *Leah and Rachel*, 296–97; William Penn, *Some Account of the Province of Pennsylvania*, 203–6.

3. For a similar argument, see Moogk, "Reluctant Exiles," 464–65.

4. Wallerstein, *Modern World-System*, 100–107, defines peripheries. Pocock, "British History," 605–13, 617–19, and "Limits and Divisions," 318–19, 325–29, and Bailyn and Morgan, introduction, 1–17, link England and colonies in Britain and America to a single history. Canny, "Origins of Empire," 1–32, shows the limitations of the connections. David Hackett Fischer, *Albion's Seed*, presents four test cases of cultural continuities.

5. Bailyn, *Peopling of British North America*, 26–28, 40–41; Games, *Migration*, 1, 13–17, 60; Souden, "Migrants and the Population Structure," 133–49, and "'Rogues, Whores, and Vagabonds,'" 23–38; Clark and Souden, *Migration and Society*, 11–48, esp. 23–26, 36–37; Ruggles, "Migration, Marriage, and Mortality," 508–10. Roger Thompson, *Mobility and Migration*, esp. 34–41, 49–51, 62–65, 73, 80–81, 92–100, 106–12, 120–25, 162–63, 171–82, 224–25, stresses the persistence of New England immigrants far too much, given the lack of data on most who left East Anglia.

6. Kenneth R. Andrews, *Trade, Plunder, and Settlement*, 9–17, chaps. 10, 14; Robert M. Bliss, *Revolution and Empire*, chap. 2; Wesley Frank Craven, "Early Settlements," 25–42; Cressy, *Coming Over*, chap. 5; Simmons, "Americana," esp. 366–70; Games, *Migration*, 18, 20, 23, 30, 62–66, 72.

7. *Oxford English Dictionary*, "adventurer"; Quinn, *North America*, chaps. 13–14, 16–18, summarizes sixteenth-century English ventures; for adventurers' self-image and the financing of the voyages, see Shammas, "Elizabethan Gentleman," chaps. 3–4, and "English Commercial Development," 151–62. On gold, see Fuller, *Voyages in Print*, 58, 64–65, 70, 73, 79–80, 85–86.

8. Louis B. Wright, *Dream of Prosperity*; Howard Mumford Jones, "Colonial Impulse," 152–56; Lefler, "Promotional Literature," 3–17; Richard Beale Davis, *Intellectual Life*, 1:13–52; Fuller, *Voyages in Print*, 13, 33, 50–51. For lists of commodities, see Hariot, *Briefe and True Report*, 325–68; John Smith, *Works* (*Map of Virginia*, 1612), 1:151–59; (*Description of New England*, 1616), 1:330–38, 348; (*Generall Historie*, 1624), 2:76–78, 89, 108–13, 300–301; *New-Englands Plantation*, 7–10.

9. Jack P. Greene, *Intellectual Construction*, 38–44; Kupperman, "Beehive," 270–88; James David Taylor, "'Base Commoditie,'" 73–89; Fuller, *Voyages in Print*, 27, 87–90; Shammas, "English Commercial Development," 157–67; Horn, *Adapting to a New World*, 3–7, 128, 130, and "Tobacco Colonies," 173–74; Newell, *From Dependency to Independence*, 30–33; Roper, "Unraveling of an Anglo-American Utopia," 278–81; Seed, *Ceremonies of Possession*, 26–30 (quote on 27).

10. Knapp, "Elizabethan Tobacco," esp. 26–37, 52, reviews the literature. See also Sarah Augusta Dickson, *Panacea or Precious Bane*, 98, 101, 129, 131–35, 155–59, 168–69, 182–83, 186–92, 196, 203–5; Horn, *Adapting to a New World*, 131, 141–42. Primary sources include Hariot, *Briefe and True Report*, 344–45; John Smith, *Works* (*Generall Historie*, 1624), 2:77, 256, 285 (quote), 287, 327; (*True Travels*, 1630), 3:215–18; (*Advertisements*, 1631), 3:271, 274; Beaumont, *Poems*, 275–321, esp. 286, 315–16.

11. Robert Brenner, *Merchants and Revolution*, chap. 2; Rabb, *Enterprise and Empire*, chap. 1 (61–62 for privateering); Carville Earle, "Pioneers of Providence," 479–81; Peter Bowden, "Agricultural Prices," 674–94.

12. On English colonization in Ireland, see Ohlmeyer, "'Civilizinge of Those Rude Partes,'" 124–46; Canny, *Kingdom and Colony*, chap. 3; Canny, "Ideology of English Colonization," 577–95; Canny "English Migration," 64–75; Raymond Gillespie, *Colonial*

Ulster, esp. chaps. 2–4; MacCarthy-Morrogh, *Munster Plantation*, chaps. 2, 4–7. On military activity, see Richard W. Stewart, "'Irish Road,'" and Thomas Garden Barnes, "Deputies," 59–60. Quotes from John White, *Planters Plea*, 15–16.

13. Rabb, *Enterprise and Empire*, esp. chap. 1 (best analysis); McCusker, review of *Enterprise and Empire*, 16–18; Robert Brenner, *Merchants and Revolution*, chap. 3; Hansen, *Dorchester Group*, 17–19, 22–24. Rabb believes that the investors he could not identify were petty London merchants. I have conservatively allocated 60 percent of the unknowns to the "merchant" category, and the rest in the same way as the knowns. Then I presumed the same ratio of gentry-merchant investment for all companies as for the Virginia Company.

14. Wesley Frank Craven, *Dissolution of the Virginia Company*, 149–50, 183–84; William Robert Scott, *Constitution and Finance*, 2:89–107, 246–48, 266–88, 306–14; Robert Brenner, *Merchants and Revolution*, 97; Rose-Throup, *Massachusetts Bay Company*, 84–93.

15. Robert Brenner, *Merchants and Revolution*, 93–102; Robert C. Johnson, "Lotteries," 259–92; Wesley Frank Craven, *Dissolution of the Virginia Company*; William Robert Scott, *Constitution and Finance*, 2:246–59, 266–89.

16. William Robert Scott, *Constitution and Finance*, 2:306–14, 246–48, 266–88; Wesley Frank Craven, *Dissolution of the Virginia Company*, chap. 7; Banks, *Winthrop Fleet*, 24–32; Rose-Throup, *Massachusetts Bay Company*, chap. 16; Robbins, "Massachusetts Bay Company," 83–98; Hansen, *Dorchester Group*, 26–27.

17. Howard Mumford Jones, "Colonial Impulse," 131–32 (quote); Games, "Venturers, Vagrants, and Vessels of Glory," 79–119, esp. 94, 101–2, 110–13; Lemay, *"New England's Annoyances,"* chap. 1 (quotes on 22, 27); Jonson, Chapman, and Marston, *Eastward Ho!*, 17–19, 138–40 (quote); Kathleen M. Brown, *Good Wives*, 53–54.

18. Cressy, *Coming Over*, chap. 9 (the best analysis); Robert C. Johnson, "Transportation of Vagrant Children," 137–51 (quote on 143); Horn, *Adapting to a New World*, 63. For examples, see Demos, *Remarkable Providences*, 46–53 (Frethorne quotes on 47, 49).

19. Firth, *American Garland*, 25–40, 51–59 (quotes on 32, 34, 35–37, 51); Morison, *Builders of the Bay Colony*, 45–46, 62–63, 384–86; Rollins, *Pepys Ballads*, 38–40; Hansen, *Dorchester Group*, 46–48. Firth (*American Garland*, xxvii–xxviii) dates "Maydens of London" to the late 1650s because of a mention of war with Spain, 1655–60. But since there was no mention of the 1622 massacre, I would date the ballad in 1621. Tensions with Spain were high in the early 1620s, and the two countries fought a war in 1625 (Hirst, *Authority and Conflict*, 126–30, 139, 142–44; Thomas Garden Barnes, "Deputies," 59–61). Moreover, there would have been no reason to send shiploads of unattached women in the 1650s, when many families paid their own passage to the Chesapeake (Ransome, "Wives for Virginia"; Horn, "'To Parts beyond the Seas,'" 93, 104–9; David Hackett Fischer, *Albion's Seed*, 214; Menard, "British Migration," 120–21). For the date of "West-Country Man's Voyage," see Lemay, *"New England's Annoyances,"* 31–34.

20. Games, "Venturers, Vagrants, and Vessels of Glory," chap. 2, esp. 79–80, emphasizes deception too strongly. Other analyses of the strategies of seventeenth-century promoters include Lemay, *American Dream*, chap. 4, and *"New England's Annoyances,"* chap. 2; Canny, "'To Establish a Common Wealthe,'" 215–17; and Kane, "Early Pennsylvania Promotion Literature," 150–68.

21. Edward Waterhouse, *Declaration*, 544–50; Firth, *American Garland*, xxi–xxii, 9–16; "Good Newes from Virginia," 351–58 (quotes on 353, 355–57).

22. *New-Englands Plantation*, 10, 12; Lemay, *"New England's Annoyances,"* 88–89.

23. Robert Johnson, *Nova Britannia*, 19, 21; John Smith, *Works* (*Description of New England*), 1:343, 346, 348; John White, *Planters Plea*, 9–10 (quote), 13–15, 16, 19 (quote),

27–33. Eburne, *Plain Pathway to Plantations* (1624), written to support a Newfoundland colony, is the longest and most systematic of these responses.

24. Menard and Carr, "Lords Baltimore," 177–83; Billings, "Sir William Berkeley," esp. 340; Gary B. Nash, "Free Society of Traders," 153–73; Corcoran, "Penn and His Purchasers."

25. Dunn, "Penny Wise and Pound Foolish," 37–48; Bronner and Fraser, *Penn's Published Writings*, 37–39, 264–76, 282–84, 298–309, 321–23, 327–29; Kane, "Early Pennsylvania Promotion Literature," 144–68; William Penn, *Some Account of the Province of Pennsylvania*, 202–10 (quotes on 203–4, 209–10), and *Letter*, 224–44.

26. This conclusion is drawn from examination of the dates of literature cited in Lefler, "Promotional Literature," 3–18, and works printed in Force, *Tracts*, and Arber, *Story of the Pilgrim Fathers*. For movement between colonies, see Hatfield, "Reconceiving Virginia," chap. 3.

27. On early Virginia trade, see Robert Brenner, *Merchants and Revolution*, 120–40, 146–48, 180, 184–93. On servant recruitment, see Galenson, "Rise and Fall of Indentured Servitude," 1–13, and *White Servitude*, 97–102; Games, *Migration*, 76–79, and "Venturers, Vagrants, and Vessels of Glory," 109–19; and Horn, "Servant Emigration," 87–94, and *Adapting to a New World*, 79–80. On wages, see Everitt, "Farm Labourers," 436–38, and Peter Bowden, "Agricultural Prices," 598–601, and "Statistical Appendix," 864–65.

28. Galenson, *White Servitude*; Abbot Emerson Smith, *Colonists in Bondage*; Games, *Migration*, 73–74, 88–98. See Prologue for servants-in-husbandry.

29. Bridenbaugh, *Vexed and Troubled Englishmen*, 411–22 (fanciful but probably accurate account of recruiting); Galenson, *White Servitude*, chap. 7; Abbot Emerson Smith, *Colonists in Bondage*, chap. 4. Galenson's evidence begins in the 1660s, but promotional pamphlets suggest knowledge of colonies was widespread by the 1630s. For the counterargument (which I do not fully accept) that servants had virtually no bargaining power, see Horn, *Adapting to a New World*, 65–68.

30. John Hammond, *Leah and Rachel*, 284–89, 292–95 (quotes); John Smith, *Works* ("True Travels," 1630), 3:216–18 (similar argument); Lemay, *Men of Letters*, 28–47, esp. 35, 38–43. On life expectancy and the condition of servants, see Carr and Menard, "Immigration and Opportunity," and Walsh and Menard, "Death in the Chesapeake," 211–27.

31. Canny, "English Migration," 63–65; Gemery, "Emigration from the British Isles," 196–98, 215, and "Markets for Migrants," 34–35; Menard, "British Migration," 102–5; Horn, *Adapting to a New World*, 24–25, 137; Dunn, *Sugar and Slaves*, 55; Delâge, *Bitter Feast*, 241–50; Morner, "Spanish Migration," 738, 750–53, 759, 767; Borah, "Mixing of Population," 707–22, esp. 707–8; Boyd-Bowman, "Patterns of Spanish Emigration," 580–604; Moogk, "Reluctant Exiles," 463–64, 497–505; Boleda, "Les Migrations au Canada," 34–35.

32. The debate has centered around New England. See Virginia DeJohn Anderson, "Migrants and Motives"; Virginia DeJohn Anderson, "Religion, the Common Thread," 418–24; Virginia DeJohn Anderson, *New England's Generation*, chap. 1; Cressy, *Coming Over*, chap. 3; David Grayson Allen, "Matrix of Motivation." For a general overview, see Quinn, *Explorers and Colonies*, 151–78. On travel times, see Canny, "Origins of Empire," 10 (quote), 26–27.

33. Carville Earle, *Geographical Inquiry*, 80–82; Menard, "British Migration," 120–21.

34. Eburne, *Plain Pathway to Plantations*, 90 (quote); Shipps, "Puritan Emigration" (quote on 89); Wrigley and Schofield, *Population History*, 185–87, 528.

35. Zuckerman, *Almost Chosen People*, 36 (quote); Bailyn, "Politics and Social Structure," 93–94; Ransome, "Wives for Virginia," 7; Gottfried, "First Depression," 655–59; Cressy, *Coming Over*, chap. 8; Games, *Migration*, 34, 44, 194–98, 202–4, 234, 236–38, and

"Venturers, Vagrants, and Vessels of Glory," 185, 203, 218, 239, 262–73, 414; David Grayson Allen, "'*Vaccuum Domicilium*,'" 3; Delbanco, *Puritan Ordeal*, chap. 6; Gragg, "Puritans in Paradise"; Sachse, "Migration"; Stout, "Morphology of Remigration"; Roger Thompson, *Mobility and Migration*, 27, 39, 41–42, 52.

36. Migration theory informs these comments on motivation. See Standing, "Migration"; Greenwood, "Research in Internal Migration"; MacDonald and MacDonald, "Chain Migration."

37. Louis B. Wright, *Dream of Prosperity*, chaps. 1–2; Quinn, *Explorers and Colonies*, 151–58; Billings, *Old Dominion*, 27–28 (quote on 27).

38. Horn, "Servant Emigration," 75–87; Horn, "Adapting to a New World," 139–45; Horn, *Adapting to a New World*, 62–66, 69–76, 80; Souden, "'Rogues, Whores, and Vagabonds,'" 24–38; Galenson, *White Servitude*, chap. 5; Wareing, "Migration to London," 357–71; Games, *Migration*, 72.

39. Bradford, *Plymouth Plantation*, 441–43; Banks, *Winthrop Fleet*, 53, 59, 61, 68–70, 88; "Partial List of the Families Who Resided in Bucks County"; "Partial List of the Families Who Arrived at Philadelphia"; Salinger, *"To Serve Well and Faithfully,"* 18–30.

40. Carr and Walsh, "Planter's Wife," 546, 550–51; Galenson, *White Servitude*, 23–31; Lorena S. Walsh, "Servitude and Opportunity," 113–14.

41. Cressy, *Coming Over*, chap. 4; Billings, *Old Dominion*, 15–17; *Relation of Maryland* (1635), 92–97; Horn, "Adapting to a New World," 141–43; Wrightson, *English Society*, 25–34; Everitt, "Farm Labourers," 419–22; Overton, "English Probate Inventories," 209; Roger Thompson, *Mobility and Migration*, 113; Games, *Migration*, 64–65.

42. David Hackett Fischer, *Albion's Seed*, 212–25; Horn, *Adapting to a New World*, 19, 53–57, and "'To Parts beyond the Seas,'" 111–17; Quitt, "Immigrant Origins of the Virginia Gentry," 629–39; Kupperman, "Founding Years," 109.

43. Compare Horn, *Adapting to a New World*, 59–61, to Quitt, "Immigrant Origins of the Virginia Gentry," 629–48. On Barbados, see Hatfield, "Reconceiving Virginia," 135–42, 148–49.

44. Horn, *Adapting to a New World*, 55–58; Bossy, "Reluctant Colonists," 158–64; John Graham, "Meetinghouse and Chapel," 245–74; David W. Jordan, *Foundations of Representative Government*, 13, 25–26, 41, 67–68 (quote), 78–80, 85–86; Perry Miller, *Errand into the Wilderness*, chap. 4; Babette M. Levy, *Early Puritanism*, chaps. 2, 4; Jon Butler, "Two 1642 Letters"; Kenneth L. Carroll, "Quakerism."

45. On the debate over the Puritan's errand and clerical motivations, see Perry Miller, *Errand into the Wilderness*, chap. 1; Bozeman, *To Live Ancient Lives*, chap. 3; Richard Waterhouse, "Reluctant Emigrants"; Stout, "Morphology of Remigration," 159–63; Tyack, "Humbler Puritans"; Foster, *Long Argument*, 23, 26–27; Banks, "Religious 'Persecution'" (denies persecution but provides evidence of clerical beliefs that it existed); and Roger Thompson, *Mobility and Migration*, 20–23, 232 (reviews economic motivations).

46. David Hackett Fischer, *Albion's Seed*, 18–24; Virginia DeJohn Anderson, *New England's Generation*, 37–46, and "Religion, the Common Thread," 420–24; Games, *Migration*, 137–41; Morison, *Builders of the Bay Colony*, 379–84; Bridenbaugh, *Vexed and Troubled Englishmen*, 448–67; Clap, *Memoirs*, 5–7, 26–27 (quotes); Selement and Woolley, *Shephard's Confessions*; Selement, "Meeting of Elite and Popular Minds."

47. McGiffert, *God's Plot*, 55–57 (first quote); Cressy, *Coming Over*, 87–98; David Grayson Allen, *In English Ways*, esp. chap. 6, and "'Both Englands,'" 60–61; Selement and Woolley, *Shephard's Confessions*, 65–66 (quote), 78–79, 89–90, 97 (quote).

48. John White, *Planters Plea*, 36 (quote); Homans, "Puritans and the Clothing Industry," 525–28; Crouse, "Causes of the Great Migration," 8–14; Breen and Foster, "Moving to the New World," 200–206; David Grayson Allen, *In English Ways*, chap. 6; Cressy,

Coming Over, 88–97; Demos, *Remarkable Providences*, 59–69; Roger Thompson, *Mobility and Migration*, 95–96.

49. Salerno, "Social Background," 31–44, 48–52; Banks, "English Sources," 368–70; Innes, *Labor in a New Land*, 5–11, 124–25.

50. Horle, *Quakers and the English Legal System*, esp. 46–55, 65–100, 138–49, 266–67, 279–84; Barry Levy, *Quakers and the American Family*, chap. 1, 110–19, 138; David Hackett Fischer, *Albion's Seed*, 420–51.

51. For different interpretations of immigration and settlement, see David Hackett Fischer, *Albion's Seed*, and Meinig, *Shaping of America*, 1:91–160.

52. Games, *Migration*, 24–25; Horn, *Adapting to a New World*, 36–38; Eltis and Engerman, "Was the Slave Trade Dominated by Men?," 242–45; Bailyn, *Voyagers to the West*, chap. 5, for the idea of a dual migration system.

53. Bailyn, *Peopling of British North America*, 26–28, 40–41; Souden, "Migrants and the Population Structure," 133–49, and " 'Rogues, Whores, and Vagabonds,' " 23–38; Clark and Souden, *Migration and Society*, 11–48, esp. 23–26, 36–37; Ruggles, "Migration, Marriage, and Mortality," 508–10; Wrigley and Schofield, *Population History*, 168–69, 182–87; Menard, "British Migration," 101–2; Arber, *Story of the Pilgrim Fathers*, 355–59; Bridenbaugh, *Vexed and Troubled Englishmen*, 409–10; Galenson, *White Servitude*, 216–17; Gemery, "Emigration from the British Isles," 215–16.

54. Quinn, *Set Fair for Roanoke*, 87–92, 264–67, 296–97, and *New American World*, 287–88, 321–22; William S. Powell, "Roanoke Colonists," 206–9, 213–26; Smith and Wilson, *North Carolina Women*, 20–21.

55. Despite disputes over numbers, no one would deny the thrust of this paragraph. See John Smith, *Works*, 3:140–42, 160–62, 190–91; Edmund S. Morgan, *American Slavery*, 83–87, 108–30; Diamond, "From Organization to Society," esp. 462–64; Ransome, " 'Shipt for Virginia,' " 447–52; Horn, "Tobacco Colonies," 177; Games, *Migration*, 47, 50–51; Bernhard, "Men, Women, and Children"; Camfield, "Can or Two of Worms," 650–62; Hecht, "Virginia Muster," 70–80; Canny, "Permissive Frontier," 25–27, 29.

56. Bernhard, "Men, Women, and Children," 614–18; Games, *Migration*, 24–25; Ransome, "Wives for Virginia" (quotes on 4, 7); Kupperman, "Founding Years," 105–7.

57. Menard, "British Migration," 100–105, 115–19; Menard, "Immigration to the Chesapeake"; Menard, "From Servants to Slaves"; Wesley Frank Craven, *White, Red, and Black*, chap. 1; Edmund S. Morgan, "Headrights and Head Counts"; Grubb and Stitt, "Liverpool Emigrant Servant Trade."

58. Menard, "British Migration," 125–27, and "Immigrants and Their Increase," 95–97; Lorena S. Walsh, "Servitude and Opportunity," 112–15, 129; Souden, " 'Rogues, Whores, and Vagabonds' "; Horn, *Adapting to a New World*, 20–22, 31–47, and "Servant Emigration"; Games, *Migration*, 50–51, 82–83.

59. Compare Mildred Campbell ("Social Origins," "Rebuttal," and "Reply"), who argued that indentured servants came mostly from the middling classes, with Galenson (*White Servitude*, chap. 3; " 'Middling People' or 'Common Sort'?"; and "Social Origins"), who contended that many were laborers. Galenson's inferences fit the evidence better than Campbell's.

60. See Horn, *Adapting to a New World*, 36–38; Menard, "From Servant to Freeholder," 51–63; Carr and Walsh, "Planter's Wife," 550–52, 562; and Chapter 5, below.

61. Menard, "British Migration," 117–21; Clemens, *Atlantic Economy*, 48–51; Horn, " 'To Parts beyond the Seas' "; Horn *Adapting to a New World*, 25–31, 39–40; Horn "Cavalier Culture?" (free immigrants came from middling classes); David Hackett Fischer, *Albion's Seed*, 207–46, and "*Albion* and the Critics," 277–89 (gentry dominance

among free immigrants); Hatfield, "Reconceiving Virginia," chaps. 3–4. The political consequences of migration are debated in the works of Fischer and Horn, in Bailyn, "Politics and Social Structure," 98–115, and in Reavis, "Maryland Gentry."

62. Bailyn, *New England Merchants*, chaps. 1–2; Heyrman, *Commerce and Culture*, 29–37, 209–20; Vickers, "Work and Life," 83–117; Christine Alice Young, *From "Good Order" to Glorious Revolution*, chap. 1; Charles E. Clark, *Eastern Frontier*, chap. 1; Charles F. Carroll, *Timber Economy*, chap. 3.

63. Moller, "Sex Composition," 115; Arber, *Story of the Pilgrim Fathers*, 359–80; Ames, *May-flower and Her Log*, 166–95; Thomas Boylston Adams, "Bad News," 133–35; Bradford, *Plymouth Plantation*, 103–10; Fussell, "Social and Agrarian Background," 185.

64. Historians have fiercely debated the origins of Massachusetts immigrants, agreeing on neither the definition of East Anglia nor the number that came from that region. See David Hackett Fischer, *Albion's Seed*, 31–42, and "*Albion* and the Critics," 263–77; Virginia DeJohn Anderson, "Migrants and Motives," 356–64, and "Origins of New England Culture"; Games, "Venturers, Vagrants, and Vessels of Glory," 70–76, and *Migration*, 28–30, 59; Roger Thompson, *Mobility and Migration*, 22–23.

65. Banks, *Winthrop Fleet*, 57–99; Games, *Migration*, 24–27; Virginia DeJohn Anderson, "Migrants and Motives," 336–54; David Hackett Fischer, *Albion's Seed*, 25–27; Archer, "New England Mosaic," 478–81; Cressy, *Coming Over*, 68–70; Roger Thompson, *Mobility and Migration*, 24–25.

66. Hansen, *Dorchester Group*, 32 (quote); Roger Thompson, "Social Cohesion" and *Mobility and Migration*, chap. 8; Games, *Migration*, 53–57, 59.

67. David Hackett Fischer, *Albion's Seed*, 30–31; Games, *Migration*, 51–52; Breen and Foster, "Moving to the New World," 196–99; Roger Thompson, *Mobility and Migration*, 26–27 (has only one-third of artisans listed in the cloth trade).

68. Cressy, *Coming Over*, 52–63; Virginia DeJohn Anderson, *New England's Generation*, 24–25, 223; Games, *Migration*, 51–52; Vickers, "Working the Fields," 56–57, 69, and "Work and Life," 84–91; Salerno, "Social Background," 33; Innes, *Labor in a New Land*, 9–10; Galenson, *White Servitude*, 82–86, 214–16; Moller, "Sex Composition," 118; Dunn, "Servants and Slaves," 160.

69. Perry, *Formation of a Society*, 160–62; Torrence, *Old Somerset*, 275–334, esp. 279–85; Kenneth L. Carroll, "Quakerism," 174–75; Hatfield, "Reconceiving Virginia," 190–93, 203–4, 208; Lindley S. Butler, "Early Settlement," 24–28; Jackson Turner Main, *Society and Economy*, chap. 1; Hansen, *Dorchester Group*, chap. 4; Kross, *Evolution of an American Town*, 31–34; Pomfret, *Province of East New Jersey*, chap. 3; Rindler, "Migration from the New Haven Colony," chaps. 1, 5–6.

70. Jackson Turner Main, *Society and Economy*, 4–10; Rindler, "Migration from the New Haven Colony," chaps. 1, 5; Bissell, "One Generation to Another," 79–82; David Grayson Allen, "'Both Englands,'" 63–69; Battis, *Saints and Sectaries*, 247–48, 261, 269–70, 274, 279, 299–328; Charles M. Andrews, *Colonial Period*, vol. 2, chaps. 1, 3, 5.

71. For South Carolina data, see Dunn, "English Sugar Islands" and "Barbados Census," 26–30 (6.4 percent of 593 people leaving Barbados in 1679 went to Carolina); Richard Waterhouse, "England, the Caribbean, and the Settlement of Carolina"; Jack P. Greene, *Imperatives*, 68–77; Peter Wood, *Black Majority*, chap. 1; Menard, "Africanization," 85–93, 104–5, 108; Greene and Harrington, *American Population*, 172.

72. Menard, "Africanization"; Peter Wood, *Black Majority*, chaps. 1, 4–5; Abbot Emerson Smith, *Colonists in Bondage*, 3–20.

73. Vann, "Quakerism"; David Hackett Fischer, *Albion's Seed*, 420–33, 438–51, and "*Albion* and the Critics," 289–94; Barry Levy, "Quakers, the Delaware Valley, and North

Midlands Emigration"; Schwartz, "*Mixed Multitude*," 66–67, 75–77; Stephanie Grauman Wolf, *Urban Village*, 38–39, 43; Lemon, *Best Poor Man's Country*, 49–55.

74. Gary B. Nash, *Quakers and Politics*, 49–56; David Hackett Fischer, *Albion's Seed*, 434–38, and "*Albion* and the Critics," 293–94; Barry Levy, "Quakers, the Delaware Valley, and North Midlands Emigration"; Salinger, "*To Serve Well and Faithfully*," 18–33; Dunn and Dunn, *Papers of William Penn*, 2:630–64. The data reported here are calculated from 701 names on immigrant lists ("Partial List of the Families Who Resided in Bucks County" and "Partial List of the Families Who Arrived at Philadelphia"), with duplicates eliminated.

75. These colonies are usually considered as a group. See Van den Boogaart, "Servant Migration"; Schwartz, "Society and Culture," 99–116, esp. 103, 115–16; Pomfret, *Province of West New Jersey*, chaps. 2–3, esp. 28–30; Rink, *Holland on the Hudson*, chap. 6; David Steven Cohen, "How Dutch Were the Dutch?"; Dennis, *Cultivating a Landscape of Peace*, 133–38; Bonomi, *Factious People*, 18–24; Goodfriend, *Before the Melting Pot*, 61–62; Wacker, "Dutch Culture Area," 1–10. I have relied most heavily on Van den Boogaart's population estimates.

76. New England social historians (Edmund S. Morgan, *Puritan Family*; Rutman, *Winthrop's Boston*; Lockridge, *New England Town*; Greven, *Four Generations*) presume the communal and familial basis of society, a presumption problematic among Chesapeake scholars (contrast Edmund S. Morgan, *American Slavery*, with Rutman and Rutman, *Place in Time: Middlesex*). Contrast as well Kulikoff, *Making of the American Yeoman Class*, chap. 1 (communal differences), with Breen, "Creative Adaptions," esp. 215–21, and David Hackett Fischer, *Albion's Seed* (cultural continuities).

77. Firth, *American Garland*, 15 (quote); Fuller, *Voyages in Print*, 45, 49 (example of empty America); Kupperman, "Beehive," 275–78, 283–85.

78. Vickers, "Competency and Competition"; Davis and Mintz, *Boisterous Sea of Liberty*, 52 (quote).

79. Breen and Foster, "Moving to the New World," 214–16.

CHAPTER TWO

1. Demos, *Remarkable Providences*, 46–53; Ludlum, *Early American Winters*, 13; Emerson, *Letters from New England*, 64–66 (another, more annotated version of the Pond letter), 110–12, 138–42, 214–15.

2. Carr, "Emigration," 272–82; Horn, "Adapting to a New World," 151–64.

3. Merrell, "'Customes of Our Countrey'"; Axtell, *After Columbus*, 226–43; Washburn, "Dispossessing the Indians"; Jennings, *Invasion of America*, chap. 5.

4. Representative environmental histories include Cronon, *Changes in the Land*; Merchant, *Ecological Revolutions*; Silver, *New Face on the Countryside*.

5. For definitions of landscape, see Meinig, "Beholding Eye," and Peirce Lewis, "Learning from Looking," 249–61. On early modern English landscape, see Hoskins, *Making of the English Landscape*, chap. 5 (classic account); Gregory King, *Two Tracts*, 35–37; J. N. L. Baker, "England in the Seventeenth Century," 386–429; Emery, "England circa 1600"; Everitt, "Marketing of Agricultural Produce," 466–502; Thirsk, "English Rural Communities," 40–61; Harley, "Meaning and Ambiguity," maps following 33; Eden, "Three Elizabethan Estate Surveyors," maps following 72; Harvey, "English Estate Maps," 27–34, 43, 48, 51–53.

6. M. J. Bowden, "Invention of American Tradition," 3–10 (quotes on 5, 7); Fritzell, "Wilderness and the Garden," 16–22; Heimert, "Puritanism, the Wilderness, and the Frontier" (quote on 365); Clap, *Memoirs*, 7–8 (quotes); McJimsey, "Topographic Terms,"

168–69, 262, 273, 276, 412; Denevan, "Pristine Myth," 380–81. Bowden dismisses such language as an invention; I believe that it was the lived experience of the first settlers.

7. Hoskins, *Making of the English Landscape*, 137–43; Emery, "England circa 1600," 265, 272–75, 283–84; J. N. L. Baker, "England in the Seventeenth Century," 394–99, 421–23; Albion, *Forests and Sea Power*, chap. 3; Roger Burrow Manning, *Hunters and Poachers*, chap. 5; Whitney, *Coastal Wilderness*, 122–26; Gregory King, *Two Tracts*, 35; Charles F. Carroll, *Timber Economy*, chap. 1.

8. Heimert, "Puritanism, Wilderness, and the Frontier," 362 (quote); Roland M. Harper, "Changes in the Forest Area," 442; Kellogg, "Forests," 52–53; Clawson, "Forests in the Long Sweep of American History," 1168–69; Parr, "Maryland Forests," 408; Wacker, "New Jersey Forests," 485; Burnett, "South Carolina Forests," 613; Dean, "Virginia Forests," 671. Penna, *Nature's Bounty*, 16, reports roughly 80 percent in the Northeast and 60 percent in the Southeast today.

9. Charles F. Carroll, *Timber Economy*, 33–37; Hawes, "New England's Forests," 210–15; Michael Williams, *Americans and Their Forests*, 45–47; Whitney, *Coastal Wilderness*, 57–59, 118–19; Cronon, *Changes in the Land*, 25–33; Thompson and Smith, "Forest Primeval," 257–62; Wacker and Clemens, *Land Use*, 38–40; Silver, *New Face on the Countryside*, 15–19; Emily W. B. Russell, "Indian-Set Fires"; Peter A. Thomas, "Understanding Indian-White Relations," 4; Day, *In Search of New England's Past*, 27–31, 330–34; Calvin Martin, "Fire and Forest Structure," 24; Peter C. Stewart, "Man and the Swamp," 57–59; Roundtree and Davidson, *Eastern Shore Indians*, 17–19; Wacker, "Human Exploitation," 3–6; Mackenthun, *Metaphors of Dispossession*, 265 (Winslow quote); Potter and Waselkov, " 'Whereby We Shall Enjoy Their Cultivated Places.' "

10. Semple, *American History*, chap. 2; Meinig, *Shaping of America*, 1:233–35; Ralph H. Brown, *Historical Geography*, 19–22; Arthur Pierce Middleton, *Tobacco Coast*, chap. 2; Carville Earle, *Evolution of a Tidewater Settlement System*, 143–45.

11. Penna, *Nature's Bounty*, 86–91; Delâge, *Bitter Feast*, 37–39; Parish, "Female Opossum," 475–89, 502–5; Canup, *Out of the Wilderness*, 23–25.

12. Bernard G. Hoffman, "Ancient Tribes," 3, 11–15, 20, 27–37; Feest, "Virginia Algonquians," 254–55, and "Nanticoke," 241; Clark and Roundtree, "Powhatans and the Maryland Mainland," 114–15; Horn, "Tobacco Colonies," 174; Howard S. Russell, *Indian New England*, 200; Sherburne F. Cook, *Indian Population*, 1–3; Peter A. Thomas, "Cultural Change," 132–35.

13. There has been a vigorous debate over Indian use of the forests. See Maxwell, "Use and Abuse of Forests"; Day, *In Search of New England's Past*, chap. 1; Calvin Martin, "Fire and Forest Structure," 23–26, 38–42; Emily W. B. Russell, "Indian-Set Fires"; Michael Williams, *Americans and Their Forests*, 32–49; Whitney, *Coastal Wilderness*, 107–20; Dennis, *Cultivating a Landscape of Peace*, 32–36; Wacker and Clemens, *Land Use*, 38–39; Delâge, *Bitter Feast*, 282–83. I think that Russell's conservative arguments (that Indian fires were not as widespread as earlier scholars believed) fits the evidence best.

14. Roundtree, "Powhatans and Other Woodland Indians as Travelers"; Tanner, "Land and Water Communication"; Hatfield, "Reconceiving Virginia," 17–21, 25, 29–30, 36–47; Dunbar, *History of Travel*, 18–21; Howard S. Russell, *Indian New England*, 199–205; Lane, *From Indian Trail to Iron Horse*, 15–18, 32–33; Wacker, *Land and People*, 59, 70, 111–12, and *Musconetcong Valley*, 25–28; Edward Graham Roberts, "Roads of Virginia," map following 111, 122–27; Goodwin, *Cherokees in Transition*, 87–92.

15. Canup, *Out of the Wilderness*, 60–64; Barley, "Rural Housing," 734–66; Nabokov and Easton, *Native American Architecture*, 52–60, 76–88; Willoughby, "Houses and Gardens," 115–20; Kevin A. McBride, "Historical Archeology," 102; Wallace, *Death and Rebirth*, 22–25; Stephen R. Potter, *Commoners*, 24–32, 96–97; Roundtree and Davidson,

Eastern Shore Indians, 33–34, 37; Roundtree, "Powhatan Indian Women," 12; Grumet, *Historic Contact*, 65, 126, 145, 168–69, 309, 314; Starna, "Pequots," 37–38; Hulton, *America 1585*, 27, 62, 67, 177–79; Merrill K. Bennett, "Food Economy," 374–75.

16. Spurr and Barnes, *Forest Ecology*, 443–48, 557–70; Bridenbaugh, "Yankee Use and Abuse of the Forest"; Michael Williams, *Americans and Their Forests*, chap. 3 (good on forests, untrustworthy on settlement systems); McCullough, *Landscape of Community*, 14–35; Whitney, *Coastal Wilderness*, 132–34; James David Taylor, "'Base Commoditie,'" 79–82; Grumet, *Historic Contact*, 447.

17. McCusker and Menard, *Economy of British America*, 314–15; Cox et al., *This Well-Wooded Land*, chap. 2; Lewis Cecil Gray, *History of Agriculture*, 1:151–60; G. Melvin Herndon, "Forest Products," 130–35; Howard S. Russell, *Long, Deep Furrow*, 48, 59, 64–65, 170–74; Charles F. Carroll, *Timber Economy*, chaps. 5–6; Hobbes, "Beginnings of Lumbering"; Hawes, "New England's Forests," 216–19; Albion, *Forests and Sea Power*, chap. 6; Penna, *Nature's Bounty*, 23–25; Wennersten, "Soil Miners," 167–70.

18. Reynolds, *Fuel Wood Used in the United States*, 1–3, 8–15; Michael Williams, *Americans and Their Forests*, 78–81; Whitney, *Coastal Wilderness*, 152–55, 213–14; McCullough, *Landscape of Community*, 28–30; Howard S. Russell, *Long, Deep Furrow*, 170–73; Cox et al., *This Well-Wooded Land*, 36–47; Roland M. Harper, "Changes in the Forest Area," 447; Kammen, "Maryland in 1699," 367 (first quote); Wayne D. Rasmussen, "Wood on the Farm," 15–20 (second quote, 15); Grettler, "Environmental Change," 198–201, 207–8; Carville Earle, *Evolution of a Tidewater Settlement System*, 24–34; Kulikoff, *Tobacco and Slaves*, 47–49; Wennersten, "Soil Miners," 169–70.

19. J. N. L. Baker, "Climate of England" (quote on 421); Manley, "Central England Temperatures," 393, 399–403; Lamb, *Changing Climate*, chap. 7; Andrew B. Appleby, "Epidemics and Famine," 656–60; M. L. Parry, *Climatic Change*, 55, 58–62, 81–85, 102–5, 112–21, 162–67.

20. Sauer, *Selected Essays*, 16–18, 21–27; Lamb, *Changing Climate*, 2:568–69; Kupperman, "Puzzle of American Climate," 1262–67; Kupperman, "Fear of Hot Climates," 213–19; Kupperman, "Founding Years," 106–7; Chaplin, "Climate and Southern Pessimism," 62, 65–67; Carville Earle "Pioneers of Providence," 486.

21. Whitney, *Coastal Wilderness*, 48–50; Fritts, Lofgren, and Gordon, "Variations in Climate," 33–34, 40; Fritts and Lough, "Estimate of Average Annual Temperature," 210–11, 219; Lamb, *Changing Climate*, 577; Manley, "Central England Temperatures," 393. The comparisons are between central England temperatures, 1660–64, and Philadelphia temperatures, 1740–49 (the earliest available), with the warmest and coldest seasons excluded.

22. Bark, "History of American Droughts," 12–14; Stahle et al., "Lost Colony and Jamestown Droughts"; Kelso, Luccketti, and Straube, *Jamestown Rediscovery*, 22, 28; Stahle and Cleveland, "Reconstruction and Analysis of Spring Rainfall," 1954–56; Stephen R. Potter, *Commoners*, 38, 154; Chaplin, "Climate and Southern Pessimism," 60–67 (quote on 65); Wennersten, "Soil Miners," 170–71.

23. Ludlum, *Early American Winters*, 2–37, reprints seventeenth-century sources. For analysis and data, see Baron, "Historical Climate Records," 82–86; Robert D. Mitchell, "Colonial Origins," 97; Peter A. Thomas, "Understanding Indian-White Relations," 8; Kupperman, "Climate and Mastery of the Wilderness," 3–13; Kupperman, "Puzzle of American Climate," 1268–73; Kupperman, "Fear of Hot Climates," 219–23; Lemay, "*New England's Annoyances*," 18; Landsberg, Yu, and Huang, "Preliminary Reconstruction," 14–15; Donald W. Linebaugh, "'All the Annoyances,'" 7, 15; Richard Brooke, "Thermometrical Account," 58–82; Lining, "Extract of Two Letters," 507; Merrens and Terry, "Dying in Paradise," 540; Philip D. Morgan, *Slave Counterpoint*, 33.

24. Kupperman, "Puzzle of American Climate," 1269–70, 1277–87, and "Fear of Hot Climates," 225–27, 229–30 (quote), 232–35; Chaplin, "Climate and Southern Pessimism," 70–73, 76; Peter A. Thomas, "Understanding Indian-White Relations," 1–8; Canup, *Out of the Wilderness*, 15, 17; Dunbar, *History of Travel*, 49–52; Beverley, *Present State of Virginia*, 297 (quote); Dunn, "English Sugar Islands," 91; Kenneth Thompson, "Forests and Climate Change," 46–48.

25. Wrigley and Schofield, *Population History*, 165–70, 176–79, 249–52, 305, 384–93, 667–93; Ruggles, "Migration, Marriage, and Mortality," 514–22; Robert V. Wells, "Population of England's Colonies," 94–96; Graunt, *Natural and Political Observations*, 19–25, 45–48; Wrightson and Levine, "Death in Whickham," 131, 135, 139–50, 155–56; Childs, *Malaria*, 125–28.

26. Kupperman, "Apathy and Death"; Carville Earle, *Geographical Inquiry*, chap. 1; David Hackett Fischer, *Albion's Seed*, 111–12.

27. Rutman and Rutman, *Small Worlds*, chap. 9, is the most compelling work on malaria, but see Merrens and Terry, "Dying in Paradise," 540–50; Childs, *Malaria*, 16–19, 131, 149–50, 153–54, 218–64; and Peter Wood, *Black Majority*, 63–76. For the debate over life expectancy in the Chesapeake area, centering on possible improvement among native-born whites in the eighteenth century, see Walsh and Menard, "Death in the Chesapeake"; Rutman and Rutman, *Place in Time: Explicatus*, chap. 4; Kulikoff, *Tobacco and Slaves*, 60–63; Daniel S. Levy, "Life Expectancies"; and Robert V. Wells, "Population of England's Colonies," 94–97.

28. David Hackett Fischer, *Albion's Seed*, 111–12; Duffy, *Epidemics in Colonial America*, 42–69, 115–28, 187–94, 206–7; Vinovskis, "Mortality Rates," 195–202; Robert V. Wells, "Population of England's Colonies," 94–97.

29. Juricek, "American Usage of the Word 'Frontier,'" 10–15; Delâge, *Bitter Feast*, 255–56; Slotkin, *Regeneration through Violence*, 69–93 (quote from Philip Vincent's 1638 account of the Pequot War on 73); Crosby, *Columbian Exchange*, chap. 3, and *Ecological Imperialism*, chaps. 7–8.

30. Edward Waterhouse, *Declaration*, 551 (quote), 554–55, reprinted in John Smith, *Works* (*Generall Historie*, 1624), 2:294.

31. I have borrowed extensively from Merrell's fine essay "'Customes of Our Countrey,'" 117–37; see also Richard White, *Middle Ground*, esp. ix–xv, 50–52, and Dennis, *Cultivating a Landscape of Peace*, 168–72.

32. Oberg, "Indians and Englishmen"; Mackenthun, *Metaphors of Dispossession*, 154–57; Gleach, *Powhatan's World*, 100–101; Roundtree, "Powhatan Indian Women," 4, 7, 9, 14–15.

33. Quitt, "Trade and Acculturation"; Barker, "Powhatan's Pursestrings," 61–75; Mackenthun, *Metaphors of Dispossession*, 263, 275; Gleach, *Powhatan's World*, 125–26; Roundtree and Davidson, *Eastern Shore Indians*, 35–36; Stephen R. Potter, *Commoners*, 41, 170, 182–84; Kathleen M. Brown, *Good Wives*, 51; Kingsbury, *Records of the Virginia Company*, 3:166; 4:172–73, 186; John Smith, *Works* (*Generall Historie*, 1624), 2:120, 262, 327.

34. Friis, *Series of Population Maps*, endpiece; Gleach, *Powhatan's World*, 134, 150, 171; *England in America*; Rowlandson, *Sovereignty and Goodness of God*, 12.

35. There has been an acrimonious debate over the number of Indians alive before contact (John D. Daniels, "Indian Population," esp. 300–301), but our concern is the number who interacted with colonists. See Ubelaker, "North American Indian Population," 170–73, and "Human Biology," 61–66; Brenda J. Baker, "Pilgrim's Progress," 39–41; Trigger and Swagerty, "Entertaining Strangers," 361–66; Pfeiffer and Fairgrieve, "Evidence from Ossuaries," 49–51; Crosby, "Virgin Soil Epidemics," esp. 290–91; Calloway, *New Worlds for All*, 34–39; Thornton, Miller, and Warren, "American Indian Popula-

tion"; Ramenofsky, *Vectors of Death*, 96–100; Denevan, "Native American Populations," xviii–xxi, xxvii–xxix; Jennings, *Invasion of America*, chap. 2; Sherburne F. Cook, *Indian Population*, chaps. 3, 4, conclusion, and "Significance of Disease"; Cave, *Pequot War*, 43, 47; Snow and Lanphear, "European Contact," esp. 24; Snow, "Disease and Population Decline"; Salwen, "Indians of Southern New England," 169; Carlson, Armelagos, and Magennis, "Impact of Disease," 148; Delâge, *Bitter Feast*, 44, 86–88; Grumet, *Historic Contact*, 154, 215, 232, 246, 289, 368; McCary, *Indians in Seventeenth-Century Virginia*, 1–10; Quitt, "Trade and Acculturation," 241–42; Kulikoff, *Tobacco and Slaves*, 28–29; Feest, "Nanticoke," 242, and "Virginia Algonquians," 258; Fausz, "Fighting 'Fire' with Firearms," 36, 38, 44, and " 'Abundance of Blood Shed,' " 10; Sugrue, "Peopling and Depeopling," 11–13.

36. Cave, *Pequot War*, 103–4 (first quote); Games, *Migration*, 212 (second quote); Horn, *Adapting to a New World*, 359 (third quote); Rowlandson, *Sovereignty and Goodness of God*, 3, 27–28. On Pidgins, see Leechman and Hall, "American Indian Pidgin English"; Goddard, "Some Early Examples"; and Hamell, "Wampum," 41–42. See also n. 37.

37. Merrell, " 'Customes of Our Countrey,' " 126–31; Feister, "Linguistic Communication," 25–38; Calloway, *New Worlds for All*, 172–77; Lepore *Name of War*, 28–33 (much too pessimistic about the willingness of the English to learn a few Indian words); Kupperman, "Presentment of Civility," 199; Lurie, "Indian Cultural Adjustment," 47; Carr, Menard, and Walsh, *Robert Cole's World*, 7, 142, 258–59, 262, 315; Fausz, "Middlemen in Peace and War," 61–63; Axtell, *European and the Indian*, 286–87; Smits, " 'We Are Not to Grow Wild,' " 11; Grumet, *Historic Contact*, 237, 255; Dennis, *Cultivating a Landscape of Peace*, 166–67; Hagedorn, "Brokers of Understanding," 381–83, 386–87, 406–7; Canny, "England's New World," 160–63; Cogley, *John Eliot's Mission*, 24–25, 57, 71, 121; Bragdon, "Gender as a Social Category," 583. For Indian phrase books, see John Smith, *Works* (*Map of Virginia*, 1612), 1:136–39; William Wood, *New England's Prospect*, 117–24; Roger Williams, *Key into the Language*; Lawson, *New Voyage*, 225–30; and Day, *In Search of New England's Past*, 99 (William Pynchon's manuscript words for months).

38. Merrell, " 'Customes of Our Countrey,' " 132–42, and "Cultural Continuity," 558–59; Lurie, "Indian Cultural Adjustment," 47; Roundtree, "Powhatans and the English," 185; Witthoft, *Green Corn*, 7–10, 12, 18–19; Kupperman, *Settling with the Indians*, chap. 2; William Wood, *New England's Prospect*, 7 (editor's introduction), pt. 2; Peter A. Thomas, "Maelstrom of Change," 477–78; Grumet, *Historic Contact*, 68, 126, 237; Joseph H. Smith, *Colonial Justice*, 55–56, 122–24, 208, 217, 223–25, 243, 262–63, 268–71, 274–75, 281–82; Thorndale, "Virginia Census," 168.

39. Barbour, "Earliest Reconnaissance," 32, 37–39, 41, 44; Carlson, Armelagos, and Magennis, "Impact of Disease," 142, 144; Stephen R. Potter, *Commoners*, 164–65, 180–82, 204–7, 210–19; Axtell, *European and the Indian*, 288–91, and *Beyond 1492*, chap. 5; William N. Fenton, *Great Law*, 268; Grumet, *Historic Contact*, 66, 76, 79, 91, 97, 136, 143, 145, 149, 168, 182, 192, 220, 223, 241, 257, 261–62, 274, 309–12, 347, 362–64, 366, 393, 397, 413, and "Sunksquaws, Shamans, and Tradeswomen," 54–60; Calloway, *New Worlds for All*, 45–50, and *Downland Encounters*, 21; Richter, *Ordeal of the Longhouse*, chap. 4; Elise Brenner, "Sociopolitical Implications," 147–74; Turnbaugh, "Assessing the Significance of European Goods," 137–57; Brenda J. Baker, "Pilgrim's Progress," 37, 39; Lurie, "Indian Cultural Adjustment," 38; Trigger, "Early Native American Responses," 1207–11; Peter A. Thomas, "Cultural Change," 142–49; Van Dongen, "Inexhaustible Kettle," 143, 159; James Homer Williams, "Great Doggs," 255–56; Roundtree and Davidson, *Eastern Shore Indians*, 55–56, 75, 138–40.

40. Axtell, *After Columbus*, 201–4; Kingsbury, *Records of the Virginia Company*, 3:20, 165, 170–71, 612–13, 4:167–68, 275–76; Quitt, "Trade and Acculturation," 228; Roundtree

and Davidson, *Eastern Shore Indians*, 54–56; Roundtree, "Powhatan Indian Women," 11; Merrell, "Cultural Continuity," 558–60; Smits, " 'Abominable Mixture,' " 175 (Smith quote); Fausz, "Fighting 'Fire' with Firearms," 42; Lurie, "Indian Cultural Adjustment," 47–48; Gleach, *Powhatan's World*, 165, 167, 172–73; McCarthy, "Influence of 'Legal Habit,' " 50–53; William N. Fenton, *Great Law*, 273–74; Hening, *Statutes at Large*, 2:20, 185.

41. Smits, " 'We Are Not to Grow Wild,' " 14; Calloway, *New Worlds for All*, 49; O'Brien, *Dispossession by Degrees*, 44–45, 69; Peter A. Thomas, "Maelstrom of Change," 172–73; Strong, "Wyandanch," 59; Malone, *Skulking Way*, 60–61; William N. Fenton, *Great Law*, 273–74; Francis, "Beads," 55; James Homer Williams, "Religion and Culture," 10–14, 22, 31; Dennis, *Cultivating a Landscape of Peace*, 165; Noble and Cronin, *Records*, 2:23 (quote).

42. M. L. Brown, *Firearms*, 127, 151–58, 281–89; Axtell, *Beyond 1492*, 140–42; Fausz, "Fighting 'Fire' with Firearms," 33–50; Shea, *Virginia Militia*, 57–58; Kingsbury, *Records of the Virginia Company*, 3:74, 93; Malone, "Changing Military Technology," 48–53, 61–62, and *Skulking Way*, chaps. 3–5; Kawashima, *Puritan Justice*, 81–82; Marshall, " 'Melancholy People,' " 415–16; Delâge, *Bitter Feast*, 133–35, 159–60; Wacker, *Land and People*, 84; Weslager, *Nanticoke Indians*, 57, 73; Jacobs and Shattuck, "Beavers for Drink," 99; Noble and Cronin, *Records*, 2:5, 27, 71, 101, 131; Joseph H. Smith, *Colonial Justice*, 208; Hening, *Statutes at Large*, 1:219, 382, 441, 525, 2:39, 215, 336.

43. Mancall, *Deadly Medicine*, 11–14, 26, 42–51, 68–70, 80–81, 93–96, 103–10; Axtell, *Beyond 1492*, 142–43; Kawashima, *Puritan Justice*, 79–85; Wacker, *Land and People*, 79–80, 84, 102; Sugrue, "Peopling and Depeopling," 25–26; Weslager, *Nanticoke Indians*, 60–61, 73; McWilliams, "Brewing Beer," 555; Noble and Cronin, *Records*, 2:33, 68, 105–6; Marshall, " 'Melancholy People,' " 413–15; Jacobs and Shattuck, "Beavers for Drink," 99–101; James Homer Williams, "Religion and Culture," 18–19; Joseph H. Smith, *Colonial Justice*, 122, 235, 262–63, 271, 281–82; Hening, *Statutes at Large*, 3:468.

44. Ceci, "Value of Wampum," "First Fiscal Crisis," and "Native Wampum"; Weeden, *Indian Money*, 1–33; Speck, "Functions of Wampum," 18–22, 40–49, 57–60; William N. Fenton, *Great Law*, chap. 16; Hamell, "Wampum," 42–50; Francis, "Beads," 54–56, 61–62; James Homer Williams, "Religion and Culture," 26–27; J. H. C. King, *First Peoples*, 47–50; Kevin A. McBride, "Source and Mother of the Fur Trade," 32, 35, 38–43; Kelso, *Jamestown Rediscovery I*, 4, 17–18; Gleach, *Powhatan's World*, 57–59; Mary W. Herman, "Wampum as a Money," 21–26; Tooker, "League of the Iroquois," 422–24; Delâge, *Bitter Feast*, 141; Cave, *Pequot War*, 51–54; Innes, *Labor in a New Land*, 454–58; Peter A. Thomas, "Maelstrom of Change," 181–82; Salisbury, *Manitou and Providence*, 150–52; Fausz, "Present at the 'Creation,' " 16–18.

45. Phillips with Smurr, *Fur Trade*, vol. 1, chaps. 7, 9, 21; Fausz, " 'To Draw Thither Trade of Beavers' " and "Present at the 'Creation,' " 16–18; Jennings, "Indian Fur Trade," 414–16; Salisbury, *Manitou and Providence*, 145–46, 152–60; Mavor and Dix, *Manitou*, 146–47; Peter A. Thomas, "Maelstrom of Change," 155–202, 261–333; Innes, *Labor in a New Land*, 3–6, 29–34; Merchant, *Ecological Revolutions*, 273–74.

46. Bachman, *Peltries or Plantations*, 78–79, 86; Jacobs and Shattuck, "Beavers for Drink," 101, 104–5; Merwick, *Possessing Albany*, 10–12, 33–34, 38, 89–99, 198; Dennis, *Cultivating a Landscape of Peace*, 121, 124–26, 137–38, 142–43, 157–64; Delâge, *Bitter Feast*, 103–10, 114–15.

47. Axtell, *European and the Indian*, 295–98; Waselkov, "Indian Maps," 292–341, esp. 292–96.

48. On Indian agriculture, see Merrill K. Bennett, "Food Economy"; Hurt, *Indian Agriculture*, chap. 3; G. Melvin Herndon, "Indian Agriculture"; William S. Fowler, "Agricultural Tools"; and Dennis, *Cultivating a Landscape of Peace*, 26–32. On colonists' bor-

rowing of Indian agricultural techniques, see Axtell, *European and the Indian*, 292–96; Mood, "Winthrop on Indian Corn" (quote on 125); and Delâge, *Bitter Feast*, 282–83, 319.

49. Dennis, *Cultivating a Landscape of Peace*, 26 (quote).

50. On fertilizer, Ceci, "Fish Fertilizer," arguing for European origins, is mostly persuasive, but for a restatement of the traditional view that Indians taught colonists about fish fertilizer, see Nanepashemet, "It Smells Fishy to Me." On establishing English fields, see Calloway, *New Worlds for All*, 54–56; Field, " 'Peculiar Manuerance,' " 16–20; and Romani, " 'Our English *Clover Grass*.' "

51. Spear, " 'They Need Wives,' " 35–49; James Homer Williams, "Religion and Culture," 14–15; Rothschild, "Social Distance," 190–92; Dennis, *Cultivating a Landscape of Peace*, 164–65; Smits, " 'We Are Not to Grow Wild,' " 1–31; Mackenthun, *Metaphors of Dispossession*, 277–78; Noble and Cronin, *Records*, 2:19 (quote); Lepore, *Name of War*, 123; Demos, *Unredeemed Captive*, 98–100, 104, 116, 142, 154–56.

52. Quitt, "Trade and Acculturation," 236–38; Smits, " 'Abominable Mixture' "; Murrin, " 'Things Fearful to Name,' " 11–13; Barker, "Powhatan's Pursestrings," 68; Vaughan, *Roots of American Racism*, 117; Merrell, *Indians' New World*, 30–31, 63–64, 86–87, 169; Jordan and Kaups, *American Backwoods Frontier*, 87–92; Godbeer, "Eroticizing the Middle Ground," 91–106; Braund, *Deerskins and Duffels*, 83–86.

53. Berry, "Marginal Groups," 290–92; Blu, *Lumbee Problem*, 36–43; Porter, "Behind the Frontier," 42–45; Harte, "Social Origins"; Roundtree and Davidson, *Eastern Shore Indians*, 163–64, 170, 184, 188; Mandell, *Behind the Frontier*, 165, 173, 182–96.

54. Axtell, *European and the Indian*, 168–206, 275–84; Vaughan and Richter, "Crossing the Cultural Divide," 46–99; Demos, *Unredeemed Captive*, 79–84; Rowlandson, *Sovereignty and Goodness of God*, 76, 79, 81–87, 92–93, 95, 97, 101, 104–6, 110; Gyles, *Memoirs*, 12–23, 93, 101–9, 113.

55. On wild men, see James Homer Williams, "Religion and Culture," 9. On Indian women's roles, see Roundtree, "Powhatan Indian Women," 7–20; Bragdon, "Gender as a Social Category," 576–78; and Virginia DeJohn Anderson, "King Philip's Herds," 607. On nakedness and unkemptness, see Kupperman, "Presentment of Civility," 200–204; Lepore, *Name of War*, 80, 84–86, 93; and Kathleen M. Brown, *Good Wives*, 50, 57–59, 65–66. On Satan, see Cave, *Pequot War*, 11, 16, 18, 22–23, 75, and "Who Killed John Stone?," 519–20, and Fausz, " 'Abundance of Blood Shed,' " 32, 35–36.

56. Merrell, *Indians' New World*, chap. 5, esp. 181–87; Paul A. Robinson, "Lost Opportunities," 25–28; Gary B. Nash, "Image of the Indian," 213–14; McCarthy, "Influence of 'Legal Habit,' " 46–52; Peter A. Thomas, "Maelstrom of Change," 170–72, 312–14; Dennis, *Cultivating a Landscape of Peace*, 124–26, 171–72, 177–78; Gleach, *Powhatan's World*, 54–55, 125–26; William N. Fenton, *Great Law*, 274–75, 388; Segal and Stineback, *Puritans, Indians*, 59–63; John Smith, *Works* ("Advertisements," 1631), 3:276.

57. Quitt, "Trade and Acculturation," 230–32, 244–58, tells this complex story.

58. Saltonstall, *Present State*, 2d p.

59. Virginia DeJohn Anderson, "King Philip's Herds," 602–4, 607–12, 620–24; Roundtree and Davidson, *Eastern Shore Indians*, 72; Vaughan, *Roots of American Racism*, 203–4; Noble and Cronin, *Records*, 2:11, 95; Kingsbury, *Records of the Virginia Company*, 4:138; Horn, *Adapting to a New World*, 358–59; LaFantasie, *Correspondence of Roger Williams*, 1:189–96; Joseph H. Smith, *Colonial Justice*, 123–24, 217, 223–24, 243, 268–70, 274–75; Segal and Stineback, *Puritans, Indians*, 61; Marshall, " 'Melancholy People,' " 407–10; James Homer Williams, "Great Doggs," 251–61; Mandell, "Waban of Natick," 174 (quote); Strong, "Wyandanch," 57–58, 61, 64–68.

60. Cronon, *Changes in the Land*, chap. 4 (the best analysis); Keary, "Retelling the History," 261–67, 276–78; James Tully, "Aboriginal Property," 159–65; Wallace, "Political

Organization," 311–19; Kroeber, "Nature of the Land-Holding Group," 303–6, 313; Sutton, *Indian Land Tenure*, 25–27; Snow, "Wabanaki 'Family Hunting Territories,'" 1143–48; Bishop, "Territoriality," 37–45; Springer, "American Indians and the Law of Real Property," 39–46; McCarthy, "Influence of 'Legal Habit,'" 56–58; Sugrue, "Peopling and Depeopling," 14–23. Vaughan, *New England Frontier*, 104–16, dissents, seeing Indian and settler conceptions of land use similar.

61. Keary, "Retelling the History," 264–73, 276–79.

62. Quitt, "Trade and Acculturation," 234–35; Cronon, *Changes in the Land*, chap. 4; Jennings, *Invasion of America*, chap. 8; Kawashima, *Puritan Justice*, chap. 2; Rowlandson, *Sovereignty and Goodness of God*, 16–18; Kevin A. McBride, "Historical Archeology," 105–7; Springer, "American Indians and Law of Real Property," 32–39; Axtell, *After Columbus*, 60–62, and *Beyond 1492*, 111–13; Peter A. Thomas, "Cultural Change," 149–55; Vaughan, *Roots of American Racism*, 118–19; Dorothy V. Jones, "British Colonial Indian Treaties," 190; Grumet, "Suscaneman," 116–32; Emerson W. Baker, "'A Scratch with a Bear's Paw,'" 237–56; Jennings, "Brother Miquon"; Marshall Becker, "Okehocking Band," 47–68; Sugrue, "Peopling and Depeopling," 14–23; Wacker, *Land and People*, 90–102; Roundtree and Davidson, *Eastern Shore Indians*, 64–67; Noble and Cronin, *Records*, 2:40, 83; Hening, *Statutes at Large*, 2:13–14, 34–35, 139, 141, 154, 351–52, 3:465–66; Stephen R. Potter, *Commoners*, 195–98 (quote on 195).

63. James Tully, "Aboriginal Property," 159–65; Pagden, "Struggle for Legitimacy," 42–46; Jennings, *Invasion of America*, chap. 5; Mackenthun, *Metaphors of Dispossession*, 194–95, 258–59, 265–68, 272; Moynihan, "Patent and the Indians," 9–11; Gary B. Nash, "Image of the Indian," 209–10; McCarthy, "Influence of 'Legal Habit,'" 54–55; Sugrue, "Peopling and Depeopling," 23–24; Delâge, *Bitter Feast*, 285; Segal and Stineback, *Puritans, Indians*, 50–51.

64. On criminal activity, see Vaughan, *Roots of American Racism*, 203–10. For contrasting studies of white-Indian warfare, see Vaughan, *New England Frontier*, xxviii–xxxiii, lix–lxii, chaps. 5, 12, and *Roots of American Racism*, chap. 8; Jennings, *Invasion of America*, chaps. 9, 11–13, 17–18. For summaries of Indian wars, see Leach, "Colonial Indian Wars," 131–42, and Washburn, "Seventeenth-Century Indian Wars."

65. Steele, *Warpaths*, chap. 3; Gary B. Nash, "Image of the Indian," 213–19; Roundtree, "Powhatans and the English," 183–94; Fausz, "'Abundance of Blood Shed'"; Fausz, "Indians, Colonialism, and the Conquest of Cant," 148–50; Fausz, "Patterns of Anglo-Indian Aggression," 235–53; Fausz, "Merging and Emerging Worlds," 75–79; Fausz, "Missing Women"; Edward Waterhouse, *Declaration*, 551, 554–56, 565–71; Kingsbury, *Records of the Virginia Company*, 3:614, 652–53, 672, 678–79, 704–7, 4:9, 71, 221–22, 235, 250, 450–51, 507–8; Gleach, *Powhatan's World*, 78–79, 130 (quote), 148–83; Stephen R. Potter, *Commoners*, 185–92; Vaughan, *Roots of American Racism*, 107, 109–14, 119–25; John Smith, *Works* (*Generall Historie*, 1624), 2:301–2, 327.

66. For a summary, see Steele, *Warpaths*, 91–99, 115–17. The Pequot War remains very controversial. For alternative views, see Vaughan, *New England Frontier*, chaps. 5–6; Jennings, *Invasion of America*, chaps. 13–15; Katz, "Pequot War Reconsidered"; Cave, *Pequot War*, chaps. 1–4, and "Who Killed John Stone"; and Karr, "'Why Should You Be So Furious?,'" esp. 894–907. On the Dutch wars, see Delâge, *Bitter Feast*, 287–90; Grumet, *Historic Contact*, 219, 255; and James Homer Williams, "Great Doggs," 254, 256.

67. Lepore, *Name of War*, 94 (first quote), 72 (third quote); LaFantasie, *Correspondence of Roger Williams*, 2:717–28 (second quotes on 722–23) (the only copy of this letter is from an eighteenth-century transcript, but the editors consider that it is probably authentic); Drake, "Restraining Atrocity," 48.

68. Steele, *Warpaths*, 99–109; Gary B. Nash, *Red, White, and Black*, 118–23; Jennings,

Invasion of America, chaps. 17–18; Leach, *Flintlock and Tomahawk*, chaps. 3–5, 7, 11–13; Lepore, *Name of War*, chap. 3, and 132, 175, 177; Puglisi, *Puritans Besieged*, chaps. 4–5, 7; Drake, "Restraining Atrocity," 3–32; Pulsipher, "Massacre at Hurtleberry Hill," 459, 465, 469–70, 476; Demos, *Unredeemed Captive*, 11–13, 18–30, 38–39, 87–91, 135–37, 171; Slotkin and Folsom, *So Dreadfull a Judgment*, 25–35, map between 45 and 46, 46–52; Mathews, *Expansion of New England*, chap. 3; Melvoin, *New England Outpost*, chaps. 4, 8; Haefeli and Sweeney, "Revisiting *the Redeemed Captive*," 14, 18–19, 22–23, 25, 29, 33; Grumet, *Historic Contact*, 76–77, 80, 93; Rowlandson, *Sovereignty and Goodness of God*, 1, 22–25, 64, 68–71, 123; Saltonstall, *Present State*, 3d–5th, 12th–13th, 16th–17th pp.; Saltonstall, *Continuation*, 5–7, 14–15; Saltonstall, *New and Further Narrative*, 2–14; Increase Mather, *Brief History*, 89–93, 99–101, 107–28, 142; LaFantasie, *Correspondence of Roger Williams*, 2:701–3, 712–27 (quote on 716); Rowlandson, *Sovereignty and Goodness of God*, 125 (second quote), 128–29.

69. Gary B. Nash, *Red, White, and Black*, 123–28; Washburn, *Governor and the Rebel*, chaps. 2–3, and "Webb's Interpretation," 348; Webb, *1676*, 3–4, 16, 21–25; Fausz, "'Engaged in Enterprises,'" 118–19; Davis and Mintz, *Boisterous Sea of Liberty*, 93–94.

70. Gary B. Nash, *Red, White, and Black*, 130–43; Perdue, *Native Carolinians*, 29–31; Haan, "'Trade Do's Not Flourish as Formerly,'" 341–58; Crane, *Southern Frontier*, 158–86.

71. Carp, "Early American Military History," 265–72; Shy, *People Numerous and Armed*, chap. 2; Cress, *Citizens in Arms*, 3–8; Steele, *Warpaths*, 102; Axtell, *After Columbus*, 205–8, 218–19; Lauber, *Indian Slavery*, 141; Selesky, *War and Society*, 9–10, 21–25; Shea, *Virginia Militia*, 31–38, 62–3, 71; Leach, "Military System," 350–61, and *Flintlock and Tomahawk*, 11–13, 51, 102–5, 123–24, 127, 183–88, 205; Puglisi, *Puritans Besieged*, 59–60; Drake, "Restraining Atrocity," 46; Gary B. Nash, *Red, White, and Black*, chap. 6; Richard R. Johnson, "Search for a Usable Indian," 628–31, 639–40.

72. J. H. C. King, *First Peoples*, 46; Merrell, "Cultural Continuity," 560–70; Grumet, *Historic Contact*, 237; Plane, "Putting a Face on Colonization," 149–56.

73. Merrell, "'Customes of Our Countrey,'" 137–56; Shea, *Virginia Militia*, 68–69; Roundtree, "Powhatans and the English," 196–203; Roundtree and Davidson, *Eastern Shore Indians*, 101–4, 110–14, 125–32, 136–37, 142–52, 168, 170–73, 208–11; Porter, "Nanticoke Indians," 141–45.

74. Mandell, *Behind the Frontier*, chaps. 2–4, and "'To Live More Like My Christian English Neighbors'"; O'Brien, *Dispossession by Degrees*, 74–87, 99–114, 154–55; Puglisi, *Puritans Besieged*, chap. 3; Lepore, *Name of War*, 137–43; Drake, "Restraining Atrocity," 44–46; Pulsipher, "Massacre at Hurtleberry Hill," 466–68; Kawashima, "Legal Origins"; Rowlandson, *Sovereignty and Goodness of God*, 120–21.

75. Horn, "Tobacco Colonies," 175–76.

76. Goodkin, *Historical Collections*, 151–52, 184–85, 188, 195–98, 201; Lepore, *Name of War*, 32–42, 86; Virginia DeJohn Anderson, "King Philip's Herds," 601–2, 605–6, 613–19; O'Brien, *Dispossession by Degrees*, 45, 47, 93–94, 96, 152; Mandell, *Behind the Frontier*, 61, 75, 93–95; Mandell, "Waban of Natick," 174–75, 180–83, 186–89; Mandell, "'To Live More Like My English Christian Neighbors,'" 557–59, 567, 572–75; Grumet, *Historic Contact*, 65, 145, 157, 172, 208, 346; Bragdon, "Material Culture," 128–31; Kevin A. McBride "'Ancient and Crazie,'" 66, 72–73, and "Historical Archeology," 108–9, 113, 115; Marshall, "'Melancholy People,'" 420, 422–23; James Homer Williams, "Great Doggs," 262; Roundtree and Davidson, *Eastern Shore Indians*, 112, 137; Stephen R. Potter, *Commoners*, 224–25.

77. See sources in n. 76 and O'Brien, *Dispossession by Degrees*, 33–39, 42, 59–62, 66–67, 73–74; Vaughan and Richter, "Crossing the Cultural Divide," 25–46; Ronda, "'We Are Well As We Are,'" esp. 78–81, and (in contrast) "Generations of Faith"; Cogley, *John*

Eliot's Mission, chaps. 5–7; Mandell, "Waban of Natick," 166–91, and *Behind the Frontier*, 36–38, 49–51, 58, 60, 96–97; Lepore, *Name of War*, 95, 142; Pulsipher, "Massacre at Hurtleberry Hill," 462–64, 469–75, 479–86.

78. Lepore, *Name of War*, 142 (quote).

79. Axtell, *Beyond 1492*, 113–15; Goodkin, *Historical Collections*, 188, 210; Sainsbury, "Indian Labor"; Rowlandson, *Sovereignty and Goodness of God*, 141–44, 147–49; Marshall, " 'Melancholy People,' " 411–12; Salisbury, "Indians and Colonists," 83–85; O'Brien, *Dispossession by Degrees*, 45, 70; Cogley, *John Eliot's Mission*, 167; Saltonstall, *Present State*, 7th p.; Mandell, *Behind the Frontier*, 26–29, 61–62, 70; Ruth Wallace Herndon, "Right to a Name," 434–42; Roundtree and Davidson, *Eastern Shore Indians*, 75–77; Mandell, "Waban of Natick," 174–75; Kawashima, "Indian Servitude," 403–6.

80. Lauber, *Indian Slavery*, chaps. 4–12; Lepore, *Name of War*, 150–51, 153–64, 170; Kawashima, "Indian Servitude," 403–6; Games, *Migration*, 210; Rowlandson, *Sovereignty and Goodness of God*, 37, 125, 145–47; Drake, "Restraining Atrocity," 53–55; Peter Wood, "Indian Servitude" and *Black Majority*, 38–40; Grinde, "Native American Slavery," 38–41; Braund, "Creek Indians," 605–7; Menard, "Africanization of the Lowcountry Labor Force," 104–5; W. Stitt Robinson, "Legal Status of the Indian," 254–56; Perdue, *Cherokee Women*, 66–69.

81. Merrell, *Indians' New World*, 171–72 (the Indians' wilderness); Tanner, *Atlas of Great Lakes Indian History*, 2, 42–45; Hauptman, "Refugee Havens," 129–30, 133; Porter, "Nanticoke Indians," 145–47; Grumet, *Historic Contact*, 76, 236, 239–40, 262–63, 273, 384; Landy, "Tuscarora among the Iroquois," 518–20; Merrell, "Cultural Continuity," 568–69; Stephen R. Potter, *Commoners*, 193–95; Marshall Becker, "Okehocking Band," 64–68; Weslager, *Nanticoke Indians*, chap. 10; Sugrue, "Peopling and Depeopling," 28–31.

82. Peter Wood, "Changing Population," 38–51; Hatley, *Dividing Paths*, chap. 4.

83. On contact in the seventeenth century, see above; Drake, "Restraining Atrocity," 51–52. On contact in the eighteenth century, see Braund, *Deerskins and Duffels*, chaps. 3, 5; Perdue, *Cherokee Women*, chap. 3; and Michael P. Morris, *Bringing of Wonder*, chap. 5. On eighteenth-century Indian wars, see Gary B. Nash, *Red, White, and Black*, chaps. 10–11; Leach, "Colonial Indian Wars," 139–42; and Chapter 3, below.

84. Vaughn and Clark, *Puritans among the Indians*, 1–28; Richard Beale Davis, *Intellectual Life*, 1:215–19; Axtell, *European and the Indian*, 298–300; Gary B. Nash, "Image of the Indian," 222–30; Alan Taylor, *Liberty Men and Great Proprietors*, chap. 7.

85. Seed, "Taking Possession," 186–89, 193–99, and *Ceremonies of Possession*, 16–23, 28–29, 38–39 (exaggerates significance of actual fences and gardens in earliest settlements); Jonathan R. T. Hughes, *Social Control*, chap. 5; Viola Florence Barnes, "Land Tenure"; Marshall Harris, *Origins of the Land Tenure System*, chaps. 5–10.

86. Marshall Harris, *Origins of the Land Tenure System*, remains the only full-length study, but see Edward T. Price, *Dividing the Land*; Kim, *Landlord and Tenant*, chap. 1; John Frederick Martin, *Profits in the Wilderness*; W. Stitt Robinson, *Mother Earth*; and Innes, *Creating the Commonwealth*, 80–83, 212.

87. Marshall Harris, *Origins of the Land Tenure System*, chap. 13; Edmund S. Morgan, *American Slavery*, 93–4, 171–3, 219; Billings, "Transfer of English Law," 237–38; Edward T. Price, *Dividing the Land*, 14, 15, 92–93, 97.

88. Marshall Harris, *Origins of the Land Tenure System*, 208–36; Edward T. Price, *Dividing the Land*, 107–10; Abbot Emerson Smith, "Indentured Servant"; Garrett Power, "Parceling Out Land," 454–55; Menard, "From Servant to Freeholder," 48–51; Clowse, *Economic Beginnings*, 45–46, 76–77, 156; Delâge, *Bitter Feast*, 309–10.

89. John Frederick Martin, *Profits in the Wilderness*, esp. 149–61, is the best account, but David Grayson Allen, *In English Ways*, 32–36, 61–66, 109–16, 125–31; David Hackett

Fischer, *Albion's Seed*, 166–74; Garvan, *Architecture and Town Planning*, chap. 3; Rife, "Land Tenure," 54–57; Kross, *Evolution of an American Town*, 34–36; and Edward T. Price, *Dividing the Land*, 29–32, 35, 37–39, 42–47, 57, 59, 61–62, 217–18, 250–51, challenge his insistence on profits, inequity, and individualism in land allocation.

90. Rife, "Land Tenure," 64–73; Rink, *Holland on the Hudson*, chap. 4, esp. 115; Kim, *Landlord and Tenant*, 12–28, 129–38, 157–80, 235–39, 250–53; Menard and Carr, "Lords Baltimore," 177–78, 186–98; Stiverson, *Poverty in a Land of Plenty*, 1–11; Edward T. Price, *Dividing the Land*, 133–34, 213–17.

91. Marshall Harris, *Origins of the Land Tenure System*, 237–54; Bidwell and Falconer, *History of Agriculture*, 58–60; John Frederick Martin, *Profits in the Wilderness*, 11–28, 327; Theodore Lewis, "Land Speculation"; Lewis Cecil Gray, *History of Agriculture*, 1:386–95; Carville Earle, *Evolution of a Tidewater Settlement System*, 193–202; Garrett Power, "Parceling Out Land," 454–55; Clowse, *Economic Beginnings*, 77–78, 96–99; Wacker, *Land and People*, 258–60; Gregg, "Land Policy," 157; Gary B. Nash, *Quakers and Politics*, 15–17, 89–97; Schweitzer, *Custom and Contract*, 89–96; Edward T. Price, *Dividing the Land*, 257–58, 262; Corcoran, "Penn and His Purchasers"; Dunn and Dunn, *Papers of William Penn*, 2:630–64.

92. Sarah S. Hughes, *Surveyors and Statesmen*, chaps. 4–6, is the best work, but see Edward T. Price, *Dividing the Land*, 6–7, 9, 13–15, 21, 37–38, 50–53, 56, 89, 95, 117, 119–23, 127, 249; Kain and Baigent, *Cadastral Map*, 265, 269–76, 285–89; Ford, *Colonial Precedents*, chap. 1; Carville Earle, *Evolution of a Tidewater Settlement System*, 182–93; Watson, "Quitrent System," 195; Garrison, *Landscape and Material Life*, 19–23; Candee, "Land Surveys," 10–14, 20–33 (quote on 11); Konig, "Community Custom," 142–43; Garvin, "Range Township"; Wacker, *Land and People*, 254, 296, 364–72; Schweitzer, *Custom and Contract*, 98–101; and Gregg, "Land Policy," 155–57.

93. See n. 63, above. On testimony of bounds, see Prince George's County Land Records, M: 434, Q: 468 (two cases quoted), M: 27–28, Q: 471, NN: 238, 258. For a Massachusetts case, see *Mansfield v. Newall* (1682) in Botein, *Early American Law*, 107. See also Seiler, "Land Processioning," esp. 416–23; Noble and Cronin, *Records*, 2:45–46.

94. Gildrie, *Salem, Massachusetts*, 60, 72–73; Vickers, *Farmers and Fishermen*, 21–22, 68–70; Jackson Turner Main, *Society and Economy*, 68–69, 99; Lorena S. Walsh, "Staying Put or Getting Out," 94–97; Games, *Migration*, 105–9; Kevin Kelly, *Economic and Social Development*, 110–13; Clemens, *Atlantic Economy*, 103; Simler, "Tenancy," 554–55; David Hackett Fischer, *Albion's Seed*, 166–67, 374–75 (Fischer misreads data in Kelly to come to incorrect conclusions about Surry County).

95. V. J. Wyckoff, "Size of Plantations"; Carr, Menard, and Walsh, *Robert Cole's World*, 24–25; Perry, *Formation of a Society*, chap. 2, esp. 56, 62–63; Horn, *Adapting to a New World*, 169; Mancall, "Landholding," 658; Wertenbaker, *Planters of Colonial Virginia*, 182–247 (1704–5 rent roll); Bergstrom and Kelly, "County around the Towns," 8, 21–23; Edward T. Price, *Dividing the Land*, 357–58. "Typical" and "average" refer to listed or computed median acreages, rounded to the nearest 25 acres.

96. John Raymond Hall, "Three Rank System"; David Grayson Allen, *In English Ways*, 32, 111, 129; Gildrie, *Salem, Massachusetts*, 56–65; Christine Alice Young, *From "Good Order" to Glorious Revolution*, 239; Lockridge, "Land, Population, and the Evolution of New England Society," 64–67 (Dedham and second Watertown data, means rather than medians reported); Wacker and Clemens, *Land Use*, 90–92.

97. Roger Thompson, *Mobility and Migration*, chap. 9; Games, *Migration*, 164–69, 172–80, 187–88; Friedeburg, "Social and Geographical Mobility," 383–86.

98. On New England, see Sumner Chilton Powell, *Puritan Village*, plate 11, 229–33; Greven, *Four Generations*, chap. 3, esp. 46, 58; Jackson Turner Main, *Society and Econ-*

omy, 68; and Bissell, "One Generation to Another," 98. On Long Island, see E. B. O'Callaghan, *Documentary History*, 2:411–542; Kross, *Evolution of an American Town*, 32–38; Breen, *Imagining the Past*, 155–63; Bonomi, *Factious People*, 90; and McAnear, "Livingston's Reasons against a Land Tax," 65. On other median holdings, see T. Burke, *Mohawk Frontier*, 35 (Rensselaerwyck, Dutch Schenectady, 53), and Rindler, "Migration from the New Haven Colony," 271 (Brandford, Connecticut, 1663, 11–20).

99. See Chapter 1, above, and Kulikoff, "Internal Migration," 334–36. Case studies include Bosworth, "Those Who Moved," 11–16; Mathews, *Expansion of New England*, chaps. 2–5; Crandall, "New England's Second Great Migration," 351–52; Archer, "New England Mosaic," 485–86; Virginia DeJohn Anderson, "Migration, Kinship," 269–87, and *New England's Generation*, 103–24; Games, *Migration*, 164–72; Cole, "Family, Settlement, and Migration"; Adams and Kasakoff, "Migration and the Family" and "Migration at Marriage"; Joseph Sutherlund Wood, *New England Village*, 22–23, 34–36, 42–43; and Cave, *Pequot War*, 87–88, 93, 95–98 (quote on 87).

100. Games, *Migration*, 143–50, 174, 227–29.

101. See Chapter 1, above, and Kulikoff, "Internal Migration," 334–36. Case studies include Kevin Kelly, " 'In Dispers'd Country Plantations' "; Robert D. Mitchell, "American Origins," 409–11; Roundtree and Davidson, *Eastern Shore Indians*, 100; Horn, "Moving On in the New World," 172–73, 176–91, 195–200, and "Tobacco Colonies," 179–80; Menard, "From Servant to Freeholder"; Lorena S. Walsh, "Staying Put or Getting Out" (quote on 92–93); Rutman and Rutman, " 'More True and Perfect Lists,' " 67–73; and Carr and Menard, "Immigration and Opportunity."

102. Richard B. Morris, *Studies in the History of American Law*, 73–74, 80–92; Jonathan R. T. Hughes, *Social Control*, chap. 5; Innes, *Creating the Commonwealth*, 212–15; Edward T. Price, *Dividing the Land*, 16; Gary B. Nash, *Quakers and Politics*, 89–92.

103. Bond, *Quit-Rent System*, is the standard work, but see also Watson, "Quitrent System."

104. Vickers, *Farmers and Fishermen*, 77–82, 95–96, 103, 135–37; Jackson Turner Main, *Society and Economy*, 68; E. B. O'Callaghan, *Documentary History*, 2:441–511 (1675–76 and 1683 tax lists, counting tenants as those without land but with livestock and those just paying the poll tax as nonhouseholders); Kim, *Landlord and Tenant*, ix–x (2-million-acre estimate), 130 (quote), and "New Look," 583 (7-million-acre estimate); Bonomi, *Factious People*, 195; Innes, *Labor in a New Land*, chap. 3, esp. 48–49; Christine Alice Young, *From "Good Order" to Glorious Revolution*, 176–82.

105. Lorena S. Walsh, "Historian as Census Taker," 256–58, and "Land, Landlord, and Leaseholder," 374–79, 382; Carr, Menard, and Walsh, *Robert Cole's World*, 26–28; Kevin Kelly, *Economic and Social Development*, 110–13, and " 'In Dispers'd Country Plantations,' " 185–91; V. J. Wyckoff, "Land Prices," 82–88 (land prices rose from seventeen to forty pounds of tobacco, or 135 percent, but since tobacco prices declined, the growth in real terms was lower, from 1.8 to 2.8 shillings per acre, or 55 percent; see Menard, "Tobacco Industry," 158–60).

106. Grosart, *Poems*, 286–87, 315–16 (first quotes); Gleach, *Powhatan's World*, 122; Barker, "Powhatan's Pursestrings," 74; Percy, *Observations*, 20 (second quote); Axtell, *After Columbus*, 193; John Smith, *Works* (*Map of Virginia*), 1:140–41, 185–89, and (*Description of New England*) 319–21; Barbour, "Earliest Reconnaissance," 285–302.

107. Keary, "Retelling the History," 270; Kelso, *Jamestown Rediscovery II*, 20–21; Schmidt, "Mapping an Empire," esp. 556–61, 573–75.

108. On English names, see David Hackett Fischer, *Albion's Seed*, 36–38; Homans, "Puritans and the Clothing Industry," 521; Demos, *Little Commonwealth*, 11; P. M. G. Harris, "Social Origins," 236–37; Emily J. Salmon, *Hornbook of Virginia History*, 103–20;

Mackenthun, *Metaphors of Dispossession*, 162, 320; Zemsky, "Nationalism in the American Place-Name," 20; Leighly, "New England Town Names," 156–64; and Krim, "Acculturation of the New England Landscape," 87. Three Virginia counties that replicate English county names could have been named to honor nobility taking the same name. On Dutch names, see Schmidt, "Mapping an Empire," 551, 557, 562, 567. The evidence here suggests that naming was far more than the "merely symbolic" activity that Seed ("Taking Possession," 199) finds.

109. Increase Mather, *Brief History*, 93; Demos, *Unredeemed Captive*, 7; Day, *In Search of New England's Past*, chaps. 6, 23; Keary, "Retelling the History," 270; Rowlandson, *Sovereignty and Goodness of God*, 15; Zinkin, "Surviving Indian Town Names"; Mary R. Miller, "Place Names," 11–14; Krim, "Acculturation of the New England Landscape," 80–83, 87–88; Green and Green, "Place-Name Dialects," 243–49; tract map attached to Heinton, *Prince George's Heritage*.

110. Clarence E. Tyler, "Topographical Terms"; McJimsey, "Topographic Terms," 16–23, 168–69, 384–85; Hawes, "New England's Forests," 209–10; Sands, "Maine Place Names," 41–42.

111. Robb, "Names of Grants"; Kelso, *Kingsmill*, 34–35; tract map attached to Heinton, *Prince George's Heritage*; Mary R. Miller, "Place Names," 18–20; Zinkin, "Names of Estates."

112. William Wood, *New England's Prospect*, chaps. 4–10 (quotes on 33–34, 36, 57–58).

113. On household formation, see Chapter 5, below; Fox-Genovese, *Within the Plantation Household*, chap. 1; and Stilgoe, *Common Landscape*, 43–73. On fences, see Seed, *Ceremonies of Possession*, 21–23.

114. This conclusion addresses the knotty issues of cultural continuity and American distinctiveness. While accepting data found in David Hackett Fischer, *Albion's Seed*, I give colonial conditions more weight. But if English (and later European) immigrants made a new world, they nonetheless borrowed from the region of migration, thus challenging Kammen's brilliant but misguided reassertion of exceptionalism in "Problem of American Exceptionalism."

115. Eugene Green, "Naming and Mapping," 83–84; Richard B. Morris, *Studies in the History of American Law*, 76–103; Konig, "Community Custom," 137–48; Billings, "Transfer of English Law," 237–38; Alston and Shapiro, "Inheritance Laws across Colonies," 277–83; Keim, "Primogeniture and Entail." Brewer, "Entailing Aristocracy," argues that a large majority of Virginia's eighteenth-century land (but only a quarter by the end of the seventeenth century) was covered by entail; but while providing a needed corrective to Keim, she greatly overestimates the percentage by ignoring the land of smallholders.

116. Carson et al., "Impermanent Architecture," 135–78; Gloria L. Main, *Tobacco Colony*, chap. 4; Neiman, "Domestic Architecture," 305–7; Upton, "Power of Things," 268–69; Lounsbury, "Development of Domestic Architecture," 46–51, 60; Cummings, *Framed Houses*, chaps. 1–3, 6, and 212–32; St. George, " 'Set Thine House in Order,' " 166–68; Lepore, *Name of War*, 76–77; Steinitz, "Rethinking Geographical Approaches"; Jordan and Kaups, *American Backwoods Frontier*, chap. 6; Michel, " 'In a Manner and Fashion Suitable to Their Degree,' " 29–65.

117. See n. 116; David Hackett Fischer, *Albion's Seed*, 62–68, 264–74, 475–81; Upton, "Traditional Timber Framing," 35–61; Donald W. Linebaugh, " 'All the Annoyances' "; Sobel, *World They Made Together*, 112–21; Stiverson, *Poverty in a Land of Plenty*, 64–65.

118. Hoffer, *Law and People*, chap. 1; Innes, *Creating the Commonwealth*, 196–97, 210–12; Chafee, "Colonial Courts" (quotes on 57); Howe, "Sources and Nature of Law," 1–15, esp. 13–14; Haskins, *Law and Authority*, chaps. 1, 10; G. B. Warden, "Law Reform";

Roetger, "Enforcing New Haven's Bylaws"; Billings, "Transfer of English Law" and "Law and Culture," 340–48.

119. Billings, "Transfer of English Law" and "Growth of Political Institutions," 225–34; Konig, " 'Dale's Laws' "; Carr, "Development of the Maryland Orphans' Court."

120. See Chap. 5, below, and Kulikoff, "Households and Markets" and *Making of the American Yeoman Class*, esp. chap. 1.

CHAPTER THREE

1. Crèvecoeur, *Letters from an American Farmer*, 54. Notwithstanding paeans to American classlessness and blindness to American poverty and diversity, his portrayal of land and yeoman identity rings true. The best short analysis of the colonial land system is Mancall, "Landholding." See also Echeverria, *Mirage in the West*, 147–55, and David M. Robinson, "Community and Utopia," 17–31. For requirements to improve land to retain it, see Hart, "Eminent Domain and Constitutionalism," 8–9, and Ely, *Guardian of Every Other Right*, 18.

2. See Prologue. For examples, see Tawney, *Agrarian Problem*, 281–310, and Hoyle, "Tenure and the Land Market" and "An Ancient and Laudable Custom."

3. Crèvecoeur, *Letters from an American Farmer*, 54, 259 (quotes); Ramsay, *History of the American Revolution*, 1:31 (quote); Macpherson, *Democratic Theory*, 124–29; Proudhon, *What Is Property?*, chap. 2, esp. 35–37; Mensch, "Colonial Origins of Liberal Property Rights," esp. 646–47, 655–60, 676–78, 681–82. For general statements, see William B. Scott, *In Pursuit of Happiness*, chap. 1, and Ely, *Guardian of Every Other Right*, chap. 1. I will deal with the political and ideological implications of this argument in *Making of the American Yeoman Class*.

4. The Germans were clients of Daniel Dulany, and the quote comes from a letter meant to entice more German immigrants to Maryland. See Cunz, *Maryland Germans*, 126 (quote), and Land, *Dulanys of Maryland*, 183–84. On eminent domain, see Ely, *Guardian of Every Other Right*, 23–25; Hart, "Eminent Domain and Constitutionalism," 4–17 (exaggerates the quantity of takings without compensation), and "Maryland Mill Act," 1–6, 11–24; and Stoebuck, "Theory of Eminent Domain," 560–62, 567 (quote from Massachusetts constitution), 579–83.

5. Terry L. Anderson, "Economic Growth in Colonial New England," 250–51; Kulikoff, "Economic Growth," 280–81; Alice Hanson Jones, *Wealth of a Nation to Be*, 98–99.

6. Colden quotes by Mensch, "Colonial Origins of Liberal Property Rights," 647.

7. Alice Hanson Jones, *Wealth of a Nation to Be*, 208–10, 225–27.

8. Crèvecoeur, *Letters from an American Farmer*, 254; David M. Robinson, "Community and Utopia," 27–31.

9. Alice Hanson Jones, *Wealth of a Nation to Be*, 208–10, 225–27. For age at land acquisition, see Kulikoff, *Tobacco and Slaves*, 54, 62–63, 85–86, 131–32; Jackson Turner Main, *Society and Economy*, 142–45, 157, 310–11; and Jedry, *World of John Cleaveland*, 63–64, 175–77. We will examine frontier landholding patterns and migration below.

10. U.S. Bureau of the Census, *Historical Statistics*, 2:1168; Fogleman, "Migrations to the Thirteen British North American Colonies," 691–93. P. M. G. Harris, "Inflation and Deflation," shows modest, if any, inflation in pre-Revolutionary America, thus meaning that land prices, reported below, represent real change.

11. The classic essay remains Lockridge, "Land, Population, and the Evolution of New England Society," but see also Edward M. Cook, "Social Behavior and Changing Values," 572–73; Gross, *Minutemen and Their World*, 78–81; Wacker, *Land and People*, 138, 142;

Main and Main, "Red Queen," 136–38. Lockridge's conception of population pressure has been persuasively challenged by Rutman and Rutman, *Small Worlds*, chap. 7; McCusker and Menard, *Economy of British America*, 104–6. For land prices, see Jedry, *World of John Cleaveland*, 60, 62; Jackson Turner Main, *Society and Economy*, 119–20; Bruce C. Daniels, *Fragmentation of New England*, 22; Kross, *Evolution of an American Town*, 208–13; Greven, *Four Generations*, 224–27 (reports falling prices after 1740, but he did not reduce prices to a common value; the exchange rate between the pound sterling and the Massachusetts pound was 628 to 100 in the 1740s, but only 133 thereafter. See P. M. G. Harris, "Inflation and Deflation," 473–74, 480–82; McCusker, *Money and Exchange*, 141–42; Terry L. Anderson, "Economic Growth in Colonial New England," 257).

12. Pruitt, "Self-Sufficiency," is the most important analysis of this problem, but see also Vickers, *Farmers and Fishermen*, chap. 5; Ryan, "Landholding, Opportunity, and Mobility," 575–76, 583–84; and Wacker and Clemens, *Land Use*, 92–97.

13. Jackson Turner Main, *Society and Economy*, 157, 201, 208, 274; Van Deventer, *Emergence of Provincial New Hampshire*, 23–25, 209, 234, 280; Edward T. Price, *Dividing the Land*, 66–75.

14. Edward M. Cook, "Social Behavior and Changing Values," 571; Gross, "Culture and Cultivation," 56; Vickers, *Farmers and Fishermen*, 224–25; Holmes, *Communities in Transition*, 18–19; Pruitt, "Agriculture and Society," chaps. 3–4, esp. 112, 149, map 8. Using the same source, Gross and Pruitt compute a slightly different percentage of landless taxpayers (Gross, 25 percent; Pruitt, 19 percent). I computed the proportion of landholders colonywide from data in Pruitt, "Agriculture and Society," 112, and *Massachusetts Tax Valuation List of 1771*, excluding Boston, Salem, and Marblehead and taxables without any property.

15. Jackson Turner Main, *Society and Economy*, 160–61; Van Deventer, *Emergence of Provincial New Hampshire*, 192–95; Ryan, "Landholding, Opportunity, and Mobility," 575–76.

16. Douglas Lamar Jones, *Village and Seaport*, 17–21; Fred Anderson, "People's Army," 510–13, and *People's Army*, 32–37; Kross, *Evolution of an American Town*, 215–16; Ryan, "Landholding, Opportunity, and Mobility," 577–78; Lender, "Enlisted Line," 113–14; Matthew C. Ward, "Army of Servants," 88–93.

17. Kulikoff, *Tobacco and Slaves*, 49, 134–36; Bergstrom, *Markets and Merchants*, 40–49; Steffen, *From Gentlemen to Townsmen*, 35–37, 74–75; Papenfuse, "Planter Behavior," 302–3; Stiverson, *Poverty in a Land of Plenty*, 144–45; Jean Butendoff Lee, *Price of Nationhood*, 272; Brown and Brown, *Virginia*, 12–13; Jackson Turner Main, "Distribution of Property," 247–48.

18. Coclanis, *Shadow of a Dream*, 69–70; Paden, "'Several & Many Grievances,'" 287–98, esp. 298; Philip D. Morgan, *Slave Counterpoint*, 43.

19. Robert E. Gallman, "Influences on the Distribution of Landholdings," 559; Jacquelyn H. Wolf, "Patents and Tithables," 267–68, 272, 275–76; Saunders, Clark, and Weeks, *Colonial and State Records*, 22:240–59 (1735 rent rolls); Morris and Morris, "Economic Conditions in North Carolina," 118–20, 128–33, table 4 following 128. Ekirch, *"Poor Carolina,"* 222–25, summarizes these data.

20. Kulikoff, *Tobacco and Slaves*, 132–33; Carville Earle, *Evolution of a Tidewater Settlement System*, 210–12; Steffen, *From Gentlemen to Townsmen*, 73–76; Clemens, *Atlantic Economy*, 231–32; Lewis Cecil Gray, *History of Agriculture*, 1:404–5; P. M. G. Harris, "Inflation and Deflation," 491–93. No one, to my knowledge, has computed a land price series for coastal areas of Virginia or the Carolinas, but see report of a price series calculated by Lorena Walsh for York County, Virginia (Brewer, "Entailing Aristocracy," 332–33), consistent with this paragraph.

21. Kulikoff, *Tobacco and Slaves*, 135; Stiverson, *Poverty in a Land of Plenty*, 144–45; Brewer, "Entailing Aristocracy," esp. 320–21 (suggests far more land was entailed than previously believed, but her argument that 70 percent of Virginia's land was entailed by 1760 is much exaggerated, ignoring land never listed in wills); Morris and Morris, "Economic Conditions in North Carolina," 118–33.

22. Lemon, *Best Poor Man's Country*, 88–92 (means read from ranges on maps); Klepp, "Five Early Pennsylvania Censuses," 492–94; Simler, "Tenancy," 549–50; Wacker, *Land and People*, 399–403; Wacker and Clemens, *Land Use*, 92–97 (sale data from newspaper ads, which exaggerate farm sizes); Lender, "Enlisted Line," 121.

23. Simler, "Tenancy," 554–57; Lemon, *Best Poor Man's Country*, 94–95; Klepp, "Five Early Pennsylvania Censuses," 492–94; Marietta, "Distribution of Wealth," 533–36.

24. For New York tenancy estimate for 1776, compare Kim, *Landlord and Tenant*, vii, with Greene and Harrington, *American Population*, 102–3. Simler, "Tenancy," 555, gives Chester County estimates. For the Chesapeake, see Stiverson, *Poverty in a Land of Plenty*, esp. 144–45, and Kulikoff, *Tobacco and Slaves*, 132–35. Since the Hudson River Valley and the Northern Neck were eighteenth-century frontier regions, I discuss tenancy in those places in more detail in the next section.

25. Lorena S. Walsh, "Land, Landlord, and Leaseholder," 386–89; Stiverson, *Poverty in a Land of Plenty*, chaps. 2, 5; Carville Earle, *Evolution of a Tidewater Settlement System*, 212–14; Simler, "Tenancy," 557–62; Lemon, *Best Poor Man's Country*, 94–96.

26. Adams and Kasakoff, "Migration and the Family," 28–38, and "Migration at Marriage," 119–29; Villaflor and Sokoloff, "Migration in Colonial America," 545–48; Susan L. Norton, "Marital Migration in Essex County," 406–18; Fred Anderson, "People's Army," 513–14; Greven, *Four Generations*, 210–14, 273–74; Edward S. Cooke Jr., *Making Furniture*, 46–47, 50–51, 70–71; Kulikoff, *Tobacco and Slaves*, 141–57. For Wrentham data, see Esther L. Friend, "Notifications and Warnings Out," 179–202, 330–46. The data show a slight rise between the 1760s and 1770s in longer distance migration: 2 of 121 came from 25+ miles in the 1760s; 36 of 430, in the 1770s.

27. Douglas Lamar Jones, "Strolling Poor" and "Poverty and Vagabondage"; Esther L. Friend, "Notifications and Warnings Out," 179–202, 330–46 (quotes on 333, 190–91, 197); Ruth Wallace Herndon, "Literacy"; Ruth Wallace Herndon, "Right to a Name," 438–47; Ruth Wallace Herndon, "Warned Out," 1–32; Salinger and Dayton, "Mapping Migration," 9–10; Cray, *Paupers and Poor Relief*, chaps. 2–3.

28. Ruth Wallace Herndon, "Warned Out," 1–32; Salinger and Dayton, "Mapping Migration," 7, 12–13, 21–28, 33, 38, 44, tables 1–5, and maps 2–3; Gary B. Nash, *Urban Crucible*, 185–86, 253, and *Race, Class, and Politics*, 185–86; Kulikoff, "Progress of Inequality," 399–404.

29. Gary B. Nash, *Urban Crucible*, 185–86, 470, and *Race, Class, and Politics*, chap. 7, esp. 181–88; Simler, "Landless Worker," 170–76, 188–98; Lemon, *Best Poor Man's Country*, 83–85, 93–94; Marietta, "Distribution of Wealth," 534–36; Calo, "From Poor Relief to the Poorhouse," 419 (quote).

30. Douglas Lamar Jones, "Strolling Poor," 44–54, and "Transformation of the Law of Poverty," 158–61, 169–71; McKinney, *Development of Local Public Services*, 48; Cray, *Paupers and Poor Relief*, chaps. 2–3, esp, 50–64, 84–99; Calo, "From Poor Relief to the Poorhouse"; Kulikoff, *Tobacco and Slaves*, 297–99; Mackey, "Operation of the Old English Poor Law," 35–39; Joan M. Jensen, *Loosening the Bonds*, 59–64.

31. Robert D. Mitchell, *Commercialism and Frontier*, 16, 18; Hofstra, "'Extension of His Majesties' Dominions,'" 1288–89; Gary B. Nash, *Red, White, and Black*, chap. 6; Michael P. Morris, *Bringing of Wonder*; Braund, *Deerskins and Duffels*, chaps. 3–6; and sources cited in Chapter 2, above, delineate Indian-white trade behind the frontier.

32. Cutcliffe, "Colonial Indian Policy," 239–46, 249–50, 254–55, 260–61 (percentages reported on 242–44 calculated for each colony, then averaged); Hinderaker, *Elusive Empires*, 22–24, 26–28, 31, 51–52, 83–85, 90; Fausz, "'Engaged in Enterprises,'" 146–47.

33. Merrell, *Into the American Woods*, esp. 32–34, 37–38, 42, 44–53, 57, 62, 68–69, 74–82, 85, 88–94, 115–22, 174, 179–83, 187, 218–19, 225, 237–39, 250–51, 282–84, 288–89, is the fullest and best account (though my definition is broader than his that focuses on diplomatic negotiation), but see also Hinderaker, *Elusive Empires*, 22–33, 46–54, 71–77, 119–21, 124–25; Fausz, "'Engaged in Enterprises,'" 124–27, 140–42; Hagedorn, "Brokers of Understanding"; Merritt, "Power of Language," 1–14; Hsiung, "Death on the Juniata"; Michelle Gillespie, "Sexual Politics"; Perdue, *Cherokee Women*, chap. 3; and Shannon, "Dressing for Success," 19–20, 26–27, 30–32.

34. Merrell, *Into the American Woods*, 68, 137–38, 142, 214–15, 227–28, 258; Hinderaker, *Elusive Empires*, 119; Colden, *Letters and Papers*, 53–55.

35. Hinderaker, *Elusive Empires*, 125–26; Merrell, *Into the American Woods*, 158–63, 169–70.

36. De Vorsey, *Indian Boundary*, 27 (Pownall quote), 30–31 (Board of Trade); Merrell, *Into the American Woods*, 295–96. See above for population pressure; the argument about land jobbers, as we shall see below, is misleading.

37. Merrell, *Into the American Woods*, 247–48 (Pisquetomen), 278; Dorothy V. Jones, *License for Empire*, 54 (Yakatastange); De Vorsey, "Indian Boundaries," 74 (second Creek quote); Fausz, "'Engaged in Enterprises,'" 130–31.

38. Volwiler, *Croghan and the Westward Movement*, 27, 70–71, 210–11 (quote); Wayland Fuller Dunaway, *Scotch-Irish*, 69–70; Hinderaker, *Elusive Empires*, 110; Buck and Buck, *Planting of Civilization in Western Pennsylvania*, 136–37; Gipson, *British Empire*, 9:425–26; Rice, *Allegheny Frontier*, 59; Leach, *Northern Colonial Frontier*, 195–96; McConnell, *Country Between*, 259–60; Ford, *Colonial Precedents*, 114–19; Nicholls, "Origins of the Virginia Southside," 14–16; Cashin, "'But Brothers It Is Our Land,'" 242, and "Sowing the Wind," 237–44; Kenneth Coleman, *Colonial Georgia*, 226–27; Anderson-Green, "New River Frontier," 421–22. Last quote is from the Proclamation of 1763 (Jack P. Greene, *Colonies to Nation*, 18).

39. Miles, "Red Man Dispossessed"; Merrell, *Indians' New World*, chap. 5; Mensch, "Colonial Origins of Liberal Property Rights," 705–8, 714–19 (New York quote on 717); Higgins, *Expansion in New York*, 29–30, 64; Jack P. Greene, *Colonies to Nation*, 16–18 (Proclamation of 1763 quote on 18); Grant, *Democracy in the Connecticut Frontier Town of Kent*, 9; Wainwright, *Croghan*, 26, 28, 273–99; Barbara Rasmussen, *Absentee Landowning*, 29–30; Dorothy V. Jones, *License for Empire*, 114–18; Abernethy, *Western Lands and the American Revolution*, 21.

40. Hatley, *Dividing Paths*, chap. 7, and Dorothy V. Jones, "British Colonial Indian Treaties" (192–94 for a list) and *License for Empire*, chaps. 3–5 (119 for another list), are the most sophisticated treatments, but see also Jennings, "Scandalous Indian Policy"; Jennings, *Ambiguous Iroquois Empire*, chaps. 16–17, 388–97; Jennings, *Empire of Fortune*, 101–8; De Vorsey, "Indian Boundaries"; and Sosin, *Opening of the West*, 11, and *Whitehall and the Wilderness*, 29–33 (quote on 31).

41. Dowd, "Panic of 1751," 531–35, 539–41, 547–52, gives an example of how rumor and fact fed mutual fears; see also Hofstra, "'Extension of His Majesties' Dominions,'" 1306–7 (quote on 1307), 1309–10; Hinderaker, *Elusive Empires*, 122, 125; Merrell, *Into the American Woods*, 185.

42. Gary B. Nash, *Red, White, and Black*, chaps. 6, 10, 11 (summary); Melvoin, *New England Outpost*, chaps. 4, 8–9, epilogue, esp. 218–20, 278; Demos, *Unredeemed Captive*, 11–13, 18–30, 38–39, 87–91, 135–37, 171; Ferling, *Struggle for a Continent*, 101–8; Leach,

Northern Colonial Frontier, 193–94; Grant, *Democracy in the Connecticut Frontier Town of Kent*, 7–8; Higgins, *Expansion in New York*, 74–75; W. Stitt Robinson, *Southern Colonial Frontier*, 107–13; Corkran, *Carolina Indian Frontier*, 21, 29, 39, 45; Meriwether, *Expansion of South Carolina*, 122–23; Thwaites and Kellogg, *Documentary History of Dunmore's War*, xvi.

43. Matthew Ward, "Fighting the 'Old Women' " (the best account); Leach, *Northern Colonial Frontier*, 200–202; Merrell, *Into the American Woods*, 227–29, 269; Matthew C. Ward, "Army of Servants," 76–78, 83; Titus, *Old Dominion at War*, 74–75 (quote), 94–95; Chester Raymond Young, "Stress of War," 254–56; Anderson-Green, "New River Frontier," 417–18; Ketcham, "Conscience, War, and Politics," 420–21, 428–30; Corkran, *Carolina Indian Frontier*, 54–55, 57.

44. Matthew Ward, "Fighting the 'Old Women,' " 312–13, 315–19 (second and third quotes on 312, 318); Chester Raymond Young, "Stress of War," 256–59, 264–65; Volwiler, *Croghan and the Westward Movement*, 100–101 (first quote). I follow Ward's estimates of casualties, captives, and abandoned areas here.

45. Hatley, *Dividing Paths*, chap. 10; Jennings, *Empire of Fortune*, chap. 20; Howard H. Peckham, *Pontiac and the Indian Uprising*, chap. 15; McConnell, *Country Between*, 120–24 and chap. 11, esp. 270–77; Parmenter, "Pontiac's Wars," 627–34; Rice, *Allegheny Frontier*, 80–83; Merrell, *Into the American Woods*, 285; Dowd, *Spirited Resistance*, 35–36, 45; Hinderaker, *Elusive Empires*, 154–56; Buck and Buck, *Planting of Civilization in Western Pennsylvania*, 176–77; Thwaites and Kellogg, *Documentary History of Dunmore's War*, ix–xvi, 3, 9–14, 19–20, 36–37, 56–57, 63–65, 103–4, 114–15, 134, 151, 192–93, 209–10, 227–29, 238, 373–80; Holton, "Ohio Indians," 473–74; Downes, "Dunmore's War"; Curry, "Lord Dunmore and the West," 232–40; Hoyt, "Fleming in Dunmore's War," 99–103; Cashin, " 'But Brothers It Is Our Land,' " 241–46.

46. Hinderaker, *Elusive Empires*, 136–38, 144–45, 149–50, 159–60, 162, 166; Fausz, " 'Engaged in Enterprises,' " 132–33; Merrell, *Into the American Woods*, 279–82, 286–87; Parmenter, "Pontiac's Wars," 630–38; Chester Raymond Young, "Stress of War," 252–53, 259; Dorothy V. Jones, *License for Empire*, 112–13; De Vorsey, "Indian Boundaries"; Harold E. Davis, *Fledgling Province*, 28–29.

47. Jack P. Greene, *Colonies to Nation*, 16–18, prints the proclamation (quote on 18). The impact of the proclamation and the Fort Stanwix Treaty (1768) are documented in De Vorsey, *Indian Boundary*, the most through work (see esp. 34–36); Friis, *Series of Population Maps*, end piece; Cappon, Petchenik, and Long, *Atlas of Early American History*, 14–16, 86, 92; McConnell, *Country Between*, 116–17, 188–89, 249, 272–73; Jennings, *Empire of Fortune*, 461–63; Arrell M. Gibson, "Indian Land Transfers," 211, 214; Holton, "Ohio Indians," 453–70; and Del Papa, "Royal Proclamation of 1763." For Indian villages, see Betty Anderson Smith, "Distribution of Eighteenth-Century Cherokee Settlements," 47–57, and Hauptman, "Refugee Havens," 128–39.

48. Merrell, *Into the Woods*, 130–35, 140–41.

49. Nobles, "Breaking into the Backcountry," 647–48; Kulikoff, *Tobacco and Slaves*, 149–52; Adams and Kasakoff, "Wealth and Migration"; Lemon, *Best Poor Man's Country*, 85; Doddridge, *Notes on the Settlement*, 82 (quotes).

50. Bosworth, "Those Who Moved," 16–24, 92–93, 116; Carville Earle, "Rate of Frontier Expansion," 190–94; Leach, *Northern Colonial Frontier*, chap. 8; Nobles, *Divisions throughout the Whole*, 107–10; Lemon, *Best Poor Man's Country*, 86–87; Dorwart, *Cape May County*, 25–26; Porter, "From Backcountry to County"; Robert D. Mitchell, *Commercialism and Frontier*, chap. 2; Norona, "Joshua Fry's Report," 37 (quote); Merrens, *Colonial North Carolina*, 62–74; Ramsey, *Carolina Cradle*, chap. 2.

51. For summaries, see Bailyn, *Peopling of British North America*, 18, 51–59, and *Voyag-*

ers to the West, 8–20; Villaflor and Sokoloff, "Migration in Colonial America"; Bosworth, "Those Who Moved," 25–33; Carville Earle, "Rate of Frontier Expansion," 190–93; Earle and Cao, "Frontier Closure," 166–67; Robert D. Mitchell, "Southern Backcountry," 10–11, 16–18, and *Commercialism and Frontier*, chap. 2; Fogleman, *Hopeful Journeys*, 8–11. Local studies include Leach, *Northern Colonial Frontier*, chap. 8; Nobles, *Divisions throughout the Whole*, 107–10; Fred Anderson, "People's Army," 514–15; Bellesiles, *Revolutionary Outlaws*, 33–41; Colin D. Campbell, "They Beckoned," 218–19; Lemon, *Best Poor Man's Country*, 86–87; Mancall, *Valley of Opportunity*, 88; Buck and Buck, *Planting of Civilization in Western Pennsylvania*, 140–46; Wacker, *Musconetcong Valley*, chap. 3; Dorwart, *Cape May County*, 25–26; Gragg, *Migration in Early America*; Keller, "Outlook of Rhinelanders," 105–7; Anderson-Green, "New River Frontier," 414–31; Rice, *Allegheny Frontier*, 66–70; Merrens, *Colonial North Carolina*, 62–74; Kenneth E. Lewis, *American Frontier*, chap. 3; Rachel N. Klein, *Unification of a Slave State*, 13–14.

52. Frontier white populations estimated from data in Greene and Harrington, *American Population*, 37–40, 73–79, 86–88, 102, 117, 120, 131–33, 150–55, 166–72, 187–94; R. Eugene Harper, *Transformation of Western Pennsylvania*, 12–15; Slaughter, *Whiskey Rebellion*, 66; Rachel N. Klein, *Unification of a Slave State*, 19, 250; Rice, *Allegheny Frontier*, 59–70; Peter Wood, "Changing Population," 38–39; Faragher, *Daniel Boone*, 123, with linear interpolations made where necessary and dates of settlement checked with Everton and Rasmuson, *Handy Book*. For east Florida, see Bailyn, *Voyagers to the West*, chap. 12. For West Florida and the Mississippi country, see Fabel, "An Eighteenth Colony," 669–70, and *Economy of British West Florida*, chap. 1. For Kentucky, see Faragher, *Daniel Boone*, chap. 4.

53. Lemon, *Best Poor Man's Country*, 76–78, 84–85; Wacker, *Musconetcong Valley*, 50–51; Nobles, *Divisions throughout the Whole*, 111–12, 227–28.

54. Nicholls, "Origins of the Virginia Southside," 46–50; Kulikoff, *Tobacco and Slaves*, 141–48; Robert D. Mitchell, *Commercialism and Frontier*, 47–52; Hofstra, "'Extension of His Majesties' Dominions,'" 1301–3 (quote on 1301); Anderson-Green, "New River Frontier," 416; Meriwether, *Expansion of South Carolina*, chaps. 3–16.

55. Nobles, "Breaking into the Backcountry," 647–48, and *Divisions throughout the Whole*, 110–11 (quote); Kulikoff, *Tobacco and Slaves*, 141–57; Douglas Lamar Jones, *Village and Seaport*, chaps. 4–6; Rutman and Rutman, *Small Worlds*, chap. 7; Lemon, *Best Poor Man's Country*, 85.

56. De Vorsey, "Colonial Georgia Backcountry," 23–25 (quote); Adams and Kasakoff, "Wealth and Migration"; Bosworth, "Those Who Moved," 134–36, 145–48.

57. Adams and Kasakoff, "Migration and the Family," 32–33; Gragg, *Migration in Early America*; Robert D. Mitchell, "Southern Backcountry," 23; Greven, *Four Generations*, 156–72, 210–14; Bellesiles, *Revolutionary Outlaws*, 37–41; Kulikoff, *Tobacco and Slaves*, 99; Beeman, *Evolution of the Southern Backcountry*, 22; Anderson-Green, "New River Frontier," 415–19, 422; Ramsey, *Carolina Cradle*, 21–22, 191–92; Kaylene Hughes, "Populating the Back Country," 52–57; Tillson, "Southern Backcountry," 390–91; Nicholls, "Origins of the Virginia Southside," 51–54.

58. Franklin, *Papers*, 4:228 (first quote); O'Callaghan and Fernow, *Documents*, 7:888–89 (second quote); Burnaby, *Travels*, 13, 14; William S. Powell, "Tryon's Book on North Carolina," 410–11; Robert E. Brown, *Middle Class Democracy*, chap. 1 (second Franklin quote on 9; English officer quote on 10–11); Ramsay, *History of the American Revolution*, 1:30–31. I do not accept the egalitarian and democratic implications Brown infers from the quotes he amasses.

59. Hofstra, "'Extension of His Majesties' Dominions,'" brilliantly delineates these

policy contradictions, but see also Rawson, "Anglo-American Frontier Settlement," 99–101, 104–7, 110, 112, 116.

60. Daniell, *Colonial New Hampshire*, 148; Ekirch, *"Poor Carolina,"* 127–43; Thornton W. Mitchell, "Granville District"; Ackerman, *South Carolina Colonial Land Policies*, 81–114; Sirmans, *Colonial South Carolina*, 31–33; Weir, *Colonial South Carolina*, 148–49; Kaylene Hughes, "Populating the Back Country," 4–19, 26–30, 33–34, 72–80, 155; Ready, "Land Tenure," 354–56; Abbot, *Royal Governors of Georgia*, 14–18, 72–73; Fabel, *Economy of British West Florida*, 7–8.

61. Rawson, "Anglo-American Frontier Settlement," 93–115; Hofstra, "'Extension of His Majesties' Dominions,'" 1282–83, 1287, 1291–92, 1298–1303, 1307–11; Richard L. Morton, *Colonial Virginia*, 539–40.

62. Bond, *Quit-Rent System*, is the standard source, but see also Richard L. Morton, *Colonial Virginia*, 539–40; Nicholls, "Origins of the Virginia Southside," 16, 35; Ready, "Land Tenure," 336; Nelson, *William Tryon*, 21–23; Lefler and Powell, *Colonial North Carolina*, 153, 229; Kulikoff, *Making of the American Yeoman Class*, chap. 6.

63. Giddens, "Land Policies and Administration," 144–47, 154–59; Nicholls, "Origins of the Virginia Southside," chaps. 2–3; Robert D. Mitchell, *Commercialism and Frontier*, 65; Richard L. Morton, *Colonial Virginia*, 539–40; Ackerman, *South Carolina Colonial Land Policies*, 106–8.

64. The most thorough analysis remains Abernethy, *Western Lands and the American Revolution*, chaps. 1, 3, 8–10, but see also Marshall Harris, *Origins of the Land Tenure System*, chap. 17; Livermore, *Early American Land Companies*, chap. 4; Gipson, *British Empire*, vol. 9, chap. 13; George E. Lewis, *Indiana Company*, 45–50, 60–63, 297–304; Kenneth P. Bailey, *Ohio Company of Virginia*, 35–60, 85–100, 233–49; Alfred P. James, *Ohio Company*, chaps. 4, 9–11, and 296–97; and Egnal, *Mighty Empire*, 94–100. These works stress the politics of the companies; only Holton, "Ohio Indians," 454–55, 469, 475, connects them to settlers.

65. Akagi, *Town Proprietors of the New England Colonies*, chaps. 6–8; Kershaw, *Kennebeck Proprietors*, esp. 37, 62–63, 66–67, 72–74; Alan Taylor, *Liberty Men and Great Proprietors*, 11–14, 24–28; Bellesiles, *Revolutionary Outlaws*, 29–32, 41–47, 91–104; Nelson, *William Tryon*, 101–8; Philip J. Schwarz, *Jarring Interests*, 168–74; Looney, "Wentworth's Land Grant Policy"; Mark, *Agrarian Conflicts*, 21–23, 164–89; Greven, *Four Generations*, 127–30, 224–27; Jedry, *World of John Cleaveland*, 59–62; Grant, *Democracy in the Connecticut Frontier Town of Kent*, chaps. 2–4; Jackson Turner Main, *Society and Economy*, 52, 58–59, 206–7.

66. Porter, "From Backcountry to County," 342–47; Land, *Dulanys of Maryland*, chap. 11; Richard L. Morton, *Colonial Virginia*, chap. 13; Billings, Selby, and Tate, *Colonial Virginia*, 208–9; Edward T. Price, *Dividing the Land*, 142–43, 152–53, 157; Ayres, "Albemarle County," 45–47, 68, 70; McCleskey, "Rich Land, Poor Prospects," 468–77, and "Shadow Land," 56–57; Tillson, "Southern Backcountry," 401; Robert D. Mitchell, *Commercialism and Frontier*, 65, 68, 72–84; Hofstra, "Land Policy," 115–16, and "'Extension of His Majesties' Dominions,'" 1311; Beeman, *Evolution of the Southern Backcountry*, 31–33; Kulikoff, *Tobacco and Slaves*, 133, 155.

67. Frederick R. Black, "West Jersey Society," 381–82; Trace, "West Jersey Society," 270, 272; Volwiler, *Croghan and the Western Movement*, 233–59; Beardslee, "Ostego Frontier," 240–49; Ekirch, *"Poor Carolina,"* 131–32; Ackerman, *South Carolina Land Policies*, 54–57, 75, 94–114; Sirmans, *Colonial South Carolina*, 170–77; Kaylene Hughes, "Populating the Back Country," 39.

68. Kenneth Coleman, *Colonial Georgia*, 121–26, 205–9; Baine, "New Perspectives on

Debtors"; Randall M. Miller, "Failure of the Colony of Georgia," 5–17; Cadle, *Georgia Land Surveying History*, 12–16, 29–38; Ready, "Land Tenure"; Paul S. Taylor, *Georgia Plan*, 98–101, 124–29, 135–38, 235–36, 276–77, 291–96, 304–10; Caldwell, "Women Landholders"; Ver Steeg, *Origins of a Southern Mosaic*, 76–77, 94; Gallay, *Formation of a Planter Elite*, chap. 4, and 171–83.

69. Kershaw, "Question of Orthodoxy"; Louis Morton, *Robert Carter of Nomini Hall*, 11, 14–21; Land, *Dulanys of Maryland*, chaps. 11–12; Porter, "From Backcountry to County," 345–48; Purvis, "Were Colonial American Politics Deferential?," 12–16, 27–29.

70. Ellis et al., *History of New York State*, 71–74; Bonomi, *Factious People*, 186–88; Kim, *Landlord and Tenant*, frontispiece; Mark, *Agrarian Conflicts*, chap. 1; Milton M. Klein, *Politics of Diversity*, 114; Colin D. Campbell, "They Beckoned," 219–23. Edith M. Fox, *Land Speculation*, examines the speculative holdings of one public official, and Higgins, *Expansion in New York*, chaps. 6–8, details many of the smaller patents.

71. Ellis et al., *History of New York State*, 71–74; Bonomi, *Factious People*, 180–210; E. B. O'Callaghan, *Documentary History*, 2:384 (first quote); William Smith, *History of the Province of New York*, 1:222, 225–26 (second quote); Kim, *Landlord and Tenant*, 36–37.

72. Greene and Harrington, *American Population*, 96–101; La Potin, "Minisink Grant," 43–48; Kim, *Landlord and Tenant*, chap. 5; Milton M. Klein, *Politics of Diversity*, 17–20; Kierner, *Traders and Gentlefolk*, 43, 66–67; Higgins, *Expansion in New York*, chaps. 6–8, esp. 64–65; Edith M. Fox, *Land Speculation*, 41–44.

73. Kim, *Landlord and Tenant*, vii, chaps. 5–6; Humphrey, "Agrarian Rioting in Albany County," chap. 1; Countryman, *People in Revolution*, 17–23; Bonomi, *Factious People*, 191–96; Kierner, *Traders and Gentlefolk*, 92–98; Lynd, *Anti-Federalism in Dutchess County*, 38–41.

74. Willard F. Bliss, "Rise of Tenancy in Virginia"; Jackson Turner Main, "Distribution of Property," 248; Hofstra, "Land Policy," 120–22; Mancall, *Valley of Opportunity*, 105–10; R. Eugene Harper, *Transformation of Western Pennsylvania*, 59–69 (the evidence of high rents he gives for the 1790s fits poorly with his description of developmental leases); Wacker, *Musconetcong Valley*, 34–35, 65–66; Stiverson, *Poverty in a Land of Plenty*, chaps. 1–2.

75. Ford, *Colonial Precedents*, chap. 7; Philip J. Schwarz, *Jarring Interests*, chaps. 6–13 (explains the border disputes that stood behind insistence of New Yorkers that they owned their lands); Franz, *Paxton*, 87–88 (Logan quote), 110–11; Fullerton, "Squatters," 168–73; Wacker, *Land and People*, 136–38, 341–44, 359, 361; Trace, "West Jersey Society," 270–71; Burger, "'Our Unhappy Purchase,'" 109; Frederick R. Black, "West Jersey Society," 386; Kessell, "Germans," 93, 95; Cadle, *Georgia Land Surveying History*, 36–37; Ackerman, *South Carolina Land Policies*, 52–54; Rachel N. Klein, *Unification of a Slave State*, 54–61, 311–15; Tillson, *Gentry and Common Folk*, 59–61 (second quote on 59), 68; Kulikoff, *Making of the American Yeoman Class*, chap. 6.

76. Thwaites and Kellogg, *Documentary History of Dunmore's War*, 371–72 (second quote); Kulikoff, *Tobacco and Slaves*, 156 (Lunenburg); Alfred F. Young, *Democratic Republicans*, 585–88; Mancall, *Valley of Opportunity*, 182–83, 242–43; Wacker, "New Jersey Tax-Ratable List," 39–44.

77. Doddridge, *Notes on the Settlement*, 81 (quote), 85; Fullerton, "Squatters," 169; Leyburn, *Scotch-Irish*, 205; Alan Taylor, *Liberty Men and Great Proprietors*, 28; Ford, *Colonial Precedents*, 130–31; Stevenson Whitcomb Fletcher, *Pennsylvania Agriculture*, 23–24. For fair play, see George D. Wolf, "Politics of Fair Play"; Linn, "Indian Land and Its Fair-Play Settlers"; and Mancall, *Valley of Opportunity*, 123, 126. Bogue, "Iowa Claims Clubs," is the best examination of that topic.

78. Slaughter, *Whiskey Rebellion*, 64–67; Thwaites and Kellogg, *Documentary History*

of Dunmore's War, 345 (Dunmore quote); Kulikoff, *Tobacco and Slaves*, 92–93, 158–60; Franz, *Paxton*, chap. 5; Mancall, *Valley of Opportunity*, 243.

79. McCleskey, "Shadow Land," 59–63; Hofstra, "Land, Ethnicity, and Community," 432–43, and "Ethnicity and Community," 64–65; MacMaster, "Religion," 89–93.

80. Kulikoff, *Tobacco and Slaves*, 133, 155–56 (Piedmont Virginia, 1760s–1780s, landownership, land prices); True, "Land Transactions," 211–17, 241, 243, 247, 250–51 (Louisa County, Virginia, land prices); Ayres, "Albemarle County," 45, 69; Wacker, "New Jersey Tax-Ratable List," 39–43; Alfred F. Young, *Democratic Republicans*, 585–88; Mancall, *Valley of Opportunity*, 182–83, 242–43; Alan Taylor, *Liberty Men and Great Proprietors*, 256; Lemon, *Best Poor Man's County*, 92–96 (Lancaster, 1781); Nicholls, "Origins of the Virginia Southside," chaps. 2–3; Morris and Morris, "Economic Conditions in North Carolina," 119; Rachel N. Klein, *Unification of a Slave State*, 24.

81. Edward T. Price, *Dividing the Land*, 175–76; Kulikoff, *Tobacco and Slaves*, 156 (Piedmont Virginia); Ayres, "Albemarle County," 45, 68–70; Tillson, *Gentry and Common Folk*, 12–14, 167; McCleskey, "Rich Land, Poor Prospects," 470–71; Robert D. Mitchell, *Commercialism and Frontier*, 68–71 (Frederick, Virginia); Ackerman, *South Carolina Land Policies*, 114; Rachel N. Klein, *Unification of a Slave State*, 24; Ekirch, *"Poor Carolina,"* 132; Weir, *Colonial South Carolina*, 210–11. Tillson, McCleskey, and Mitchell present somewhat different medians for Augusta County); Alan Taylor, *Liberty Men and Great Proprietors*, 256 (Maine); Mancall, *Valley of Opportunity*, 180–82, 242–43; Franz, *Paxton*, 191–207; W. Stitt Robinson, *Southern Colonial Frontier*, 155 (Frederick, Maryland).

82. Edward T. Price, *Dividing the Land*, 67–76, 144–50, 153–56, 163–72, 262–63, 268–76, 361–63; Hofstra, "'Extension of His Majesties' Dominions,'" 1305–6 (quotes on 1306).

CHAPTER FOUR

1. Archdeacon, *Becoming American*, 13–22 (a good short analysis of eighteenth-century immigration and ethnicity); Horn, "British Diaspora," 35–49; Fogleman, "Migrations to the Thirteen British North American Colonies," 698 (307,400); and Gemery, "European Emigration," 311 (336,800), are the best estimates of European emigration. See also Schlenter, "'English Is Swallowing Up Their Language'" (first quotes on 206–7); Klepp and Smith, *Infortunate*, 8–13, 17–32, 50–52 (quote), 110–11 (quote); E. R. R. Green, "Ulster Emigrants' Letters," 91–92.

2. Blum, *End of the Old Order*, chap. 9, brilliantly evokes the condition of Europe's early modern peasantry, but see also van Leeuwen, "Logic of Charity," esp. 600–607, and Post, *Food Shortage*, chap. 6. For Arthur Young, see *Tour of Ireland*, vol. 2, chap. 6, and *Travels in France*, 341–48.

3. De Vries, *European Urbanization*, 36–37 (population); Saalfeld, "Struggle to Survive"; Malthus, *Essay on the Principle of Population* (1793), 93–94 (quote), and *Essay on the Principle of Population* (1803), 1:1, 16–17 (quotes), 226–27, 250–51, 282–84, 288 (quote), 291–92 (quote), 359.

4. Horn, "British Diaspora," 51 (quote).

5. Moch, *Moving Europeans*, 30–40, 70–76, and Flinn, *European Demographic System*, chap. 5, are the best summaries.

6. Rural industry has excited vast debate. See Gutmann, *Toward the Modern Economy*, chaps. 4–5; Mendels, "Proto-industrialization"; Kriedte, Medick, and Schlumbohm, "Proto-industrialization Revisited"; D. C. Coleman, *Myth, History, and the Industrial Revolution*, chap. 5; Mager, "Proto-industrialization." For England, see Pawson, *Early Industrial Revolution*, chaps. 2, 4–5; Cunningham, "Employment and Unemployment of

Children," 115–33; Valenze, "Art of Women," 142–57; Charles Wilson, *England's Apprenticeship*, chap. 14; Corfield, *Impact of English Towns*, chaps. 1–2, 5–7; Chambers, "Population Change," 103–4, 107, 111–13; and Wrigley, *People, Cities, and Wealth*, 166. For other places, see J. P. Shaw, "New Industries"; Durie, "Scottish Linen Industry," 88–94; Kisch, *From Domestic Manufacture to Industrial Revolution*, chaps. 2–4; and Braun, "Early Industrialization."

7. De Vries, *European Urbanization* (data on 119, 270–71); Wrigley, *People, Cities, and Wealth*, chaps. 6–7; Moch, *Moving Europeans*, 43–58, 93–99. The debate between Sharlin, "Natural Decrease," and Finlay, "Natural Decrease," on migration and natural decrease is informative. For case studies, see Peter A. Clark, "Migrants in the City," 273–75, 281–86; Corfield, *Impact of English Towns*, chaps. 2, 5–7 (quote on 101); Houston and Withers, "Population Mobility in Scotland," 288–89, 292–93; and Hochstadt, "Migration in Preindustrial Germany," 199–209, 214.

8. Jan Lucassen, *Migrant Labour* and "Netherlands," 166–81; Leo Lucassen, "Blind Spot," 209–27; Moch, *Moving Europeans*, 76–88; Fontaine, "Family Cycles, Peddling, and Society," 62–64; Moogk, "Mannon's Fellow Exiles," 256–57; Louis Michael Cullen, "Irish Diaspora," 131–35, 139–40; Perrenoud, "Migrations en Suisse," 253–54; Viazzo, *Upland Communities*, 162–77, 249–52; Hochstadt, "Migration in Preindustrial Germany," 210–13; Whyte, "Population Mobility in Early-Modern Scotland," 54–58; Houston and Withers, "Population Mobility in Scotland," 287–89; Ruth-Ann Harris, "Seasonal Migration," 363–64, 368 (quote 363); Casway, "Irish Women Overseas," 114–17. For England, see below.

9. Houston and Withers, "Population Mobility in Scotland," 297; Mayhew, *Rural Settlement*, chap. 4, esp. 121–22, 150–67; references in n. 66, below; Poussou, "Mouvements Migratoires," 55; Head, "L'Émigration des Régions Préalpines," 190–92.

10. Jon Butler, *Huguenots in America*, chaps. 1–2 (estimates on 23–27); Gwynn, *Huguenot Heritage*, chaps. 1–2, 4–5 (estimate on 35); Moch, *Moving Europeans*, 28; Moogk, "Mannon's Fellow Exiles," 242; Jan Lucassen, "Netherlands," 157. Since the numbers found in these sources are in much dispute, I have averaged them.

11. Kettner, *Development of American Citizenship*, chaps. 4–5, is the most important work, but see also Winkel, "Naturalization in Colonial New Jersey." On anti-Catholicism, see Cogliano, "Nil Desperandum," 183–89; Bosworth, "Anti-Catholicism as a Political Tool"; Madden, "Catholics in Colonial South Carolina," 30–35; Casino, "Anti-Popery in Colonial Pennsylvania"; Kerby A. Miller, *Emigrants and Exiles*, 142–47; and Metzger, *Catholics and the American Revolution*, chaps. 1, 5. Schwartz, "*Mixed Multitude*," 104–5, 150–52, 240–42, dissents somewhat.

12. There is no general agreement on numbers. Fogleman, "Migrations to the Thirteen British North American Colonies," revised downward earlier guesses. I accept somewhat larger estimates: Canny, "English Migration," 64; Horn, "British Diaspora," 30–32; Jon Butler, *Huguenots in America*, 46–48; Jan Lucassen, "Netherlands," 178–79; Smout, Landsman, and Devine, "Scottish Emigration," 86–90, 93–99; Louis Michael Cullen, "Irish Diaspora," 115–16, 127–30, 138–40; Fertig, "Transatlantic Migration," 196–203; Wokeck, *Trade in Strangers*, 44–47; and Schelbert, "On Becoming an Emigrant," 461, 463.

13. Deane and Cole, *British Economic Growth*, 18–21 and chap. 2; Crafts, *British Economic Growth*, chap. 2; Overton, *Agricultural Revolution*, chap. 3; Shammas, *Preindustrial Consumer*, chaps. 4–6, 8, and "Decline of Textile Prices," 487–502; Hatcher, "Labor, Leisure, and Economic Thought," 79, 92–94, 98–101; Timmins, *Made in Lancashire*, 16–17, 50–52, 159; Eversley, "Home Market," 207–59; McKendrick, "Commercialization," 9–33; Mingay, "Agricultural Depression"; John, "Agricultural Productivity,"

19–27; Wrigley and Schofield, *Population History*, 207–15, 412–35; Gilboy, "Cost of Living"; Tucker, "Real Wages of Artisans in London"; L. D. Schwarz, "Standard of Living"; Snell, *Annals of the Labouring Poor*, chap. 1; Robert C. Allen, *Enclosure and the Yeoman*, 254–57, 294–96; Petersen, *Bread*, 6–7, 18–23, 33–35, 58–59, 132, 189–95, 241; Razzell, "Growth of Population," 752–69; Wrigley, "Explaining the Rise in Marital Fertility," 438–42, 452–60.

14. E. P. Thompson, *Customs in Common*, chap. 3; Turner, "Common Property and Property in Common" and *English Parliamentary Enclosure*, 66–76; Neeson, *Commoners*, chaps. 2, 6; Humphries, "Enclosures, Common Rights, and Women," 19–41; Gregory Clark, "Commons Sense," 97–99 (challenges utility of commons, but his aggregate evidence ignores regional differences); Beckett, "Hammonds Revisited," 53–57; Wordie, "Chronology of English Enclosure," 502; Yelling, *Common Field and Enclosure*, 101–3, 227–31; Rogers, "Industrialisation and the Local Community"; Souden, "Movers and Stayers," 24–25; Tranter, "Population and Social Structure"; Tate, *Domesday of English Enclosure Acts*, 56; Chambers, "Population Change," 101–3.

15. Louis Michael Cullen, "Irish Economic History"; "Economic Development, 1750–1800," 166–69; and *Economic History*, 60–66, deny the traditional bleak picture of Ireland's eighteenth-century economy, but the author might be too optimistic. See Conrad Gill, *Rise of the Irish Linen Industry*, 10–11, 21–29, 34–43, 75–79, 83–91, 125–31, 138–60, 341–43; William H. Crawford, "Women in the Domestic Linen Industry," 255–63, and *Domestic Industry in Ireland*, 3–7, 24–40; Gribbon, "Irish Linen Board," 82–84; Collins, "Proto-Industrialization," 129–31; Cullen and Smout, "Economic Growth," 6–7; Arthur Young, *Tour of Ireland*, 1:124–28, 150; Donald E. Jordan, *Land and Popular Politics*, 59–62; Kirkham, "Economic Diversification," 64–65, 70–71 (quote on 70).

16. Louis Michael Cullen, *Emergence of Modern Ireland*, 42–44, and "Catholic Social Classes," 57–71; Nuala Cullen, "Women and the Preparation of Food," 268; Maureen Wall, *Catholic Ireland*, 73–92; David Dickson, "Catholics and Trade" and "Property and Social Structure"; Whelan, "Catholic Community," 130–48; Keller, "Ulster Scots Emigration," 76; William H. Crawford, "Landlord-Tenant Relations"; William H. Crawford, "Social Structure of Ulster," 118–19; William H. Crawford, "Ulster as a Mirror," 62–66; Roebuck, "Rent Movement," 90–96; Arthur Young, *Tour of Ireland*, 2:256–57 (tea).

17. Louis Michael Cullen, *Economic History*, 52–58; Connell, *Population of Ireland*, 268–69; David Dickson, "Place of Dublin," 186–88; Ó Gráda, *Ireland*, 26, 32–35; Francis G. James, "Irish Colonial Trade"; Shepherd and Walton, *Shipping, Maritime Trade, and Economic Development*, 110–12; Thomas P. Power, *Land, Politics, and Society*, 21–35; Donald E. Jordan, *Land and Popular Politics*, 39–52.

18. Arthur Young, *Tour of Ireland*, 1:35 (quote), 59, 161, 2:43–46; Connell, *Population of Ireland*, 122–35, 147–49, 155–59; Louis Michael Cullen, *Emergence of Modern Ireland*, chap. 7, and "Irish History without the Potato"; Nuala Cullen, "Women and Preparation of Food," 265–70; David Dickson, "Gap in Famines," 104; Mokyr and Ó Gráda, "Height of Irishmen and Englishmen," 88–92; Ó Gráda, *Ireland*, 19; Macafee, "Demographic History of Ulster," 47–50.

19. Flinn, *Scottish Population History*, pts. 3–4, esp. 164–200, 216–22, 241–50, 284–89; Handley, *Scottish Farming*, 9–18, 28–37; Houston, "Demographic Regime," 12–14; Houston, "Age at Marriage"; Houston, *Population History*, 28–30, 37–38, 48–49; Tyson, "Famine in Aberdeenshire"; Hamilton, *Economic History of Scotland*, 18, 395; Mitchison, "Webster Revisited," 62–64; Lenman, *Economic History of Modern Scotland*, 45–47; Smout, Landsman, and Devine, "Scottish Emigration," 85–90; Biegańska, "Scots in Poland." The early eighteenth-century figure is based on eighteenth-century guesses and the baptism/burial ratio (Flinn, *Scottish Population History*, 242–43).

20. Houston, *Population History*, 31–32, and *Social Change*, 18; Devine, "Urbanisation," 27–35, and *Transformation of Rural Scotland*, 252; Smout, "Scottish Economy," 56–59; Turnock, *Historical Geography*, 64–65; Whyte, *Agriculture and Society*, 8–9; W. Iain Stevenson, "Geography of the Clyde Tobacco Trade," 28.

21. Devine, *Tobacco Lords*, chaps. 3–4; Devine, "Colonial Trades"; Devine, "Colonial Commerce and the Scottish Economy"; W. Iain Stevenson, "Geography of the Clyde Tobacco Trade"; U.S. Bureau of Census, *Historical Statistics*, 2:1190; Hamilton, *Economic History of Scotland*, 184–212, 255–67; Shepherd and Walton, *Shipping, Maritime Trade, and Economic Development*, 38–40; Jacob M. Price, "Rise of Glasgow" and "Economic Growth of the Chesapeake."

22. Durie, *Scottish Linen Industry*, esp. 8–9, 13, 22–31, 38–54, 65–67, 74–81, 88–91, 158–60, is the standard source, but see also Smout, "Scottish Economy," 62–65; Hamilton, *Economic History of Scotland*, chaps. 5–6; Cullen and Smout, "Economic Growth," 7; Turnock, *Historical Geography*, 52–62; and Hamilton, *Life and Labour*, xxxiv–xli, 137–73.

23. Smout, *History of the Scottish People*, chap. 5; Handley, *Scottish Farming*, chaps. 2, 8, esp. 37–52; Whyte, *Agriculture and Society*, chaps. 1, 3, and "Agriculture in Aberdeenshire," 11–15, 24–30; Whittington and Brett, "Locational Decision-Making," esp. 38–39; Dodgshon, "Toward an Understanding" and "Strategies of Farming"; Alexander Fenton, *Northern Isles*, chaps. 5–6, 10; Callander, *Pattern of Landownership*, chap. 7; Landsman, *Scotland*, chap. 3; Turnock, *Historical Geography*, 29–30; Francis J. Shaw, *Northern and Western Islands*, chaps. 6–8; Clapham, "Agricultural Change," 146–48.

24. Kussmaul, *Servants in Husbandry*, 104, 114; Humphries, "Enclosures, Common Rights, and Women," 25–41; Holderness, " 'Open' and 'Close' Parishes," 128–29; Hammond and Hammond, *Village Labourer*, chaps. 1, 4–5 (still the most persuasive account); Robert C. Allen, *Enclosure and the Yeoman*, 28–36, 43–45, 72–75, 235–55, 287; Yelling, *Common Field and Enclosure*, 11–16, 54–58, 63–69, 101–3, 227–31; Turner, *English Parliamentary Enclosure*, chaps. 3, 7; Mingay, *Enclosure and the Small Farmer*, 15–32; Armstrong, "Labour I," 672–74, 721–26; Hoskins, *Midland Peasant*, chaps. 8–10; Charlesworth, *Atlas of Rural Protest*, 44, 48–50, 56–62, 68–70, 83–94; Neeson, *Commoners*, 200–204 and chap. 9; E. P. Thompson, *Customs in Common*, chaps. 4–5.

25. For protests, see n. 24. Komlos, "Malthusian Episode" and "Secular Trend," 123, 128–42; Fogel, "Nutrition and the Decline of Mortality," 486–96; Mokyr and Ó Gráda, "Height of Irishmen and Englishmen"; and Floud, Wachter, and Gregory, *Height, Health, and History*, chaps. 3–4, 7, esp. 140–48, 166–70, 175, 197, find an increase in height.

26. Peter King, "Customary Rights" and "Gleaners," 131–37; Snell, *Annals of the Labouring Poor*, chaps. 3–4; Solar, "Poor Relief"; Slack, *English Poor Law*, 26–39, 53–56; Boyer, *Economic History of the English Poor Law*, 1, 24, 29–35, 38–41; Huzel, "Labourer and the Poor Law," 760–66, 771–72; Stapleton, "Inherited Poverty," 342–55; Holderness, " 'Open' and 'Close' Parishes," 127–29.

27. Mokyr and Ó Gráda, "New Developments," 474–79, summarizes the debate over population and marriage ages, but see Connell, *Population of Ireland*, 4–5, 22–25, 52–53, 58–59; Houston, *Population History*, 30–31, 38–39; Clarkson, "Irish Population Revisited," 17–18, 22–35; Daultrey, Dickson, and Ó Gráda, "Eighteenth-Century Irish Population," 620–21; Ó Gráda, *Ireland*, 9–13; Macafee and Morgan, "Population in Ulster," 59–63; Macafee, "Demographic History of Ulster," 42–46; Ruth-Ann Harris, "Seasonal Migration," 368 (for land); Collins, "Proto-Industrialization," 127–37 (for linen and population).

28. Louis Michael Cullen, "Economic Development, 1750–1800," 167–79, and *Economic History*, 45–47, 78–80; William H. Crawford, "Ulster as Mirror," 63–64; R. J.

Dickson, *Ulster Emigration*, 45–47, 69–74; E. R. R. Green, "'Strange Humours'" (quotes on 117–18); Bigger, *Ulster Land War*, 18–19; Roebuck, "Rent Movement," 86–89; William H. Crawford, "Landlord-Tenant Relations," 13–21; Arthur Young, *Tour of Ireland*, 1:201–2; Lyne, "Land Tenure," esp. 40–45 (Kerry).

29. Cullen, Smout, and Gibson, "Wages and Comparative Development," 106–11; Arthur Young, *Tour of Ireland*, 2:35–58 (quote on 40), 1:59 (second quote), 78–79, 142, 190 (fourth quote), 288 (third quote), 290, 299; Donald E. Jordan, *Land and Popular Politics*, 54–57; John Harwood Andrews, "Land and People, c. 1685," 465–66, and "Land and People, c. 1780," 241–44; McDowell, "Ireland in 1800," 670–72; David Dickson, "Property and Social Structure," 133; Beames, *Peasants and Power*, 6–7; Thomas P. Power, *Land, Politics, and Society*, 185–88, 202; Louis Michael Cullen, *Economic History*, 81–82.

30. David Dickson, "In Search of the Old Irish Poor Law," 149–55; McDowell, "Ireland in 1800," 676–77; Solar, "Poor Relief," 10; Post, *Food Shortage*, 174–78; Louis Michael Cullen, "Economic Development, 1750–1800," 168–70; Doyle, *Ireland, Irishmen, and Revolutionary America*, 61.

31. Post, *Food Shortage*, esp. 32, 34, 37–38, 58, 86–110, 124–25, 219–20, 243–46 264–66 (best examination of the Europewide 1740–41 famine); E. Margaret Crawford, "Wilde's Table of Famines," 11–15; David Dickson, "Gap in Famines," 96–111; Mokyr and Ó Gráda, "New Developments," 476, 486–87; McCracken, "Social Structure and Social Life," 32–34 (quote on 33); Louis Michael Cullen, "Economic Development, 1691–1750," 146–50, and "Irish Diaspora," 128–31; Daultrey, Dickson, and Ó Gráda, "Eighteenth-Century Irish Population," 622–27.

32. Smout, *History of the Scottish People*, chaps. 6, 12; Callander, *Pattern of Landownership*, chaps. 3–4, 7; Landsman, *Scotland*, chap. 1; Whyte, *Agriculture and Society*, 30–31, 37–51; Whyte and Whyte, "Continuity and Change"; Clapham, "Agricultural Change," 150–57; Mitchison, *Life in Scotland*, 62–69; Dodgshon, "Origins of Traditional Field Systems"; Devine, *Transformation of Rural Scotland*, chaps. 1–2; Malcolm Gray, "Social Impact," 53–57.

33. Devine, "Social Responses"; Smout, "Scottish Economy," 46–49, and *History of the Scottish People*, 291–301, 324–31; Whyte, *Agriculture and Society*, 102–3, 116–33, and "Agriculture in Aberdeenshire," 15–18; R. H. Campbell, "Scottish Improvers," 204–12; Turner, *Enclosures in Britain*, 28–32, 73; Caird, "Reshaped Agricultural Landscape," 204–9; Third, "Changing Landscape"; Ian H. Adams, "Economic Process" and "Agricultural Revolution"; Handley, *Scottish Farming*, 198–202, and *Agricultural Revolution*, chap. 1; Dodgshon, "Removal of Runrig"; Landsman, *Scotland*, chap. 3; Malcolm Gray, "Social Impact," 57–59.

34. Whyte, *Agriculture and Society*, 34–36; Alex J. S. Gibson, "Proletarianization?"; Devine, *Transformation of Rural Scotland*, 12–14, and "Introduction," 1–2; Houston, "Women in the Economy," 120–26; Hamilton, *Life and Labour*, 60–63; Malcolm Gray, "Farm Workers," 10–14, and "Social Impact," 54–55; Orr, "Farm Servants and Farm Labour," 30, 34, 40–42; Robson, "Border Farm Worker," 72–75, 84; Howatson, "Grain Harvesting," 126–29; Petersen, *Bread*, 132.

35. Smout, *History of the Scottish People*, 332–42; Malcolm Gray, *Highland Economy*, 11–30; Parker, *Scottish Highlanders*, 26–28; Richards, *Highland Clearances*, vol. 2, chaps. 2–4; Francis J. Shaw, *Northern and Western Islands*, chaps. 4–5; Withers, *Gaelic*, 104–6; Richards and Clough, *Cromartie*, chaps. 1, 3–5; McLean, *People of Glengarry*, chap. 2.

36. Malcolm Gray, *Highland Economy*, 24–41, 250–53; Richards, *Highland Clearances*, vol. 1, chaps. 2–3; Richards and Clough, *Cromartie*, 38–39, 72–74, 77 (quote), 90; Adam, "Eighteenth-Century Highland Landlords," 8–10; Youngson, *After the Forty-five*, 38–39; Whyte, *Agriculture and Society*, 83–86.

37. Adam, "Eighteenth-Century Highland Landlords"; Whyte, *Agriculture and Society*, 234–42; Smout, "Scottish Economy," 54–57, and *History of the Scottish People*, 342–60; Youngson, *After the Forty-five*, chaps. 2, 4–5; Richards, *Highland Clearances*, vol. 1, chaps. 6–7; Malcolm Gray, *Highland Economy*, chaps. 2, 3; Devine, *Clanship to Crofters' War*, chap. 3; Handley, *Scottish Farming*, chap. 10; Alexander Fenton, *Northern Isles*, chaps. 6, 7, 11; Richards and Clough, *Cromartie*, chaps. 5–8; Parker, *Scottish Highlanders*, 28–31.

38. Gibson and Smout, *Prices, Food, and Wages*, chaps. 8–9, is the basic source, but see Treble, "Standard of Living," 194–99; Alex J. S. Gibson, "Proletarianization," 357–89; Smout, "Scottish Economy," 61–62; Hamilton, *Life and Labour*, xxiv–xxix, 55–56; Turnock, *Historical Geography*, 43.

39. Gibson and Smout, *Prices, Food, and Wages*, 11–13, 53, 78–80, 133, 137–41, 144, 166–67, 171–74, 179–84, 193–200, 228–43, 248–50, 275–99, 337–64, is the best work; see also Hamilton, *Life and Labour*, xix–xx; Mitchison, *Life in Scotland*, 70–73; Handley, *Scottish Farming*, 78–80; Post, *Food Shortage*, 93–96, 123–24, 146–52, 220–21; Durie, *Scottish Linen Industry*, 22, 65, 76; and Houston, *Social Change*, 253–61.

40. Mitchison, "Making of the Old Scottish Poor Law," 70–90; Mitchison, "Who Were the Poor?," 140–44: Mitchison, "North and South," 208–20; Cage, "Making of the Old Scottish Poor Law" and *Scottish Poor Law*, chaps. 1–3; Lindsay, *Scottish Poor Law*, 29–30, 38, 48, 52, 67, 76, 81–85, 104–5, 117–18, 125–29, 141; Houston, *Social Change*, chap. 4, and 320–31; Tyson, "Famine in Aberdeenshire," 32–40; Post, *Food Shortage*, 62–63, 71–75, 93–96, 146–52, 220–21; Flinn, *Scottish Population History*, 218–19, 225, 247; Gibson and Smout, *Prices, Food, and Wages*, 172–73.

41. Kießling, "Markets and Marketing," 150–58; Wunder, "Peasant Organization and Class Conflict," 93–94; Robisheaux, *Rural Society*, 147–62; Harnish, "Peasants and Markets," 46–58; Slicher van Bath, *Agrarian History*, 308–10; Redlich, "Eighteenth-Century German Guide for Investors," 98–103; Sabean, *Property, Production, and Family*, 46, 259–75, 300–304, 355–64, 373–91; William Robert Lee, *Population Growth*, 184–86, 405; Gagliardo, *From Pariah to Patriot*, 46–47; Brakensiek, "Agrarian Individualism," 153, 157.

42. C. T. Smith, *Historical Geography*, 196–99, 229–33, 264–68, 274–76; Tom Scott, "Economic Landscapes," 6–9; Wunder, "Peasant Organization," 93–94; James J. Sheehan, *German History*, 91–92; Bruford, *Germany in the Eighteenth Century*, 114–17; Sabean, *Property, Production, and Family*, 52, 55; Blum, *End of the Old Order*, 264–67; Gagliardo, *From Pariah to Patriot*, 8–9; Mayhew, *Rural Settlement*, 143, 169–79; Grab, "Enlightened Despotism," 49–50; Brakensiek, "Agrarian Individualism," 138–41, 144–48; Kriedte, *Peasants, Landlords, and Merchant Capitalists*, 112–13.

43. Scribner, "Communities," 295–309; Knodel, *Demographic Behavior*, 144–52; Imhof, "Structure of Reproduction," 25; Medick, "Village Spinning Bees" (quote on 332); Berdahl, "Garve on the German Peasantry," 91–92, 95–96; Robisheaux, *Rural Society*, 70–79, 259–60; Braun, "Early Industrialization," 296, 299.

44. Rabb, *Struggle for Stability*, 76–77, 119–23, and "Effects of the Thirty Years' War"; Kamen, "Economic and Social Consequences"; Wedgwood, *Thirty Years War*, 146–48, 207, 213–16, 255–57, 352–53, 511–17; Pounds, *Historical Geography*, 74–77; Mayhew, *Rural Settlement*, 120–21, 142–43; Robisheaux, *Rural Society*, 70–79, 201–8, 211–16, 223–26, 243–56, 259–60; Theibault, *German Villages in Crisis*, chaps. 5–6; Abel, *Agricultural Fluctuations*, 180–84; Sabean, *Power in the Blood*, 8–9, and *Property, Production, and Family*, 40–41; Gagliardo, *From Pariah to Patriot*, 4–6. For the population of post-1871 Germany and Switzerland, see Moller, introduction, 5–6; De Vries, "European Urbanization," 36; James J. Sheehan, *German History*, 74–76; and Kellenbenz, with Schawacht, Schneider, and Peters, "Germany," 191–92.

45. Blum, *End of the Old Order*, 29–30, 52–54, 61, 75, 308–9; Nichtweiss, "Second Serfdom," 115–27, 134–35; Harnish, "Peasants and Markets," 44–45, 48–50; James J. Sheehan, *German History*, 79; Bruford, *Germany in the Eighteenth Century*, 108–12; Gagliardo, *From Pariah to Patriot*, 11–15, 77–80, 88; Blickle, "From Subsistence to Property"; Sabean, *Property, Production, and Family*, 19–21; Berkner, "Inheritance, Land Tenure, and Peasant Family Structure," 76–77; Fogleman, *Hopeful Journeys*, 47–51, 56.

46. Roeber, *Palatinates, Liberty, and Property*, 47–58, and "In German Ways?," 754–55; Tom Scott, "Economic Landscapes," 5–7; Sabean, *Property, Production, and Family*, 1, 4, 13–17, 21–22, 51–54, 442–52; Kellenbenz, with Schawacht, Schneider, and Peters, "Germany," 196–98; William Robert Lee, *Population Growth*, 110–16, 139, 142; Berkner, "Inheritance, Land Tenure, and Peasant Family Structure," 79–84.

47. Berkner, "Peasant Household Organization," 61–63; Knodel, *Demographic Behavior*, 119–36.

48. Tom Scott, "Economic Landscapes," 5–7; Schlumbohm, "Land-Family Bond," 461–67, and "From Peasant Society to Class Society," 185–89, 192–95; Berkner, "Peasant Household Organization," 62–63, and "Inheritance, Land Tenure, and Peasant Family Structure," 80–81; Peter Kei Taylor, *Indentured to Liberty*, 63, 70–74; Slicher van Bath, "Agriculture," 128–29; Eric R. Wolf, "Inheritance of Land," 108–10; William Robert Lee, "Germany," 145; Berkner and Mendels, "Inheritance Systems," 213, 219–20; Melton, *Absolutism*, 123–26; Blum, *End of the Old Order*, 105–6.

49. Berkner, "Peasant Household Organization," 57–67, and "Inheritance, Land Tenure, and Peasant Family Structure," 80, 84–95; Schlumbohm, "From Peasant Society to Class Society," 189–94; Knodel, *Demographic Behavior*, 121–35. On illegitimacy, see William Robert Lee, "Bastardy and the Socioeconomic Structure," 416–21, and the debate between Shorter, "Bastardy," and William Robert Lee, "Bastardy: Reply."

50. Adelmann, "Structural Change," 82–87; Melton, *Absolutism*, chap. 5 (quote on 128); Pounds, *Historical Geography*, 235–39; Sabean, *Property, Production, and Family*, 49, 55, 156; Schlumbohm, "From Peasant Society to Class Society," 189–91; Braun, "Early Industrialization," 292, 296; Kellenbenz, "Rural Industries," 58–72; Kisch, "Textile Industries," 544–47, 551–52, 555–57, and *From Domestic Manufacture to Industrial Revolution*, 18, 24, 28–31, 37, 56–60, 66–67, 70, 115–18, 122–23, 130–31, 178; Kriedte, "Demographic and Economic Rhythms," 266–67; Keller, "From the Rhineland to the Virginia Frontier," 485–93; Wolfram Fisher, "Rural Industrialization," 163–65.

51. Kriedte, *Peasants, Landlords, and Merchant Capitalists*, 133–34; Pfister, "Work Roles and Family Structure," 83–105; Adelmann, "Structural Change," 82–87; Braun, "Early Industrialization," 296–301.

52. De Vries, *European Urbanization*, 26, 29–32, 36, 39, 58–59, 115, 272–73; Koellmann, "Population of Barmen," 588–93; Kisch, *From Domestic Manufacture to Industrial Revolution*, 78–79; Kriedte, "Demographic and Economic Rhythms," 260–67.

53. Blum, *End of the Old Order*, 185–89; Abel, *Agricultural Fluctuations*, 142–43, 252–59; Kellenbenz, with Schawacht, Schneider, and Peters, "Germany," 197, 216–19; Pounds, *Historical Geography*, 178–79; Morley, *Robinson in Germany*, 22, 44–45 (quote on 44), 56–57, 88, 109; James J. Sheehan, *German History*, 77, 91; Slicher van Bath, "Agriculture," 84; Jefferson, *Papers*, 13:13–14, 22, 26–27 (quotes); Dirlmeier and Fouquet, "Diet and Consumption," 86–89, 94–95, 98–100; Bruford, *Germany in the Eighteenth Century*, 121 (von Lowe quote); Gagliardo, *From Pariah to Patriot*, 28–30.

54. Gagliardo, *From Pariah to Patriot*, 46, 102–4; Schleunes, *Schooling and Society*, 20–26; Blum, *End of the Old Order*, 190–91; James J. Sheehan, *German History*, 88–89; Melton, *Absolutism*, chap. 1; William Robert Lee, *Population Growth*, 318–21; Ulbricht, "World of a Beggar," esp. 165–67. For a dissent, see Abel, *Agricultural Fluctuations*, 216–17.

55. Post, *Food Shortage*, 66, 70, 73, 75, 84–85, 88–89, 104–6, 111–12, 130–32, 193–200, 224, 251–56, 266–67, 279.

56. Schofield, "Age-Specific Mobility"; Souden, "Movers and Stayers," 23–25; Lawrence J. White, "Enclosures and Population Movements"; Wareing, "Migration to London," 360–77, and "Geographical Distribution of the Recruitment of Apprentices," 243–49; Holderness, "Personal Mobility" (Yorkshire); Stephen King, "Migrants on the Margin"; Timmins, *Made in Lancashire*, 35–37, 70–71.

57. The emigration estimate, higher than Fogleman, "Migrations to the Thirteen British North American Colonies," 698, 708 (41,100), and Canny, "English Migration," 58–59 (50,000), adds Ekirch's (*Bound for America*, chaps. 1–2, 7, esp. 22–25, 50–51 [convicts]) and Dunn's ("Servants and Slaves," 159 [indentured servants]) estimates and applies the ratio of free and indentured emigrants from the 1770s (Bailyn, *Voyagers to the West*, 168–69) to earlier periods.

58. Galenson, *White Servitude*, 24–28, 51–64, 102–13, 134–40; Bailyn, *Voyagers to the West*, chaps. 5–9; McCusker, "View from British North America," 692–93; Grubb, "Fatherless and Friendless," 85–86, 91–106 (quote on 91); Horn, "British Diaspora," 35–36. For Moraley, see Klepp and Smith, *Infortunate*, 11–12, 96–97.

59. Winchester, "Ministers, Merchants, and Migrants"; Bailyn, *Voyagers to the West*, 138–40, 161–64, 170–71, 210–11, 361–90, 401–26.

60. John Harwood Andrews, "Land and People, c. 1865," 454–63 (quote on 454); Louis Michael Cullen, "Irish Diaspora," 118–19, and *Emergence of Modern Ireland*, 42–43, 55–56; Smout, Landsman, and Devine, "Scottish Emigration," 78–80, 85, 87–88; Canny, "English Migration," 62–68, and "Marginal Kingdom," 40–53, 62–66; Myers, *Immigration of the Irish Quakers*, 8–9; Gailey, "Scots Element," 1–9.

61. Doyle, *Ireland, Irishmen, and Revolutionary America*, 61–63; Casway, "Irish Women Overseas," 114–17; Myers, *Immigration of the Irish Quakers*, 81–83; Lockhart, "Quakers and Emigration," esp. 83–92; Arthur Young, *Tour of Ireland*, 124, 153; Ekirch, *Bound for America*, 24–25, 47, 83–85; Wayland Fuller Dunaway, *Scotch-Irish*, 28–33; McCracken, "Ecclesiastical Structure," 99–104; Brady and Corish, *Church under the Penal Code*, 1 (percent Catholic).

62. Estimates of immigrants from Ireland vary greatly: see Fogleman, "Migrations to the Thirteen British North American Colonies," 698, 704–7 (108,600); Truxes, *Irish-American Trade*, 129–30 (175,000 to 200,000); Doyle, *Ireland, Irishmen, and Revolutionary America*, 53, 65, 70–71 (250,000); R. J. Dickson, *Ulster Emigration*, 23, 34, 50, 55, 57–61, 64–65 (115,000 from Ulster). Wokeck, *Trade in Strangers*, 169–76, 188–92, 201–2, dramatically reduces the number of emigrants to 51,676, but the author's data cover only the Delaware Valley ports and may well underrecord even those arrivals. I have placed the number roughly halfway between the estimates of Fogleman and Truxes. See also Kirkham, "Ulster Emigration," 75–97; R. J. Dickson, *Ulster Emigration*, 45–47; Arthur Young, *Tour of Ireland*, 1:161 (first quote); Kerby A. Miller, *Emigrants and Exiles*, 155–56 (second quote); De Wolfe, *Discoveries of America*, 5–6.

63. Houston and Withers, "Population Mobility in Scotland," 285–97; Houston, "Geographical Mobility in Scotland"; Lovett, Whyte, and Whyte, "Poisson Regression Analysis"; Whyte and Whyte, "Geographical Mobility of Women"; Malcolm Gray, "Migration in the Rural Lowlands," 104–14; Withers, "Highland, Lowland Migration," 76–79, and *Gaelic*, 182–89, 192–96; Devine, "Highland Migration," 137–41.

64. I applied Fogleman's temporal distribution in "Migrations to the Thirteen British North American Colonies," 698, 707–8, to Smout, Landsman, and Devine's aggregate totals in "Scottish Emigration," 90–111; for transported criminals, see Ekirch, "Transportation of Scottish Criminals."

65. Parker, *Scottish Highlanders*, 35; Adams and Somerville, *Cargoes of Hope and Despair*, 23; Ian Charles Cargill Graham, *Colonists from Scotland*, 25–28; Bumsted, "Highland Emigration," 511–27; De Wolfe, *Discoveries of America*, 5–6.

66. Smout, Landsman, and Devine, "Scottish Emigration," 90–111; Adams and Somerville, *Cargoes of Hope and Despair*, 19–23; Ian Charles Cargill Graham, *Colonists from Scotland*, chaps. 2–4, 6; Fingerhut, "Assimilation of Immigrants," 306–22; Landsman, *Scotland*, 271–73; Youngson, *After the Forty-five*, 41–46; Ekirch, "Transportation of Scottish Criminals," 366–74; Dobson, *Scottish Emigration*, 6–7, 89–98, 110–21; Adam, "Highland Emigration of 1770"; Devine, *Clanship to Crofters' War*, 177–84.

67. Bailyn, *Voyagers to the West*, 147–54, 158–60, 183–85, 510–19; De Wolfe, *Discoveries of America*, 23–24; Flinn, *Scottish Population History*, 212, 218–19, 231–32; Adam, "Highland Emigration of 1770"; Richards, *Highland Clearances*, vol. 2, chap. 7; Devine, *Clanship to Crofters' War*, 181–84; Richards and Clough, *Cromartie*, 90–92; Bumsted, *People's Clearance*, chaps. 1–3, 229 (downplays economic problems).

68. Faust, "Swiss Emigration," 24–39; Fogleman, *Hopeful Journeys*, 19–21, and "Review Essay," 319–23; Blanning, *Reform and Revolution*, 89; Fenske, "International Migration," 339–40; Roeber, "German-American Concepts of Property," 152–53, and " 'Origin of Whatever Is Not English,' " 243–44; Fertig, "Transatlantic Migration," 204–6; Wokeck, *Trade in Strangers*, xxv, 9–11, 22–23.

69. Hochstadt, "Migration in Preindustrial Germany," 195–224; Fertig, "Transatlantic Migration," 194–96, 203–9; Pounds, *Historical Geography*, 100; Kriedte, "Demographic and Economic Rhythms," 275–76; Ulbricht, "World of a Beggar," 158–65, 171–72, 181–82; Peter Kei Taylor, *Indentured to Liberty*, 24–26, 62, 66–70; Berkner, "Peasant Household Organization," 67.

70. Mayhew, *Rural Settlement*, 121–22, 163–67; Fenske, "International Migration"; Fertig, "Transatlantic Migration," 196, 202–3; James J. Sheehan, *German History*, 86–87; Fogleman, *Hopeful Journeys*, 25–31; Wokeck, *Trade in Strangers*, 18–20, and "Harnessing the Lure," 208–11; Schelbert, "On Becoming an Emigrant," 461, 463; Huggett, *Land Question*, 92–93; W. O. Henderson, *Studies in the Economic Policies of Frederick the Great*, 126–30, 134–35; Kirchner, *Commercial Relations*, chaps. 9–10; Faust, "Swiss Emigration," 26, 43.

71. Fertig, "Transatlantic Migration," 196, 202–3; Schelbert, "On Becoming an Emigrant," 461, 463; Fogleman, *Hopeful Journeys*, 31–32, and "Review Essay," 319; Wokeck, *Trade in Strangers*, 2–3, 6, 37, 44–47. I have used Wokeck's totals, subtracting the 1,300 who arrived in Louisiana but including the 2,400 who went to Nova Scotia.

72. Fertig, "Transatlantic Migration," 222–26; Wokeck, *Trade in Strangers*, 37–58; Faust, "Swiss Emigration," 42–44; Fenske, "International Migration," 342.

73. On recruiters, see below; Fertig, "Transatlantic Migration," 220–35; Dyck, "Mennonite Motivation for Emigration," 3 (quote); Wokeck, *Trade in Strangers*, 29; Grubb, "German Immigration," 429–33; Braun, "Early Industrialization," 300–305; and Statt, *Foreigners and Englishmen*, 135–36.

74. Schwartz, *"Mixed Multitude,"* 23–24, 84–85, 92–93, 202–3 (quote on 202; emphasis eliminated); Abbot Emerson Smith, *Colonists in Bondage*, 55; Wokeck, *Trade in Strangers*, 25–26, 61.

75. Ackerman, *South Carolina Colonial Land Policies*, chaps. 5, 8; Kaylene Hughes, "Populating the Back Country," chap. 1; Migliazzo, "Tarnished Legacy."

76. Paul S. Taylor, *Georgia Plan*, chap. 3; Parker, *Scottish Highlanders*, chap. 3; Ready, "Economic History of Colonial Georgia," 23–31; Baine, "Oglethorpe and the Early Promotional Literature"; Oglethorpe, *Publications of James Oglethorpe*, xiii–xx, 159–66, 200–240, 245–51; Randall M. Miller, "Failure of the Colony of Georgia," 2–4; George Fenwick Jones, *Georgia Dutch*, 17–21; E. R. R. Green, "Queensborough Township."

77. R. J. Dickson, *Ulster Emigration*, chap. 8; Purcell, "Irish Contribution," 598–99; Schwartz, *"Mixed Multitude,"* 84–85, 246–47, 258–60; Richards and Clough, *Cromartie*, 91–92; Roeber, "'Origin of Whatever Is Not English,'" 244–45; Selig, "Emigration, Fraud, Humanitarianism."

78. Jack P. Greene, *Intellectual Construction*, chaps. 3–4, is a sophisticated analysis of visions of eighteenth-century America. On promotional literature, see Gallay, *Voices of the Old South*, chap. 3; Bailyn, *Voyagers to the West*, 432–37, 504; Wokeck, *Trade in Strangers*, 26–27; De Wolfe, *Discoveries of America*, 7, 108–21.

79. Wokeck, *Trade in Strangers*, 27–34, and "Harnessing the Lure," 219–22; Roeber, "In German Ways?," 760–61; Fogleman, *Hopeful Journeys*, 33–34, 60, 71–72; O'Reilly, "Conceptualizing America," 110–15 (points to similar Newlanders who returned from Eastern Europe); Parsons, *Pennsylvania Dutch*, 44–45; Mittelberger, *Journey*, 9, 26–31; Cunz, *Maryland Germans*, 126–27; Myers, *Immigration of the Irish Quakers*, 59–61; Parkhill, "Philadelphia Here I Come," 119–23, 125, 132.

80. DeWolfe, *Discoveries of America* (reprints and analyzes these letters); Myers, *Immigration of the Irish Quakers*, 69–80 (first quotes on 71–74); Lockhart, "Quakers and Emigration,"; Paul S. Taylor, *Georgia Plan*, 76–77, 230 (second and third quotes); Fogleman, "From Slaves, Convicts, and Servants to Free Passengers," 53–55 (points out that letter writers did see servitude as an obstacle to prosperity).

81. De Wolfe, *Discoveries of America*, 1, 8–10, 15–26, 95–100, 104, 113, 116–17, 127, 143–44, 154–57, 163, 172–74, 177–81, 184–86, 194–95, 210–11, 217; Wokeck, *Trade in Strangers*, 29–30; Fogleman, *Hopeful Journeys*, 4, 125.

82. Roeber, *Palatines, Liberty, and Property*, 96–101; Roeber, "German-American Concepts of Property," 152–53; Roeber, "'Origin of Whatever Is Not English,'" 243–44; Grubb, "Redemptioner Immigration"; Grubb, "Immigrant Servant Labor," 249–51, 255–64; Grubb, "Market Structure," 37–38; Wokeck, *Trade in Strangers*, 119–21, 127, 149–63, 187–88; Moltmann, "Migration of German Redemptioners," 106–13; Abbot Emerson Smith, *Colonists in Bondage*, 20–22; Ekirch, *Bound for America*, 21–27; Brophy, "Law and Indentured Servitude," 85–89; Fogleman, *Hopeful Journeys*, 73–74.

83. Bailyn, *Voyagers to the West*, chap. 9; Klepp and Smith, *Infortunate*, 50–52 (quote); Truxes, *Irish-American Trade*, chap. 7 (quote on 131); Grubb, "Market Structure," esp. 36–39; Wokeck, *Trade in Strangers*, 59–92, 199–201; Abbot Emerson Smith, *Colonists in Bondage*, 59–61; Fertig, "Transatlantic Migration," 229–30; Selig, "Emigration, Fraud, Humanitarianism," 5.

84. Bailyn, *Voyagers to the West*, 313–23; Grubb, "Market Structure," 45–47, and "Morbidity and Mortality"; Wokeck, *Trade in Strangers*, 53–54, 78–80, 115–16, 130–37, 204–7 (tends to moderate this dismal picture).

85. Bailyn, *Voyagers to the West*, chap. 10; Kenneth Morgan, "Convict Trade," 217–22; Mittelberger, *Journey*, xiii–xvi, 4, 17–19 (quotes); Wokeck, *Trade in Strangers*, 98–101, 137–41, 156–63, and "Harnessing the Lure," 217–18; Brophy, "Law and Indentured Servitude," 87–95.

86. Grubb, "Immigrant Servant Labor," 257–67; Brophy, "Law and Indentured Servitude," 92–93; Bailyn, *Voyagers to the West*, 344–52; Kenneth Morgan, "Convict Trade," 218–19. On ethnic composition, see Purvis, "Pennsylvania Dutch," 86–89, and "Patterns of Ethnic Settlement," 115–20.

87. Myers, *Immigration of the Irish Quakers*, 95–102; Landsman, *Scotland*, 176, 208–9, 218, 224. See below for the Palatine and Georgia migrants.

88. Knittle, *Early-Eighteenth-Century Palatine Emigration*, chaps. 1–7 (religion on 8), 274–82; Statt, *Foreigners and Englishmen*, chaps. 5–6; Dickinson, "Poor Palatines"; Savory, "Huguenot-Palatine Settlements," 122–30; McCracken, "Social Structure and Social

Life," 41–42; Roeber, *Palatines, Liberty, and Property*, 8–16; Henry Z. Jones, *Palatine Families*, i–xvi; Otterness, "New York Naval Stores Project"; McGregor, "Cultural Adaptation."

89. Knittle, *Early-Eighteenth-Century Palatine Emigration*, 98–110, 147–49; Merrens, *Colonial North Carolina*, 21–22; Lefler and Powell, *Colonial North Carolina*, 60–72; E. B. O'Callaghan, *Documentary History*, 3:559, 568; Otterness, "New York Naval Stores Project," 138–39.

90. O'Callaghan, *Documentary History*, 3:542–53, 559–80, 583–607, 707–26 (quote on 707–8); O'Callaghan and Fernow, *Documents*, 5:211, 214–15, 553–55; Knittle, *Early-Eighteenth-Century Palatine Emigration*, chaps. 5–8; Roeber, *Palatines, Liberty, and Property*, 8–16; Otterness, "New York Naval Stores Project"; McGregor, "Cultural Adaptation."

91. On the debate over the origins of the Georgia colonists and trustees' goals, see Kenneth Coleman, *Colonial Georgia*, chap. 2; Kenneth Coleman, "Southern Frontier"; Kenneth Coleman, "Rebuttal"; Saye, *New Viewpoints*, chap. 1, and "Genesis of Georgia"; Ready, "Georgia Concept," 157–72, and "Georgia Trustees," 269–72; Paul S. Taylor, *Georgia Plan*, chaps. 1, 3, 7, and 299–321; Phinizy Spalding, "Oglethorpe's Quest," 67–76; Jack P. Greene, *Imperatives*, chap. 5; Betty Wood, "Oglethorpe, Race, and Slavery," 66–79 (second quote on 72); Oglethorpe, *Publications of James Oglethorpe*, xi–xxii (first quote on xvi), 163–66, and *Some Account of the Design*, 11–14, 22–49 (third quote on 23).

92. Baine, "New Perspectives on Debtors"; Paul S. Taylor, *Georgia Plan*, chaps. 2–3, 6–7 (quote on 230), and 300–306; Parker, *Scottish Highlanders*, 20–23, 42–43; Betty Wood, "Georgia Malcontents," 266, 271–72; Bolzius, "Bolzius Answers a Questionnaire," 219–20; Phinizy Spalding, *Oglethorpe in America*, chaps. 5, 7–9; Ready, "Economic History of Colonial Georgia," 105–13, esp. 113.

93. Paul S. Taylor, *Georgia Plan*, chaps. 3–8, and 300–306; Ver Steeg, *Origins of a Southern Mosaic*, chap. 3; Randall M. Miller, "Failure of Georgia," 8–15; Ready, "Georgia Trustees"; Betty Wood, "Georgia Malcontents"; Chesnutt, "South Carolina's Expansion," chaps. 1–3; Gray and Wood, "Transition from Indentured to Involuntary Servitude."

94. Wokeck, *Trade in Strangers*, 44–46; Selig, "Emigration, Fraud, Humanitarianism"; Knittle, *Early-Eighteenth-Century Palatinate Emigration*, chap. 8; Purcell, "Irish Contribution," 113; Fingerhut "Assimilation of Immigrants," 168–71; Kirkham, "Ulster Emigration," 81–84; De Wolfe, *Discoveries of America*, 118.

95. Robert D. Mitchell, *Commercialism and Frontier*, 34–35, 54–55; Fogleman, *Hopeful Journeys*, 93–99, and "Women on the Trail"; Ramsey, *Carolina Cradle*, 87–93, 117–29, 211–16; Rosenberger, "Migrations of the Pennsylvania Germans," 320–35, 59–62; Roeber, "In German Ways?," 765, and *Palatines, Liberty, and Property*, 118–19; Thorp, *Moravian Community*, 11, 42–48, 50–51.

96. See Chapter 3 for land acquisition; Wokeck, *Trade in Strangers*, 47–54, 61. On Baden-Durlach and Kraichtal, see Häberlein, "German Migrants"; Fogleman, *Hopeful Journeys*, 37, 60–65, 74–76; and De Wolfe, *Discoveries of America*, 32–36, 173. For other examples of wealth of emigrants, see Roeber, *Palatines, Liberty, and Property*, 101–13. Following Mittelberger, *Journey*, 17, I assume an exchange rate of 6 florins to £1 sterling; I have averaged 1758 and 1771 Pennsylvania tax list data and German property Häberlein compares to holdings in Germany.

CHAPTER FIVE

1. *American Husbandry*, 48–50, 52, 110, 131–32, 135, 176–77, 179, 305, 510, 512.

2. Robert E. Gallman, "Can We Build National Accounts?," 24, 26; Bushman, "Markets and Composite Farms," 352–53, 363–74.

3. Fox-Genovese, *Within the Plantation Household*, chap. 1; Ulrich, "Housewife and Gadder" and *Good Wives*, chaps. 1–2; Carr and Walsh, "Planter's Wife."

4. Edward Johnson, *Wonder-Working Providence*, 96, 179, 196, 249. Rutman and Rutman, *Small Worlds*, 94, 105, called attention to these passages. Studies of these towns—Sumner Chilton Powell, *Puritan Village*; Lockridge, *New England Town*; Greven, *Four Generations*; and David Grayson Allen, *In English Ways*, 117–49—ignore markets, implicitly presuming the practice of self-sufficient and subsistence agriculture.

5. For a distinction between marketplaces and the market, see Agnew, "Threshold of Exchange," and Rothenberg, *From Market-Places to a Market Economy*, 4–7, 20–21, 95–99. On risk-averse behavior, see Gavin Wright, *Political Economy*, 62–74.

6. For definition of household, see Fox-Genovese, *Within the Plantation Household*, chap. 1; Mary Beth Norton, *Founding Mothers*, 17–18 (family defined the same household used here); Wilk and Netting, "Households," 5–19 (defines households by functions of "production, distribution, transmission, reproduction, and coresidence" [5; emphasis deleted]); Shammas, "Anglo-American Household Government," 103–9 (definition in terms of authority of father/husband). On competency and the market, see Vickers, "Competency and Competition" and "Northern Colonies," 213, 216, 223. For case studies of farm household organization, see Merchant, *Ecological Revolutions*, pt. 2; Christopher Clark, *Roots of Rural Capitalism*; Joan M. Jensen, *Loosening the Bonds*; and Carr, Menard, and Walsh, *Robert Cole's World*.

7. McCusker and Menard, *Economy of British America*, chap. 14; Lewis Cecil Gray, *History of Agriculture*, vol. 1, chaps. 2–3, 11; Bidwell and Falconer, *History of Agriculture*, chap. 4; Bushman, "Markets and Composite Farms," 359–60.

8. Bidwell and Falconer, *History of Agriculture*, 46–48; Jonathan R. T. Hughes, *Social Control*, 27–30; Lemon, "Colonial America," 132–36; Joseph Sutherlund Wood, *New England Village*, 93–96; Rutman, *Winthrop's Boston*, 181; McNealy, "Bristol," 500–502; Friedmann, "Victualing Colonial Boston"; McWilliams, "Brewing Beer," 543–57; Little, "Men on Top?," 135–38; Matson, *Merchants and Empire*, 30–31, 50–51, 95–100, 110–12; Rosen, *Courts and Commerce*, 20–22.

9. Bassett, "Virginia Planter and the London Merchant," 555–58; Christine Daniels, " 'WANTED,' " 749; Pleasants and Scisco, *County Court of Charles County*, lviii–lix, 416, 466–76, 516–18; Rainbolt, "Absence of Towns"; Bruce, *Economic History*, 2:388–91; Rutman and Rutman, *Place in Time: Middlesex*, chap. 7; Jacob M. Price, "Rise of Glasgow," 194–99; Kulikoff, *Tobacco and Slaves*, 123–27; Thorp, "Doing Business."

10. This regional perspective permeates eighteenth-century accounts (*American Husbandry*) as well as recent interpretations (McCusker and Menard, *Economy of British America*, chaps. 5–9; Jack P. Greene, *Pursuits of Happiness*; David Haclett Fischer, *Albion's Seed*). On merchant networks, see Bailyn, "Communications and Trade." On New York negotiations, see Matson, *Merchants and Empire*, 99, 102, 106–7, 115.

11. The vast literature on tobacco marketing includes Price and Clemens, "Revolution of Scale," 1–37; Jacob M. Price, "Merchants and Planters," 11–15; Menard, "Tobacco Industry," 127–42; Bassett, "Virginia Planter and the London Merchant"; Lewis Cecil Gray, "Market Surplus Problem," 13–21, and *History of Agriculture*, vol. 1, chap. 11.

12. Rutman, *Husbandmen of Plymouth*, 13–23, and Rutman and Rutman, *Small Worlds*, chap. 6, are the best analyses, but see Gottfried, "First Depression," 671–78; Garrison, "Farm Dynamics," 1–3; Bailyn, "Communications and Trade," 383–84; Bushman, *From Puritan to Yankee*, 26–32; Bridenbaugh, *Fat Mutton*, chap. 5; William Davis Miller, "Narragansett Planters," 71–106; Romani, "Pettaquamscut Purchase"; Vickers, "Working the Fields," 69; Vickers, *Farmers and Fishermen*, 47–48; Vickers, "Northern

Colonies," 219–22; McMahon, "Comfortable Subsistence," 52–53; and Friedmann, "Victualing Colonial Boston."

13. McCusker and Menard, *Economy of British America*, chaps. 5–9; Egnal, "Economic Development," 200–207; Bidwell and Falconer, *History of Agriculture*, 89–96, 132–44; Pruitt, "Self-Sufficiency and the Agricultural Economy"; Bruce C. Daniels, *Fragmentation of New England*, 3–13; Edward S. Cooke Jr., *Making Furniture*, 72–74; Wermuth, " 'To Market, To Market,' " chap. 2; Otto, *Southern Frontiers*, chap. 4; Nobles, "Breaking into the Backcountry," 656–62; Robert D. Mitchell, *Commercialism and Frontier*, chap. 5; Merrens, *Colonial North Carolina*, chap. 6.

14. For import and export data, see U.S. Bureau of the Census, *Historical Statistics*, 2:1176–84. For interpretations, see Shepherd and Walton, *Shipping, Maritime Trade, and Economic Development*, 41–44; Brinley Thomas, "Atlantic Economy of the Eighteenth Century," 29–31; Egnal, *New World Economies*, chaps. 4–6, and "Economic Development," 199–216; Sachs, "Agricultural Conditions," 284–86; and S. D. Smith, "Market for Manufacturers." For prices, see Shammas, "Decline of Textile Prices." For regional economies, see McCusker and Menard, *Economy of British America*, chaps. 5–9.

15. Rothenberg, *From Market-Places to a Market Economy*, 95–101; Pruitt, "Self-Sufficiency," 338–63; Klingaman, "Food Surpluses," 557–66; Bidwell and Falconer, *History of Agriculture*, 142–44; Gross, "Culture and Cultivation," 44–48, 57–58; Lockridge, "Land, Population, and the Evolution of New England Society," 66–74; Schumaker, "Northern Farmer," 28–29 (quote); Bruce C. Daniels, *Fragmentation of New England*, 4–9; Vickers, *Farmers and Fishermen*, 211–15.

16. McMahon, "Comfortable Subsistence," is the most important study, but see Howard S. Russell, *Long, Deep Furrow*, chaps. 14–16; Schumaker, "Northern Farmer," 2–3, 22–25; and Jedry, *World of John Cleaveland*, 58–70.

17. Mid-eighteenth-century urban population estimated at 27,200 from data in Kulikoff, "Progress of Inequality," 393; Bruce C. Daniels, *Connecticut Town*, 197; Robert V. Wells, *Population of the British Colonies*, 70, 97–99; and Greene and Harrington, *American Population*, 23. Massachusetts, New Hampshire, and Rhode Island towns (except Boston) chosen on the basis of the high commercial indexes reported in Edward M. Cook, *Fathers of the Towns*, 200–211. Populations were then multiplied by .28 (the proportion living in town centers in New Haven and Hartford).

18. Shepherd and Walton, *Shipping, Maritime Trade, and Economic Development*, 211–12, 217, 220, 223–24; Klingaman, "Food Surpluses," 557–66; Thomas Reed Lewis, "From Suffield to Saybrook," 107–10; Friedmann, "Victualing Colonial Boston"; Baker and Izard, "New England Farmers," 33–36; Garrison, "Farm Dynamics," 5–7; Schumaker, "Northern Farmer," 20, 31–33; Gross, "Culture and Cultivation," 47–48; Newell, *From Dependency to Independence*, 247, 252–53; Vickers, *Farmers and Fishermen*, 207–10.

19. Bruce C. Daniels, *Connecticut Town*, chap. 6 and 197; Carville Earle, *Geographical Inquiry*, 158–61; Thomas Reed Lewis, "From Suffield to Saybrook," 208–15; Rothenberg, *From Market-Places to a Market Economy*, 82–95; Edward M. Cook, *Fathers of the Towns*, chap. 7; Joseph Sutherlund Wood, *New England Village*, chap. 2.

20. Pelizzon, "Grain Flour Commodity Chains"; Sachs, "Agricultural Conditions," 274–85; Doerflinger, *Vigorous Spirit of Enterprise*, 93, 101–16; Lemon, *Best Poor Man's Country*, 23, 27–29; Egnal, *New World Economies*, 50–54, 60–62, 67; Arthur Jensen, *Maritime Commerce*, 7–8, 45–46, 60, 71, 292–93; Stevenson Whitcomb Fletcher, *Pennsylvania Agriculture*, 282–83; Shepherd and Walton, *Shipping, Maritime Trade, and Economic Development*, 64, 98–100, 168, 170, 212, 217, 221, 224.

21. Nettels, *Money Supply*, 119–24; Lemon, *Best Poor Man's Country*, chap. 5; Schweit-

zer, *Custom and Contract*, 67–71; Doerflinger, "Farmers and Dry Goods"; Dauer, "Philadelphia's Hinterland Economy," 1–33, and "Philadelphia's Intraregional Transportation System"; Stephanie Grauman Wolf, *Urban Village*, chap. 3.

22. Lemon, *Best Poor Man's Country*, 123–27, 195–217; Billy G. Smith, *"Lower Sort,"* 95–104, 204–7; Klepp, "Demography," 94–95, 104–5; Jerome H. Wood, *Conestoga Crossroads*, 47–50. To estimate urban food needs, I used Smith's laborer diet and population data in Smith, in Klepp, and in Lemon, presuming that men made up a quarter of the population; women (two-thirds of male consumption), a quarter; and children (one-half of adult male consumption), a half.

23. Klingaman, "Food Surpluses," 561–63; Harrington, *New York Merchant*, 171–72, 214–16; Armour, *Merchants of Albany*, 174–77, 228–37; Matson, *Merchants and Empire*, 143, 197, 199–201, 227, 251–53; Kierner, *Traders and Gentlefolk*, 65, 79, 82–83, 91–92; Kim, *Landlord and Tenant*, 27–28, 190–95, 215, 232–33; David Steven Cohen, *Dutch-American Farm*, chap. 4; Hedrick, *History of Agriculture*, 67–74; Rossano, "Launching Prosperity," 32–35; Schumaker, "Northern Farmer," 9, 117, 119; Fabend, *Dutch Family*, 80–85; Wacker, *Musconetcong Valley*, 55–63.

24. Compare Robert Gough, "Myth of the 'Middle Colonies,' " 394–95, 404–5, 408–12; Gary B. Nash, *Urban Crucible*, 102, 313–15, 407–9; and Jacob M. Price, "Economic Function," 156–60. On merchants, see Countryman, "Uses of Capital," 15–22; Matson, *Merchants and Empire*, 219–22, 228–34, 238–39, 262–63, 278; McCusker, "Sources of Investment Capital," 153–55; and Doerflinger, *Vigorous Spirit of Enterprise*, 157–64. On New York fur trade, see Thomas Elliot Norton, *Fur Trade*, esp. chaps. 7, 10, and Harrington, *New York Merchant*, 232–37.

25. Carr, "Diversification," 342–75; Lorena S. Walsh, "Summing the Parts," 56–76, and "Plantation Management," 393–403; Russo, "Self-Sufficiency"; Clemens, *Atlantic Economy*, chap. 6; John W. Tyler, "Foster Cunliffe," 260–66; Harold B. Gill Jr., "Wheat Culture," 380–82; Klingaman, "Significance of Grain," 270–74, and "Food Surpluses," 557–63; Kulikoff, *Tobacco and Slaves*, 100–104, 120–22; Menard, "Economic and Social Development," 265–68; Lorena S. Walsh, "Provisioning Tidewater Towns," second section.

26. U.S. Bureau of the Census, *Historical Statistics*, 2:1190–91; Egnal, *New World Economies*, chap. 5; Price and Clemens, "Revolution of Scale," 19–33; Jacob M. Price, "Rise of Glasgow," 179–99; Jacob M. Price, "Buchanan and Simpson," 9–29; Jacob M. Price, "Virginia-London Consignment," 82–93; Jacob M. Price, "Merchants and Planters," 25–31; Jacob M. Price, *Capital and Credit*, 124–30; John W. Tyler, "Foster Cunliffe," 246–79; J. H. Soltow, "Scottish Traders."

27. Robert D. Mitchell, "Metropolitan Chesapeake," 109–15, 121–23; Kulikoff, *Tobacco and Slaves*, 104–7, 122–27, 223–29; Carville Earle, *Geographical Inquiry*, chap. 2, and *Evolution of a Tidewater Settlement System*, 78–100; Brune, "Changing Spatial Organization," 74–84; John W. Tyler, "Foster Cunliffe," 255–56; Jacob M. Price, "Rise of Glasgow," 191–97; Clarence P. Gould, "Economic Causes"; Preisser, "Alexandria"; Siener, "Charles Yates," 409–21; Christine Daniels, " 'WANTED,' " 749–51, 754, 760.

28. The best work is Charles J. Farmer, *In the Absence of Towns*, chaps. 2–6, but see also Robert D. Mitchell, "Metropolitan Chesapeake," 113; Wellenreuther, "Urbanization in the Colonial South," 663–67; and Nicholls, "Competition, Credit, and Crisis," 276, 280.

29. Terry G. Jordan, *North American Cattle-Ranching Frontiers*, 109–20, 170–78; Lewis Cecil Gray, *History of Agriculture*, 1:209–12; Otto, "Livestock-Raising in Early South Carolina"; Mart Stewart, " 'Whether Wast, Deodand, or Stray' "; R. C. Nash, "Urbanization," 6–7.

30. Coclanis, *Shadow of a Dream*, 77–84, 97–98, 106–10; Egnal, *New World Economies*,

chap. 6; Menard, "Economic and Social Development," 276–78, 281–85; R. C. Nash, "South Carolina and the Atlantic Economy" and "Urbanization," 5–8, 19; Richard Waterhouse, "Economic Growth," 207–17; Philip D. Morgan, *Slave Counterpoint*, 47–48; Harold E. Davis, *Fledgling Province*, 69–72, 123–24, 158–59; Peter Wood, *Black Majority*, 144–52; Statom, "Negro Slavery in Eighteenth-Century Georgia," 165–78; Clowse, *Economic Beginnings*, 256–58, and *Measuring Charleston's Overseas Commerce*, 44–45, 47; Alice Hanson Jones, *American Colonial Wealth*, 3:1506–7, 1569–70, 1588–89; Bridenbaugh, *Cities in Revolt*, 217. The estimate of the population of small planters in the Lowcountry presumes that four whites lived in each household and that the Lowcountry Georgia planter population was half that of South Carolina.

31. Kessel, "Germans," 93–100; Robert D. Mitchell, *Commercialism and Frontier*, chaps. 5–7; Hofstra and Mitchell, "Town and Country," 619–39; MacMaster, "Cattle Trade," 127–45; Klingaman, "Significance of Grain," 271–74, and "Food Surpluses," 557–63; Kulikoff, *Tobacco and Slaves*, 120–22.

32. Otto, *Southern Frontiers*, 52–57; Merrens, *Colonial North Carolina*, 115–19, 134–41, 144, 155–72; Thorp, *Moravian Community*, 40–41, 107–47; Johanna Miller Lewis, *Artisans*, 70–71; Rachel N. Klein, *Unification of a Slave State*, chap. 1; Chaplin, *Anxious Pursuit*, 289–97, 332–38; Ernst and Merrens, " 'Camden's Turrets,' " 557–68 (they would dispute this interpretation).

33. Matthew Patten, *Diary*, 5–38, 132–84, 294–353. On Patten, see Charles E. Clark, *Eastern Frontier*, 239–44, and Thomas C. Thompson, "Life Course." On Bedford, see Edward M. Cook, *Fathers of the Towns*, 147, 252.

34. Mary Beth Norton, *Founding Mothers*, 207–10, 445–46; John Frederick Martin, *Profits in the Wilderness*, chap. 1; Innes, *Labor in a New Land*, chaps. 3–4 and 199–227; Kierner, *Traders and Gentlefolk*, 65–66, 71–74, 78–79, 83, 91–96; Gloria L. Main, *Tobacco Colony*, 70–71, 80–82, 92; Land, *Dulanys of Maryland*, chap. 11. Kierner would not accept this interpretation.

35. Carville Earle, "Rate of Frontier Expansion," 190–95; Earle and Cao, "Frontier Closure," 166–68; Menard, "Tobacco Industry," 110–16, 129–42; P. M. G. Harris, "Social Origins," 220–38, 257–64.

36. See sources in n. 14; Jacob M. Price, *Capital and Credit*, 125–26 (Henderson quote); Egnal, *New World Economies*, 92–96; Sheridan, "British Credit Crisis," 163–70; Shammas, *Preindustrial Consumer*, 267–72, 275; Charles J. Farmer, *In the Absence of Towns*, 174–83; Steffen, *From Gentlemen to Townsmen*, chap. 5; Tommy R. Thompson, "Personal Indebtedness," 14–16; Coclanis, "Retailing," 2–5, and *Shadow of a Dream*, 103, 147–49; Thorp, "Doing Business," 387–408, and *Moravian Community*, chap. 5.

37. Egnal, "Pennsylvania Economy," 64–66; Gwyn, "Private Credit," 277–85, 288–89; Sweeney, "Gentleman Farmers," 66–73; Nobles, "Rise of Merchants," 7–12; Johanna Miller Lewis, *Artisans*, 71; Rosen, *Courts and Commerce*, 42–44; Snydaker, "Kinship and Community," 56; Wacker and Clemens, *Land Use*, 275–77; Rona Stephanie Weiss, "Development of the Market Economy," 161–70.

38. Rothenberg, "Role of Mortgage Credit," 16–22; Newell, *From Dependency to Independence*, 179–80; Thayer, "Land Bank"; Schweitzer, *Custom and Contract*, chaps. 4–5 (data on 148–49, with missing years 1731–35 excluded from the average); Kross, *Evolution of an American Town*, 222–23; Matson, *Merchants and Empire*, 246–48; Kemmerer, "Colonial Loan Office," 867–74; John L. Brooke, *Heart of the Commonwealth*, 55–65.

39. Rothenberg, "Role of Mortgage Credit," 1, 8–16; Rosen, *Courts and Commerce*, 44–47, 146–48; Nicholls, "Competition, Credit, and Crisis," 283–85. New York debt figures were reduced using McCusker, *Money and Exchange*, 164–65.

40. Vickers, "Working the Fields," 56–61, and *Farmers and Fishermen*, 45–46, 49, 60–

64; Lorena S. Walsh, "Community Networks," 218–36 (quote on 235–36); Carr, "Diversification," 351–53, 364–67, 374–76; Carr and Walsh, "Inventories," 95–96.

41. On the term "borrowing system," see Faragher, *Sugar Creek*, 133–36. For alternative visions of local exchange and commerce, see Merrill, "Cash Is Good to Eat," esp. 54–61, and "Survey of the Debate," 9–10; Merrill and Wilentz, *Key of Liberty*, 11–14, 197–98; Henretta, "Families and Farms," 15–20; Rothenberg, *From Market-Places to a Market Economy*, chap. 2; and Vickers, *Farmers and Fishermen*, 237–47.

42. Crèvecoeur, *Letters from an American Farmer*, 103–4, 277–78, 281–82; Kulikoff, *Tobacco and Slaves*, 185; Maryland Court of Appeals, TDM#1, 83–85 (Evans); Wacker and Clemens, *Land Use*, 1 (Beatty quote), 7, 9–10, 15.

43. Gloria L. Main, "Gender, Work, and Wages," 52–54, 58; Ulrich, "Housewife and Gadder," 27–32; Ulrich, *Midwife's Tale*, 72–90 (quote on 73), 152, 169–71; Ulrich, "Wheels, Looms, and the Gender Division of Labor," 17–19; Hood, "Textile Manufacture," chaps. 3–4; Kulikoff, *Agrarian Origins*, 31; Newell, *From Dependency to Independence*, 99–100; Johanna Miller Lewis, *Artisans*, chap. 6; Joan M. Jensen, *Loosening the Bonds*, chaps. 5–7.

44. We know little about rural artisans. Bridenbaugh, *Colonial Craftsman*, chaps. 1–2, the only general work, is out-of-date and often inaccurate. The best works are local studies of Maryland (Russo, "Self-Sufficiency," esp. 425–32; Christine Daniels, "'WANTED'" and "From Father to Son," 3–16), Connecticut (Edward S. Cooke Jr., *Making Furniture*, esp. chaps. 1–4), and backcountry North Carolina (Johanna Miller Lewis, *Artisans*, esp. chaps. 2–6). See also Stephanie Grauman Wolf, *As Various as Their Land*, 176–79; Alice Hanson Jones, *Wealth of a Nation to Be*, 225–27, 231–32. On growing population density, see Kulikoff, *Tobacco and Slaves*, 48–49, 206, and Wacker, *Land and People*, 138.

45. Alice Hanson Jones, *Wealth of a Nation to Be*, 104; Philip D. Morgan, *Slave Counterpoint*, 54; Wacker and Clemens, *Land Use*, 286; Russo, "Self-Sufficiency," 411–14; Johanna Miller Lewis, *Artisans*, 23, 48.

46. Rorabaugh, *Craft Apprentice*, 9–10; Rothenberg, *From Market-Places to a Market Economy*, 117–18; Vickers, *Farmers and Fishermen*, 253–58; Newell, *From Dependency to Independence*, 97–99, 248–49; Edward S. Cooke Jr., *Making Furniture*, 14, 45, 77; McCoy, *Elusive Republic*, 63–66; Coxe, *View of the United States of America*, 442–43 (quote).

47. On kin networks, see Daniel Scott Smith, "'All in Some Degree Related,'" 44–68; Kulikoff, *Tobacco and Slaves*, chap. 6; and Perry, *Society on Virginia's Eastern Shore*, 70–77. On divorce witnesses, see Cott, "Eighteenth-Century Family," 24–26. On marriage migration, see Susan L. Norton, "Marital Migration," 419–28; Adams and Kasakoff, "Migration at Marriage," 119–23; and Ulrich and Stabler, "'Girling of It,'" 27–29. On interethnic exchange, see Kross, *Evolution of an American Town*, 218–22; Barry Levy, *Quakers and the American Family*, 184, 312; Laura A. Becker, "Diversity," 209–10; Kessel, "Germans," 97–101; Marietta, *Reformation of American Quakerism*, 7, 23–25; and Snydaker, "Kinship and Community," 56. Burnard, "Associational Networks," 17–45, and Crowley, "Importance of Kinship," 562, 565–77, argue that kinship was unimportant in economic exchange.

48. Mutch, "Yeoman and Merchant," 291–97; Nobles, "Rise of Merchants," 5–12; Coe and Coe, "Mid-Eighteenth-Century Food and Drink"; Rosen, *Courts and Commerce*, 35–40, 47–51, 76–78; Derven, "Wholesome, Toothsome, and Diverse"; Newell, *From Dependency to Independence*, 95–97; Jacob M. Price, "American Port Towns," 138–40; Lemon, *Best Poor Man's Country*, 125–40; Shammas, *Preindustrial Consumer*, 267–83; Charles J. Farmer, *In the Absence of Towns*, 162–68; Nicholls, "Competition, Credit, and Crisis," 278–79; Wermuth, "'To Market, To Market,'" 74–75.

49. Nicholls, "Competition, Credit, and Crisis," 281–82.

50. On peddlers, see Breen, "Empire of Goods," 467–68, 494–95; Robert D. Mitchell, "Metropolitan Chesapeake," 110; Schumacher, "Northern Farmer," 87; and Matson, *Merchants and Empire*, 131–32. On cottagers, see Simler, "Landless Worker." On transients and the poor, see Chapter 3, above; Douglas Lamar Jones, "Strolling Poor," 28–54; Kulikoff, "Progress of Inequality," 398–404; and Mackey, "Operation of the English Old Poor Law," 36–40.

51. Berlin, *Many Thousands Gone*, 35–36, 57, 113, 119, 134, 157–58, 165–66; Lorena S. Walsh, "Slave Society," 191–92; Schlotterbeck, "Internal Economy of Slavery," 173–77; Philip D. Morgan, *Slave Counterpoint*, 115, 250–53, 304–7, 310–15, 360–72; Olwell, *Masters, Slaves, and Subjects*, 141–58, 167–78; Betty Wood, *Women's Work*, 55–58, 64–65, 69–71, 82–86, 91–92, 94.

52. Alan Tully, *William Penn's Legacy*, 65, reports the house raising. I examine the idea of rural independence and yeoman ideology in *Making of the American Yeoman Class*, chap. 6. On the relation between independence and local gift exchange, see Mutch, "Yeoman and Merchant," 296–97, and Pruitt, "Self-Sufficiency," 354–55.

53. Perkins, *American Public Finance*, chap. 3; Hoffer, *Law and People*, 50–55; Bruce H. Mann, *Neighbors and Strangers*, chap. 2; Rosen, *Courts and Commerce*, 80–91; Rothenberg, *From Market-Places to a Market Economy*, 117–38; Marietta, *Reformation of American Quakerism*, 23–26; Charles J. Farmer, *In the Absence of Towns*, 180–83; Shammas, *Preindustrial Consumer*, 269–72; Land, *Bases of the Plantation Economy*, 112–14; Peter Coleman, *Debtors and Creditors*, 40–42, 45–48 (at most, 27 percent of Massachusetts creditors sued for assignment of property during 1764–67 were farmers), 76–79, 86–88, 95–96, 106–14, 130–35, 141–45, 163–69, 179–85, 192–98, 203, 207–10, 215–19, 228–30, 233–34; Feer, "Imprisonment for Debt"; Tommy R. Thompson, "Debtors, Creditors, and the General Assembly in Maryland" and "Personal Indebtedness," 16–25; Nicholls, "Competition, Credit, and Crisis," 273–74, 277–79, 281.

54. Sheridan, "British Credit Crisis," 173–77; Jacob M. Price, *Capital and Credit*, 130–37; Kulikoff, *Tobacco and Slaves*, 129–30; Egnal, "Pennsylvania Economy," 99–104, and *New World Economies*, 74–76; Peter Coleman, *Debtors and Creditors*, 167–69; Nicholls, "Competition, Credit, and Crisis," 283–89; Rosen, *Courts and Commerce*, 52–54, 65, 155–56.

55. These comments grow out of my reading of the conflicting views of Loehr, "Self-Sufficiency"; Merrill, "Cash Is Good to Eat"; Henretta, "Families and Farms"; Pruitt, "Self-Sufficiency"; Shammas, "How Self-Sufficient Was Early America?"; McCusker and Menard, *Economy of British America*, chaps. 1, 4, 14; Weiman, "Families, Farms, and Rural Society"; and Thomas C. Thompson, "Life Course," 135–45. I have staked out a position in *Agrarian Origins*, chaps. 1–2.

56. Bradford, *Of Plymouth Plantation*, 120–21 (quotes), 361–64. Mary Beth Norton, *Founding Mothers*, 7–10, brilliantly interprets this passage.

57. Paul E. Johnson, "Modernization of Patch"; Fox-Genovese, *Within the Plantation Household*, chap. 1 (best rendition of the household economy); Kulikoff, *Tobacco and Slaves*, chap. 5; Stephanie Grauman Wolf, *As Various as Their Land*, 17–21, 30–37, 43–47; Thomas C. Thompson, "Life Course"; Henretta, "Families and Farms"; Breen, "Back to Sweat and Toil," 244–55 (critiques Henretta). On women's place in the household, see Mary Beth Norton, *Liberty's Daughters*, 9–20; Marylynn Salmon, *Women and the Law of Property*, chap. 3; and Lisa Wilson, *Life after Death*, chap. 4.

58. See sources in n. 47, and Sen, "Gender and Cooperative Conflicts," 123–49. On England, see Prologue, above.

59. Robert V. Wells, "Population of England's Colonies," 88–91; Archer, "New England

Mosaic," 489–91; Carr and Menard, "Immigration and Opportunity," 218–27; Treckel, " 'To Comfort the Heart,' " 141–44; Kathleen M. Brown, *Good Wives*, 83–84.

60. Franklin, *Papers*, 3:438–41, 4:225–34 (quotes on 228, 232; written in 1751 and published in 1755); Robert V. Wells, "Population of England's Colonies," 85–86.

61. Hajnal, "European Marriage Patterns," 101–12, 130–40, and "Two Kinds of Pre-Industrial Household Formation System," 68–72; Robert V. Wells, "Population of England's Colonies," 88–91; Levine, *Reproducing Families*, 76–78, 12–22, 131–32; Prologue, above. Evidence for declining ages of females at first marriage and higher levels of prenuptial pregnancy and bastardy dates from the eighteenth century.

62. For analytical statements, see Daniel Scott Smith, "American Family and Demographic Patterns," 394–96; Menard, "Early American Population History," 358–59; and Robert V. Wells, "Population of England's Colonies," 88–91. For data, see Robert V. Wells, "Quaker Marriage Patterns," 417–36; Archer, "New England Mosaic," 489–92; Higgs and Stettler, "Colonial New England Demography," 284–86; Kulikoff, *Tobacco and Slaves*, 49–58; David Hackett Fischer, *Albion's Seed*, 487, 674–75; James Matthew Gallman, "Relative Ages" and "Determinants of Age at Marriage," 177–82; Smith and Wilson, *North Carolina Women*, 23–24; Linzer, "Population and Society," 24–25; Marc Harris, "People of Concord," 88–90; and Roger C. Henderson "Demographic Patterns," 355–58. Moravian Bethlehem, Pennsylvania, with its communal control stands as the only exception (Smaby, *Transformation of Moravian Bethlehem*, 73–75).

63. On premarital sex, see Berkin and Horowitz, *Women's Voices*, 17, and Gundersen, *To Be Useful to the World*, 43–45. The classic essay on bridal pregnancy remains Smith and Hindus, "Premarital Pregnancy," 537–64, but see also Gladwin, "Tobacco and Sex," 61–72, and Mary Beth Norton, *Founding Mothers*, 66–72, 335–42, 424–25, 466–67, 469. On bastardy, see Robert V. Wells, "Illegitimacy," 354–55; Daniel Scott Smith, "Long Cycle," 371–75; Carr and Walsh, "Planter's Wife," 548–50, and Kathleen M. Brown, *Good Wives*, 92–94, 187–99, 203.

64. Domar, "Causes of Slavery and Serfdom"; Daniel Scott Smith, "Malthusian-Frontier Interpretation"; Little, "Men on Top?," 126–27.

65. Robert V. Wells, "Population of England's Colonies," 91–93, and "Family Size"; David Hackett Fischer, *Albion's Seed*, 71, 277, 483; Kulikoff, *Tobacco and Slaves*, 33–34, 54–61, 173–75, 186–87; Gundersen, "Double Bonds of Race and Sex," 361–63; Rutman and Rutman, *Place in Time: Explicatus*, chap. 5; Roger C. Henderson, "Demographic Patterns," 369–74; Daniel Scott Smith, "Demographic History," 177–83; Greven, "Average Size"; Withey, "Household Structure."

66. Controversy besets the familial composition of households. Laslett, "Introduction," 23–44, 61–62; Laslett, *Family Life and Illicit Love*, 19–25; and Laslett, "Stem Family," assert that extended families were rare. Berkner, using a life-cycle approach in "Stem Family," argues that they were common. The best work, centering on the nineteenth century, supports Berkner (Ruggles, "Transformation" and *Prolonged Connections*, 37–39, 223–33, and Lee and Gjerde, "Comparative Household Morphology," 89–95). For the colonies, compare Demos, "Demography and Psychology," 561–62, and "Families in Colonial Bristol," 44–45, with Waters, "Family, Inheritance, and Migration," and Kulikoff, *Tobacco and Slaves*, 173–75. On houses and families, see Flaherty, *Privacy in Colonial New England*, 46–52, and Felt, "Statistics of Population in Massachusetts," 148–57.

67. Mary Beth Norton, *Founding Mothers*, 7–13, 21–22, 38–42, 48, 51–53, 58–65, 96–111, 118–19, 140–44, and "Evolution of White Woman's Experience," 597–605; Shammas, "Anglo-American Household Government," 115–28. Daniel Scott Smith, "Curious History," shows the ideological origins of the connection between nuclear structure and the imputation of individualism to household members.

68. In addition to the references in the preceding note, see Folbre, "Patriarchy in Colonial New England," and Barry Levy, "Girls and Boys," 295–97 (disputes this argument, showing larger than expected numbers of children, at least half in eighteenth-century agrarian New England, who had lost their fathers by age fourteen).

69. Shammas, "Anglo-American Household Government," 115–28; Joan M. Jensen, *Loosening the Bonds*, chap. 2; Mary Beth Norton, *Founding Mothers*, 12–14; Carr and Walsh, "Planter's Wife"; Kulikoff, *Tobacco and Slaves*, 165–74; Philip D. Morgan, *Slave Counterpoint*, 273–96.

70. Richard B. Morris, *Studies in the History of American Law*, chap. 3; Marylynn Salmon, *Women and the Law of Property*, esp. chaps. 1–2, 5–7; Marylynn Salmon, "Women and Property"; Marylynn Salmon, "Equality or Submersion," 92–111; Mary Beth Norton, *Founding Mothers*, 57–62, 72–78, 83–89, 155–56, 427; Dayton, *Women before the Bar*, chap. 1; Rosen, *Courts and Commerce*, 117–24; Gundersen and Gampel, "Married Women's Legal Status"; Gampel, "Planter's Wife Revisited," 20–35; Little, "Men on Top?" For examples of prenuptial agreement, see Berkin and Horowitz, *Women's Voices*, 81–83. Morris and Gundersen and Gampel argue for fuller rights of married women than the evidence supports. Fox-Genovese, *Within the Plantation Household*, chap. 1, and Coontz, *Social Origins of Private Life*, chaps. 1, 3, come closest to my interpretation.

71. Scott and Wishy, *America's Families*, 86–89 (Wadsworth quotes; emphasis removed). See also Berkin and Horowitz, *Women's Voices*, 54–60; Kathleen M. Brown, *Good Wives*, 340–41.

72. Examples of conflicting analyses of the wife's status in the family can be found in Gundersen, *To Be Useful to the World*, 51–53; Ulrich, *Good Wives*, chap. 2; Lisa Wilson, *Life after Death*, chap. 4; Rosen, *Courts and Commerce*, 100–104; Mary Beth Norton, *Founding Mothers*, 84–85, 138–39; and Kulikoff, *Tobacco and Slaves*, 179–80, 188–93.

73. Henretta, "Families and Farms," 21–28, examines northern lineal ideology and practice. On naming patterns, see Rutman and Rutman, *Place in Time: Explicatus*, chap. 7; Kulikoff, *Tobacco and Slaves*, 241, 245–50; Tebbenhoff, "Tacit Rules"; Daniel Scott Smith, "Child-Naming Practices"; David Hackett Fischer, "Forenames and the Family," 219–31; Waters, "Family, Inheritance, and Migration," 66–67, and "Naming and Kinship," 171–75; and Gloria L. Main, "Naming Children," 1–16, and "English Family in America," 148–49, 158–59.

74. David Hackett Fischer, *Albion's Seed*, 284–85, 485–90; Mary Beth Norton, *Founding Mothers*, 62–66, 108–9; Farber, *Guardians of Virtue*, 124–36; Kulikoff, *Tobacco and Slaves*, 252–55; Reid, "Church Membership," 400–401, 404–12; Daniel Scott Smith, "Parental Power," 422–26. Burnard, "Associational Networks," argues that kinship was unimportant in the marriages of Maryland gentlemen.

75. For this formulation, see Janeway, *Powers of the Weak*, chaps. 11–12, and Fox-Genovese, "Property and Patriarchy." Mary Beth Norton, *Founding Mothers*, esp. 6–14, 17, 38–41, 48–49, 59–62, 74, 293–304, insists that early New Englanders not only espoused patriarchal theory but that they successfully linked state and family. The truncated hierarchy (no king, few lords, no parliament), in my view, renders her argument unpersuasive.

76. Koehler, *Search for Power*, 137–42, 160; Mary Beth Norton, *Founding Mothers*, 28–29, 34, 42, 52–53, 79 (first quote), 426; Gundersen, *To Be Useful to the World*, 124–25; Roger Thompson, *Sex in Middlesex*, chaps. 7–9; Pleck, *Domestic Tyranny*, 19, 21, 24, 27, 29; Berkin and Horowitz, *Women's Voices*, 85–87; Lisa Wilson, *Ye Heart of a Man*, 90–97; Merrill D. Smith, *Breaking the Bonds*, 105 (stale bread); Sheldon S. Cohen, " 'To Parts of the World Unknown,' " 285; Lacey, "World of Hannah Heaton," 294–95 (quotes); Kulikoff, *Tobacco and Slaves*, 181–83 (Pattison quotes); Spruill, *Women's Life*, 167–70.

77. Koehler, *Search for Power*, 142–46, 153–56; Mary Beth Norton, *Founding Mothers*, 79–81, 426, and *Liberty's Daughters*, 48; Lerner, *Female Experience*, 75–76 (quoting Pennsylvania *Chronicle*, August 1767); Merrill D. Smith, *Breaking the Bonds*, 106–7, 139–41, 149–50; Wawrzyczek, "Women of Accomack," 10–25; Kathleen M. Brown, *Good Wives*, 336–38.

78. Roderick Phillips, *Putting Asunder*, 134–53, 241–55; Gundersen, *To Be Useful to the World*, 49–51; Mary Beth Norton, *Founding Mothers*, 89–94; Marylynn Salmon, *Women and the Law of Property*, chap. 4; Koehler, *Search for Power*, chap. 5; Sheldon S. Cohen, " 'To Parts of the World Unknown,' " 275–90, esp. 277, 287, 290; Cott, "Eighteenth-Century Family," esp. 31, and "Divorce" (587, for occupations; I assume that all artisans, merchants, and tradesmen were known, and I distributed the unknowns among the other groups); Thomas Weiss, "U.S. Labor Force Estimates," 22, 37, 51 (60 percent of Massachusetts workers in agriculture); Dayton, *Women before the Bar*, chap. 3; Moynihan, "Four Centuries of Connecticut Women," 39–40 (second quote); Merrill D. Smith, *Breaking the Bonds*, 16–19, 65–66; Kulikoff, *Tobacco and Slaves*, 180–83 (first quote).

79. On family strategy, see Chayanov, *Theory of the Peasant Economy* (esp. 53–90), and Tilly and Scott, *Women, Work, and Family*, chaps. 1–3. For debates swirling around the concept, see Moch et al., "Family Strategy."

80. On household consumption, see below. Studies based on probate documents can rarely distinguish dowries from inheritances, but see Ditz, *Property and Kinship*, 112–15; Nylander, "Provision for Daughters," 12–27; Folbre, "Patriarchy in Colonial New England," 8–9; Barbara McLean Ward, "Women's Property," 77–78; and Jean Butendoff Lee, "Land and Labor," 330–31.

81. This interpretation is based on studies of farmer and planter probate inventories and tax lists and wills. See Alice Hanson Jones, *Wealth of a Nation to Be*, 90, 96–99; Jackson Turner Main, *Society and Economy*, 77–78, 203, 212–13, 215, 218, 237–38; Wacker, "New Jersey Tax-Ratable List," 39–44; Carr and Menard, "Immigration and Opportunity," 217–20; Carr, "Inheritance," 158–60; Kulikoff, *Tobacco and Slaves*, 51–52; Henry M. Miller, "Archaeological Perspective," 177–81; Lorena S. Walsh, " 'Till Death Us Do Part,' " 143, 147. Carr, Menard, and Walsh, *Robert Cole's World*, 38, 45–50, 74, is a first-rate case study.

82. Auwers, "Fathers, Sons, and Wealth," 144–45; Jackson Turner Main, *Society and Economy*, chap. 6; Carr, "Inheritance," 158–60, 202–3; Jean Butendoff Lee, "Land and Labor," 314–30; Kulikoff, *Tobacco and Slaves*, 51–52; Menard, "Population, Economy, and Society," 79; Main and Main, "Economic Growth," 35–37; Alice Hanson Jones, *Wealth of a Nation to Be*, 90–91, 96–99, 109–10, 204–5, 224–29, 270.

83. Daniel Scott Smith, "Perspective on Demographic Methods," 455–58, 463–66; Demos, *Past, Present, and Personal*, 124–27, 135, 137; Waters, "Patrimony," 154–59; Jackson Turner Main, *Society and Economy*, 69–70, 109–10, 123–25, 136, 201–3, 229–30, and "Standards of Living"; Gloria L. Main, *Tobacco Colony*, 267–73; Menard, Harris, and Carr, "Opportunity and Inequality," 178–81; Alice Hanson Jones, *Wealth of a Nation to Be*, 166–69, 220–30, 238, 381–88; Williamson and Lindert, *American Inequality*, 25–30.

84. On inequality, see Alice Hanson Jones, *Wealth of a Nation to Be*, chaps. 6–8; Kulikoff, "Growth and Welfare," 360–61; David Hackett Fischer, *Albion's Seed*, 168–71, 375–76, 566–67; Ritchie, *Duke's Province*, 131–38; and Lemon and Nash, "Distribution of Wealth." Williamson and Lindert, *American Inequality*, chap. 2, admits inequality grew in some places but argues that "*aggregate* colonial inequality was stable at low levels (11)." Shammas, "New Look," 419–20, 429–31, reworks Jones's data, arguing that Jones underestimated the wealth of the top 1 percent. Osberg and Siddiq, "Inequality," 148–54, pushes the percentage held by the top tenth to as high as an implausible 84. On eco-

nomic growth *and* decline, see Alice Hanson Jones, *Wealth of a Nation to Be*, 72–79, 304–5; Main and Main, "Economic Growth," 35–39; Terry L. Anderson, "Economic Growth in Colonial New England"; Ball, "Dynamics of Population," esp. 635; Ball and Walton, "Agricultural Productivity Change"; Menard, "Comment"; and Kulikoff, "Economic Growth."

85. Nearly all work is on the Chesapeake region (Rutman and Rutman, *Small Worlds*, chap. 10; Lorena S. Walsh, " 'Till Death Us Do Part,' " 135–37, 143–46; Carr, "Development of Maryland's Orphans' Court"; Kulikoff, *Tobacco and Slaves*, 168–74), but for other regions (with lower death rates), see David Hackett Fischer, *Growing Old in America*, 27–28, 275–76; Calhoun, *Social History of the American Family*, 1:172–73, 295, 307–12; Demos, *Little Commonwealth*, 115–16, 120–22; Barry Levy, *Quakers and the American Family*, 197–98; Watson, "Orphanage"; and Crowley, "Family Relations," 42–44.

86. On the status of widows, see Mary Beth Norton, *Founding Mothers*, 139–40, 144–46, 164. On inheritance, see Shammas, "Early American Women," 141–49; Shammas, Salmon, and Dahlin, *Inheritance in America*, chaps. 1–2, esp. 30–36, 51–55; Lisa Wilson, *Life after Death*, 104–10, 120–21; Deborah Mathias Gough, "Further Look at Widows," 829–33; Ditz, *Property and Kinship*, 125–37; Gloria L. Main, "Widows," 80–86; Kulikoff, *Tobacco and Slaves*, 189–93; Carr, "Inheritance," 158–86; and Crowley, "Family Relations," 44–47.

87. On women's property, see Mary Beth Norton, *Founding Mothers*, 147–49, 157–60, 438; Shammas, "Early American Women," 137–47; Alice Hanson Jones, "Wealth of Women," 248–56; Narrett, "Men's Wills," 108–12; and Wulf, " 'As We Are All Single,' " 1–9, 15–23, 27, 31–32, appendix, tables 1–2. On remarriage and widowhood, see Daniel Scott Smith, "Female Householding," 88–93 (number and age of widow householders); Grigg, "Remarriage" (evidence mostly urban); Faragher, "Old Women," 18–28; Keyssar, "Widowhood"; Daniel Scott Smith, "Population, Family, and Society," 281–84, and "Inheritance," 55–57; Ricketson, "To Be Young," 113–27, esp. 125–26; and Rutman and Rutman, *Place in Time: Explicatus*, 66–69. Lisa Wilson, *Life after Death*, chap. 4, challenges this interpretation.

88. Shammas, "Early American Women," 137–47; Alice Hanson Jones, "Wealth of Women," 248–56; Ditz, *Property and Kinship*, 125–37; Gloria L. Main, "Widows," 80–86; Rosen, *Courts and Commerce*, 103–4; Carr, "Inheritance," 158–86; Crowley, "Family Relations," 44–47; Keyssar, "Widowhood"; Kulikoff, *Tobacco and Slaves*, 192–93 (quote).

89. Examples are in Greven, *Four Generations*, chaps. 3, 6, 8; Vickers, *Farmers and Fishermen*, 222–25; Waters, "Family, Inheritance, and Migration," 71–84; Daniel Scott Smith, "Parental Power"; Lisa Wilson, *Ye Heart of a Man*, 35; Daniel Blake Smith, *Inside the Great House*, 242–48; and Kulikoff, *Tobacco and Slaves*, 193–202.

90. Shammas, Salmon, and Dahlin, *Inheritance in America*, 30–33, 42–51; Carr, "Inheritance," 169–70, 180–81; Jean Butendoff Lee, "Land and Labor," 314–29; Kulikoff, *Tobacco and Slaves*, 200–203; Keim, "Primogeniture and Entail," 551–57; Brewer, "Entailing Aristocracy," 311–46; Ditz, *Property and Kinship*, 61–73; Crowley, "Family Relations," 47–50; Keller, "Outlook of Rhinelanders," 101–2.

91. Shammas, Salmon, and Dahlin, *Inheritance in America*, 30–33, 42–48; Gloria L. Main, "Widows," 79; Carr, "Inheritance," 158–60; Jean Butendoff Lee, "Land and Labor," 330–38; Conger, " 'If Widow,' " 253–55 (suggests sons elsewhere in Maryland received more slaves); Crowley, "Family Relations," 48–55.

92. See Chapter 4. The classic essay remains Lockridge, "Land, Population, and the Evolution of New England Society," but Rutman and Rutman, *Small Worlds*, chap. 7, and Main and Main, "Red Queen," 130–46, are more persuasive.

93. On seventeenth-century houses, see nn. 116–17 in Chapter 2. On house sizes and room use, see Lee Soltow, *Distribution of Wealth*, 51–55, 61–65, and "Housing Characteristics"; Flaherty, *Privacy in Colonial New England*, 38–44; Lanier, "Ethnic Perceptions," 11–29; Steinitz, "Rethinking Geographical Approaches"; Donald W. Linebaugh, "All the Annoyances"; Upton, "Vernacular Domestic Architecture"; Gary Wheller Stone, "Artifacts Are Not Enough," 71–72; Camille Wells, "Planter's Prospect," 5–22; Michel, " 'In a Manner and Fashion Suitable to Their Degree,' " 29–65; Bernard L. Herman, *Architecture and Rural Life*, chap. 2, and "Home and Hearth"; Menard, "Economic and Social Development," 254; Stiverson, *Poverty in a Land of Plenty*, chap. 3; and Barnett, "Tobacco, Planters, Tenants, and Slaves," 13–16.

94. Shammas, *Preindustrial Consumer*, 84–100, 169–88; Shammas, "Domestic Environment," 14–19; Shammas, "Changes in English and Anglo-American Consumption," 177–78, 183–202; Carr and Walsh, "Inventories"; Carr and Walsh, "Standard of Living"; Carr and Walsh, "Changing Lifestyles," 59–166; Ann Smart Martin, "Role of Pewter," 11–19, and "Creamware Revolution," 169–84, esp. 174–75; Gloria L. Main, "Standard of Living"; Main and Main, "Economic Growth," 36–45; Edward S. Cooke Jr., *Making Furniture*, chap. 5; Breen, "Meaning of Things," 254–58; De Vries, "Purchasing Power," 98–104; R. C. Nash, "Urbanization," 15; Rosen, *Courts and Commerce*, 21–29; Wacker and Clemens, *Land Use*, 275–77, 290.

95. Billy G. Smith, "Comment," makes a similar point, and the empirical data of studies cited in the preceding and succeeding notes sustain it. On growth of gentility, see Trautman, "Dress in Seventeenth-Century Cambridge," 51–63 (little difference in dress by wealth); Bushman, *Refinement of America*, chaps. 1, 3–4, 6 (esp. 182–86); and Kierner, "Hospitality," 449–60, 465–68. The evidence, in my view, does not support Breen's arguments for the democratization of consumption nor the political implications he draws from it ("Empire of Goods," 485–98; "Meaning of Things"; " 'Baubles of Britain,' " 77–104). On the gentry class, see Kulikoff, *Making of the American Yeoman Class*.

96. Edward S. Cooke Jr., *Making Furniture*, 100–105; Ann Smart Martin, "Frontier Boys," 71–100, and "Common People and the Local Store," 44–49; Gary Wheller Stone, "Artifacts Are Not Enough," 73–76; Wacker and Clemens, *Land Use*, 280–81.

97. Despite excellent work on slavery and servitude, family labor and wage labor are understudied. Surveys include Dunn, "Servants and Slaves"; Rediker, " 'Good Hands' "; and Galenson, "Labor Market Behavior," 84–95.

98. On the growing season, see Chapter 2; Bidwell and Falconer, *History of Agriculture*, 10–11, 90–92; Vickers, *Farmers and Fishermen*, 50–51; Hensley, "Time, Work, Social Context," 531–55; Thomas C. Thompson, "Life Course," 145–49; Wacker and Clemens, *Land Use*, 8–11, 229–56; Barbara Clark Smith, *After the Revolution*, 55–58; Carr and Menard, "Land, Labor, and Economies of Scale," 414–16; Breen, *Tobacco Culture*, 46–58; David Hackett Fischer, *Albion's Seed*, 164–65, 370–72, 565, 744 (backcountry spring marriages); Robert V. Wells, "Marriage Seasonals," 299–307; Cressy, "Seasonality"; Rutman, Weatherall, and Rutman, "Rhythms of Life," 36–38, 42–43 (pre-Lent peak); and Kulikoff, *Tobacco and Slaves*, 256, 383–84.

99. Little is known about the division of labor on seventeenth-century farms, but see Gloria L. Main, "Gender, Work, and Wages," 52–54; Kathleen M. Brown, *Good Wives*, 85–87, 101; Carr and Walsh, "Planter's Wife," 547, 561–64; and Henry M. Miller, "Archaeological Perspective," 179–85.

100. On male farm work, see Gloria L. Main, "Gender, Work, and Wages," 54–55; Thomas C. Thompson, "Life Course," 145–49; Barbara Clark Smith, *After the Revolution*, 55–58; and Carr and Menard, "Land, Labor, and Economies of Scale," 414–16. On

land use, see Carville Earle, *Geographical Inquiry*, chap. 7, esp. 258–62, 276–85 (system of field rotation innovative), and Avery Odells Craven, *Soil Exhaustion*, chaps. 1–2.

101. Berkin and Horowitz, *Women's Voices*, 112–13.

102. Scholten, *Childbearing*, 52–54 (quote on 53 from a Pennsylvania advertisement for female household help); Mary Beth Norton, *Liberty's Daughters*, 9–15; Gundersen, *To Be Useful to the World*, 60–62; Boydston, *Home and Work*, 13–14; Ulrich, *Good Wives*, chap. 2, and *Midwife's Tale*, 310–14, 323–29; Gloria L. Main, "Gender, Work, and Wages," 55–57, 63–64; McMahon, "Comfortable Subsistence," 28–29, 38; Shammas, "How Self-Sufficient Was Early America?," 255–59; Oakes, "Ticklish Business," 201–4; Derven, "Wholesome, Toothsome, and Diverse," 50–51; Joan M. Jensen, *Loosening the Bonds*, chaps. 2–3; Kulikoff, *Tobacco and Slaves*, 174–83, 217–31.

103. Gloria L. Main, "Gender, Work, and Wages," 54–55; Mary Beth Norton, *Liberty's Daughters*, 13–14; Bidwell and Falconer, *History of Agriculture*, 116; Stevenson Whitcomb Fletcher, *Pennsylvania Agriculture*, 108; Rush, *Essays*, 232 (quote); Simler and Clemens, "Rural Labor," 130–34; Lacey, "Women in the Era of the American Revolution," 539; Kathleen M. Brown, *Good Wives*, 295–96, 335; Kulikoff, *Tobacco and Slaves*, 165.

104. Lorena S. Walsh, "Consumer Behavior," 218–21, 231–34, 238–43, summarizes current knowledge. See also McMahon, "Provisions Laid Up" and "Comfortable Subsistence" (the most important works); Cowan, *More Work for Mother*, 20–25; Matthei, *Economic History*, 40–44; Ulrich, *Good Wives*, 19–24; Lemon, "Household Consumption," 61–66; Gloria L. Main, *Tobacco Colony*, 169–71, 191–92, 219–20; Henry M. Miller, "Archaeological Perspective," 176–99; Kierner, *Beyond the Household*, 15; Smith and Wilson, *North Carolina Women*, 28–29.

105. Tryon, *Household Manufactures*, chaps. 2–3, and Alice Morse Earle, *Home Life in Colonial Days*, chaps. 8–10, are classic studies but exaggerate home cloth production. Correctives include Ouellette, "Divine Production and Collective Endeavor"; Ulrich, "Wheels, Looms, and the Gender Division of Labor," 3–18, 37–38; Little, "Men on Top?," 126–27; Hood, "Industrial Opportunism," 139–44; Hood, "Gender Division of Labor"; Hood, "Material World of Cloth"; Hood, "Textile Manufacture," chaps. 2–4; Henretta, "War for Independence," 58–68; Edward S. Cooke, *Making Furniture*, 15–17; Rosen, *Courts and Commerce*, 26, 102; Matson, *Merchants and Empire*, 248–49, 253–55; Wacker and Clemens, *Land Use*, 286; Boydston, *Home and Work*, 12–14; S. D. Smith, "Market for Manufacturers," 696–702; A. L. Macdonald, *No Idle Hands*, chap. 1; Beales, "Slavish and Other Female Work," 52–53; Kulikoff, *Tobacco and Slaves*, 179–80, 185; Philip D. Morgan, *Slave Counterpoint*, 246–48; and Kierner, *Beyond the Household*, 18–19.

106. Hood, "Material World of Cloth," 43–57, 61–62; Ulrich, "Wheels, Looms, and the Gender Division of Labor," 6.

107. For summaries, see Scholten, *Childbearing*, chaps. 1–3, and n. 57, above, for fertility and child spacing. For implications of heavy childbearing on women's lives, see Robert V. Wells, "Demographic Change," 273–82, and David Hackett Fischer, *Growing Old in America*, 56–57, 278–79. On nursing, see Treckel, "Breastfeeding"; Beales, "Nursing and Weaning," 53–60, 63; and Lorena S. Walsh, "'Till Death Us Do Part,'" 141–42. See Waltzer, "Period of Ambivalence," 367 (1790 Sedgwick quote). I have been influenced by Folbre ("Of Patriarchy Born," 263–73, and "Patriarchy and Capitalism," chap. 5).

108. Little is known about child labor, but see Edith Abbott, "Early History," 16–22 (quote on 16); Bremner et al., *Children and Youth*, 103–5; Alice Morse Earle, *Home Life in Colonial Days*, chap. 11; Lystad, *At Home*, 31–33; Demos, *Little Commonwealth*, 140–42; Beales, "Parkman's Farm Workers," 124–25; Vickers, *Farmers and Fishermen*, 64–71, 221–22; Edmund S. Morgan, *Puritan Family*, 65–79; Mary Beth Norton, *Liberty's Daughters*,

12–13; Towner, *Past Imperfect*, 44–47; Gloria L. Main, "Gender, Work, and Wages," 55; Folbre, "Wealth of Patriarchs," 217–19; Thomas C. Thompson, "Life Course," 150–52; Frost, *Quaker Family*, 83; Lorena S. Walsh, " 'Till Death Us Do Part,' " 146–47, and "Experiences and Status of Women," 6–8; Kulikoff, *Tobacco and Slaves*, 185–86; and Smith and Wilson, *North Carolina Women*, 27. Midwestern data from 1860 on the value of children's farm labor in Craig, *To Sow One Acre More*, 79–86, can be used as a crude proxy for the colonies.

109. Education literature emphasizes rich or poor children and adult literacy, but inferences can be drawn from Vinovskis, "Family and Schooling," 22–25; Gundersen, *To Be Useful to the World*, 80–83; Cremin, *American Education*, chap. 17; Hiner, "Cry of Sodom," 9–13; Lockridge, *Literacy*; Beales, "Studying Literacy," 91–102; Auwers, "Reading the Marks"; Perlmann and Shirley, "When Did New England Women Acquire Literacy?"; Grubb, "Educational Choice," 366–73; Kessel, " 'Mighty Fortress,' " 376–80; Frost, *Quaker Family*, 95–96, 99–102, 108–18; Alan Tully, "Literacy Levels" (lower levels in rural Pennsylvania); Kulikoff, *Tobacco and Slaves*, 193–99; Robert E. Gallman, "Changes in the Level of Literacy"; and Watson, "Women in Colonial North Carolina," 12–14.

110. Marylynn Salmon, "Family Management"; Barry Levy, *Quakers and the American Family*, 128–31, 193–230 (quotes on 193, 224); Frost, *Quaker Family*, 85–87; Kierner, *Beyond the Household*, 28–30; Kulikoff, *Tobacco and Slaves*, 184–85, 193–95. Bloch, "American Feminine Ideals," assesses the postrevolutionary origins of domesticity and child nurture, and Kerber, *Toward an Intellectual History*, 159–99, brilliantly delineates historians' uses of the ideology of separate spheres.

111. Dunn, "Servants and Slaves," 185–88; Galenson, "Labor Market Behavior," 54–62, 77–81; Ball and Walton, "Agricultural Productivity Change," 108–9 (corrected by Menard, "Comment," 119–21); Simler and Clemens, " 'Best Poor Man's Country,' " 246–61; Menard, "From Servants to Slaves," 368–71; Kulikoff, *Tobacco and Slaves*, 85–88, 133–37; Peter Wood, *Black Majority*, 160–61.

112. Barry Levy, "Girls and Boys"; Vickers, *Farmers and Fishermen*, 55–60, and "Working the Fields," 54–58; Towner, *Past Imperfect*, 44–45, 53; Quimby, *Apprenticeship*, 38–39, 105–10; Demos, "Families in Colonial Bristol," 46. A tiny handful of small farmers may have employed Indian servants (especially children after King Philip's War), but most went to connected, richer men. See Rowlandson, *Sovereignty and Goodness of God*, 50, 141–44, 147–49.

113. Vickers, *Farmers and Fishermen*, 229–32; Masur, "Slavery in Eighteenth-Century Rhode Island," esp. 142–43; Lorenzo Johnson Greene, *Negro in Colonial New England*, chaps. 3–4; Pierson, *Black Yankees*, chap. 2; Rosivach, "Agricultural Slavery."

114. McManus, *Black Bondage*, 38–42, 175–77, 199–214; Thomas J. Davis, "New York's Long Black Line"; Kross, *Evolution of an American Town*, xiv–xv, 90–93, 109, 155–56, 230–33, 248–49; Alan Tully, "Patterns of Slaveholding"; Soderlund, *Quakers and Slavery*, chap. 3, and "Black Importation," 147–51; Nash and Soderlund, *Freedom by Degrees*, chap. 1; Grubb, "Servant Auction Records," 165–67; Salinger, *"To Serve Well and Faithfully,"* 68–71; Berlin, *Many Thousands Gone*, 181.

115. Galenson, "Labor Market Behavior," 79–81; Philip D. Morgan, *Slave Counterpoint*, 1–2, 175, 351–52, 515–16, and "Black Society," 93–98; Menard, "From Servants to Slaves," 369, 385; Menard, "Slave Demography," 284–85, 303; Menard, "Africanization"; Menard, "Economic and Social Development," 278; Lorena S. Walsh, "Servitude and Opportunity," 118–22; Kevin Kelly, *Economic and Social Development*, 77–80; U.S. Bureau of the Census, *Century of Population Growth*, 185; Robert V. Wells, *Population of the British Colonies*, 148–51; Kay and Cary, *Slavery in North Carolina*, 16, 23–24, 48–50, 229–31, 238;

Gundersen, *To Be Useful to the World*, 60; Kulikoff, *Tobacco and Slaves*, 136–37, 153–54; Sarah S. Hughes, "Slaves for Hire"; Jacob M. Price, "Merchants and Planters," 7; Christine Daniels, "Gresham's Laws," 224; Peter Wood, *Black Majority*, 152, 160–63; Olwell, *Masters, Slaves, and Subjects*, 158–66; Rachel N. Klein, *Unification of a Slave State*, 20–28; De Wolfe, *Discoveries of America*, 163.

116. Jacob M. Price, "Merchants and Planters," 17–24.

117. Almost all we know about slave work revolves around slave labor on the largest units, but hints can be found in Berlin, *Many Thousands Gone*, 55–56, 135–36, 146–48, 181–82; Hodges, *Slavery and Freedom*, 44–50; Kathleen M. Brown, *Good Wives*, 115; Philip D. Morgan, *Slave Counterpoint*, 5, 7, 48–52, 122, 129, 139–41, 144, 147–75, 182–83, 193, 197–98, 249; Peter Wood, *Black Majority*, 28–36, 55–62; Carney, "Rice Milling"; Kulikoff, *Tobacco and Slaves*, 397–402; Menard, "Economic and Social Development," 280; Lorena S. Walsh, "Summing the Parts," 69; Christine Daniels, "Gresham's Laws," 222–24; and Wacker and Clemens, *Land Use*, 13, 99.

118. Philip D. Morgan, *Slave Counterpoint*, 6, 8–9, 13–14, 36, 39–41, 104–8, 144, 187–88, 300–302; Kulikoff, *Tobacco and Slaves*, 330–31, 337–39, 411; Hodges, *Slavery and Freedom*, 47–49. The important issues of slave community, farm life, and religion—central to recent literature on colonial slavery—are beyond the scope of this volume.

119. Berkin and Horowitz, *Women's Voices*, 103–6 (Chesapeake quotes), 115–16, 119 (1780 quote); Davis and Mintz, *Boisterous Sea of Liberty*, 66–67.

120. A small sampling of the literature on servant and slave resistance includes Berlin, *Many Thousands Gone*, 11, 32–33, 67, 98, 106, 115–16, 118, 120–21, 150–51; Edmund S. Morgan, *American Slavery*, 126–29, 246, 308; Mary Beth Norton, *Founding Mothers*, 45–46, 103, 115–16, 120–26, 130, 132–36, 433, 435; Kathleen M. Brown, *Good Wives*, 102, 351–52, 354–55; Kulikoff, *Tobacco and Slaves*, 328–29, 343–45, 376, 379–80; Mullin, *Flight and Rebellion*, chaps. 2–3; Philip J. Schwarz, *Twice Condemned*, chaps. 3–6; and Peter Wood, *Black Majority*, chaps. 9, 11–12.

121. Galenson, "Labor Market Behavior," 84–95, and "Settlement and Growth," 137, 159, 166–68; Simler, "Hired Labor"; Rothenberg, *From Market-Places to a Market Economy*, chap. 6; Vickers, *Farmers and Fishermen*, 26–28, 45–46, 52–54, 232–37 (quote on 45); Gloria L. Main, "Gender, Work, and Wages"; Beales, "Parkman's Farm Workers"; Hensley, "Time, Work, and Social Context," 542–51; Wacker and Clemens, *Land Use*, 18–20, 103–9; Christine Daniels, " 'Getting His [or Her] Livelyhood,' " 129–30, 147, 149–57, 160–61; Hodges, *Slavery and Freedom*, 47. See above for labor in the borrowing system.

122. Christine Daniels, " 'Getting His [or Her] Livelyhood,' " 128–48, and "Gresham's Laws," 212–22.

123. Mary Beth Norton, *Founding Mothers*, 84, 87 (quotes on 87); Norton does not connect these examples to cottagers.

124. Simler, "Hired Labor" (good summary); Simler, "Tenancy in Colonial Pennsylvania," 546–50; Simler, "She Came to Work," 2–20; Simler and Clemens, " 'Best Poor Man's Country,' " 239–55; Simler and Clemens, "Rural Labor"; Berkin and Horowitz, *Women's Voices*, 117–18 (an account book segment, not identified as cottagers, but clearly working in that way).

125. De Wolfe, *Discoveries of America*, 144.

126. Egnal, *New World Economies*, 42–44; Alice Hanson Jones, *Wealth of a Nation to Be*, 77–82; Kulikoff, "Economic Growth," 277–80; Carr and Menard, "Wealth and Welfare," 96–101, 108–11, 118–20; Menard, "Economic and Social Development," 256–59, 268–70; Steckel, "Nutritional Status," 37–45; Main and Main, "Red Queen." Mancall and Weiss, "Was Economic Growth Likely?," 17–33, dissent, but their controlled conjectures

presume no change in key economic variables, a highly implausible result, and exclude home manufacture, which clearly jumped in the pre-Revolutionary decades (a growth they deny). On diet, see Derven, "Wholesome, Toothsome, and Diverse," 48, 63.

EPILOGUE

1. Buel and Buel, *Way of Duty*, chaps. 4–7 (tells Mary Silliman's story); Fennelly, *Connecticut Women*, 56 (quote). This chapter is based on my *The Revolution the Farmers Made*, in progress.

2. Nettels, *Emergence of a National Economy*, chap. 1; McCusker and Menard, *Economy of British America*, 361–63; Carrington, "American Revolution," 823–41, esp. 840–41; Jacob M. Price, *France and the Chesapeake*, 2:682–83, and "New Time Series," 319, 323; Lewis Cecil Gray, *History of Agriculture*, 2:576–78, 589–94.

3. Baugh, "Politics of British Naval Failure," esp. 224, 230–35; McCusker and Menard, *Economy of British America*, 361–63; Albion and Pope, *Sea Lanes in Wartime*, 37–40, 63; Gruber, *Howe Brothers*, 80–81, 102–3, 136, 140–41, 150–51, 201–4, 210, 264–65, 291–93; Doerflinger, *Vigorous Spirit of Enterprise*, 210–11; Lewis Cecil Gray, *History of Agriculture*, 2:579–81.

4. Jacob M. Price, *France and the Chesapeake*, 2:682, 687, 700–727; Kulikoff, *Tobacco and Slaves*, 157–58; Lewis Cecil Gray, *History of Agriculture*, 2:589–92; Barbara Clark Smith, *After the Revolution*, 121; Customs 16/1 (1768–72 tobacco exports); Bergstrom, *Markets and Merchants*, 150 (1773 exports); "Tobacco Exports" (postwar hogsheads).

5. Parry and Sherlock, *Short History of the West Indies*, 137–38; East, *Business Enterprise*, 150–52, 154, 167, 169, 174–77; Bezanson et al., *Prices and Inflation*, 51, 89–91, 95, 255–57; Lewis Cecil Gray, *History of Agriculture*, 2:578–79, 582–83; Goldenberg, "Virginia Ports," 318, 324; Cooper, "Trial and Triumph," 292–94, 299–302; Gilbert, "Baltimore's Flour Trade," chap. 3.

6. See n. 7 for size of armies. On diet, see U.S. Bureau of the Census, *Historical Statistics*, 2:1175; Frey, *British Soldier*, 32–33; Cash, *Medical Men*, 26–27, 82; and Huston, *Logistics of Liberty*, 126. For livestock weights, see Howard S. Russell, *Long, Deep Furrow*, 153, 160, and Lewis Cecil Gray, *History of Agriculture*, 205. On wagons and horses, see Shy, "Logistical Crisis," 165–67. On exports, see Shepherd and Walton, *Shipping, Maritime Trade and Economic Development*, 220–26.

7. On Continentals, see Lesser, *Sinews of Independence*, xxi–xxxii, 2–211. On the French, see Christopher Ward, *War of the Revolution*, 2:586, 611, 690, 870, 881–82, 885–86, and Kennett, *French Forces*, 20–21, 41–42, 66–75. On the British, see Bowler, *Logistics*, chaps. 2–3; Atwood, *Hessians*, 257; and Katcher, *Encyclopedia*, 135–41.

8. Jeremy Black, *War for America*, 50; Conway, *War of American Independence*, 163; Bowler, "Logistics and Operations," 57–71, and *Logistics*, chaps. 2–3; Bernard Mason, "Entrepreneurial Activity," 195, 209–10; Schlebecker, "Agricultural Markets and Marketing," 24–35; Whitfield, "Initial Settling," 7–13; Syrett, *Shipping and the American War*, chaps. 7–9 and 249–50; Kennett, *French Forces*, 41, 66–75.

9. E. James Ferguson, *Power of the Purse*, chap. 4; Shy, *People Numerous and Armed*, 228, 317; Buel, *Dear Liberty*, 199 (quote); Carp, *To Starve the Army*, 64–65, 90, 93–94, 106–7, 109, 186; Risch, *Supplying Washington's Army*, 126, 223–24; Schlebecker, "Agricultural Markets and Marketing," 23–24; Jackson Turner Main, *Sovereign States*, 265–66; Thibaut, *This Fatal Crisis*, 35–36; Lemon, *Best Poor Man's Country*, 5–6, 225.

10. Bodle, *Vortex of Small Fortunes*, 91–92, 145, 177, 179–80, 249, 253, 287–88, 295, 376; Van Tyne, *Loyalists in the American Revolution*, 205–6, 327–29; Buel, *Dear Liberty*, 257–58; Bowler, *Logistics*, 47, 70, 72–74; Bernard Mason, "Entrepreneurial Activity," 193–94,

209–10; Frederic Gregory Mather, *Refugees of 1776*, 934–40, 962–63, 965–70; Ryden, *Letters to and from Caesar Rodney*, 195–96, 200–201, 241, 244–45; Lemon, *Best Poor Man's Country*, 5–6, 225.

11. Wermuth, "'To Market, to Market,'" 128–35; Schlebecker, "Agricultural Markets and Marketing," 28–35; Thibaut, *This Fatal Crisis*, 78, 131–32, 146, 155, 158, 538; Shy, "Logistical Crisis," 169–71; Meyer, "Evolution of the Interpretation of Economic Life," 36–38; Pinkett, "Maryland as a Source of Food Supplies," 159; Robert D. Mitchell, "Agricultural Change," esp. 121–24, 128–29.

12. Royster, *Revolutionary People at War*, 192–93, 296–97; Huston, *Logistics of Liberty*, 128–29; Thibaut, *This Fatal Crisis*, 105, 107, 128; Nelson, "American Soldier," 48–49; Bodle, *Vortex of Small Fortunes*, 157, 353; Fitzpatrick, *Writings of George Washington*, 10:327 (first quote), 12:479; Joseph Plumb Martin, *Private Yankee Doodle*, 182–88 (second quote on 186); Prince et al., *Papers of William Livingston*, 2:175.

13. Joseph Plumb Martin, *Private Yankee Doodle*, 175 (quote); Kim, "Continental Army and the American People," 463–64.

14. Calomiris, "Depreciation of the Continental," 54–59; McCusker, "How Much Is That in Real Money?," 351–58; Bezanson et al., *Prices and Inflation*, 10, 54–55, 80–93, 316, 325–26, 345; Behrens, *Paper Money*, chaps. 6–7; Cometti, "Inflation in Revolutionary Maryland"; Rothenberg, "Price Index for Rural Massachusetts," 982–83, 988; P. M. G. Harris, "Inflation and Deflation," 479, 492, 501.

15. Wermuth, "'To Market, To Market,'" 114–15; Christopher Moore, *Loyalists*, 118 (first quote); Harrell, *Loyalism in Virginia*, 79–82 (second quote on 79); Cometti, "Inflation in Revolutionary Maryland," 231–33; Nevins, *American States*, 478–92.

16. Eric Foner, *Paine and Revolutionary America*, 147, 161–62, 174, 186; Bezanson et al., *Prices and Inflation*, 88–89, 93; Bezanson, "Inflation and Controls," 15–19; Buel, "Sampson Shorn," 156–59; Kenneth Scott, "Price Control in New England," 443–73; Gwyn, "Impact of British Military Spending," 80–85, 98; Conway, *War of American Independence*, 163–64; Kim, "Limits of Politicization," 883–84; Calhoon, *Loyalists in Revolutionary America*, 326–27, 375, 390.

17. Lord, *War over Walloomscoick*, 116–19, 125–53, 156; Page, "Economic Structure," 74–75; Doblin and Lynn, *Specht Journal*, 103; Wacker and Clemens, *Land Use*, 143, 147–48, 178–82, 186, 197, 205–11, 217–24; Bruegel, "Uncertainty, Pluriactivity, and Neighborhood Exchange," 247–50; Lorena S. Walsh, "Plantation Management," 394–400, and "Rural African-Americans," 334; Chaplin, "Creating a Cotton South," 182–83.

18. On prewar servants, see Chapter 5, above, and Nash and Soderlund, *Freedom by Degrees*, 34. On wartime immigration, see Bailyn, *Voyagers to the West*, 52–57, 98–101, and Adams and Somerville, *Cargoes of Hope and Despair*, 135–36, 140–41, 209–12. On ex-servants at war, see Papenfuse and Stiverson, "General Smallwood's Recruits," 125–26; Conway, *War of American Independence*, 165; Mayer, *Belonging to the Army*, 58, 79; William Miller, "Effects of the American Revolution," 138–41; Richard B. Morris, *Government and Labor*, 291–93; Neimeyer, *America Goes to War*, 47–48; Alexander, "How Maryland Tried to Raise Her Continental Quotas," 192; and Knouff, "'Arduous Service,'" 46–48.

19. Kulikoff, *Tobacco and Slaves*, 66–67, 136–37, 154, 405–7, and *Agrarian Origins*, 232–33, 239; Philip D. Morgan, "Black Society," 84–85, 93–95; Hodges, *Slavery and Freedom*, 47–50; Nash and Soderlund, *Freedom by Degrees*, 32–40; Soderlund, *Quakers and Slavery*, 57–61, 71–75; Rachel N. Klein, *Unification of a Slave State*, 20–22; Kay and Cary, *Slavery in North Carolina*, 228–31; Deyle, "Irony of Liberty," 40, 44, 59; Du Bois, *Suppression of the African Slave Trade*, 40–49; Quarles, *Negro in the American Revolution*, 40–42. For more data, see Chapter 5 at nn. 114–17.

20. Hodges and Brown, "*Pretends to Be Free*," xxx–xxxiii; Mullin, *Flight and Rebellion*, 124–30; Kay and Cary, *Slavery in North Carolina*, chap. 5, and 257–68; Hodges, *Black Loyalist Directory*, 217–23; Gutman, *Black Family in Slavery and Freedom*, 241–43; Frey, *Water from the Rock*, 92–93; Rachel N. Klein, *Unification of a Slave State*, 105–7. Kulikoff, *Agrarian Origins*, 232, and *Tobacco and Slaves*, 418–19, report and justify the numbers.

21. Conway, *War of American Independence*, 30 (good summary of data); Shy, *People Numerous and Armed*, 248–49, 337; Paul H. Smith, "American Loyalists," 266–67, 271–77, and *Loyalists and Redcoats*, 60–61, 77; Wallace Brown, *Good Americans*, 97–98; Neimeyer, *America Goes to War*, 16–17.

22. Selesky, *Demographic Survey*, tabs. 10–65 and 145–69; Royster, *Revolutionary People at War*, 296, 426; Isaac W. Hammond, *State of New Hampshire*, 1:76–77, 107–17, 175–76, 4:3–12, 14. On class and immigrants, see Neimeyer, *America Goes to War*, 15–25; Lender, "Social Structure," 30–31; Papenfuse and Stiverson, "General Smallwood's Recruits," 121–25, 34–43, 47–64; and Delaware Archives Commission, *Delaware Archives*, 1:67–68.

23. Kulikoff, *Agrarian Origins*, 168–71; Isaac W. Hammond, *State of New Hampshire*, 3:756–57; Gross, *Minutemen and Their World*, 142; Showman, Conrad, and Chase, *Papers of Nathanael Greene*, 9:501; Abbot, Twohig, et al., *Papers of George Washington*, 5:209, 251, 333, 349, 444, 667–68, 6:275; Prince et al., *Papers of William Livingston*, 1:78–79, 95, 98, 101.

24. Quarles, *Negro in the American Revolution*, 26–32, chaps. 4–6; Neimeyer, *America Goes to War*, 72–85; Philip S. Foner, *Blacks in the American Revolution*, 67–69 (5,000 estimate); David O. White, *Connecticut's Black Soldiers*, 20–35, 56–63; Frey, *Water from the Rock*, 91–92, 97, 100–101, 116, 123–27, 138, 140, 163–64, 170–71; Crow, *Black Experience*, 69–81; Holton, " 'Rebel against Rebel,' " 161, 174–75, 182; McDonnell, "Other Loyalists," 13–16; James W. St. G. Walker, "Blacks as American Loyalists," 57–59; Hodges, *Slavery and Freedom*, 97–105; Mayer, *Belonging to the Army*, 166–70.

25. Blumenthal, *Women Camp Followers*, 15–54, 60–95; Kopperman, "British High Command"; Potter-MacKinnon, *While the Women Only Wept*, 45–47; Mayer, *Belonging to the Army*, 123, 126, 130–34, 139–45, 154, 156, 158–59; DePauw, "Women in Combat"; Kerber, *Toward an Intellectual History*, 70–74; Gundersen, *To Be Useful to the World*, 164–67; Elizabeth Evans, "Heroines All," 24, 26; Chalou, "Women in the American Revolution," 84–85.

26. Mary Beth Norton, *Liberty's Daughters*, 9–20, 195–99, 202–8, 214–26; Ellet, *Women of the American Revolution*, 2:166; 3:128, 132–33, 266, 276–77, 294; Kerber, *Toward an Intellectual History*, 63–68, 78–79, 84; Mary Beth Norton, "Eighteenth-Century American Women" (underestimates women's farm management); Potter-MacKinnon, *While the Women Only Wept*, 65, 70–71; Schultz, "Daughters of Liberty," 146.

27. Neimeyer, *America Goes to War*, 137–39; Bowman, *Morale*, 68–92; Neagles, *Summer Soldiers*, 33–37; Kulikoff, *Agrarian Origins*, 161; Goldenberg, Nelson, and Fletcher, "Revolutionary Ranks," 185–86, 189; Alexander, "Desertion and Its Punishment," 381–97, esp. 389–90, and "Footnote on Deserters."

28. Dann, *Revolution Remembered*, 170–71, 228–29, 311, 330, 399–400; Kulikoff, *Agrarian Origins*, 159–60; Lender, "Social Structure," 33; Baller, "Kinship and Culture," 298–99; Alexander, "Service by Substitute"; Knouff, " 'Arduous Service,' " 49–50, 59; Jonathan Smith, "How Massachusetts Raised Her Troops," 366–68.

29. Prince et al., *Papers of William Livingston*, 1:59 (first quote), 78–79, 2:223–24; Kwasny, *Washington's Partisan War*, 59–60; Baller, "Kinship and Culture," 298; Hodges, *Slavery and Freedom*, 101; MacMaster, Horst, and Ulle, *Conscience in Crisis*, 279, 292, 317,

329–30 (second quote on 330); Kulikoff, *Agrarian Origins*, 160 (third quote), 170; Tillson, *Gentry and Common Folk*, 88.

30. Isaac W. Hammond, *State of New Hampshire*, 4:77; Dann, *Revolution Remembered*, 262, 264, 267 (first quote), 271, 273, 279, 306, 310; Faragher, *Daniel Boone*, 146–47, 182, 186, 193; Knouff, " 'Arduous Service,' " 58–59 (second quote on 59); Ellet, *Women of the American Revolution*, 2:264–65; Underwood, "Indian and Tory Raids," 206.

31. Simler, "Landless Worker," 192–96; Carp, *To Starve the Army*, 64; Van Tyne, *Loyalists in the American Revolution*, 233–35; Lacey, "Women in the Era of the American Revolution," 539 (quote); Jefferson, *Papers*, 5:464, 587.

32. Atwood, *Hessians*, 198–200; Baxter, *Documentary History of Maine*, 15:267; Blumenthal, *Women Camp Followers*, 31–32; Leamon, *Revolution Downeast*, 143; Metzger, *Prisoner in the American Revolution*, 122, 128; Showman, Conrad, et al., *Papers of Nathanael Greene*, 9:441.

33. Mayer, *Belonging to the Army*, 218–19; John W. Jackson, *Valley Forge*, 81; Pingeon, *Blacks in the Revolutionary Era*, 20 (quote); Showman, Conrad, et al., *Papers of Nathanael Greene*, 2:322, 3:173, 175, 209, 230–31, 245, 297–98, 345, 363–64, 417–19, 450, 463, 475, 4:33, 135, 280, 311, 449, 5:308.

34. This will be documented in my forthcoming book on the Revolution. For examples, see Pinkett, "Maryland as a Source of Food Supplies," 169–70; Bezanson et al., *Prices and Inflation*, 84, 88–89, 134; Bodle, *Vortex of Small Fortunes*, 80–81, 145, 179, 269, 283, 285 (quote); Risch, *Supplying Washington's Army*, 112–13, 122–23, 210–11; Barbara Clark Smith, "Food Rioters"; Bruegel, "Uncertainty, Pluriactivity, and Neighborhood Exchange," 250–55 (New York quote on 251); Countryman, *People in Revolution*, 182–83; Buel, *Dear Liberty*, 159–65, 169; Leamon, *Revolution Downeast*, 136–43; Keith Mason, "Localism, Evangelicalism and Loyalism," 32–36; Showman, Conrad, et al., *Papers of Nathanael Greene*, 7:102 (quote); Ronald Hoffman, " 'Disaffected' in the Revolutionary South," 283–84; Faragher, *Daniel Boone*, 152–54; Crow, "Liberty Men and Loyalists," 137–38; and Tarleton, *History of the Campaigns*, 158, 160, 229, 279, 309.

35. Ryden, *Letters to and from Caesar Rodney*, 358; Jean Butendoff Lee, *Price of Nationhood*, 168–69; Barbara Clark Smith, "Food Rioters," 25; Cometti, "Women in the American Revolution," 329–32; Fennelly, *Connecticut Women*, 54–55; Leamon, *Revolution Downeast*, 148; Isaac W. Hammond, *State of New Hampshire*, 3:558–60, 592–94, 598–99, 620–21, 623–24, 640–41, 661, 667, 704–5, 766, 779–81, 788–89, 800–802, 813, 825, 874.

36. Lewis Cecil Gray, *History of Agriculture*, 2:580, 582; Buel, *Dear Liberty*, 168; Wermuth, " 'To Market, To Market,' " 119–20; Huston, *Logistics of Liberty*, 82–83; Gilbert, "Baltimore's Flour Trade," 41–43; Ryden, *Letters to and from Caesar Rodney*, 379–80; Pinkett, "Maryland as a Source of Food Supplies," 165–66; Emory G. Evans, *Nelson of Yorktown*, 113; Selby, *Revolution in Virginia*, 178; Bezanson et al., *Prices and Inflation*, 30, 38, 85, 91, 128–30, 134.

37. Dann, *Revolution Remembered*, 31, 33 (quote).

38. Kim, "Limits of Politicization," 868–89 (the essential study of Westchester); Hufeland, *Westchester County*, 368 (Thatcher); Dwight, *Travels*, 3:345–46. I have rearranged both quotes to tell the story more sharply.

39. Kim, "Limits of Politicization," 878–81 (quote on 878); Hufeland, *Westchester County*, 94–96, 154–67, 277, 295; Crary, "DeLancey's Cowboys," 17–18.

40. Mayer, *Belonging to the Army*, 25; Dann, *Revolution Remembered*, 80; Hufeland, *Westchester County*, 366.

41. Bodle, "Learning to Live with War," 1–50; Kim, "Limits of Politicization," 868–89; Conway, "To Subdue America" and " 'Greatest Mischief Complain'd Of' "; and astute

comments in Shy, *People Numerous and Armed*, 232–34, 236, are the only studies of military-civilian relations.

42. Examples include Rankin, "Officer out of His Time," 336–41, 346–47; Conway, "To Subdue America," 383, 400; Showman, Conrad, et al., *Papers of Nathanael Greene*, 5:451 (first quote); Fitzpatrick, *Writings of George Washington*, 7:46 (second quote).

43. Instances are scattered throughout the Washington papers; see Abbot, Twohig, and Chase, *Papers of George Washington*, vols. 1 and 5.

44. Ibid., 2:409; Showman, Conrad, et al., *Papers of Nathanael Greene*, 1:243–44.

45. Bowler, *Logistics*, chap. 2; Tiedmann, "Patriots by Default," 42–49; Rivers, "Gray's Observations," 141; Ellet, *Women of the American Revolution*, 3:210; Tarleton, *History of the Campaigns*, 381; Frey, *Water from the Rock*, 146–47; Showman, Conrad, et al., *Papers of Nathanael Greene*, 9:543. See below for Valley Forge.

46. Buel, "Sampson Shorn," 152–53, makes a similar point; see also Risch, *Supplying Washington's Army*, 103–4; Fitzpatrick, *Writings of George Washington*, 10:267 (first quote), 9:275 (second quote).

47. Shy, "Logistical Crisis," 167–70; Bodle, *Vortex of Small Fortunes*, 87–89, 92, 107–8, 144–46, 173–74, 250, 269, 273, 283, 286–88, 377, 389, 392; Thibaut, *This Fatal Crisis*, 35, 40, 47, 60, 104, 114–16, 118, 121–22, 127–29, 145, 162–63, 174, 199, 211, 213, 231, 241; John W. Jackson, *Valley Forge*, 39, 61, 80–81, 85–87, 99, 144–47; Risch, *Supplying Washington's Army*, 103–5, 109, 209–10, 212–15 (first quote on 215); Kashatus, "Mutual Sufferance"; Fitzpatrick, *Writings of George Washington*, 10:342 (second quote).

48. One must piece this story together from the Washington and the Greene papers. See Fitzpatrick, *Writings of George Washington*, vols. 17–19 (quotes on 17:347, 439, 450), and Showman, Conrad, et al., *Papers of Nathanael Greene*, vol 5.

49. Risch, *Supplying Washington's Army*, 230–31, 238–41, 249; Carp, *To Starve the Army*, 97–98, 180–86; Van Dusen, "Connecticut," 18–19, 22; Pinkett, "Maryland as a Source of Food Supplies," 162–63; Rich, "Prudence or Tyranny," 61–65; Selby, *Revolution in Virginia*, 248, 253–54, 259, 261, 284; Hufeland, *Westchester County*, 328; Crow, "Liberty Men and Loyalists," 162–64, 166–67.

50. For examples, see Abbot, Twohig, and Chase, *Papers of George Washington*, vol. 6, and Fitzpatrick, *Writings of George Washington*, vols. 9, 12, 15, 17. For the British, see Conway, "'Greatest Mischief Complain'd Of,'" 385–89, and "To Subdue America," 384–405.

51. For examples, see Jameson, "Subsistence for Middle States Militia," 122–34; Joseph Plumb Martin, *Private Yankee Doodle*, 28, 55, 104–5, 108–9, 174–75, 274–75, 284; Ellet, *Women of the American Revolution*, 2:229–30, 236–36, 298; Isaac W. Hammond, *State of New Hampshire*, vol. 1 and 2:346–417 (marching to units); Pennypacker, *Valley Forge Orderly Book*, 37–38, 42–43, 54, 201, 297–98 (quotes on 37, 297); Thibaut, *This Fatal Crisis*, 104, 128; Fitzpatrick, *Writings of George Washington*, 8:469, 9:199–200, 10:205–7, 322, 11:312, 12:132, 17:331, 460, 18:73; Christopher Moore, *Loyalists*, 72–74, 117–18; James Kirby Martin, "'Most Undisciplined, Profligate Crew,'" 131–32; Kashatus, "Mutual Sufferance," 173–74; and Trussell, *Birthplace of an Army*, 62–63.

52. Atwood, *Hessians*, chap. 8; Conway, "'Greatest Mischief Complain'd Of,'" 376–85; Calhoon, *Loyalists in Revolutionary America*, 361–62; Bowler, *Logistics*, 80; Gruber, *Howe Brothers*, 242; Wallace Brown, *Good Americans*, 118–20; Van Tyne, *Loyalists in the American Revolution*, 247–48; Shy, *People Numerous and Armed*, 225–26, 334; Gerlach, *New Jersey in the American Revolution*, 296–97, 299–300, 393–94.

53. Ronald Hoffman, "'Disaffected' in the Revolutionary South"; Crow, "Liberty Men and Loyalists" and "What Price Loyalism?," 216–17, 224–27; Underwood, "Indian and Tory Raids," 210–13, 218; Delaney, "Outer Banks," 3, 12, 14–15; Fallaw and Stoer, "Old

Dominion under Fire," 446–52; Frey, *Water from the Rock*, 150–52; Jean Butendoff Lee, *Price of Nationhood*, 144–53.

54. McDevitt, *Connecticut Attacked*, 30–31, 37–43, 46–48, 56–57, 59–61; Fennelly, *Connecticut Women*, 55–58; Kwasny, *Washington's Partisan War*, 23, 120–21, 231–32, 240, 284–85, 307–8.

55. Hufeland, *Westchester County*, 201–5, gives some examples.

56. Frank Moore, *Diary of the American Revolution*, 210–16 (quotes); Fitzpatrick, *Writings of George Washington*, 8:341; Amandus Johnson, *Journal and Biography of Nicholas Collin*, 247–48; Calhoon, *Loyalists in Revolutionary America*, 360–63; Shy, *People Numerous and Armed*, 206; Lobdell, "Six Generals Gather Forage," 35, 37, 39–43; Leiby, *Revolutionary War in the Hackensack Valley*, 84–88.

57. Ronald Hoffman, " 'Disaffected' in the Revolutionary South," 290–98; Rachel N. Klein, *Unification of a Slave State*, 95–99, 100–107; Crow, "Liberty Men and Loyalists," 130, 137, 141, and "What Price Loyalism?," 222 (second quote); Gary D. Olson, "Thomas Brown," 8–9, 188–92, 195; Alan S. Brown, "Simpson's Reports," 515–16, 519 (first quote); Cashin, *King's Ranger*, 42, 44, 79–80, 249–52; Buchanan, *Road to Guilford*, 105–6, 124, 238–39, 242, 247–48; Nadelhaft, *Disorders of War*, 56–63, 66; Rivers, "Gray's Observations," 145, 148–49, 152–53; Frey, *Water from the Rock*, 105–6, 129–30, 132–37.

58. Crow, "Liberty Men and Loyalists," 151, 159–61, 168–69, 173; Lindley S. Butler, "David Fanning's Militia"; Massey, "British Expedition," 387–88, 392–94, 399–405; Crittenden, *Commerce of North Carolina*, 152–54; Showman, Conrad, et al., *Papers of Nathanael Greene*, 7:82, 8:199–200 (quote), 9:11–12 (quote); Clyde R. Ferguson, "Carolina and Georgia Patriot and Loyalist Militia," 192.

59. The key essay remains Cometti, "Depredations in Virginia"; see also Keith Mason, "Localism, Evangelicalism, and Loyalism," 23–24, 46–54; Ronald Hoffman, "Popularizing the Revolution," 136–37; Frey, *Water from the Rock*, 158–60, 167–68; and Tarleton, *History of the Campaigns*, 335–36, 344–45, 402–3.

60. Charles Knowles Bolton, *Private Soldier*, 214–15; Underwood, "Indian and Tory Raids," 198–99 (quote on 199); Ekirch, "Whig Authority and Public Order," 111; Crow, "Liberty Men and Loyalists," 140.

61. Ellet, *Women of the American Revolution*, 3:180–82, 218–19, 226, 376; Underwood, "Indian and Tory Raids," 212–13; Mary Beth Norton, *Liberty's Daughters*, 202–4; Kerber, *Women of the Republic*, 46–47; Frank Moore, *Diary of the American Revolution*, 217, 378, 380, 388; Fennelly, *Connecticut Women*, 56–57; Brownmiller, *Against Our Will*, 22, 122–27; Chalou, "Women in the American Revolution," 83–84; Ekirch, "Whig Authority and Public Order," 107; Force, *American Archives*, 3:1188, 1376.

62. Lender and Martin, *Citizen Soldier*, 54, 59.

63. Calloway, *American Revolution in Indian Country* (the most comprehensive work); Graymont, *Iroquois in the American Revolution*; O'Donnell, *Southern Indians*; Horsman, "Image of the Indian," 6–8; Kashatus, "Wyoming Massacre," 117; Fitzpatrick, *Writings of George Washington*, 16:375; Knouff, " 'Arduous Service,' " 59–60, 62–70, and "Common Soldiers in Total War," 1, 7, 9–17, 26–27.

64. Graymont, *Iroquois in the American Revolution*, chaps. 6–9; Calloway, *American Revolution in Indian Country*, chap. 4; Christopher Ward, *War of the Revolution*, vol. 2, chap. 56; Commager and Morris, *Spirit of Seventy-six*, 2:1005–11, 1028–31; Fryer, *King's Men*, 92–99, 102–7, 114–16, 139, 143–46, 159–68.

65. Calloway, *American Revolution in Indian Country*, 14, 51–53, 99, 130, 135–37, 139; Graymont, *Iroquois in the American Revolution*, 206, 213, 215, 218–24; Joseph R. Fischer, *Well-Executed Failure*, 3–4, 58, 82–86, 122–23, 127; Fryer, *King's Men*, 147–59; Potter-MacKinnon, *While the Women Only Wept*, 68–69; Siebert, "Loyalist and Six Nation

Indians," 83, 86; Huston, *Logistics of Liberty*, 231–33; Commager and Morris, *Spirit of Seventy-six*, 2:1016–21.

66. Calloway, *American Revolution in Indian Country*, 39, 49, 124–25 (quote), 167; O'Donnell, *Southern Indians*, 42, 76; Graymont, *Iroquois in the American Revolution*, 232–33; Ousterhout, *State Divided*, 242; Ellet, *Women of the American Revolution*, 2:181–83; Kashatus, "Wyoming Massacre," 115, 117; Bernard W. Sheehan, " 'Famous Hair Buyer General,' " 12–14; Olmstead, *Blackcoats*, 32–33, 54–56, and *Zeisberger*, 330–35; Richard White, *Middle Ground*, 389–90; Fryer, *King's Men*, 145–46, 170; Faragher, *Daniel Boone*, 210; Cashin, " 'But Brothers It Is Our Land,' " 266–67; E. Raymond Evans, "Last Battle," 33; Knouff, " 'Arduous Service,' " 63–68, and "Common Soldiers in Total War," 9–10, 16, 18–21, 28–29.

67. Ellet, *Women of the American Revolution*, 2:257–58; Fryer, *King's Men*, 330. Potter-MacKinnon, *While the Women Only Wept*, 33, makes a similar point.

68. Leiby, *Revolutionary War in the Hackensack Valley*, 43; Van Tyne, *Loyalists in the American Revolution*, 43, 46, 51–52, 57–59, 146–47, 215, 217–19, 239–41; Alfred F. Young, "Women of Boston," 206–7; Stark, *Loyalists of Massachusetts*, 58; Robert Ernst, "Tory-eye View," 377–79, 388–93; Ellen Gibson Wilson, *Loyal Blacks*, 63–67; Mary Beth Norton, *British-Americans*, 32–40; Rosenwaike, *Population History*, 15; Siebert, "Dispersion of the American Tories," 185–87; Potter-MacKinnon, *While the Women Only Wept*, 74–75, 87–93, 114–18.

69. Troxler, "Refuge, Resistance, and Reward," 563–74, 580–89; Troxler, "Origins of the Rawdon," 63, 70; Troxler, "Loyalist Life," 73, 79; McCowen, *British Occupation*, 148–50; Samuel Cole Williams, *Tennessee*, 100–101; Siebert, *Legacy of the American Revolution*, 14–22; Wallace Brown, "American Loyalists in Jamaica," 121–23, 133; Kozy, "Tories Transplanted," 18–33; J. Leitch Wright, *Florida in the American Revolution*, 21–23, 103–5, 126–29, 132–33, 138–39; Starr, *Tories, Dons, and Rebels*, 48–50, 228–31, 234–37; Frey, *Water from the Rock*, 182–88.

70. Baxter, *Documentary History of Maine*, 17:112, 334–35, 350–51, 361, 18:113–14, 221–22, 19:58, 399–400, 455; Alfred F. Young, "Women of Boston," 206–7; Kerber, *Toward an Intellectual History*, 66; Cash, *Medical Men*, 95, 111; Willard, *Letters*, 113, 175–76 (quotes on 176); Gross, *Minutemen and Their World*, 135, 222; Frederic Gregory Mather, *Refugees of 1776*, 166–72, 187–89, 192–93; Fennelly, *Connecticut Women*, 56–57; Tiedmann, *Reluctant Revolutionaries*, 233–34, 240–41; Rosenwaike, *Population History*, 14–15; Mary Beth Norton, *British-Americans*, 32–33; Leiby, *Revolutionary War in the Hackensack Valley*, 49.

71. Barbara Clark Smith,, *After the Revolution*, 119; Hast, *Loyalism in Revolutionary Virginia*, 48, 58–59; Selby, *Revolution in Virginia*, 80–84; Gary D. Olson, "Thomas Brown," 187–88; Searcy, "British Occupation," 187; Frey, *Water from the Rock*, 112–13.

72. Stilwell, *Migration from Vermont*, 90–91; Underwood, "Indian and Tory Raids," 200, 204–5; Gundersen, *To Be Useful to the World*, 156–57; Wendy Martin, "Women and the American Revolution," 323; Boyle, "From Saratoga to Valley Forge," 239 (first quote); Fryer, *King's Men*, 116 (second quote).

73. Gundersen, *To Be Useful to the World*, 156–57; Hinderaker, *Elusive Empires*, 223–24; Samuel Cole Williams, *Tennessee*, 43–44; Faragher, *Daniel Boone*, 152; Buchanan, *Road to Guilford*, 122–24, 343, 348–49; Ellet, *Women of the American Revolution*, 1:238–40, 280–81; Selby, *Revolution in Virginia*, 202 (quote); Dann, *Revolution Remembered*, 57, 307, 360–61; Showman, Conrad, et al., *Papers of Nathanael Greene*, 7:252.

74. On the British, see Bowler, *Logistics*, 88–89; Van Tyne, *Loyalists in the American Revolution*, 248–49; Tiedmann, "Patriots by Default," 61–62; McCowen, *British Occupation*, 61; and Tarleton, *History of the Campaigns*, 71–72, 89, 186–89. On Patriots, see Kerber, *Women of the Republic*, 50–51, chap. 4; Ousterhout, *State Divided*, 173; Gun-

dersen, *To Be Useful to the World*, 160–61; Potter-MacKinnon, *While the Women Only Wept*, 52–54, 61–62, 84–87; and Janice Potter, "Patriarchy and Paternalism," 15.

75. Van Tyne, *Loyalists in the American Revolution*, 269–81; Stark, *Loyalists of Massachusetts*, 60, 136–44; Christopher Moore, *Loyalists*, 116; Jonathan Smith, "Toryism in Worcester County," 21–23, 26–27; Hufeland, *Westchester County*, 445; Fryer, *King's Men*, 123; John D. McBride, "Virginia War Effort," 238–41; Harrell, "North Carolina Loyalists," 581–90; Nadelhaft, *Disorders of War*, 80–85; Calhoon, *Loyalist Perception*, 166; Weir, "*Last of the American Freemen*," 141–42.

76. Lower, *Colony to Nation*, 119–23; Brown and Senior, *Victorious in Defeat*, 28–30; Wilfred Campbell, *Report on Manuscript Lists*, 8, 15–16, 20; Rawlyk, "Loyalist Military Settlement," 100; Robert S. Allen, "Loyalist Military Settlement," 92; Capon et al., *Atlas of Early American History*, 59, 127–28; Siebert, "Dispersion of the American Tories," 188–89, 194–96; Mary Beth Norton, *British-Americans*, 8–9; Robert R. Palmer, *Age of the Democratic Revolution*, 1:188–89 (French comparison). On social composition, see Condon, *Envy of the American States*, 85; Potter-MacKinnon, *While the Women Only Wept*, 12–25; Janice Potter, "Patriarchy and Paternalism," 4–6; and Wallace Brown, *King's Friends*, 81–83, 87, 260–67, and "American Farmer."

77. On farm destruction, see Chastellux, *Travels in North America*, 1:130; Castiglioni, *Viaggio*, 261; Tiedmann, "Patriots by Default," 39; Frey, *Water from the Rock*, 206–8; and Nadelhaft, *Disorders of War*, 64. On economy, see Davis and Engerman, "Economy of British North America," 21; McCusker and Menard, *Economy of British America*, 63, 369–71; Bjork, "Weaning of the American Economy"; Merrill Jensen, *New Nation*, 187–93; Lewis Cecil Gray, *History of Agriculture*, 2:596–99; and Harrell, *Loyalism in Virginia*, 118–21. I follow McCusker and Menard's ideas, rather than the rosier views of Jensen and Bjork. On income, see McCusker and Menard, *Economy of British America*, 373–74, and Alice Hanson Jones, *Wealth of a Nation to Be*, 81, inflated to 1994 dollars (rounded to the nearest hundred) using U.S. Bureau of the Census, *Statistical Abstract*, 492.

78. U.S. Bureau of the Census, *Historical Statistics*, 2:1189, 1192; Seybert, *Statistical Annals*, 100–103; Frey, *Water from the Rock*, 208–9; Nadelhaft, *Disorders of War*, 134, 151–53, 254–55; Lewis Cecil Gray, *History of Agriculture*, 2:606; Chaplin, "Creating a Cotton South," 187–96; Bruchey, *Cotton and the Growth of the American Economy*, 14–19; Kulikoff, *Agrarian Origins*, 233–40.

79. Lorena S. Walsh, "Plantation Management," 395–401; Lorena S. Walsh, "Land, Landlord, and Leaseholder," 385, 390; Lorena S. Walsh, *From Calabar to Carter's Grove*, 130; Sarah S. Hughes, "Slaves for Hire"; Bjork, *Stagnation and Growth*, 92; Coxe, *View of the United States of America*, 414, 417; Gilbert, "Baltimore's Flour Trade," 172; Lewis Cecil Gray, *History of Agriculture*, 2:605; Jacob M. Price, *France and the Chesapeake*, 2:730–32; Frey, *Water from the Rock*, 219–20; "Tobacco Exports."

80. Bjork, *Stagnation and Growth*, 54–58, 83; Coxe, *View of the United States of America*, 414, 417.

81. Coatsworth, "American Trade," 245–51; Shepherd and Walton, "Economic Change," 406–7; Carrington, "American Revolution," 843–44; Oakes, "Ticklish Business," 210–11. The data do not allow direct comparisons, for pre- and postwar data on bread and grain are listed differently.

82. Wacker and Clemens, *Land Use*, 1–30, 204–14; Christopher Clark, *Roots of Rural Capitalism*, 29–31, 34–38, 66–71, 76–80.

83. Bruegel, "Uncertainty, Pluriactivity, and Neighborhood Exchange," 272. See Chapter 5, above, for details on the origins of this system.

84. There is a vast literature on these issues. For a range of views, see Kulikoff, *Agrarian Origins*, chap. 4; Gordon Wood, *Radicalism of the American Revolution*, chap. 8; Rothen-

berg, *From Market-Places to a Market Economy*, chaps. 4–8; Christopher Clark, *Roots of Rural Capitalism*, chaps. 2–3; Simler, "Landless Worker," 175–77, 190–97; Simler and Clemens, " 'Best Poor Man's Country,' " 239–43; and Nash and Soderlund, *Freedom by Degrees*, 182–93.

85. Richard White, *Middle Ground*, chaps. 10–11, is the most subtle analysis. Other fine works include Horsman, "American Indian Policy" and "Image of the Indian," 7–10; Merrell, "Declarations of Independence," 197–210, esp. 206; Graymont, *Iroquois in the American Revolution* chap. 10; and Kenneth Coleman, "Federal Indian Relations," 436–58.

86. Gemery, "European Emigration," 286, 304–13; Grabbe, "European Immigration," 190–201; James Kelly, "Resumption of Emigration," 66–76; Grubb, "Immigrant Servant Labor," 262–72, and "German Immigration," 431–34; Fogleman, "From Slaves, Convicts, and Servants to Free Passengers," 44, 60–65, 73–75; Darlington, "Peopling the Post-Revolutionary New York Frontier," 371–73. The lower Grabbe estimate comes from shipping records; the higher Gemery estimate is a net migration figure computed from census totals.

87. Onuf, "Settlers, Settlements, and New States," 179; Ryan, "Landholding, Opportunity, and Mobility," 577–91 (quote on 588); Cox, " 'Touch of Kentucky News' "; Marks, "Rage for Kentucky," 112, 116, 121.

88. Ellis, "Yankee Invasion," 6–8; Darlington, "Peopling the Post-Revolutionary New York Frontier," 348–49, 357, 370–71; Hinderaker, *Elusive Empires*, 242, 245; William Wyckoff, *Developer's Frontier*, 106–7; Mathews, *Expansion of New England*, 141, 154–55, 157, 160–61; Eslinger, "Migration and Kinship"; Thomas W. Spalding, "Maryland Catholic Diaspora," 163–69; Marks, "Rage for Kentucky," 112–12, 119–26; Ranck, "Travelling Church"; Peskin, "Restless Generation," 316; Reid, "Church Membership," 399–400, 406–9.

89. Mathews, *Expansion of New England*, 129–37, 142, 146, 151, 157, 159, 174–79; Stilwell, *Migration from Vermont*, 91, 95–103, 116–21; Margaret Walsh, "Women's Place," 241; Darlington, "Peopling the Post-Revolutionary New York Frontier," esp. 349–55, 358; Gross, *Minutemen and Their World*, 177–80, 233; Ellis, "Yankee Invasion," 4–9; Whitaker, *Mississippi Question*, 8; Richard White, *Middle Ground*, 418; Samuel Cole Williams, *Tennessee*, 59–60; Marks, "Rage for Kentucky," 121; Cox, " 'Touch of Kentucky News,' " 219; Peskin, "Restless Generation," 311–27; Sellers, "Common Soldier," 158–60; Martinac, " 'Unsettled Disposition,' " chaps. 3–4.

90. Kulikoff, *Agrarian Origins*, 239; Greene and Harrington, *American Population*, 7–8, 75–87, 103–4, 117, 182, 192–4; U.S. Bureau of the Census, *Returns of the Whole Numbers*; Blodget, *Economica*, 60; Christopher Clark, *Roots of Rural Capitalism*, 61–63.

91. Wallace Brown, *Good Americans*, 177, and *King's Friends*, 136, 160, 168, 214; Troxler, "Origins of the Rawdon," 70–71; Donovan, " 'Taking Leave,' " 137; Zeichner, "Loyalist Problem," 295–302; Riccards, "Patriots and Plunderers," 19, 24–25; Lambert, "Confiscation of Loyalist Property," 92–94; Nadelhaft, *Disorders of War*, 78; Coker, "Case of James Nassau Colleton"; Peter M. Mitchell, "Loyalist Property," 209–20.

92. Stiverson, *Poverty in a Land of Plenty*, 112–17, 121–34.

93. Nettels, *Emergence of a National Economy*, 142; Humphrey, "Agrarian Rioting in Albany County," 238–43; Crary, "Forfeited Loyalist Lands"; Lynd, *Class Conflict*, 53–55, 59. Reubens, "Pre-emptive Rights," 433–55, paints a more optimistic picture than the evidence warrants.

94. Lutz, "Land Grants," 221–31, and "State's Concern," 317–18; Jerry A. O'Callaghan, "War Veteran," 164–65, 168; Lee Soltow, *Distribution of Wealth*, 155–60; Shy, *People Numerous and Armed*, 252; Hitz, "Georgia Bounty Land Grants," 338, 344; Ellen F.

Wilson, "Gaining Title to the Land," 69; Frey, *Water from the Rock*, 213–14; Vivian, "Military Land Bounties," 238; Richter, "Onas and the Long Knives," 25–26.

95. Nettels, *Emergence of a National Economy*, 143–46; Lutz, "Land Grants," 229–30; Tatter, "Land Policy," 178–82; Lewis Cecil Gray, *History of Agriculture*, 2:625–26; Hitz, "Georgia Bounty Land Grants," 339, 341, 344–45; Gates, *History of Public Land Law Development*, 67; Aron, *How the West Was Lost*, 75–76; Picht, "American Squatter," 73.

96. Nettels, *Emergence of a National Economy*, chaps. 7–8, remains the best analysis. Case studies include Alan Taylor, *Liberty Men and Great Proprietors*; William Wyckoff, *Developer's Frontier*; R. Eugene Harper, *Transformation of Western Pennsylvania*, chap. 4; Aron, *How the West Was Lost*, chap. 3; and Abernethy, *South in the New Nation*, chaps. 4, 6.

97. Wilma A. Dunaway, *First American Frontier*, 61–65; Lee Soltow, "Land Inequality," 382; Rice, *Allegheny Frontier*, 136–49.

98. There is first-rate literature on these speculators and developers, beginning with Hedrick, *History of Agriculture*, 45–60; Chazanof, *Joseph Ellicott*, chaps. 1–5; William Wyckoff, *Developer's Frontier*, chaps. 1–4; Siles, "Pioneering in Genessee Country," 36–65; Alfred F. Young, *Democratic Republicans*, 232–43, 252, 265, 498–505; and Schein, "Unofficial Proprietors" and "Farming the Frontier," 21–23.

99. There is a vast literature on the Northwest Ordinance and the early federal land system. Gates, *History of Public Land Law Development*, chaps. 4–5, 7, provides the best summary. Other important works are Hibbard, *History of Public Land Policies*, chaps. 1–5, and on Federalist policy, Onuf, *Statehood and Union*, chaps. 1–2, and "Settlers, Settlement, and New States," 180–87, and Cayton, *Frontier Republic*, chap. 2.

100. Hibbard, *History of Public Land Policies*, 45–55; Tagney, "Essex County Looks West," 92–101; Shannon, "Ohio Company"; Cayton, *Frontier Republic*, 18–20, 25, 27; Schein, "Farming the Frontier," 8 (military reserves).

101. Picht, "American Squatter," 72–74; Aron, *How the West Was Lost* (see 71, 80, 97, 151–52, 204–5, for tenancy and squatting). For landholding patterns, see Lee Soltow, "Inequality amidst Abundance," 135–36, and "Kentucky Wealth," 620–22; Wilma A. Dunaway, *First American Frontier*, 68–70; and Soltow and Keller, "Tenancy and Asset Holding."

AFTERWORD

1. The ideas in this afterword interpret the findings of this book and introduce ideas that will be developed in my forthcoming work, *Making of the American Yeoman Class*. But see also Kulikoff, *Agrarian Origins*, esp. chaps. 1–2, and Bushman, "Massachusetts Farmers and the Revolution."

2. Kulikoff, *Making of the American Yeoman Class*, chap. 5; Richard Maxwell Brown, *South Carolina Regulators*; McConville, *Those Daring Disturbers of the Public Peace*.

3. Slaughter, *Whiskey Rebellion*; Gross, *In Debt to Shays*.

4. Lee and Lalli, "Population," 26–32; George Rogers Taylor, "Comment," 39–46; Thomas Weiss, "U.S. Labor Force Estimates," 37, 50.

5. Soltow and Keller, "Rural Pennsylvania in 1800," 35–38.

6. Merrill Peterson, *Jefferson*, 512, 1138; Kulikoff, *Agrarian Origins*, chaps. 3, 7.

bibliography

ABBREVIATIONS

AgH	*Agricultural History*
EcHR	*Economic History Review*, 2d ser.
JEH	*Journal of Economic History*
JIH	*Journal of Interdisciplinary History*
MdHM	*Maryland Historical Magazine*
NEQ	*New England Quarterly*
P&P	*Past and Present*
VMHB	*Virginia Magazine of History and Biography*
WMQ	*William and Mary Quarterly*, 3d ser.

Abbot, W. W. *The Royal Governors of Georgia, 1754–1775*. Chapel Hill, N.C., 1959.

Abbot, W. W., Dorothy Twohig, and Philander D. Chase, eds. *The Papers of George Washington. Revolutionary War Series*. 7 vols. to date. Charlottesville, Va., 1985–97.

Abbott, Edith. "A Study of the Early History of Child Labor in America." *American Journal of Sociology* 14 (1908): 15–37.

Abbott, Mary. *Family Ties: English Families, 1540–1920*. London, 1993.

Abel, Wilhelm. *Agricultural Fluctuations in Europe from the Thirteenth to the Twentieth Centuries*. Translated by Olive Ordish. New York, 1980.

Abernethy, Thomas Perkins. *The South in the New Nation, 1789–1819*. Baton Rouge, La., 1961.

——. *Western Lands and the American Revolution*. New York, 1937.

Ackerman, Robert K. *South Carolina Colonial Land Policies*. Columbia, S.C., 1977.

Adam, Margaret I. "Eighteenth-Century Highland Landlords and the Poverty Problem." *Scottish Historical Review* 19 (1921–22): 1–20, 161–79.

——. "The Highland Emigration of 1770." *Scottish Historical Review* 16 (1919): 280–93.

Adams, Ian H. "The Agricultural Revolution in Scotland: Contributions to the Debate." *Area* 10 (1978): 198–203.

——. "Economic Process and the Scottish Land Surveyor." *Imago Mundi* 27 (1975): 13–18.

Adams, Ian H., and Meredyth Somerville. *Cargoes of Hope and Despair: Scottish Emigration to North America, 1603–1803*. Edinburgh, 1993.

Adams, John H., and Alice Bee Kasakoff. "Migration and the Family in Colonial New England: The View from Genealogies." *Journal of Family History* 9 (1984): 24–43.

——. "Migration at Marriage in Colonial New England: A Comparison of Rates

Derived from Genealogies with Rates from Vital Records." In *Genealogical Demography*, edited by Bennett Dyke and Warren Morrill, 115–38. New York, 1980.
——. "Wealth and Migration in Massachusetts and Maine, 1771–1798." *JEH* 45 (1985): 363–68.
Adams, Thomas Boylston. "Bad News from Virginia." *VMHB* 74 (1966): 131–40.
Adelmann, Gerhard. "Structural Change in the Rhenish Linen and Cotton Trades at the Outset of Industrialization." In *Essays in European Economic History, 1789–1914*, edited by F. Crouzet, W. H. Chaloner, and W. M. Stern, 82–97. New York, 1969.
Agnew, Jean Christophe. "The Threshold of Exchange: Speculations upon the Market." *Radical History Review*, no. 21 (1979): 99–118.
Akagi, Roy Hidemichi. *The Town Proprietors of the New England Colonies: A Study of Their Development, Organization, Activities, and Controversies, 1620–1770*. Philadelphia, 1924.
Albion, Robert Greenhalgh. *Forests and Sea Power: The Timber Problem of the Royal Navy, 1652–1862*. Cambridge, Mass., 1926.
Albion, Robert Greenhalgh, and Jennie Barnes Pope. *Sea Lanes in Wartime: The American Experience, 1775–1942*. New York, 1942.
Alexander, Arthur J. "Desertion and Its Punishment in Revolutionary Virginia." *WMQ* 3 (1946): 383–97.
——. "A Footnote on Deserters from the Virginia Forces during the American Revolution." *VMHB* 55 (1947): 137–46.
——. "How Maryland Tried to Raise Her Continental Quotas." *MdHM* 42 (1947): 184–96.
——. "Service by Substitute in the Militia of Northampton and Lancaster Counties (Pennsylvania) during the Revolution." *Military Affairs* 9 (1945): 278–82.
Allen, David Grayson. " 'Both Englands.' " In Hall and Allen, *Seventeenth-Century New England*, 55–82.
——. *In English Ways: The Movement of Societies and the Transferal of English Local Law and Custom to Massachusetts Bay in the Seventeenth Century*. Chapel Hill, N.C., 1981.
——. "The Matrix of Motivation." *NEQ* 59 (1986): 408–18.
——. " '*Vaccuum Domicilium*: The Social and Cultural Landscape of Seventeenth-Century New England.' " In Fairbanks and Trent, *New England Begins*, 1:1–10.
Allen, Robert C. *Enclosure and the Yeoman*. Oxford, 1992.
Allen, Robert S. "Loyalist Military Settlement in Quebec." In Robert S. Allen, *Loyal Americans*, 91–96.
——, ed. *The Loyal Americans: The Military Role of the Loyalist Provincial Corps and Their Settlement in British North America, 1775–1784*. Ottawa, Canada, 1983.
Alston, Lee J., and Morton Owen Schapiro. "Inheritance Laws across Colonies: Causes and Consequences." *JEH* 44 (1984): 277–88.
Altman, Ida, and James Horn, eds. *"To Make America": European Emigration in the Early Modern Period*. Berkeley, Calif., 1991.
American Husbandry. 1775. Edited by Harry J. Carman. New York, 1939.
Ames, Azel. *The May-flower and Her Log Chiefly from Original Sources*. Boston, 1901.
Amussen, Susan Dwyer. *An Ordered Society: Gender and Class in Early Modern England*. Oxford, 1988.
Anderson, Fred. *A People's Army: Massachusetts Soldiers and Society in the Seven Years' War*. Chapel Hill, N.C., 1984.
——. "A People's Army: Provincial Military Service in Massachusetts during the Seven Years' War." *WMQ* 40 (1983): 499–527.
Anderson, Terry L. "Economic Growth in Colonial New England: 'Statistical Renaissance.' " *JEH* 39 (1979): 243–58.

Anderson, Virginia DeJohn. "King Philip's Herds: Indians, Colonists, and the Problem of Livestock in Early New England." *WMQ* 51 (1994): 601–24.
——. "Migrants and Motives: Religion and the Settlement of New England, 1630–1640." *NEQ* 58 (1985): 339–83.
——. "Migration, Kinship, and the Integration of Colonial New England Society: Three Generations of the Danforth Family." In Taylor and Crandall, *Generations and Change*, 269–90.
——. *New England's Generation: The Great Migration and the Formation of Society and Culture in the Seventeenth Century*. Cambridge, 1991.
——. "The Origins of New England Culture." *WMQ* 48 (1991): 231–37.
——. "Religion, the Common Thread." *NEQ* 59 (1986): 418–24.
Anderson-Green, Paula Hathaway. "The New River Frontier Settlement on the Virginia–North Carolina Border, 1760–1820." *VMHB* 86 (1978): 413–31.
Andrews, Charles M. *The Colonial Period in American History*. 4 vols. New Haven, Conn., 1934–38.
Andrews, John Harwood. "Land and People, c. 1685." In *A New History of Ireland*, edited by Theodore William Moody, F. X. Martin, and F. J. Byrne. Vol. 3, *Early Modern Ireland, 1534–1691*, 454–508. Oxford, 1976.
——. "Land and People, c. 1780." In Moody and Vaughan, *New History of Ireland*, 236–64.
Andrews, Kenneth R. *Trade, Plunder, and Settlement: Maritime Enterprise and the Genesis of the British Empire, 1480–1630*. Cambridge, 1984.
Andrews, Kenneth R., N. P. Canny, and P. E. H. Hair, eds. *The Westward Enterprise: English Activities in Ireland, the Atlantic, and America, 1480–1650*. Detroit, 1979.
Appleby, Andrew B. "Epidemics and Famine in the Little Ice Age." *JIH* 10 (1980): 643–64.
Appleby, Joyce Oldham. *Economic Thought and Ideology in Seventeenth-Century England*. Princeton, N.J., 1978.
Arber, Edward. *The Story of the Pilgrim Fathers, 1606–1623 A.D.; as told by Themselves, their Friends, and their Enemies*. London, 1897.
Archdeacon, Thomas J. *Becoming American: An Ethnic History*. New York, 1983.
Archer, Richard. "New England Mosaic: A Demographic Analysis for the Seventeenth Century." *WMQ* 47 (1990): 477–502.
Armour, David Arthur. *The Merchants of Albany, New York, 1686–1760*. New York, 1986.
Armstrong, W. A. "Labour I: Rural Population Growth, Systems of Employment, and Incomes." In Mingay, *Agrarian History*, 641–728.
Aron, Stephen. *How the West Was Lost: The Transformation of Kentucky from Daniel Boone to Henry Clay*. Baltimore, 1996.
Aston, T. H., and C. H. E. Philpin, eds. *The Brenner Debate: Agrarian Class Structure and Economic Development in Pre-Industrial Europe*. Cambridge, 1985.
Aston, T. H., P. R. Cross, Christopher Dyer, and Joan Thirsk, eds. *Social Relations and Ideas: Essays in Honour of R. H. Hilton*. Cambridge, 1983.
Atwood, Rodney. *The Hessians: Mercenaries from Hessen-Kassel in the American Revolution*. Cambridge, 1980.
Ault, Warren O. *Open-Field Husbandry and the Village Community: A Study in Agrarian By-Laws in Medieval England*. *Transactions of the American Philosophical Society*, n.s., 55 (1965).
Auwers, Linda. "Fathers, Sons, and Wealth in Colonial Windsor, Connecticut." *Journal of Family History* 3 (1978): 136–49.
——. "Reading the Marks of the Past: Exploring Female Literacy in Colonial Windsor, Connecticut." *Historical Methods* 13 (1980): 204–14.

Axtell, James. *After Columbus: Essays in the Ethnohistory of Colonial North America*. New York, 1988.
——. *Beyond 1492: Encounters in Colonial North America*. New York, 1992.
——. *The European and the Indian: Essays in the Ethnohistory of Colonial North America*. New York, 1981.
Aylmer, G. E. "The Meaning and Definition of 'Property' in Seventeenth-Century England." *P&P*, no. 86 (1980): 87–97.
Ayres, S. Edward. "Albemarle County, 1744–1770: An Economic, Political, and Social Analysis." *Magazine of Albemarle County History* 25 (1966–67): 37–72.
Bachman, Van Cleaf. *Peltries or Plantations: The Economic Policies of the Dutch West India Company in New Netherland, 1623–1639*. Baltimore, 1969.
Bailey, Kenneth P. *The Ohio Company of Virginia and the Westward Movement, 1748–1792: A Chapter in the History of the Colonial Frontier*. Glendale, Calif., 1939.
Bailey, Mark. "Peasant Welfare in England, 1290–1348." *EcHR* 51 (1998): 223–51.
Bailyn, Bernard. "The Challenge of Modern Historiography." *American Historical Review* 87 (1982): 1–24.
——. "Communications and Trade: The Atlantic in the Seventeenth Century." *JEH* 13 (1953): 378–87.
——. *The New England Merchants in the Seventeenth Century*. Cambridge, Mass., 1955.
——. *The Peopling of British North America: An Introduction*. New York, 1986.
——. "Politics and Social Structure in Virginia." In James Morton Smith, *Seventeenth-Century America*, 90–115.
——. *Voyagers to the West: A Passage in the Peopling of America on the Eve of the Revolution*. New York, 1986.
Bailyn, Bernard, and Philip D. Morgan. Introduction to Bailyn and Morgan, *Strangers within the Realm*, 1–31.
——, eds. *Strangers within the Realm: Cultural Margins of the First British Empire*. Chapel Hill, N.C., 1991.
Baine, Rodney M. "James Oglethorpe and the Early Promotional Literature for Georgia." *WMQ* 45 (1988): 100–106.
——. "New Perspectives on Debtors in Colonial Georgia." *Georgia Historical Quarterly* 77 (1993): 2–19.
Baker, Andrew H., and Holly V. Izard. "New England Farmers and the Marketplace, 1780–1865: A Case Study." *AgH* 65 (1991): 29–52.
Baker, Brenda J. "Pilgrim's Progress and Praying Indians: The Biocultural Consequences of Contact in Southern New England." In Larsen and Milner, *In the Wake of Contact*, 35–45.
Baker, Emerson W. " 'A Scratch with a Bear's Paw': Anglo-Indian Land Deeds in Early Maine." *Ethnohistory* 36 (1989): 235–56.
Baker, J. N. L. "The Climate of England in the Seventeenth Century." *Quarterly Journal of the Royal Meteorological Society* 58 (1932): 421–36.
——. "England in the Seventeenth Century." In *An Historical Geography of England before A.D. 1800*, edited by H. C. Darby, 387–443. Cambridge, 1936.
Ball, Duane E. "Dynamics of Population and Wealth in Eighteenth-Century Chester County, Pennsylvania." *JIH* 6 (1976): 621–44.
Ball, Duane E., and Gary M. Walton. "Agricultural Productivity Change in Eighteenth-Century Pennsylvania." *JEH* 36 (1976): 102–17.
Baller, Bill. "Kinship and Culture in the Mobilization of Colonial Massachusetts." *Historian* 57 (1995): 291–302.
Banks, Charles Edward. "English Sources of Emigration to the New England Colonies

in the Seventeenth Century." *Massachusetts Historical Society, Proceedings* 60 (1927): 366–73.
——. "Religious 'Persecution' as a Factor in Emigration to New England, 1630–1640." *Massachusetts Historical Society, Proceedings* 63 (1929–30): 136–51.
——. *The Winthrop Fleet of 1630: An Account of the Vessels, the Voyage, and Passengers and Their English Homes from Original Authorities*. Boston, 1930.
Barbour, Philip L. "The Earliest Reconnaissance of the Chesapeake Bay Area." *VMHB* 79 (1971): 280–302; 80 (1972): 21–51.
Bark, L. Dean. "History of American Droughts." In *North American Droughts*, edited by Norman J. Rosenberg, 9–23. AAAS Selected Symposia Series, no. 15. Boulder, Colo., 1978.
Barker, Alex W. "Powhatan's Pursestrings: On the Meaning of Surplus in a Seventeenth-Century Algonkian Chiefdom." In *Lords of the Southeast: Social Inequality and the Native Elites of Southeastern North America*, edited by Alex W. Barker and Timothy R. Pauketat, 63–79. American Anthropological Association, Archaeological Papers, no. 3 (1992).
Barley, M. W. "Rural Housing in England." In Thirsk, *Agrarian History*, 696–766.
Barnes, Thomas Garden. "Deputies Not Principals, Lieutenants Not Captains: The Institutional Failure of Lieutenancy in the 1620s." In Fissel, *War and Government*, 58–86.
Barnes, Viola Florence. "Land Tenure in English Colonial Charters of the Seventeenth Century." In *Essays in Colonial History*, 4–40.
Barnett, Todd H. "Tobacco, Planters, Tenants, and Slaves on Maryland's Piedmont: A Portrait of Montgomery County in 1783." Paper, McNeil Center for Early American Studies, September 1992.
Baron, W. R. "Historical Climate Records from the Northeastern United States, 1640 to 1900." In *Climate since A.D. 1500*, edited by Raymond S. Bailey, 74–91. Rev. ed. London, 1995.
Bassett, John Spencer. "The Relation between the Virginia Planter and the London Merchant." *American Historical Association, Annual Report for 1901* 1 (1902): 553–75.
Battis, Emery. *Saints and Sectaries: Anne Hutchinson and the Antinomian Controversy in the Massachusetts Bay Colony*. Chapel Hill, N.C., 1962.
Baugh, Daniel A. "The Politics of British Naval Failure, 1775–1777." *American Neptune* 52 (1992): 221–46.
Baxter, James Phinney, ed. *Documentary History of the State of Maine*. 24 vols. Portland, Maine, 1896–1916.
Beales, Ross W., Jr. "Nursing and Weaning in an Eighteenth-Century New England Household." In Benes, *Families and Children*, 48–63.
——. "The Reverend Ebenezer Parkman's Farm Workers, Westborough, Massachusetts, 1726–1782." *American Antiquarian Society, Proceedings* 99 (1989): 121–49.
——. "Slavish and Other Female Work in the Parkman Household, Westborough, Massachusetts, 1724–1782." In Benes, *House and Home*, 48–57.
——. "Studying Literacy at the Community Level: A Research Note." *JIH* 9 (1978): 93–102.
Beardslee, G. William. "An Ostego Frontier Experience, 1770–1795." *New York History* 79 (1998): 233–54.
Beaudry, Mary C., ed. *Documentary Archaeology in the New World*. Cambridge, 1988.
Becker, Laura A. "Diversity and Its Significance in an Eighteenth-Century Pennsylvania Town." In *Friends and Neighbors: Group Life in America's First Plural Society*, edited by Michael Zuckerman, 196–221. Philadelphia, 1982.

Becker, Marshall. "The Okehocking Band of Lenape: Cultural Continuities and Accommodation in Southeastern Pennsylvania." In Porter, *Strategies for Survival*, 43–84.
Beckett, J. V. "The Disappearance of the Cottager and the Squatter from the English Countryside: The Hammonds Revisited." In *Land, Labour, and Agriculture, 1700–1920: Essays for Gordon Mingay*, edited by B. A. Holderness and Michael Turner, 49–67. London, 1991.
Beeman, Richard R. *The Evolution of the Southern Backcountry: A Case Study of Lunenburg County, Virginia, 1746–1832*. Philadelphia, 1984.
Behrens, Kathryn L. *Paper Money in Maryland, 1727–1789*. Ser. 41, no. 1, *Johns Hopkins University Studies in Historical and Political Science*. Baltimore, 1923.
Beier, A. L. *Masterless Men: The Vagrancy Problem in England, 1560–1640*. London, 1985.
——. "Poverty and Progress in Early Modern England." In *The First Modern Society: Essays in English History in Honour of Lawrence Stone*, edited by A. L. Beier, David Cannadine, and James M. Rosenheim, 201–39. Cambridge, 1989.
Bellesiles, Michael A. *Revolutionary Outlaws: Ethan Allen and the Struggles for Independence on the Early American Frontier*. Charlottesville, Va., 1993.
Ben-Amos, Ilana Krausman. *Adolescence and Youth in Early Modern England*. New Haven, Conn., 1994.
Bender, Thomas. "Wholes and Parts: The Need for Synthesis in American History." *Journal of American History* 73 (1986): 120–36.
Benes, Peter, ed. *Algonkians of New England: Past and Present*. Vol. 16 of *The Dublin Seminar for New England Folklife*. Boston, 1993.
——. *Early American Probate Inventories*. Vol. 12 of *The Dublin Seminar for New England Folklife*. Boston, 1989.
——. *Families and Children*. Vol. 10 of *The Dublin Seminar for New England Folklife*. Boston, 1987.
——. *Foodways in the Northeast*. Vol. 7 of *The Dublin Seminar for New England Folklife*. Boston, 1984.
——. *House and Home*. Vol. 13 of *The Dublin Seminar for New England Folklife*. Boston, 1990.
——. *New England Prospect: Maps, Place Names, and the Historical Landscape*. Vol. 5 of *The Dublin Seminar for New England Folklife*. Boston, 1980.
——. *Plants and People*. Vol. 20 of *The Dublin Seminar for New England Folklife*. Boston, 1996.
Bennett, Judith M. *Ale, Beer, and Brewsters in England: Women's Work in a Changing World, 1300–1600*. New York, 1996.
——. "Medieval Peasant Marriage: An Examination of Marriage License Fines in Liber Gersumarum." In Raftis, *Pathways to Medieval Peasants*, 193–246.
——. *Women in the Medieval English Countryside: Gender and Household in Brigstock before the Plague*. New York, 1987.
Bennett, Merrill K. "The Food Economy of the New England Indians, 1605–75." *Journal of Political Economy* 63 (1955): 369–97.
Berdahl, Robert M. "Christian Garve on the German Peasantry." *Peasant Studies* 8 (1979): 86–102.
Beresford, Maurice. "A Review of Historical Research (to 1968)." In *Deserted Medieval Villages*, edited by Maurice Beresford and John G. Hurst, 3–75. London, 1971.
Berger, Ronald M. "The Development of Retail Trade in Provincial England, ca. 1550–1700." *JEH* 40 (1980): 123–28.

Bergstrom, Peter V. *Markets and Merchants: Economic Diversification in Colonial Virginia, 1700–1775*. New York, 1985.

Bergstrom, Peter V., and Kevin P. Kelly. "The Country around the Towns: Society, Demography, and the Rural Economy of York County, Virginia, 1695–1705." Manuscript draft in possession of author, 1984.

Berkin, Carol, and Leslie Horowitz, eds. *Women's Voices, Women's Lives: Documents in Early American History*. Boston, 1998.

Berkner, Lutz K. "Inheritance, Land Tenure, and Peasant Family Structure: A German Regional Example." In Goody, Thirsk, and Thompson, *Family and Inheritance*, 71–95.

——. "Peasant Household Organization and Demographic Change in Lower Saxony, 1689–1766." In *Population Patterns in the Past*, edited by Ronald Demos Lee, 53–69. New York, 1977.

——. "The Stem Family and the Developmental Cycle of the Peasant Household: An Eighteenth-Century Austrian Example." *American Historical Review* 77 (1972): 398–418.

Berkner, Lutz K., and Franklin F. Mendels. "Inheritance Systems, Family Structure, and Demographic Patterns in Western Europe, 1700–1900." In Tilly, *Historical Studies of Changing Fertility*, 209–23.

Berlin, Ira. *Many Thousands Gone: The First Two Centuries of Slavery in North America*. Cambridge, Mass., 1998.

Bernhard, Virginia. " 'Men, Women and Children' at Jamestown: Population and Gender in Early Virginia." *Journal of Southern History* 53 (1992): 599–618.

Berry, Brewton. "Marginal Groups." In Trigger, *Northeast*, 290–95.

Beverley, Robert. *The History and Present State of Virginia*. Edited by Louis B. Wright. 1705. Reprint. Charlottesville, Va., 1947.

Bezanson, Anne. "Inflation and Controls, Pennsylvania, 1774–1779." *JEH* 8, supplement (1948): 1–20.

Bezanson, Anne, assisted by Blanch Daley, Marjorie C. Denison, and Miriam Hussey. *Prices and Inflation during the American Revolution: Pennsylvania, 1776–1790*. Philadelphia, 1951.

Biddick, Kathleen. "Missing Links: Taxable Wealth, Markets, and Stratification among Medieval English Peasants." *JIH* 17 (1987): 277–98.

Bidwell, Percy Wells, and John I. Falconer. *History of Agriculture in the Northern United States, 1620–1860*. Washington, D.C., 1925.

Biegańska, Anna. "A Note on Scots in Poland, 1550–1800." In *Scotland and Europe, 1200–1850*, edited by T. Christopher Smout, 157–65. Edinburgh, 1986.

Bigger, Francis Joseph. *The Ulster Land War of 1770 (the Hearts of Steel)*. Dublin, 1910.

Billings, Warren M. "The Growth of Political Institutions in Virginia, 1634 to 1676." *WMQ* 31 (1974): 225–42.

——. "Law and Culture in the Colonial Chesapeake Area." *Southern Studies* 17 (1978): 333–48.

——. "Sir William Berkeley and the Carolina Proprietary." *North Carolina Historical Review* 72 (1995): 329–42.

——. "The Transfer of English Law to Virginia, 1606–50." In Andrews, Canny, and Hair, *Westward Enterprise*, 215–44.

——, ed. *The Old Dominion in the Seventeenth Century: A Documentary History of Virginia, 1606–1689*. Chapel Hill, N.C., 1975.

Billings, Warren M., John E. Selby, and Thad W. Tate. *Colonial Virginia: A History*. White Plains, N.Y., 1986.

Birrell, Jean. "Common Right in the Medieval Forest: Disputes and Conflicts in the Thirteenth Century." *P&P*, no. 117 (1987): 22–49.

——. "Peasant Deer Poachers in the Medieval Forest." In Britnell and Hatcher, *Progress and Problems in Medieval England*, 68–88.

Bishop, Charles A. "Territoriality among Northeastern Algonquians." *Anthropologica*, n.s., 28 (1986): 37–63.

Bissell, Linda Auwers. "From One Generation to Another: Mobility in Seventeenth-Century Windsor, Connecticut." *WMQ* 31 (1974): 79–110.

Bjork, Gordon C. *Stagnation and Growth in the American Economy, 1784–1792*. New York, 1985.

——. "The Weaning of the American Economy: Independence, Market Changes, and Economic Development." *JEH* 24 (1964): 541–66.

Black, Frederick R. "The West Jersey Society, 1768–1784." *Pennsylvania Magazine of History and Biography* 97 (1973): 379–406.

Black, Jeremy. *The War for America: The Fight for Independence, 1775–1783*. New York, 1991.

Blanchard, Ian. "Industrial Employment and the Rural Land Market, 1380–1520." In Richard M. Smith, *Land, Kinship, and the Life Cycle*, 227–75.

Blanning, T. C. W. *Reform and Revolution in Mainz, 1743–1803*. Cambridge, 1974.

Blethen, H. Tyler, and Curtis W. Wood Jr., eds. *Ulster and North America: Transatlantic Perspectives on the Scotch-Irish*. Tuscaloosa, Ala., 1997.

Blickle, Renate. "From Subsistence to Property: Traces of a Fundamental Change in Early Modern Bavaria." *Central European History* 25 (1992): 377–85.

Bliss, Robert M. *Revolution and Empire: English Politics and the American Colonies in the Seventeenth Century*. Manchester, England, 1990.

Bliss, Willard F. "The Rise of Tenancy in Virginia." *VMHB* 58 (1950): 427–41.

Bloch, Ruth H. "American Feminine Ideals in Transition: The Rise of the Moral Mother, 1785–1815." *Feminist Studies* 4 (1978): 101–26.

Blodget, Samuel. *Economica: A Statistical Manual for the United States of America*. Washington, D.C., 1806.

Blu, Karen I. *The Lumbee Problem: The Making of an American Indian People*. Cambridge, 1980.

Blum, Jerome. *The End of the Old Order in Rural Europe*. Princeton, N.J., 1978.

Blumenthal, Walter Hart. *Women Camp Followers of the American Revolution*. Philadelphia, 1952.

Bodle, Wayne K. "Learning to Live with War: Civilians and Revolutionary Conflict in the Delaware Valley in 1777." Paper, McNeil Center for Early American Studies, February 1995.

——. *The Vortex of Small Fortunes: The Continental Army at Valley Forge, 1777–1778*. Vol. 1 of *The Valley Forge Historical Report*. Valley Forge, Pa., 1980.

Bogue, Allan G. "The Iowa Claims Clubs: Symbol and Substance." *Mississippi Valley Historical Review* 45 (1958): 231–53.

Bois, Guy. "Against the Neo-Malthusian Orthodoxy." In Aston and Philpin, *Brenner Debate*, 107–18.

Boleda, Mario. "Les Migrations au Canada sous le régime français, 1607–1760." *Cahiers Quebecois de demographie* 13 (1984): 23–28.

Bolton, Charles Knowles. *The Private Soldier under Washington*. New York, 1902.

Bolton, J. L. *The Medieval English Economy, 1150–1500*. London, 1980.

Bolzius, Johann Martin. "Johann Martin Bolzius Answers a Questionnaire on Carolina

and Georgia." Edited by Kalus G. Leowald, Beverely Starika, and Paul S. Taylor. *WMQ* 14 (1957): 218–61; 15 (1958): 228–52.
Bond, Beverly W., Jr. *The Quit-Rent System in the American Colonies*. New Haven, Conn., 1919.
Bonfield, Lloyd. "Normative Rules and Property Transmission: Reflections on the Link between Marriage and Inheritance in Early Modern England." In *The World We Have Gained: Histories of Population and Social Structure* (Essays Presented to Peter Laslett on His Seventieth Birthday), edited by Lloyd Bonfield, Richard M. Smith, and Keith Wrightson, 155–76. London, 1986.
——. "What Did English Villagers Mean by 'Customary Law'?" In Razi and Smith, *Medieval Society and the Manor Court*, 103–16.
Bonomi, Patricia U. *A Factious People: Politics and Society in Colonial New York*. New York, 1971.
Borah, Woodrow. "The Mixing of Population." In Chiappelli, *First Images of America*, 707–22.
Bossy, John. "Reluctant Colonists: The English Catholics Confront the Atlantic." In Quinn, *Early Maryland in a Wider World*, 149–64.
Bosworth, Timothy Woody. "Anti-Catholicism as a Political Tool in Mid-Eighteenth Century Maryland." *Catholic Historical Review* 61 (1975): 539–63.
——. "Those Who Moved: Internal Migrants in America before 1840." Ph.D. diss., University of Wisconsin, 1980.
Botein, Stephen. *Early American Law and Society*. New York, 1983.
Bowden, M. J. "The Invention of American Tradition." *Journal of Historical Geography* 18 (1992): 3–26.
Bowden, Peter. "Agricultural Prices, Farm Profits, and Rents." In Thirsk, *Agrarian History*, 593–695.
——. "Statistical Appendix." In Thirsk, *Agrarian History*, 814–70.
Bowler, R. Arthur. "Logistics and Operations in the American Revolution." In *Reconsiderations on the Revolutionary War: Selected Essays*, edited by Don Higginbotham, 54–71. Westport, Conn., 1978.
——. *Logistics and the Failure of the British Army in America, 1775–1783*. Princeton, N.J., 1975.
Bowman, Allen. *The Morale of the American Revolutionary Army*. Washington, D.C., 1943.
Boyd-Bowman, Peter. "Patterns of Spanish Emigration to the Indies until 1600." *Hispanic American Historical Review* 56 (1976): 580–604.
Boydston, Jeanne. *Home and Work: Housework, Wages, and the Ideology of Labor in the Early Republic*. New York, 1990.
Boyer, George R. *An Economic History of the English Poor Law, 1750–1850*. Cambridge, 1990.
Boyle, Joseph Lee, ed. "From Saratoga to Valley Forge: The Diary of Lt. Samuel Armstrong." *Pennsylvania Magazine of History and Biography* 121 (1997): 237–70.
Bozeman, Theodore Dwight. *To Live Ancient Lives: The Primitivist Dimension in Puritanism*. Chapel Hill, N.C., 1988.
Bradford, William. *Of Plymouth Plantation, 1620–1647*. Edited by Samuel Eliot Morison. New York, 1952.
Brady, John, and Patrick J. Corish. *The Church under the Penal Code*. Vol. 4, no. 2, of *A History of Irish Catholicism*. Dublin, 1971.
Bragdon, Kathleen. "Gender as a Social Category in Native Southern New England." *Ethnohistory* 43 (1996): 573–92.

——. "The Material Culture of the Christian Indians of New England, 1650–1775." In Beaudry, *Documentary Archaeology*, 126–31.
Brakensiek, Stefan. "Agrarian Individualism in North-West Germany, 1770–1870." *German History* 12 (1994): 137–79.
Braun, Rudolf. "Early Industrialization and Demographic Change in the Canton of Zürich." In Tilly, *Historical Studies of Changing Fertility*, 289–334.
Braund, Kathryn E. Holland. "The Creek Indians, Blacks, and Slavery." *Journal of Southern History* 57 (1991): 601–36.
——. *Deerskins and Duffels: The Creek-Indian Trade with Anglo-America, 1685–1815*. Lincoln, Nebr., 1993.
Breen, Timothy Hall. "Back to Sweat and Toil: Suggestions for the Study of Agricultural Work in Early America." *Pennsylvania History* 49 (1982): 241–58.
——. " 'Baubles of Britain': The American and Consumer Revolutions of the Eighteenth Century." *P&P*, no. 119 (1988): 73–104.
——. "Creative Adaptations: Peoples and Cultures." In Greene and Pole, *Colonial British America*, 195–232.
——. "An Empire of Goods: The Anglicization of Colonial America, 1690–1776." *Journal of British Studies* 25 (1986): 467–99.
——. *Imagining the Past: East Hampton Histories*. Reading, Mass., 1989.
——. "The Meaning of Things: Interpreting the Consumer Economy in the Eighteenth Century." In Brewer and Porter, *Consumption and the World of Goods*, 249–60.
——. *Puritans and Adventurers: Change and Persistence in Early America*. New York, 1980.
——. *Tobacco Culture: The Mentality of the Great Tidewater Planters on the Eve of Revolution*. Princeton, N.J., 1985.
Breen, Timothy Hall, and Stephen Foster. "Moving to the New World: The Character of Early Massachusetts Immigration." *WMQ* 30 (1973): 189–222.
Bremer, Francis J. "The Heritage of John Winthrop: Religion along the Stour Valley, 1548–1630." *NEQ* 70 (1997): 515–47.
Bremner, Robert H., John Barnard, Tamara K. Hareven, and Robert M. Mennel, eds. *Children and Youth in America*. Vol. 1, *1600–1865*. Cambridge, Mass., 1970.
Brenner, Elise. "Sociopolitical Implications of Mortuary Ritual Remains in Seventeenth-Century Southern New England." In *The Recovery of Meaning: Historical Archaeology in the Eastern United States*, edited by Mark P. Leone and Parker B. Potter Jr., 147–81. Washington, D.C., 1988.
Brenner, Robert. "Agrarian Class Structure and Economic Development in Pre-Industrial Europe." In Aston and Philpin, *Brenner Debate*, 10–63.
——. "The Agrarian Roots of European Capitalism." In Aston and Philpin, *Brenner Debate*, 213–327.
——. *Merchants and Revolution: Commercial Change, Political Conflict, and London's Overseas Traders, 1550–1653*. Princeton, N.J., 1993.
Brewer, Holly. "Entailing Aristocracy in Colonial Virginia: 'Ancient Feudal Restraints' and Revolutionary Reform." *WMQ* 54 (1997): 307–46.
Brewer, John, and Roy Porter, eds. *Consumption and the World of Goods*. London, 1993.
Bridenbaugh, Carl. *Cities in Revolt: Urban Life in America, 1743–1776*. New York, 1955.
——. *The Colonial Craftsman*. New York, 1950.
——. *Fat Mutton and Liberty of Conscience: Society in Rhode Island, 1630–1690*. Providence, R.I., 1974.
——. *Vexed and Troubled Englishmen, 1590–1642*. New York, 1968.
——. "Yankee Use and Abuse of the Forest in the Building of New England, 1620–1660." *Massachusetts Historical Society, Proceedings* 89 (1977): 3–35.

Britnell, Richard H. *The Commercialisation of English Society, 1000–1500*. 2d ed. Manchester, England, 1996.
Britnell, Richard H., and John Hatcher, eds. *Progress and Problems in Medieval England: Essays in Honour of Edward Miller*. Cambridge, 1996.
Bronner, Edwin B., and David Fraser, eds. *William Penn's Published Writings, 1660–1726*. Vol. 5 of *The Papers of William Penn*. Edited by Richard S. Dunn and Mary Maples Dunn. Philadelphia, 1986.
Brooke, John L. *The Heart of the Commonwealth: Society and Political Culture in Worcester County, Massachusetts, 1713–1861*. Cambridge, 1989.
Brooke, Richard. "A Thermometrical Account of the Weather, for One Year, beginning September 1753. Kept in Maryland, by Mr. Richard Brooke, Physician and Surgeon in that Province." *Transactions of the Philosophical Society* 59 (1759): 58–82.
Brophy, Alfred L. "Law and Indentured Servitude in Mid-Eighteenth Century Pennsylvania." *Willamette Law Review* 28 (1991): 69–126.
Brown, Alan S., ed. "James Simpson's Reports on the Carolina Loyalists, 1779–1780." *Journal of Southern History* 21 (1955): 513–19.
Brown, Kathleen M. *Good Wives, Nasty Wenches, and Anxious Patriarchs: Gender, Race, and Power in Colonial Virginia*. Chapel Hill, N.C., 1996.
Brown, M. L. *Firearms in Colonial America: The Impact on History and Technology, 1492–1792*. Washington, D.C., 1980.
Brown, Ralph H. *Historical Geography of the United States*. New York, 1948.
Brown, Richard Maxwell. *The South Carolina Regulators*. Cambridge, Mass., 1963.
Brown, Robert E. *Middle-Class Democracy and the Revolution in Massachusetts, 1691–1780*. Ithaca, N.Y., 1955.
Brown, Robert E., and B. Katherine Brown. *Virginia, 1705–1786: Democracy or Aristocracy*. East Lansing, Mich., 1964.
Brown, Wallace. "The American Farmer during the Revolution: Rebel or Loyalist." *AgH* 42 (1968): 327–38.
——. "The American Loyalists in Jamaica." *Journal of Caribbean History* 26 (1992): 121–46.
——. *The Good Americans: The Loyalists in the American Revolution*. New York, 1969.
——. *The King's Friends: The Composition and Motives of the American Loyalist Claimants*. Providence, R.I., 1965.
Brown, Wallace, and Hereward Senior. *Victorious in Defeat: The American Loyalists in Exile*. New York, 1984.
Brownmiller, Susan. *Against Our Will: Men, Women, and Rape*. New York, 1976.
Bruce, Philip Alexander. *Economic History of Virginia in the Seventeenth Century: An Inquiry into the Material Condition of the People, based upon Original and Contemporaneous Records*. 2 vols. London, 1896.
Bruchey, Stuart, ed. *Cotton and the Growth of the American Economy, 1790–1860: Sources and Readings*. New York, 1967.
Bruegel, Martin, "Uncertainty, Pluriactivity, and Neighborhood Exchange in the Rural Hudson Valley in the late Eighteenth Century." *New York History* 77 (1996): 245–72.
Bruford, W. H. *Germany in the Eighteenth Century: The Social Background of the Literary Revival*. Cambridge, 1935.
Brune, Basel H. "The Changing Spatial Organization of Early Tobacco Marketing in the Patuxent River Basin." In Mitchell and Muller, *Geographic Perspectives on Maryland's Past*, 71–89.
Buchanan, John. *The Road to Guilford Courthouse: The American Revolution in the Carolinas*. New York, 1997.

Buck, Solon J., and Elizabeth Hawthorn Buck. *The Planting of Civilization in Western Pennsylvania*. Pittsburgh, 1939.
Buel, Joy Day, and Richard Buel Jr. *The Way of Duty: A Woman and Her Family in Revolutionary America*. New York, 1984.
Buel, Richard, Jr. *Dear Liberty: Connecticut's Mobilization for the Revolutionary War*. Middletown, Conn., 1980.
——. "Sampson Shorn: The Impact of the Revolutionary War on Estimates of the Republic's Strength." In Hoffman and Albert, *Arms and Independence*, 141–65.
Bumsted, J. M. "Highland Emigration to the Island of St. John and the Scottish Catholic Church." *Dalhousie Review* 58 (1978): 510–27.
——. *The People's Clearance: Highland Emigration to British North America*. Edinburgh, 1982.
Burger, John Strauss. " 'Our Unhappy Purchase': The West Jersey Society, Louis Morris, and Jersey Lands, 1703–36." *New Jersey History* 98 (1980): 97–115.
Burke, Thomas E. *Mohawk Frontier: The Dutch Community of Schenectady, New York, 1661–1710*. Ithaca, N.Y., 1991.
Burnaby, Rev. Andrew. *Travels through the Middle Settlements in North America in the Years 1759 and 1766 with Observations upon the State of the Colonies*. 1775. Reprint. Ithaca, N.Y., 1960.
Burnard, Trevor. "A Tangled Cousinry? Associational Networks of the Maryland Elite, 1691–1776." *Journal of Southern History* 61 (1995): 17–44.
Burnett, G. Wesley. "South Carolina Forests." In Richard C. Davis, *Encyclopedia of Forest History*, 2:613–15.
Bush, Jonathan A. " 'Take This Job and Shove It': The Rise of Free Labor." *Michigan Law Review* 91 (1993): 1382–1413.
Bushman, Richard Lyman. *From Puritan to Yankee: Character and the Social Order in Connecticut, 1690–1765*. Cambridge, Mass., 1967.
——. "Markets and Composite Farms in Early America." *WMQ* 55 (1998): 351–74.
——. "Massachusetts Farmers and the Revolution." In *Society, Freedom, and Conscience: The American Revolution in Virginia, Massachusetts, and New York*, edited by Richard M. Jellison, 77–124. New York, 1976.
——. *The Refinement of America: Persons, Houses, Cities*. New York, 1992.
Butler, Jon. *The Huguenots in America: A Refugee People in a New World Society*. Cambridge, Mass., 1983.
——, ed. "Two 1642 Letters from Virginia Puritans." *Massachusetts Historical Society, Proceedings* 84 (1972): 99–109.
Butler, Lindley S. "David Fanning's Militia: A Roving Partisan Community." In *Loyalists and Community in North America*, edited by Robert McCluer Calhoon, Timothy M. Barnes, and George A. Rawlyk, 147–57. Westport, Conn., 1994.
——. "The Early Settlement of Carolina: Virginia's Southern Frontier." *VMHB* 79 (1971): 20–28.
Cadle, Farris W. *Georgia Land Surveying History and Law*. Athens, Ga., 1991.
Cage, R. A. "The Making of the Old Scottish Poor Law." *P&P*, no. 69 (1975): 113–18.
——. *The Scottish Poor Law, 1745–1845*. Edinburgh, 1981.
Cahn, Susan. *Industry of Devotion: The Transformation of Women's Work in England, 1500–1660*. New York, 1987.
Caird, J. B. "The Reshaped Agricultural Landscape." In Parry and Slater, *Making of the Scottish Countryside*, 203–22.
Caldwell, Lee Ann. "Women Landholders of Colonial Georgia." In Jackson and Spalding, *Forty Years of Diversity*, 183–97.

Calhoon, Robert McCluer. *The Loyalist Perception and Other Essays*. Columbia, S.C., 1989.
——. *The Loyalists in Revolutionary America, 1760–1781*. New York, 1973.
Calhoun, Arthur W. *A Social History of the American Family*. Vol. 1, *The Colonial Period*. New York, 1917.
Callander, Robin Fraser. *A Pattern of Landownership in Scotland with Particular Reference to Aberdeenshire*. Finzean, Scotland, 1987.
Calloway, Colin G. *The American Revolution in Indian Country: Crisis and Diversity in Native American Communities*. Cambridge, 1995.
——. *New Worlds for All: Indians, Europeans, and the Remaking of Early America*. Baltimore, 1997.
——, ed. *Downland Encounters: Indians and Europeans in Northern New England*. Hanover, N.H., 1991.
Calo, Zachary Ryan. "From Poor Relief to the Poorhouse: The Response to Poverty in Prince George's County, Maryland, 1710–1770." *MdHM* 93 (1998): 393–427.
Calomiris, Charles W. "Institutional Failure, Monetary Security, and the Depreciation of the Continental." *JEH* 48 (1988): 47–68.
Camfield, Thomas M. "A Can or Two of Worms: Virginia Bernhard and the Historiography of Early Virginia." *Journal of Southern History* 60 (1994): 649–62.
Campbell, Bruce M. S. "A Fair Field Once Full of Folk: Agrarian Change in an Era of Population Decline, 1348–1500." *Agricultural History Review* 41 (1993): 60–70.
——. "Population Pressure, Inheritance, and the Land Market in a Fourteenth-Century Peasant Community." In Richard M. Smith, *Land, Kinship, and the Life Cycle*, 87–134.
——, ed. *Before the Black Death: Studies of the "Crisis" of the Early Fourteenth Century*. Manchester, England, 1991.
Campbell, Bruce M. S., and Mark Overton. "A New Perspective on Medieval and Early Modern Agriculture: Six Centuries of Norfolk Farming, c. 1250–c1850." *P&P*, no. 141 (1993): 38–87.
Campbell, Bruce M. S., Kenneth C. Bartley, and John P. Power. "The Demesne-Farming Systems of Post–Black Death England: A Classification." *Agricultural History Review* 44 (1996): 131–79.
Campbell, Colin D. "They Beckoned and We Came: The Settlement of Cherry Valley." *New York History* 79 (1998): 215–32.
Campbell, Mildred. *The English Yeoman under Elizabeth and the Early Stuarts*. New Haven, Conn., 1942.
——. " 'Of People Either Too Few or Too Many': The Conflict of Opinion on Population and Its Relation to Emigration." In *Conflict in Stuart England: Essays in Honour of Wallace Notestein*, edited by William Appleton Aiken and Basil Duke Henning, 171–201. New York, 1960.
——. "Rebuttal [to David Galenson]." *WMQ* 35 (1978): 526–40.
——. "Reply [to David Galenson]." *WMQ* 36 (1979): 277–86.
——. "The Social Origins of Some Early Americans." In James Morton Smith, *Seventeenth-Century America*, 63–89.
Campbell, R. H. "The Scottish Improvers and the Course of Agrarian Change in the Eighteenth Century." In Cullen and Smout, *Comparative Aspects of Scottish and Irish Economic and Social History*, 204–15.
Campbell, Wilfrid, comp. *Report on Manuscript Lists in the Archives Relating to the United Empire Loyalists with Reference to Other Sources*. Ottawa, Canada, 1909.
Candee, Richard M. "Land Surveys of William and John Godsoe of Kittery, Maine: 1689–1769." In Benes, *New England Prospect*, 9–46.

Canny, Nicholas. "England's New World and the Old, 1480s–1630s." In Canny and Low, *Origins of Empire*, 148–69.
——. "English Migration into and across the Atlantic during the Seventeenth and Eighteenth Centuries." In Canny, *Europeans on the Move*, 39–75.
——. "The Ideology of English Colonization: From Ireland to America." *WMQ* 30 (1973): 575–98.
——. *Kingdom and Colony: Ireland in the Atlantic World, 1560–1800*. Baltimore, 1988.
——. "The Marginal Kingdom: Ireland as a Problem in the First British Empire." In Bailyn and Morgan, *Strangers within the Realm*, 35–66.
——. "The Origins of Empire: An Introduction." In Canny and Low, *Origins of Empire*, 1–33.
——. "The Permissive Frontier: The Problem of Social Control in English Settlements in Ireland and Virginia, 1550–1620." In Andrews, Canny, and Hair, *Westward Enterprise*, 17–44.
——. " 'To Establish a Common Wealthe': Captain John Smith as New World Colonist." *VMHB* 96 (1988): 213–22.
——, ed. *Europeans on the Move: Studies on European Migration, 1500–1800*. Oxford, 1994.
Canny, Nicholas, and Alaine Low, eds. *The Origins of Empire: British Overseas Enterprise to the Close of the Seventeenth Century*. Vol. 1 of *The Oxford History of the British Empire*. Edited by William Roger Louis. Oxford, 1998.
Canny, Nicholas, Joseph E. Illick, Gary B. Nash, and William Pencak, eds. *Empire, Society, and Labor: Essays in Honor of Richard S. Dunn*. *Pennsylvania History* 64, special supplement (1997).
Cantor, Leonard. *The Changing English Landscape, 1400–1700*. London, 1987.
Canup, John. *Out of the Wilderness: The Emergence of American Identity in Colonial New England*. Middletown, Conn., 1990.
Cappon, Lester J., Barbara Bartz Petchenik, and John Hamilton Long, eds. *Atlas of Early American History: The Revolutionary Era, 1760–1790*. Princeton, N.J., 1976.
Carlson, Catherine C., George J. Armelagos, and Ann L. Magennis. "Impact of Disease on the Precontact and Early Historic Populations of New England and the Maritimes." In Verano and Ubelaker, *Disease and Demography*, 141–53.
Carlson, Eric Josef. *Marriage and the English Reformation*. Oxford, 1994.
Carney, Judith. "Rice Milling, Gender, and Slave Labour in Colonial South Carolina." *P&P*, no. 153 (1996): 108–34.
Carp, E. Wayne. "Early American Military History: A Review of Recent Work." *VMHB* 94 (1986): 259–84.
——. *To Starve the Army at Pleasure: Continental Army Administration and American Political Culture, 1775–1783*. Chapel Hill, N.C., 1984.
Carr, Lois Green. "The Development of the Maryland Orphans' Court, 1654–1715." In Land, Carr, and Papenfuse, *Law, Society, and Politics in Early Maryland*, 41–62.
——. "Diversification in the Colonial Chesapeake: Somerset County, Maryland, in Comparative Perspective." In Carr, Morgan, and Russo, *Colonial Chesapeake Society*, 342–88.
——. "Emigration and the Standard of Living: The Seventeenth-Century Chesapeake." *JEH* 52 (1992): 271–92.
——. "Inheritance in the Colonial Chesapeake." In Hoffman and Albert, *Women in the Age of the American Revolution*, 155–208.
Carr, Lois Green, and Russell R. Menard. "Immigration and Opportunity: The

Freedman in Early Colonial Maryland." In Tate and Ammerman, *Chesapeake in the Seventeenth Century*, 206–42.
———. "Land, Labor, and Economies of Scale in Early Maryland: Some Limits to Growth in the Chesapeake System of Husbandry." *JEH* 49 (1989): 407–18.
———. "Wealth and Welfare in Early Maryland: Evidence from St. Mary's County." *WMQ* 56 (1999): 95–120.
Carr, Lois Green, and Lorena S. Walsh. "Changing Lifestyles and Consumer Behavior in the Colonial Chesapeake." In *Of Consuming Interests: The Style of Life in the Eighteenth Century*, edited by Cary Carson, Ronald Hoffman, and Peter J. Albert, 59–166. Charlottesville, Va., 1994.
———. "Inventories and the Analysis of Wealth and Consumption Patterns in St. Mary's County, Maryland, 1658–1777." *Historical Methods* 13 (1980): 81–104.
———. "The Planter's Wife: The Experience of White Women in Seventeenth-Century Maryland." *WMQ* 34 (1977): 542–71.
———. "The Standard of Living in the Colonial Chesapeake." *WMQ* 45 (1988): 135–59.
Carr, Lois Green, Russell R. Menard, and Lorena S. Walsh. *Robert Cole's World: Agriculture and Society in Early Maryland*. Chapel Hill, N.C., 1991.
Carr, Lois Green, Philip D. Morgan, and Jean B. Russo, eds. *Colonial Chesapeake Society*. Chapel Hill, N.C., 1989.
Carrington, Selwyn H. H. "The American Revolution and the British West Indies' Economy." *JIH* 17 (1987): 823–50.
Carroll, Charles F. *The Timber Economy of Puritan New England*. Providence, R.I., 1973.
Carroll, Kenneth L. "Quakerism on the Eastern Shore of Virginia." *VMHB* 74 (1966): 170–89.
Carson, Cary, Norman F. Barka, William M. Kelso, Gary Wheeler Stone, and Dell Upton. "Impermanent Architecture in the Southern American Colonies." *Winterthur Portfolio* 16 (1981): 135–81.
Casey, Kathleen. "The Cheshire Cat: Reconstructing the Experience of Medieval Woman." In *Liberating Women's History: Theoretical and Critical Essays*, edited by Berenice A. Carroll, 224–49. Urbana, Ill., 1976.
Cash, Philip. "The Canadian Military Campaign of 1775–1776: Medical Problems and Effects of Disease." *Journal of the American Medical Association* 236, no. 1 (1976): 52–56.
———. *Medical Men at the Siege of Boston, April 1775–April 1776*. Vol. 98 of *Memoirs of the American Philosophical Society*. Philadelphia, 1973.
Cashin, Edward J. " 'But Brothers It Is Our Land We Are Talking About': Winners and Losers in the Georgia Backcountry." In Hoffman, Tate, and Albert, *Uncivil War*, 240–75.
———. *The King's Ranger: Thomas Brown and the American Revolution on the Southern Frontier*. Athens, Ga., 1989.
———. "Sowing the Wind: Governor Wright and the Georgia Backcountry on the Eve of the Revolution." In Jackson and Spalding, *Forty Years of Diversity*, 233–50.
Casino, Joseph J. "Anti-Popery in Colonial Pennsylvania." *Pennsylvania Magazine of History and Biography* 105 (1981): 279–309.
Castiglioni, Luigi. *Luigi Castiglioni's Viaggio: Travels in the United States of North America, 1785–87*. Edited and translated by Antonio Pace. Syracuse, N.Y., 1983.
Casway, Jerrold. "Irish Women Overseas, 1500–1800." In MacCurtain and O'Dowd, *Women in Early Modern Ireland*, 112–32.
Cave, Alfred A. *The Pequot War*. Amherst, Mass., 1996.

——. "Who Killed John Stone? A Note on the Origins of the Pequot War." *WMQ* 49 (1992): 509–21.

Cayton, Andrew R. L. *The Frontier Republic: Ideology and Politics in the Ohio Country, 1780–1825*. Kent, Ohio, 1986.

Ceci, Lynn. "The First Fiscal Crisis in New York." *Economic Development and Cultural Change* 28 (1980): 839–47.

——. "Fish Fertilizer: A Native North American Practice?" *Science* 188 (1975): 26–30.

——. "Native Wampum as a Peripheral Resource in the Seventeenth-Century World System." In Hauptman and Wherry, *Pequots in Southern New England*, 48–63.

——. "The Value of Wampum among the New York Iroquois: A Case Study in Artifact Analysis." *Journal of Anthropological Research* 38 (1982): 97–107.

Chafee, Zechariah, Jr. "Colonial Courts and the Common Law." In *Essays in the History of Early American Law*, edited by David H. Flaherty, 53–82. Chapel Hill, N.C., 1969.

Chalou, George C. "Women in the American Revolution: Vignettes or Profiles." In *Clio Was a Woman: Studies in the History of American Women*, edited by Mabel E. Deutrich and Virginia C. Purdy, 73–90. Washington, D.C., 1980.

Chambers, J. D. "Population Change in a Provincial Town: Nottingham, 1700–1800." In *Studies in the Industrial Revolution Presented to T. S. Ashton*, edited by L. S. Pressnell, 97–124. London, 1960.

Chaplin, Joyce E. *An Anxious Pursuit: Agricultural Innovation and Modernity in the Lower South, 1730–1815*. Chapel Hill, N.C., 1993.

——. "Climate and Southern Pessimism: The Natural History of an Idea, 1500–1800." In *The South as an American Problem*, edited by Larry J. Griffin and Don H. Doyle, 57–81. Athens, Ga., 1995.

——. "Creating a Cotton South in Georgia and South Carolina, 1760–1815." *Journal of Southern History* 57 (1991): 171–200.

Chappell, William, and J. Woodfall Webworth, eds. *The Roxburghe Ballads*. 9 vols. in 8. Hertford, England, 1872–99.

Charles, Lindsey, and Lorna Duffin, eds. *Women and Work in Pre-Industrial England*. London, 1985.

Charlesworth, Andrew, ed. *An Atlas of Rural Protest in Britain, 1548–1908*. Philadelphia, 1983.

Chartres, J. A. "The Marketing of Agricultural Produce." In *The Agrarian History of England and Wales*, edited by Joan Thirsk. Vol. 5, *1640–1750*, 406–502. Cambridge, 1985.

Chastellux, Marquis François Jean de. *Travels in North America in the Years 1780, 1781, and 1782*. Edited by Howard C. Rice Jr. 2 vols. Chapel Hill, N.C., 1963.

Chayanov, A. V. *The Theory of the Peasant Economy*. Edited by Daniel Thorner, Basile Kerblay, R. E. F. Smith, and Theodor Shanin, Madison, Wis., 1986.

Chaytor, Miranda. "Household and Kinship: Ryton in the Late Sixteenth and Early Seventeenth Centuries." *History Workshop* 10 (1980): 25–60.

Chazanof, William. *Joseph Ellicott and the Holland Land Company: The Opening of Western New York*. Syracuse, N.Y., 1970.

Chesnutt, David Rogers. "South Carolina's Expansion into Colonial Georgia, 1720–1765." Ph.D. diss., University of Georgia, 1973.

Chiappelli, Fredi, ed. *First Images of America: The Impact of the New World on the Old*. 2 vols. Berkeley, Calif., 1976.

Childs, St. Julien Ravenel. *Malaria and Colonization in the Carolina Low Country, 1526–1696*. Baltimore, 1940.

Clap, Roger. *Memoirs of Capt. Roger Clap, Relating some of GOD's remarkable*

providence to him, in bringing him into New England . . . To which is annexed, a SHORT ACCOUNT OF THE AUTHOR and his FAMILY. Boston, 1774.
Clapham, Peter. "Agricultural Change and Its Impact on Tenancy: The Evidence of Angus Rentals and Tacks, c. 1760–1850." In *Industry, Business, and Society in Scotland since 1700: Essays Presented to Professor John Butt*, edited by A. J. G. Cummings and Thomas Martin Devine, 144–63. Edinburgh, 1994.
Clark, Alice. *Working Life of Women in the Seventeenth Century*. 1919. 2d ed. Introduction by Miranda Chaytor and Jane Lewis. London, 1982.
Clark, Charles E. *The Eastern Frontier: The Settlement of Northern New England, 1610–1763*. 2d ed. Hanover, N.H., 1983.
Clark, Christopher. *The Roots of Rural Capitalism: Western Massachusetts, 1780–1860*. Ithaca, N.Y., 1990.
Clark, Elaine. "Charitable Bequests, Deathbed Land Sales, and the Manor Court in Later Medieval England." In Razi and Smith, *Medieval Society and the Manor Court*, 143–61.
——. "Debt Litigation in a Late Medieval English Vill." In Raftis, *Pathways to Medieval Peasants*, 247–79.
Clark, Gregory. "Commons Sense: Common Property Rights, Efficiency, and Institutional Change." *JEH* 58 (1998): 73–102.
——. "The Economics of Exhaustion, the Postan Thesis, and the Agricultural Revolution." *JEH* 52 (1992): 61–84.
Clark, Jonathan. "The Problem of Allegiance in Revolutionary Poughkeepsie." In Hall, Murrin, and Tate, *Saints and Revolutionaries*, 285–317.
Clark, Peter A. *The English Alehouse: A Social History, 1200–1830*. London, 1983.
——. "The Migrant in Kentish Towns, 1580–1640." In *Crisis and Order in English Towns, 1500–1700: Essays in Urban History*, edited by Peter A. Clark and Paul Slack, 117–63. Toronto, 1972.
——. "Migrants in the City: The Process of Social Adaptation in English Towns, 1500–1800." In Clark and Souden, *Migration and Society*, 267–91.
——, ed. *The Transformation of English Provincial Towns, 1600–1800*. London, 1984.
Clark, Peter, and David Souden, eds. *Migration and Society in Early Modern England*. Totowa, N.J., 1987.
Clark, Victor S. *History of Manufactures in the United States*. 2 vols. Washington, D.C., 1929.
Clark, Wayne E., and Helen C. Roundtree. "The Powhatans and the Maryland Mainland." In Roundtree, *Powhatan Foreign Relations*, 112–35.
Clarkson, L. A. "Irish Population Revisited, 1687–1821." In *Irish Population, Economy, and Society: Essays in Honour of the Late K. H. Connell*, edited by J. M. Goldstrom and L. A. Clarkson, 13–35. Oxford, 1981.
——. *The Pre-Industrial Economy in England, 1500–1750*. New York, 1972.
Clawson, Marion. "Forests in the Long Sweep of American History." *Science* 205 (1979): 1168–74.
Clay, C. G. A. *Economic Expansion and Social Change: England, 1500–1700*. 2 vols. Cambridge, 1984.
Clemens, Paul G. E. *The Atlantic Economy and Colonial Maryland's Eastern Shore: From Tobacco to Grain*. Ithaca, N.Y., 1980.
Clowse, Converse D. *Economic Beginnings in Colonial South Carolina, 1670–1730*. Columbia, S.C., 1971.
——. *Measuring Charleston's Overseas Commerce, 1717–1767: Statistics from the Port's Naval Lists*. Washington, D.C., 1981.

Coatsworth, John H. "American Trade with European Colonies in the Caribbean and South America, 1790–1812." *WMQ* 24 (1967): 243–66.
Coclanis, Peter A. "Retailing in Early South Carolina." In *Retailing: Theory and Practice for the 21th Century*, edited by Robert L. King, 1–5. Charleston, S.C., 1986.
——. *The Shadow of a Dream: Economic Life and Death in the South Carolina Low Country, 1670–1920*. New York, 1989.
Coe, Michael D., and Sophie D. Coe. "Mid-Eighteenth-Century Food and Drink on the Massachusetts Frontier." In Benes, *Foodways in the Northeast*, 39–46.
Cogley, Richard W. *John Eliot's Mission to the Indians before King Philip's War*. Cambridge, Mass., 1999.
Cogliano, Francis D. "Nil Desperandum Christo Duce: The New England Crusade against Louisbourg, 1745." *Essex Institute Historical Collections* 128 (1992): 180–207.
Cohen, Daniel A. *Pillars of Salt, Monuments of Grace: New England Crime Literature and the Origins of American Popular Culture, 1674–1860*. New York, 1993.
Cohen, David Steven. *The Dutch-American Farm*. New York, 1992.
——. "How Dutch Were the Dutch of New Netherland?" *New York History* 62 (1981): 43–60.
Cohen, Jon S., and Martin L. Weitzman. "Enclosures and Depopulation: A Marxist Analysis." In Parker and Jones, *European Peasants and Their Markets*, 161–76.
Cohen, Sheldon S. " 'To Parts of the World Unknown': The Circumstances of Divorce in Connecticut, 1750–1797." *Canadian Review of American Studies* 11 (1980): 275–93.
Coker, Kathy Roe. "The Case of James Nassau Colleton before the Commissioners of Forfeited Estates." *South Carolina Historical Magazine* 87 (1986): 106–16.
Colden, Cadwallader. *Letters and Papers of Cadwallader Colden*. Vol. 7, *1765–1775*. Collections of the New-York Historical Society for the Year 1923. New York, 1923.
Cole, Thomas R. "Family, Settlement, and Migration in Southeastern Massachusetts, 1650–1805: The Case for Regional Analysis." *New England Historical and Genealogical Register* 132 (1978): 171–85.
Coleman, D. C. *Industry in Tudor and Stuart England*. London, 1975.
——. *Myth, History, and the Industrial Revolution*. London, 1992.
Coleman, Kenneth. *Colonial Georgia: A History*. New York, 1976.
——. "Federal Indian Relations in the South, 1781–1789." *Chronicles of Oklahoma* 35 (1957–58): 435–58.
——. "Rebuttal to 'The Georgia Concept.' " *Georgia Historical Quarterly* 55 (1971): 172–76.
——. "The Southern Frontier: Georgia's Founding and the Expansion of South Carolina." *Georgia Historical Quarterly* 56 (1972): 163–74.
Coleman, Peter. *Debtors and Creditors in America: Insolvency, Imprisonment for Debt, and Bankruptcy, 1607–1900*. Madison, Wis., 1974.
Collins, Brenda. "Proto-Industrialization and Pre-Famine Emigration." *Social History* 7 (1982): 127–46.
Cometti, Elizabeth. "Depredations in Virginia during the Revolution." In *The Old Dominion: Essays for Thomas Perkins Abernethy*, edited by Darrett B. Rutman, 135–51. Charlottesville, Va., 1964.
——. "Inflation in Revolutionary Maryland." *WMQ* 8 (1951): 228–34.
——. "Women in the American Revolution." *NEQ* 20 (1947): 329–46.
Commager, Henry Steele, and Richard B. Morris, eds. *The Spirit of Seventy-six: The Story of the American Revolution as Told by Participants*. 2 vols. Indianapolis, 1958.
Condon, Ann Gorman. *The Envy of the American States: The Loyalist Dream for New Brunswick*. Frederickton, Canada, 1984.

Conger, Vivian Bruce. " 'If Widow, Both Housewife and Husband May Be': Widows' Testamentary Freedom in Colonial Massachusetts and Maryland." In *Women and Freedom in Early America*, edited by Larry D. Eldridge, 244–66. New York, 1997.
Connell, K. H. *The Population of Ireland, 1750–1845*. Oxford, 1950.
Conway, Stephen. " 'The Greatest Mischief Complain'd Of': Reflections on the Misconduct of British Soldiers in the Revolutionary War." *WMQ* 47 (1990): 370–90.
——. "To Subdue America: British Army Officers and the Conduct of the Revolutionary War." *WMQ* 43 (1986): 381–407.
——. *The War of American Independence, 1775–1783*. London, 1995.
Cook, Edward M., Jr. *The Fathers of the Towns: Leadership and Community Structure in Eighteenth-Century New England*. Baltimore, 1976.
——. "Social Behavior and Changing Values in Dedham, Massachusetts, 1700 to 1775." *WMQ* 27 (1970): 546–80.
Cook, Sherburne F. *The Indian Population of New England in the Seventeenth Century*. Vol. 12 of *University of California Publications in Anthropology*. Berkeley, Calif., 1976.
——. "The Significance of Disease in the Extinction of the New England Indians." *Human Biology* 45 (1973): 485–508.
Cooke, Edward S., Jr. *Making Furniture in Preindustrial America: The Social Economy of Newtown and Woodbury, Connecticut*. Baltimore, 1996.
Cooke, Jacob Ernest, ed. in chief. *Encyclopedia of the North American Colonies*. 3 vols. New York, 1993.
Coontz, Stephanie. *The Social Origins of Private Life: A History of American Families, 1600–1900*. London, 1988.
Cooper, Todd. "Trial and Triumph: The Impact of the Revolutionary War on Baltimore Merchants." In Eller, *Chesapeake Bay in the American Revolution*, 282–309.
Corcoran, Irma. "William Penn and His Purchasers: Problems in Paradise." *Proceedings of the American Philosophical Society* 138 (1994): 476–86.
Corfield, P. J. *The Impact of English Towns, 1700–1800*. Oxford, 1982.
Corkran, David H. *The Carolina Indian Frontier*. Columbia, S.C., 1970.
Cornwall, J. C. K. *Wealth and Society in Sixteenth-Century England*. London, 1988.
Cott, Nancy. "Divorce and the Changing Status of Women in Eighteenth-Century Massachusetts." *WMQ* 33 (1976): 586–614.
——. "Eighteenth-Century Family and Social Life Revealed in Massachusetts Divorce Records." *Journal of Social History* 10 (1976): 20–43.
Countryman, Edward. *A People in Revolution: The American Revolution and Political Society in New York, 1760–1790*. Baltimore, 1981.
——. "The Uses of Capital in Revolutionary America: The Case of the New York Loyalist Merchants." *WMQ* 49 (1992): 3–28.
Cowan, Ruth Schwartz. *More Work for Mother: The Ironies of Household Technology from the Open Hearth to the Microwave*. New York, 1983.
Cox, Richard J., ed. " 'A Touch of Kentucky News & State of Politicks': Two Letters of Levi Todd, 1784 and 1788." *Register, Kentucky Historical Society* 76 (1978): 217–22.
Cox, Thomas R., Robert S. Maxwell, Philip Drennon Thomas, and Joseph J. Malone. *This Well-Wooded Land: Americans and Their Forests from Colonial Times to the Present*. Lincoln, Nebr., 1985.
Coxe, Tench. *A View of the United States of America in a Series of Papers Written at Various Times, in the Years Between 1787 and 1794*. Philadelphia, 1794.
Crafts, N. F. R. *British Economic Growth during the Industrial Revolution*. Oxford, 1985.
Craig, Lee A. *To Sow One Acre More: Child Bearing and Productivity in the Antebellum North*. Baltimore, 1993.

Crandall, Ralph J. "New England's Second Great Migration: The First Three Generations of Settlement, 1630–1700." *New England Historical and Genealogical Register* 129 (1975): 347–60.

Crane, Verner W. *The Southern Frontier, 1670–1732*. Ann Arbor, Mich., 1929.

Crary, Catherine Snell. "Forfeited Loyalist Lands in the Western District of New York: Albany and Tryon Counties." *New York History* 35 (1954): 239–58.

——. "Guerilla Activities of Jame DeLancey's Cowboys in Westchester County: Conventional Warfare or Self-Interested Freebooting?" In *The Loyalist Americans: A Focus on Greater New York*, edited by Robert A. East and Jacob Judd, 14–24, 153–55. Tarrrytown, N.Y., 1975.

Crass, David Colin, Stephen D. Smith, Martha A. Zierden, and Richard D. Brooks, eds. *The Southern Colonial Backcountry: Interdisciplinary Perspectives on Frontier Communities*. Knoxville, Tenn., 1998.

Craven, Avery Odells. *Soil Exhaustion as a Factor in the Agricultural History of Virginia and Maryland, 1606–1860*. Urbana, Ill., 1926.

Craven, Wesley Frank. *Dissolution of the Virginia Company: The Failure of a Colonial Experiment*. New York, 1932.

——. "The Early Settlements: A European Investment of Capital and Labor." In *The Growth of the American Economy*, edited by Harold F. Williamson, 19–43. 2d ed. Englewood Cliffs, N.J., 1951.

——. *White, Red, and Black: The Seventeenth-Century Virginian*. Charlottesville, Va., 1971.

Crawford, E. Margaret, ed. *Famine: The Irish Experience, 900–1900: Subsistence Crises and Famines in Ireland*. Edinburgh, 1989.

——. "William Wilde's Table of Irish Famines, 900–1850." In E. Margaret Crawford, *Famine*, 1–30.

Crawford, William H. *Domestic Industry in Ireland: The Experience of the Linen Industry*. Dublin, 1972.

——. "Landlord-Tenant Relations in Ulster, 1609–1820." *Irish Economic and Social History* 2 (1975): 5–21.

——. "The Social Structure of Ulster in the Eighteenth Century." In Cullen and Furet, *Ireland and France*, 117–27.

——. "Ulster as a Mirror of the Two Societies." In Devine and Dickson, *Ireland and Scotland*, 60–69.

——. "Women in the Domestic Linen Industry." In MacCurtain and O'Dowd, *Women in Early Modern Ireland*, 255–64.

Cray, Robert E., Jr. *Paupers and Poor Relief in New York City and Its Rural Environs, 1700–1830*. Philadelphia, 1988.

Cremin, Lawrence A. *American Education: The Colonial Experience, 1607–1783*. New York, 1970.

Cress, Lawrence Delbert. *Citizens in Arms: The Army and Militia in American Society to the War of 1812*. Chapel Hill, N.C., 1982.

Cressy, David. *Birth, Marriage, and Death: Ritual, Religion, and the Life-Cycle in Tudor and Stuart England*. Oxford, 1997.

——. *Coming Over: Migration and Communication between England and New England in the Seventeenth Century*. Cambridge, 1987.

——. "The Seasonality of Marriage in Old and New England." *JIH* 16 (1985): 1–22.

Crèvecoeur, J. Hector St. John de. *Letters from an American Farmer and Sketches of Eighteenth-Century America*. 1782. Edited by Albert E. Stone. New York, 1981.

Crittenden, Charles Christopher. *The Commerce of North Carolina, 1763–1789*. New Haven, Conn., 1936.
Cronon, William. *Changes in the Land: Indians, Colonists, and the Ecology of New England*. New York, 1983.
Crosby, Alfred W., Jr. *The Columbian Exchange: Biological and Cultural Consequences of 1492*. Westport, Conn., 1972.
——. *Ecological Imperialism: The Biological Expansion of Europe, 900–1900*. Cambridge, 1986.
——. "Virgin Soil Epidemics as a Factor in the Aboriginal Depopulation of America." *WMQ* 33 (1976): 289–99.
Crouse, Nellis M. "Causes of the Great Migration, 1630–1640." *NEQ* 5 (1932): 3–36.
Crow, Jeffrey J. *The Black Experience in Revolutionary North Carolina*. Raleigh, N.C., 1977.
——. "Liberty Men and Loyalists: Disorder and Disaffection in Backcountry North Carolina, 1776–1783." In Hoffman, Tate, and Albert, *Uncivil War*, 125–78.
——. "What Price Loyalism? The Case of John Cruden, Commissioner of Sequestered Estates." *North Carolina Historical Review* 58 (1981): 215–33.
Crowley, John E. "Family Relations and Inheritance in Early South Carolina." *Histoire Sociale—Social History* 17 (1984): 35–57.
——. "The Importance of Kinship: Testamentary Evidence from South Carolina." *JIH* 16 (1986): 559–77.
Cullen, Louis Michael. "Catholic Social Classes under the Penal Laws." In Power and Whelan, *Endurance and Emergence*, 57–84.
——. "Economic Development, 1691–1750." In Moody and Vaughan, *New History of Ireland*, 123–58.
——. "Economic Development, 1750–1800." In Moody and Vaughan, *New History of Ireland*, 159–95.
——. *An Economic History of Ireland since 1660*. London, 1972.
——. *The Emergence of Modern Ireland, 1600–1900*. New York, 1981.
——. "The Irish Diaspora of the Seventeenth and Eighteenth Centuries." In Canny, *Europeans on the Move*, 113–49.
——. "Irish History without the Potato." *P&P*, no. 40 (1968): 72–83.
——. "Problems in the Interpretation and Revision of Eighteenth-Century Irish Economic History." *Transactions of the Royal Historical Society*, 5th ser., 17 (1967): 1–22.
Cullen, Louis Michael, and François Furet, eds. *Ireland and France, Seventeenth–Twentieth Centuries: Toward a Comparative Study of Rural History*. Ann Arbor, Mich., 1980.
Cullen, Louis Michael, and T. Christopher Smout, eds. *Comparative Aspects of Scottish and Irish Economic and Social History, 1600–1900*. Edinburgh, 1977.
——. "Economic Growth in Scotland and Ireland." In Cullen and Smout, *Comparative Aspects of Scottish and Irish Economic and Social History*, 3–18.
Cullen, Louis Michael, T. Christopher Smout, and Alex Gibson. "Wages and Comparative Development in Ireland and Scotland, 1565–1780." In Mitchison and Roebuck, *Economy and Society in Scotland and Ireland*, 105–16.
Cullen, Nuala. "Women and the Preparation of Food in Eighteenth-Century Ireland." In MacCurtain and O'Dowd, *Women in Early Modern Ireland*, 265–75.
Cummings, Abbott Lowell. *The Framed Houses of Massachusetts Bay, 1625–1725*. Cambridge, Mass., 1979.

Cunningham, Hugh. "The Employment and Unemployment of Children in England, c. 1680–1851." *P&P*, no. 126 (1990): 115–50.
Cunz, Dieter. *The Maryland Germans: A History*. Princeton, N.J., 1948.
Curry, Richard Orr. "Lord Dunmore and the West: A Re-evaluation." *West Virginia History* 19 (1958): 231–43.
Customs 16/1. Public Record Office, London. Film at Colonial Williamsburg.
Cutcliffe, Stephen H. "Colonial Indian Policy as a Measure of Rising Imperialism: New York and Pennsylvania, 1700–1755." *Western Pennsylvania Historical Magazine* 64 (1981): 237–68.
Dahlman, Carl Johan. *The Open Field System and Beyond*. Cambridge, 1980.
Daniell, Jere R. *Colonial New Hampshire: A History*. Millwood, N.Y., 1981.
Daniels, Bruce C. *The Connecticut Town: Growth and Development, 1635–1790*. Middletown, Conn., 1979.
——. *The Fragmentation of New England: Comparative Perspectives on Economic, Political, and Social Divisions in the Eighteenth Century*. New York, 1988.
Daniels, Christine. "From Father to Son: Economic Roots of Craft Dynasties in Eighteenth-Century Maryland." In *American Artisans: Crafting Social Identity, 1750–1850*, edited by Howard B. Rock, Paul A. Gilje, and Robert Asher, 3–16, 201–9. Baltimore, 1995.
——. "'Getting His [or Her] Livelyhood': Free Workers in Slave Anglo-America, 1675–1810." *AgH* 71 (1997): 125–62.
——. "Gresham's Laws: Labor Management on an Early-Eighteenth Century Chesapeake Plantation." *Journal of Southern History* 62 (1996): 205–38.
——. "'WANTED': A Blacksmith Who Understands Plantation Work': Artisans in Maryland, 1700–1800." *WMQ* 50 (1993): 743–67.
Daniels, John D. "The Indian Population of North America in 1492." *WMQ* 49 (1992): 298–320.
Dann, John C., ed. *The Revolution Remembered: Eyewitness Accounts of the War for Independence*. Chicago, 1980.
Darlington, James W. "Peopling the Post-Revolutionary New York Frontier." *New York History* 74 (1993): 341–81.
Dauer, David E. "Colonial Philadelphia's Hinterland Economy: The Wheat Supply System for the Milling Industry." Paper, McNeil Center for Early American Studies, January 1980.
——. "Colonial Philadelphia's Intraregional Transportation System: An Overview." *Working Papers from the Regional Economic History Center* 2, no. 3 (1979): 1–16.
Daultrey, Stuart, David Dickson, and Cormac Ó Gráda. "Eighteenth-Century Irish Population: New Perspectives from Old Sources." *JEH* 46 (1981): 601–28.
Davies, Kathleen M. "The Sacred Condition of Equality: How Original Were the Puritan Doctrines of Marriage?" *Social History* 2 (1977): 563–80.
Davis, David Brion, and Steven Mintz, eds. *The Boisterous Sea of Liberty: A Documentary History of America from Discovery through the Civil War*. New York, 1998.
Davis, Harold E. *The Fledgling Province: Social and Cultural Life in Colonial Georgia, 1733–1776*. Chapel Hill, N.C., 1976.
Davis, Lance, and Stanley Engerman. "The Economy of British North America: Miles Traveled, Miles to Go." *WMQ* 56 (1999): 9–22.
Davis, Richard Beale. *Intellectual Life in the Colonial South, 1585–1763*. 3 vols. Knoxville, Tenn., 1978.
Davis, Richard C., ed. *Encyclopedia of American Forest and Conservation History*. 2 vols. New York, 1983.

Davis, Thomas J. "New York's Long Black Line: A Note on the Growing Slave Population, 1626–1790." *Afro-Americans in New York Life and History* 2 (1978): 41–59.
Day, Gordon M. *In Search of New England's Past: Selected Essays of Gordon M. Day.* Edited by Michael K. Foster and William Cowan. Amherst, Mass., 1998.
Dayton, Cornelia Hughes. "Taking the Trade: Abortion and Gender Relations in an Eighteenth-Century New England Village." *WMQ* 48 (1991): 19–49.
——. *Women before the Bar: Gender, Law, and Society in Connecticut, 1639–1789.* Chapel Hill, N.C., 1995.
Deal, Laura K. "Widows and Reputation in the Diocese of Chester, England, 1560–1650." *Journal of Family History* 23 (1998): 382–92.
Dean, George W. "Virginia Forests." In Richard C. Davis, *Encyclopedia of Forest History,* 2:671–73.
Deane, Phyllis, and W. A. Cole. *British Economic Growth, 1688–1959.* Cambridge, Mass., 1969.
Delâge, Denys. *Bitter Feast: Amerindians and Europeans in Northeastern North America, 1600–64.* Vancouver, 1993.
Delaney, Norman C. "The Outer Banks of North Carolina during the Revolutionary War." *North Carolina Historical Review* 36 (1959): 1–16.
Delaware Archives Commission. *Delaware Archives.* 3 vols. Wilmington, Del., 1911–19.
Delbanco, Andrew. *The Puritan Ordeal.* Cambridge, Mass., 1989.
Del Papa, Eugene M. "The Royal Proclamation of 1763: Its Effect upon Virginia Land Companies." *VMHB* 83 (1975): 406–11.
Demos, John Putnam. "Demography and Psychology in the History Study of Family-Life: A Personal Report." In Laslett and Wall, *Household and Family in Past Time,* 561–70.
——. "Families in Colonial Bristol, Rhode Island: An Exercise in Historical Demography." *WMQ* 25 (1968): 40–57.
——. *A Little Commonwealth: Family Life in Plymouth Colony.* New York, 1970.
——. *Past, Present, and Personal: The Family and Life Course in American History.* New York, 1986.
——. *The Unredeemed Captive: A Family Story from Early America.* New York, 1994.
——, ed. *Remarkable Providences: Readings on Early American History.* 2d ed. Boston, 1991.
Denevan, William M. "Native American Populations in 1492: Recent Research and a Revised Hemispheric Estimate." In *The Native Population of the Americas in 1492,* edited by William M. Denevan, xvii–xxxviii. 2d ed. Madison, Wis., 1992.
——. "The Pristine Myth: The Landscape of the Americas in 1492." *Annals of the Association of American Geographers* 82 (1992): 369–85.
Dennis, Matthew. *Cultivating a Landscape of Peace: Iroquois-European Encounters in Seventeenth-Century America.* Ithaca, N.Y., 1993.
DePauw, Linda Grant. "Women in Combat: The Revolutionary War Experience." *Armed Forces and Society* 7 (1981): 209–26.
Derven, Daphne L. "Wholesome, Toothsome, and Diverse: Eighteenth-Century Foodways in Deerfield, Massachusetts." In Benes, *Foodways in the Northeast,* 47–63.
Devine, Thomas Martin. *Clanship to Crofters' War: The Social Transformation of the Highlands.* Manchester, England, 1994.
——. "Colonial Commerce and the Scottish Economy, c. 1730–1815." In Cullen and Smout, *Comparative Aspects of Scottish and Irish Economic and Social History,* 177–90.
——. "The Colonial Trades and Industrial Investment in Scotland, c. 1700–1815." *EcHR* 29 (1976): 1–13.

——. "Highland Migration to Lowland Scotland, 1760–1860." *Scottish Historical Review* 62 (1983): 137–49.
——. "Introduction: Scottish Farm Service in the Agricultural Revolution." In Devine, *Farm Servants and Labour*, 1–8.
——. "Social Responses to Agrarian 'Improvement': The Highland and Lowland Clearances in Scotland." In Houston and Whyte, *Scottish Society*, 148–68.
——. *The Tobacco Lords: A Study of the Tobacco Merchants of Glasgow and Their Trading Activities, c 1740–1790*. Edinburgh, 1975.
——. *The Transformation of Rural Scotland: Social Change and the Agrarian Economy, 1660–1815*. Edinburgh, 1994.
——. "Urbanisation." In Devine and Mitchison, *People and Society in Scotland*, 27–52.
——, ed. *Farm Servants and Labour in Lowland Scotland, 1770–1914*. Edinburgh, 1984.
Devine, Thomas Martin, and David Dickson, eds. *Ireland and Scotland, 1600–1850: Parallels and Contrasts in Economic and Social Development*. Edinburgh, 1983.
Devine, Thomas Martin, and Rosalind Mitchison, eds. *People and Society in Scotland*. Vol. 1, *1760–1830*. Edinburgh, 1988.
De Vorsey, Louis, Jr. "The Colonial Georgia Backcountry." In *Colonial Augusta: "Key of the Indian Country,"* edited by Edward J. Cashin, 3–26. Macon, Ga., 1986.
——. "Indian Boundaries in Colonial Georgia." *Georgia Historical Quarterly* 54 (1970): 63–78.
——. *The Indian Boundary in the Southern Colonies, 1763–1775*. Chapel Hill, N.C., 1966.
De Vries, Jan. "Between Purchasing Power and the World of Goods: Understanding the Household Economy in Early Modern Europe." In Brewer and Porter, *Consumption and the World of Goods*, 85–132.
——. *European Urbanization, 1500–1800*. Cambridge, Mass., 1984.
DeWindt, Anne Reiber. "Redefining the Peasant Community in Medieval England: The Regional Perspective." *Journal of British Studies* 26 (1987): 163–207.
De Wolfe, Barbara, ed. *Discoveries of America: Personal Accounts of British Emigrants to North America during the Revolutionary Era*. Cambridge, 1997.
Deyle, Steven. "The Irony of Liberty: The Origins of the Domestic Slave Trade." *Journal of the Early Republic* 12 (1992): 37–62.
Diamond, Sigmund. "From Organization to Society: Virginia in the Seventeenth Century." *American Journal of Sociology* 63 (1958): 457–75.
Dickinson, H. T. "The Poor Palatines and the Parties." *English Historical Review* 82 (1967): 464–85.
Dickson, David. "The Catholics and Trade in Eighteenth-Century Ireland: An Old Debate Revisited." In Power and Whelan, *Endurance and Emergence*, 85–100.
——. "The Gap in Famines: A Useful Myth?" In E. Margaret Crawford, *Famine*, 96–111.
——. "In Search of the Old Irish Poor Law." In Mitchison and Roebuck, *Economy and Society in Scotland and Ireland*, 149–59.
——. "The Place of Dublin in the Eighteenth-Century Irish Economy." In Devine and Dickson, *Ireland and Scotland*, 178–92.
——. "Property and Social Structure in Eighteenth-Century South Munster." In Cullen and Furet, *Ireland and France*, 129–38.
Dickson, R. J. *Ulster Emigration to Colonial America, 1718–1755*. London, 1966.
Dickson, Sarah Augusta. *Panacea or Precious Bane: Tobacco in Sixteenth Century Literature*. New York, 1954.
Dietz, B. "The North-East Coal Trade, 1550–1750: Measures, Markets, and the Metropolis." *Northern History* 22 (1986): 280–94.

Dirlmeier, Ulf, and Gerhard Fouquet. "Diet and Consumption." In Scribner, *Germany*, 85–111.
Ditz, Toby L. *Property and Kinship: Inheritance in Early Connecticut, 1750–1820*. Princeton, N.J. 1986.
Dobb, Maurice. *Studies in the Development of Capitalism*. New York, 1947.
Doblin, Helga, trans., and Mary C. Lynn, ed. *The Specht Journal: A Military Journal of the Burgoyne Campaign*. Westport, Conn., 1995.
Dobson, David. *Scottish Emigration to Colonial America, 1607–1885*. Athens, Ga., 1994.
Dodd, A. H. *The Character of Early Welsh Emigration to the United States*. Cardiff, Wales, 1953.
Doddridge, Joseph. *Notes on the Settlement and Indian Wars of the Western Parts of Virginia and Pennsylvania from 1763 to 1782 inclusive, together with a Review of the State of Society and Manners of the First Settlers of the Western Country*. 1824. Pittsburgh, 1912.
Dodgshon, Robert A. "Origins of the Traditional Field Systems." In Parry and Slater, *Making of the Scottish Countryside*, 69–92.
——. "Removal of the Runrig in Roxburghshire and Berwickshire, 1680–1766." *Scottish Studies* 16 (1972): 121–37.
——. "Strategies of Farming in the Western Highlands and Islands of Scotland prior to Crofting and the Clearances." *EcHR* 46 (1993): 679–701.
——. "Toward an Understanding and Definition of Runrig: The Evidence for Roxburghshire and Berwickshire." *Institute of British Geographers, Transactions*, no. 64 (1975): 15–33.
Doerflinger, Thomas M. "Farmers and Dry Goods in the Philadelphia Market Area, 1750–1800." in Hoffman et al., *Economy of Early America*, 166–95.
——. *A Vigorous Spirit of Enterprise: Merchants and Economic Development in Revolutionary Philadelphia*. Chapel Hill, N.C., 1986.
Döhla, Johann Conrad. *A Hessian Diary of the American Revolution*. Edited and translated by Bruce E. Burgoyne. Norman, Okla., 1989.
Dolan, Frances E. *Dangerous Familiars: Representations of Domestic Crime in England, 1550–1700*. Ithaca, N.Y., 1994.
Domar, Evsey. "The Causes of Slavery and Serfdom: A Hypothesis." *JEH* 30 (1970): 18–32.
Donne, John. *A Sermon Preached to the Honourable Company of the Virginia Plantation, 13 November 1622*. In *The Sermons of John Donne*, edited by George R. Potter and Evelyn M. Simpson, 4:264–82. Berkeley, Calif., 1959.
Donovan, Kenneth. " 'Taking Leave of an Ungrateful Country': The Loyalist Exile of Joel Stone." *Dalhousie Review* 64 (1984): 125–45.
Dorwart, Jeffery M. *Cape May County, New Jersey: The Making of an American Resort Community*. New Brunswick, N.J., 1992.
Dowd, Gregory Evans. "The Panic of 1751: The Significance of Rumors on the South Carolina–Cherokee Frontier." *WMQ* 53 (1996): 527–60.
——. *A Spirited Resistance: The North American Indian Struggle for Unity, 1745–1815*. Baltimore, 1992.
Downes, Randolph C. "Dunmore's War: An Interpretation." *Mississippi Valley Historical Review* 21 (1934): 311–30.
Doyle, David Noel. *Ireland, Irishmen, and Revolutionary America, 1760–1820*. Dublin, 1981.
Drake, James. "Restraining Atrocity: The Conduct of King Philip's War." *NEQ* 70 (1997): 3–32.

Du Bois, W. E. B. *The Suppression of the African Slave Trade to the United States*. New York, 1904.
Duffy, John. *Epidemics in Colonial America*. Baton Rouge, La., 1953.
Dunaway, Wayland Fuller. *The Scotch-Irish of Colonial Pennsylvania*. Chapel Hill, N.C., 1944.
Dunaway, Wilma A. *The First American Frontier: Transition to Capitalism in Southern Appalachia, 1700–1860*. Chapel Hill, N.C., 1996.
Dunbar, Seymour. *A History of Travel in America . . . .* New York, 1937.
Dunn, Richard S. "The Barbados Census of 1680: Profile of the Richest Colony in English America." *WMQ* 26 (1969): 3–30.
——. "The English Sugar Islands and the Founding of South Carolina." *South Carolina Historical Magazine* 72 (1971): 81–93.
——. "Experiments Holy and Unholy, 1630–1." In Andrews, Canny, and Hair, *Westward Enterprise*, 271–89.
——. "An Odd Couple: John Winthrop and William Penn." *Massachusetts Historical Society, Proceedings* 90 (1987): 1–24.
——. "Penny Wise and Pound Foolish: Penn as a Businessman." In Dunn and Dunn, *World of William Penn*, 37–54.
——. "Servants and Slaves: The Recruitment and Employment of Labor." In Greene and Pole, *Colonial British America*, 157–94.
——. *Sugar and Slaves: The Rise of the Planter Class in the English West Indies, 1624–1713*. Chapel Hill, N.C., 1972.
Dunn, Richard S., and Mary Maples Dunn, eds. *The Papers of William Penn*. Vol. 2, *1680–1684*. Philadelphia, 1982.
——. *The World of William Penn*. Philadelphia, 1986.
Durie, Alastair J. *The Scottish Linen Industry in the Eighteenth Century*. Edinburgh, 1979.
——. "The Scottish Linen Industry in the Eighteenth Century: Some Aspects of Expansion." In Cullen and Smout, *Comparative Aspects of Scottish and Irish Economic and Social History*, 88–99.
Durston, Christopher. *The Family in the English Revolution*. Oxford, 1989.
Dwight, Timothy. *Travels in New England and New York*. 4 vols. Edited by Barbara Miller Solomon. Cambridge, Mass., 1969.
Dyck, Cornelius J. "European Mennonite Motivation for Emigration, 1650–1750." *Pennsylvania Mennonite Heritage* 6 (1983): 2–9.
Dyer, Christopher. "English Diet in the Later Middle Ages." In Aston et al., *Social Relations and Ideas*, 191–216.
——. *Everyday Life in Medieval England*. London, 1994.
——. *Lords and Peasants in a Changing Society: The Estates of the Bishopric of Worcester, 680–1540*. Cambridge, 1980.
——. *Standards of Living in the Later Middle Ages: Social Change in England c. 1200–1520*. Cambridge, 1989.
Earle, Alice Morse. *Home Life in Colonial Days*. New York, 1898.
Earle, Carville. *The Evolution of a Tidewater Settlement System: All Hallow's Parish, Maryland, 1650–1783*. University of Chicago Department of Geography Research Paper, no. 170, 1975.
——. *Geographical Inquiry and American Historical Problems*. Stanford, Calif., 1992.
——. "Pioneers of Providence: Anglo-American Experience, 1492–1792." *Association of American Geographers, Annals* 82 (1992): 478–99.
——. "The Rate of Frontier Expansion in American History, 1650–1890." In *Lois Green Carr*, 183–204.

Earle, Carville, and Changyong Cao. "Frontier Closure and the Involution of American Society, 1840–1890." *Journal of the Early Republic* 13 (1993): 163–79.
Earle, Peter. "The Female Labour Market in London in the Late Seventeenth and Early Eighteenth Centuries." *EcHR* 42 (1989): 328–53.
East, Robert A. *Business Enterprise in the American Revolutionary Era*. New York, 1938.
Eburne, Richard. *A Plain Pathway to Plantations*. 1624. Edited by Lewis B. Wright. Ithaca, N.Y., 1962.
Echeverria, Durand. *Mirage in the West: A History of the French Image of American Society to 1815*. Princeton, N.J., 1957.
Eden, Peter. "Three Elizabethan Estate Surveyors: Peter Kempe, Thomas Clerke, and Thomas Langdon." In Sarah Tyacke, *English Map-Making*, 68–84.
Egnal, Marc. "The Economic Development of the Thirteen Continental Colonies, 1720 to 1775." *WMQ* 32 (1975): 191–222.
——. *A Mighty Empire: The Origins of the American Revolution*. Ithaca, N.Y., 1988.
——. *New World Economies: The Growth of the Thirteen Colonies and Early Canada*. New York, 1998.
——. "The Pennsylvania Economy, 1748–1762: An Analysis of Short-Run Fluctuations in the Context of Long-Run Changes in the Atlantic Trading Community." Ph.D. diss., University of Wisconsin, 1974.
Ekirch, A. Roger. *Bound for America: The Transportation of British Convicts to the Colonies, 1718–1775*. Oxford, 1987.
——. *"Poor Carolina": Politics and Society in Colonial North Carolina, 1729–1776*. Chapel Hill, N.C., 1981.
——. "The Transportation of Scottish Criminals to America during the Eighteenth Century." *Journal of British Studies* 24 (1985): 366–74.
——. "Whig Authority and Public Order in Backcountry North Carolina, 1776–1783." In Hoffman, Tate, and Albert, *Uncivil War*, 99–124.
Eley, Geoff, and William Hunt, eds. *Reviving the English Revolution: Reflections and Elaborations on the Work of Christopher Hill*. London, 1988.
Eller, Ernest McNeill, ed. *Chesapeake Bay in the American Revolution*. Centerville, Md., 1981.
Ellet, Elizabeth F. *The Women of the American Revolution*. 3 vols. Reprint of 4th (1850) ed. New York, 1969.
Elliott, Vivien Brodsky. "Single Women in the London Marriage Market: Age, Status, and Mobility, 1598–1619." In *Marriage and Society: Studies in the Social History of Marriage*, edited by R. B. Outhwaite, 81–100. New York, 1981.
Ellis, David M. "The Yankee Invasion of New York, 1783–1850." *New York History* 32 (1951): 1–17.
Ellis, David M., James A. Frost, Harold C. Syrett, and Harry J. Carman. *A History of New York State*. Ithaca, N.Y., 1967.
Eltis, David, and Stanley L. Engerman. "Was the Slave Trade Dominated by Men?" *JIH* 23 (1992): 237–58.
Ely, James W., Jr. *The Guardian of Every Other Right: A Constitutional History of Property Rights*. New York, 1992.
Emerson, Everett, ed. *Letters from New England: The Massachusetts Bay Colony, 1629–1638*. Amherst, Mass., 1976.
Emery, F. V. "England circa 1600." In *A New Historical Geography of England*, edited by H. C. Darby, 248–301. Cambridge, 1973.
Emmer, P. C., ed. *Colonialism and Migration: Indentured Labour before and after Slavery*. Dordrecht, Netherlands, 1986.

Engerman, Stanley L., and Robert E. Gallman, eds. *The Cambridge Economic History of the United States*. Vol. 1, *The Colonial Era*. Cambridge, 1996.
England in America: The Chesapeake Bay from Jamestown to St. Mary's City, 1607–1634. Map. Champlain, Va., 1996.
Erickson, Amy Louise. *Women and Property in Early Modern England*. London, 1993.
Ernst, Joseph Albert. *Money and Politics in America, 1755–1775: A Study in the Currency Act of 1764 and the Political Economy of Revolution*. Chapel Hill, N.C., 1973.
Ernst, Joseph Albert, and H. Roy Merrens. "'Camden's Turrets Pierce the Skies!': The Urban Process in the Southern Colonies during the Eighteenth Century." *WMQ* 30 (1973): 549–74.
Ernst, Robert. "A Tory-eye View of the Evacuation of New York City." *New York History* 64 (1983): 376–94.
Eslinger, Ellen. "Migration and Kinship on the Trans-Appalachian Frontier: Strode's Station, Kentucky." *Filson Club Historical Quarterly* 62 (1988): 52–66.
Essays in Colonial History Presented to Charles McLean Andrews by his Students. New Haven, Conn., 1931.
Evans, Elizabeth. "Heroines All: The Plight of Women at War in America, 1776–1778." In *Conflict at Monmouth Court House: Proceedings of a Symposium Commemorating the Two-Hundredth Anniversary of the Battle of Monmouth, April 8, 1978*, edited by Mary R. Murrin and Richard Waldron, 23–28. Trenton, N.J., 1983.
Evans, Emory G. *Thomas Nelson of Yorktown: Revolutionary Virginian*. Williamsburg, Va., 1975.
Evans, E. Raymond. "Was the Last Battle of the American Revolution Fought on Lookout Mountain?" *Journal of Cherokee Studies* 5 (1980): 30–40.
Everitt, Alan. "Farm Labourers." In Thirsk, *Agrarian History*, 396–465.
——. "The Marketing of Agricultural Produce." In Thirsk, *Agrarian History*, 466–592.
Eversley, D. E. C. "The Home Market and Economic Growth in England, 1750–1850." In *Land, Labour, and Population in the Industrial Revolution: Essays Presented to J. D. Chambers*, edited by E. L. Jones and G. E. Mingay, 206–59. New York, 1967.
Everton, George B., Sr., and Gunnar Rasmuson, eds. *The Handy Book for Genealogists*. 3d ed. Logan, Utah, 1957.
Ezell, Margaret J. M. *The Patriarch's Wife: Literary Evidence and the History of the Family*. Chapel Hill, N.C., 1987.
Fabel, Robin F. A. *The Economy of British West Florida, 1763–1783*. Tuscaloosa, Ala., 1988.
——. "An Eighteenth Colony: Dreams for Mississippi on the Eve of the Revolution." *Journal of Southern History* 59 (1993): 647–72.
Fabend, Firth Haring. *A Dutch Family in the Middle Colonies, 1660–1800*. New Brunswick, N.J., 1991.
Fairbanks, Jonathan L., and Robert F. Trent, eds. *New England Begins: The Seventeenth Century*. 3 vols. Boston, 1982.
Faith, Rosamond. "Berkshire: Fourteenth and Fifteenth Centuries." In Harvey, *Peasant Land Market*, 107–78.
Fallaw, Robert, and Marion West Stoer. "The Old Dominion under Fire: The Chesapeake Invasions, 1779–1781." In Eller, *Chesapeake Bay in the American Revolution*, 432–74.
Faragher, John Mack. *Daniel Boone: The Life and Legend of an American Pioneer*. New York, 1992.
——. "Old Women and Old Men in Seventeenth-Century Wethersfield, Connecticut." *Women's Studies* 4 (1976): 11–31.
——. *Sugar Creek: Life on the Illinois Prairie*. New Haven, Conn., 1986.

Farber, Bernard. *Guardians of Virtue: Salem Families in 1800*. New York, 1972.
Farmer, Charles J. *In the Absence of Towns: Settlement and Country Trade in Southside Virginia, 1730–1800*. Lanham, Md., 1993.
Farmer, David L. "The *Famuli* in the Later Middle Ages." In Britnell and Hatcher, *Progress and Problems in Medieval England*, 207–36.
——. "Marketing the Produce of the Countryside, 1200–1500." In Edward Miller, *Agrarian History of England and Wales*, 324–430.
——. "Prices and Wages, 1350–1500." In Edward Miller, *Agrarian History of England and Wales*, 431–525.
Faust, Albert B. "Swiss Emigration to the American Colonies in the Eighteenth Century." *American Historical Review* 22 (1916): 21–44.
Fausz, J. Frederick. "An 'Abundance of Blood Shed on Both Sides': England's First Indian War, 1609–1614." *VMHB* 98 (1990): 3–56.
——. " 'Engaged in Enterprises Pregnant with Terror': George Washington's Formative Years among the Indians." In *George Washington and the Virginia Backcountry*, edited by Warren R. Hofstra, 115–55. Madison, Wis., 1998.
——. "Fighting 'Fire' with Firearms: The Anglo-Powhatan Arms Race in Early Virginia." *American Indian Culture and Research Journal* 3 (1979): 33–50.
——. "Indians, Colonialism, and the Conquest of Cant: A Review-Essay on Anglo-Indian Relations in the Chesapeake." *VMHB* 95 (1987): 133–56.
——. "Merging and Emerging Worlds: Anglo-Indian Interest Groups and the Development of the Seventeenth-Century Chesapeake." In Carr, Morgan, and Russo, *Colonial Chesapeake Society*, 47–98.
——. "Middlemen in Peace and War: Virginia's Earliest Indian Interpreters, 1608–1632." *VMHB* 95 (1987): 41–64.
——. "The Missing Women of Martin's Hundred." *American History* 33 (March 1998): 56–60, 62.
——. "Patterns of Anglo-Indian Aggression and Accommodation along the Mid-Atlantic Coast, 1584–1634." In Fitzhugh, *Cultures in Contact*, 225–68.
——. "Present at the 'Creation': The Chesapeake World That Greeted the Maryland Colonists." *MdHM* 79 (1984): 7–20.
——. " 'To Draw Thither the Trade of Beavers': The Strategic Significance of the English Fur Trade in the Chesapeake, 1620–1660." In *"Le Castor Fait Tout": Selected Papers of the Fifth North American Fur Trade Conference*, edited by Bruce G. Trigger, Toby Morantz, and Louise Dechene, 42–71. Montreal, 1985.
Feer, Robert A. "Imprisonment for Debt in Massachusetts before 1800." *Mississippi Valley Historical Review* 48 (1961): 252–69.
Feest, Christian F. "Nanticoke and Neighboring Tribes." In Trigger, *Northeast*, 240–52.
——. "Virginia Algonquians." In Trigger, *Northeast*, 253–70.
Feister, Lois M. "Linguistic Communication between the Dutch and Indians in New Netherland, 1609–1664." *Ethnohistory* 20 (1973): 24–38.
Felt, Joseph B. "Statistics of Population in Massachusetts." *American Statistical Association, Collections* 1, pt. 2 (1845): 121–216.
Fennelly, Catherine. *Connecticut Women in the Revolutionary Era*. Connecticut Bicentennial Series, 15. Chester, Conn., 1975.
Fenske, Hans. "International Migration: Germany in the Eighteenth Century." *Central European History* 13 (1980): 332–47.
Fenton, Alexander. *The Northern Isles: Orkney and Shetland*. Edinburgh, 1978.
Fenton, William N. *The Great Law and the Longhouse: A Political History of the Iroquois Confederacy*. Norman, Okla., 1998.

Ferguson, Clyde R. "Carolina and Georgia Patriot and Loyalist Militia in Action, 1778–1783." In *The Southern Experience in the American Revolution*, edited by Jeffrey J. Crow, and Larry E. Tise, 174–99. Chapel Hill, N.C., 1978.

Ferguson, E. James. *The Power of the Purse: A History of Public Finance, 1776–1790*. Chapel Hill, N.C., 1961.

Ferling, John. *Struggle for a Continent: The Wars of Early America*. Arlington Heights, Ill., 1993.

Fertig, Georg. "Transatlantic Migration from the German-Speaking Parts of Central Europe, 1600–1800: Proportions, Structures, and Explanations." In Canny, *Europeans on the Move*, 192–235.

Field, Jonathan Beecher. " 'Peculiar Manuerance': Puritans, Indians, and the Rhetoric of Agriculture." In Benes, *Plants and People*, 12–24.

Fildes, Valerie. "Maternal Feelings Re-Assessed: Child Abandonment and Neglect in London and Westminster, 1550–1800." In *Women as Mothers in Pre-Industrial England: Essays in Memory of Dorothy McLaren*, edited by Valerie Fildes, 139–78. London, 1990.

Finch, Andrew. "*Repulsa Uxore Sua*: Marital Difficulties and Separation in the Later Middle Ages." *Continuity and Change* 8 (1993): 11–38.

Fingerhut, Eugene R. "Assimilation of Immigrants on the Frontier of New York, 1764–1776." Ph.D. diss., Columbia University, 1962.

Finlay, Roger. "Natural Decrease in Early Modern Cities." *P&P*, no. 92 (1981): 169–74.

——. *Population and Metropolis: The Demography of London, 1580–1650*. Cambridge, 1981.

Finlay, Roger, and Beatrice Shearer. "Population Growth and Suburban Development." In *London, 1500–1700: The Making of a Metropolis*, edited by A. L. Beier and Findlay, 37–59. London, 1986.

Firth, C. H., ed. *An American Garland, being a Collection of Ballads Relating to America, 1563–1759*. Oxford, 1915.

Fischer, David Hackett. "*Albion* and the Critics: Further Evidence and Reflections." *WMQ* 48 (1991): 260–308.

——. *Albion's Seed: Four British Folkways in America*. New York, 1989.

——. "Forenames and the Family in New England: An Exercise in Historical Onomastics." In Taylor and Crandall, *Generations and Change*, 215–42.

——. *Growing Old in America: The Bland-Lee Lectures at Clark University*. Expanded ed. New York, 1978.

——. *Paul Revere's Ride*. New York, 1994.

Fischer, Joseph R. *A Well-Executed Failure: The Sullivan Campaign against the Iroquois, July–September 1779*. Columbia, S.C., 1997.

Fisher, F. J. *London and the English Economy, 1500–1700*. Edited by P. J. Corfield and N. B. Harte. London, 1990.

Fisher, Wolfram. "Rural Industrialization and Population Change." *Comparative Studies in Society and History* 15 (1973): 158–70.

Fissel, Mark Charles, ed. *War and Government in Britain, 1598–1650*. Manchester, England, 1991.

Fitzhugh, William G., ed. *Cultures in Contact: The European Impact on Native American Cultural Institutions in Eastern North America, A.D. 1000–1800*. Washington, D.C., 1985.

Fitzpatrick, John C., ed. *The Writings of George Washington from the Original Manuscript Sources, 1745–1799*. 39 vols. Washington, D.C. 1931–44.

Flaherty, David H. *Privacy in Colonial New England*. Charlottesville, Va., 1967.

Fletcher, Anthony. *Gender, Sex, and Subordination in England, 1500–1800*. New Haven, Conn., 1995.
Fletcher, Stevenson Whitcomb. *Pennsylvania Agriculture and Country Life, 1640–1840*. Harrisburg, Pa., 1950.
Flinn, Michael W. *The European Demographic System, 1500–1820*. Baltimore, 1981.
——, ed. *Scottish Population History from the Seventeenth Century to the 1930s*. Cambridge, 1977.
Floud, Roderick, Kenneth Wachter, and Annabel Gregory. *Height, Health, and History: Nutritional Status in the United Kingdom, 1750–1980*. Cambridge, 1990.
Fogel, Robert William. "Nutrition and the Decline of Mortality since 1700: Some Preliminary Findings." In *Long-Term Factors in American Economic Growth*, edited by Stanley L. Engerman and Robert E. Gallman. Vol. 51, *Studies in Income and Wealth*, 439–527, 537–55. Chicago, 1986.
Fogleman, Aaron S. "From Slaves, Convicts, and Servants to Free Passengers: The Transformation of Immigration in the Era of the American Revolution." *Journal of American History* 85 (1998): 43–76.
——. *Hopeful Journeys: German Immigration, Settlement, and Political Culture in Colonial America, 1717–1775*. Philadelphia, 1996.
——. "Migrations to the Thirteen British North American Colonies: New Estimates." *JIH* 22 (1992): 691–709.
——. "Review Essay: Progress and Possibilities in Migration Studies: The Contributions of Werner Hacker to the Study of Early German Migration to Pennsylvania." *Pennsylvania History* 56 (1989): 318–29.
——. "Women on the Trail in Colonial America: A Travel Journal of German Moravians Migrating to North Carolina in 1766." *Pennsylvania History* 61 (1994): 206–34.
Folbre, Nancy. "Of Patriarchy Born: The Political Economy of Fertility Decisions." *Feminist Studies* 9 (1983): 261–84.
——. "Patriarchy and Capitalism in New England." Ph.D. diss., University of Massachusetts, 1979.
——. "Patriarchy in Colonial New England." *Review of Radical Political Economics* 12 (summer 1980): 4–13.
——. "The Wealth of Patriarchs: Deerfield, Massachusetts, 1760–1840." *JIH* 16 (1985): 199–220.
Foner, Eric. *Tom Paine and Revolutionary America*. New York, 1976.
Foner, Philip S. *Blacks in the American Revolution*. Westport, Conn., 1976.
Fontaine, Laurence. "Family Cycles, Peddling, and Society in Upper Alpine Valleys in the Eighteenth Century." In *Domestic Strategies: Work and Family in France and Italy, 1600–1800*, edited by Stuart Woolf, 43–68. Cambridge, 1991.
Foot, Michael, and Isaac Kramnick, eds. *The Thomas Paine Reader*. Harmondsworth, England, 1987.
Force, Peter, comp. *American Archives: Consisting of a Collection of Authentick Records, State Papers Debates . . . .* 5th ser. 3 vols. Washington, D.C., 1853.
——. *Tracts and Other Papers, Relating Principally to the Origin, Settlement, and Progress of the Colonies in North America, from the Discovery of the Country to the Year 1776*. 3 vols. Washington, D.C., 1836.
Ford, Amelia Clewley. *Colonial Precedents of Our National Land System as It Existed in 1800. Bulletin of the University of Wisconsin*, no. 352. Madison, Wis., 1908.
Foster, Stephen. *The Long Argument: English Puritanism and the Shaping of New England Culture, 1570–1700*. Chapel Hill, N.C., 1991.

Fowler, David H. "Connecticut's Freemen: The First Forty Years." *WMQ* 15 (1958): 312–33.
Fowler, William S. "Agricultural Tools of the Northeast." *Bulletin of the Massachusetts Archaeological Society* 15, no. 3 (1954): 41–51.
Fox, Edith M. *Land Speculation in the Mohawk Country*. Cornell Studies in American History, Literature, and Folklore, no. 3. Ithaca, N.Y., 1949.
Fox, H. S. A. "Exploitation of the Landless by Lords and Tenants in Early Medieval England." In Razi and Smith, *Medieval Society and the Manor Court*, 518–68.
Fox-Genovese, Elizabeth. "Property and Patriarchy in Classical Bourgeois Political Theory." *Radical History Review* 4 (1977): 36–59.
——. *Within the Plantation Household: Black and White Women of the Old South*. Chapel Hill, N.C., 1988.
Fox-Genovese, Elizabeth, and Eugene D. Genovese. *Fruits of Merchant Capital: Slavery and Bourgeois Property in the Rise and Expansion of Capitalism*. New York, 1983.
Foyster, Elizabeth. "A Laughing Matter? Marital Discord and Gender Control in Seventeenth-Century England." *Rural History* 4 (1993): 5–21.
Francis, Peter, Jr. "The Beads That Did Not Buy Manhattan Island." In Van Dongen et al., *One Man's Trash*, 53–70.
Franklin, Benjamin. *The Papers of Benjamin Franklin*. Edited by Leonard W. Larbaree and William B. Wilcox. 26 vols. to date. New Haven, Conn., 1959– .
Franz, George W. *Paxton: A Study of Community Structure and Mobility in the Colonial Pennsylvania Backcountry*. New York, 1989.
Frey, Sylvia R. *The British Soldier in America: A Social History of Military Life in the Revolutionary Period*. Austin, Tex., 1981.
——. *Water from the Rock: Black Resistance in a Revolutionary Age*. Princeton, N.J., 1991.
Friedeburg, Robert von. "Reformation of Manners and the Social Composition of Offenders in an East Anglian Cloth Village: Earl's Colne, Essex, 1531–1642." *Journal of British Studies* 29 (1990): 347–85.
——. "Social and Geographical Mobility in the Old World and New World Communities: Earls Colne, Ipswich, and Springfield, 1636–1685." *Journal of Social History* 29 (1995): 375–400.
Friedmann, Karen J. "Victualing Colonial Boston." *AgH* 47 (1973): 189–205.
Friend, Craig. "Merchants and Markethouses: Reflections on Moral Economy in Early Kentucky." *Journal of the Early Republic* 17 (1997): 553–74.
Friend, Esther L. "Notifications and Warnings Out: Strangers Taken into Wrentham, Massachusetts, between 1732 and 1812." *New England Historical and Genealogical Register* 141 (1987): 179–202, 330–57; 142 (1988): 56–84.
Fries, Adelaide L. "North Carolina Certificates of the Revolutionary War Period." *North Carolina Historical Review* 9 (1932): 229–41.
Friis, Herman R. *A Series of Population Maps of the Colonies and the United States, 1625–1790*. American Geographical Society, Mimeographed Publication No. 3. New York, 1940.
Fritts, Harold C., and J. M. Lough. "An Estimate of Average Annual Temperature Variations for North America, 1602 to 1961." *Climatic Change* 7 (1985): 203–24.
Fritts, Harold C., G. Robert Lofgren, and Geoffrey A. Gordon. "Variations in Climate since 1602 as Reconstructed from Tree Rings." *Quaternary Research* 12 (1979): 18–46.
Fritzell, Peter A. "The Wilderness and the Garden: Metaphors for the American Landscape." *Forest History* 12 (1968): 16–23.
Frost, J. William. *The Quaker Family in Colonial America: A Portrait of the Society of Friends*. New York, 1973.

Fryde, E. B. *Peasants and Landlords in Later Medieval England, c. 1380–1525*. Stroud, England, 1996.
Fryer, Mary Beacock. *King's Men: The Soldier Founders of Ontario*. Toronto, 1980.
Fuller, Mary C. *Voyages in Print: English Travel to America, 1576–1624*. Cambridge, 1995.
Fullerton, James N. "Squatters and Land Titles in Early Western Pennsylvania; or, An Introduction to Early Western Pennsylvania Land Titles." *Western Pennsylvania Historical Magazine* 6 (1923): 165–78.
Fussell, G. E. "Social and Agrarian Background of the Pilgrim Fathers." *AgH* 7 (1933): 183–202.
Gagliardo, John G. *From Pariah to Patriot: The Changing Image of the German Peasant, 1770–1840*. Lexington, Ky., 1969.
Gailey, Alan. "The Scots Element in North Irish Popular Culture." *Ethnologia Europaea* 8 (1975): 2–22.
Galenson, David W. "Labor Market Behavior in Colonial America: Servitude, Slavery, and Free Labor." In *Markets in History: Economic Studies of the Past*, edited by David W. Galenson, 52–96. Cambridge, 1989.
——. " 'Middling People' or 'Common Sort'? The Social Origins of Some Early Americans Reexamined." *WMQ* 35 (1978): 499–524.
——. "The Rise and Fall of Indentured Servitude in the Americas: An Economic Analysis." *JEH* 44 (1984): 1–26.
——. "The Settlement and Growth of the Colonies: Population, Labor, and Economic Development." In Engerman and Gallman, *Cambridge Economic History*, 135–207.
——. "The Social Origins of Some Early Americans: Rejoinder." *WMQ* 36 (1979): 264–77.
——. *White Servitude in Colonial America: An Economic Analysis*. Cambridge, 1981.
Gallay, Alan. *The Formation of a Planter Elite: Jonathan Bryan and the Southern Colonial Frontier*. Athens, Ga., 1989.
——, ed. *Voices of the Old South: Eyewitness Accounts, 1528–1861*. Athens, Ga., 1994.
Gallman, James Matthew. "Determinants of Age at Marriage in Colonial Perquimans County, North Carolina." *WMQ* 39 (1982): 176–91.
——. "Relative Ages of Colonial Marriages." *JIH* 14 (1984): 609–18.
Gallman, Robert E. "American Economic Growth before the Civil War: The Testimony of the Capital Stock Estimates." In Gallman and Wallis, *American Economic Growth and Standards of Living*, 79–115.
——. "Can We Build National Accounts for the Colonial Period of American History?" *WMQ* 56 (1999): 23–30.
——. "Changes in the Level of Literacy in a New Community of Early America." *JEH* 48 (1988): 567–82.
——. "Influences on the Distribution of Landholdings in Early Colonial North Carolina." *JEH* 42 (1982): 549–76.
——. "The Pace and Pattern of American Economic Growth." In *American Economic Growth: An Economist's History of the United States*, edited by Lance E. Davis, Richard A. Easterlin, and William N. Parker, 15–60. New York, 1972.
Gallman, Robert E., and John Joseph Wallis, eds. *American Economic Growth and Standards of Living before the Civil War*. Chicago, 1992.
Games, Alison F. *Migration and the Origins of the English Atlantic World*. Cambridge, Mass., 1999.
——. "Venturers, Vagrants, and Vessels of Glory: Migration from England to the Colonies under Charles I." Ph.D. diss., University of Pennsylvania, 1992.
Gampel, Gwen Victor. "The Planter's Wife Revisited: Women, Equity Law, and the

Chancery Court in Seventeenth-Century Maryland." In *Women and the Structure of Society: Selected Papers from the Fifth Berkshire Conference on the History of Women*, 20–35, 256–59. Durham, N.C., 1984.

Garrison, J. Ritchie. "Farm Dynamics and Regional Exchange: The Connecticut Valley Beef Trade, 1670–1850." *AgH* 61 (1987): 1–17.

——. *Landscape and Material Life in Franklin County, Massachusetts, 1770–1860*. Knoxville, Tenn., 1991.

Garvan, Anthony. *Architecture and Town Planning in Colonial Connecticut*. New Haven, Conn., 1951.

Garvin, James L. "The Range Township in Eighteenth-Century New Hampshire." In Benes, *New England Prospect*, 47–68.

Gates, Paul W. *History of Public Land Law Development*. Washington, D.C., 1968.

Gemery, Henry A. "Emigration from the British Isles to the New World, 1630–1700: Inferences from Colonial Populations." *Research in Economic History* 5 (1980): 179–231.

——. "European Emigration to North America, 1700–1820: Numbers and Quasi-Numbers." *Perspectives in American History*, n.s., 1 (1984): 283–342.

——. "Markets for Migrants: English Indentured Servitude and Emigration in the Seventeenth and Eighteenth Centuries." In Emmer, *Colonialism and Migration*, 33–54.

George, Charles H. "The Making of the English Bourgeoisie, 1500–1750." *Science and Society* 35 (1971): 385–414.

George, Charles H., and Katherine George. *The Protestant Mind of the English Reformation, 1570–1640*. Princeton, N.J., 1961.

George, Margaret. *Women in the First Capitalist Society: Experiences in Seventeenth-Century England*. Urbana, Ill., 1988.

Gerlach, Larry R., ed. *New Jersey in the American Revolution, 1763–1783: A Documentary History*. Trenton, N.J., 1975.

Gibson, Alex J. S. "Proletarianization? The Transition to Full-Time Labour on a Scottish Estate, 1723–1787." *Continuity and Change* 3 (1990): 357–89.

Gibson, Alex J. S., and T. Christopher Smout. *Prices, Food, and Wages in Scotland, 1550–1780*. Cambridge, 1995.

Gibson, Arrell M. "Indian Land Transfers." In Washburn, *History of Indian-White Relations*, 211–29.

Giddens, Paul H. "Land Policies and Administration in Colonial Maryland, 1753–1769." *MdHM* 28 (1933): 142–71.

Gilbert, Geoffrey N. "Baltimore's Flour Trade to the Caribbean, 1750–1815." Ph.D. diss., Johns Hopkins University, 1975.

Gilboy, E. W. "The Cost of Living and Real Wages in Eighteenth-Century England." In Arthur J. Taylor, *Standard of Living in Britain*, 1–20.

Gilchrist, David T., ed. *The Growth of the Seaport Cities, 1790–1825: Proceedings of a Conference Sponsored by the Eleutherian Mills—Hagley Foundation, March 17–19, 1966*. Charlottesville, Va., 1967.

Gildrie, Richard P. *Salem, Massachusetts, 1626–1683: A Covenant Community*. Charlottesville, Va., 1975.

Gilje, Paul A., ed. *Wages of Independence: Capitalism in the Early Republic*. Madison, Wis., 1997.

Gill, Conrad. *The Rise of the Irish Linen Industry*. Oxford, 1925.

Gill, Harold B., Jr. "Wheat Culture in Colonial Virginia." *AgH* 52 (1978): 380–94.

Gillespie, Michelle. "The Sexual Politics of Race and Gender: Mary Musgrove and the

Georgia Trustees." In *The Devil's Lane: Sex and Race in the Early South*, edited by Catherine Clinton and Michelle Gillespie, 187–201. New York, 1997.
Gillespie, Raymond. *Colonial Ulster: The Settlement of East Ulster, 1600–1641*. Cork, Ireland, 1985.
Gipson, Lawrence Henry. *The British Empire before the American Revolution*. 15 vols. New York, 1965.
Gladwin, Lee A. "Tobacco and Sex: Some Factors Affecting Non-Marital Sexual Behavior in Colonial Virginia." *Journal of Social History* 12 (1978): 57–75.
Glass, D. V., and D. E. C. Eversley, eds. *Population in History: Essays in Historical Demography*. London, 1965.
Gleach, Frederick Wright. *Powhatan's World and Colonial Virginia: A Conflict of Cultures*. Lincoln, Nebr., 1997.
Glennie, Paul. "In Search of Agrarian Capitalism: Manorial Land Markets and the Acquisition of Land in the Lea Valley, c. 1450–c. 1560." *Continuity and Change* 3 (1988): 11–40.
Godbeer, Richard. "Eroticizing the Middle Ground: Anglo-Indian Sexual Relations along the Eighteenth-Century Frontier." In Hodes, *Sex, Love, Race*, 91–111.
Goddard, Ives. "Some Early Examples of American Pidgin English from New England." *International Journal of American Linguistics* 43 (1977): 37–41.
Goheen, R. B. "Social Ideals and Social Structure: Rural Gloucestershire, 1450–1500." *Histoire Sociale—Social History* 12 (1979): 262–80.
Goldberg, Jonathan. "Fatherly Authority: The Politics of Stuart Family Images." In *Rewriting the Renaissance: The Discourse of Sexual Difference in Early Modern Europe*, edited by Margaret W. Ferguson, Maureen Quilligan, and Nancy J. Vickers, 3–32. Chicago, 1986.
Goldberg, Peter Jeremy Piers. "Female Labour, Service, and Marriage in the Late Medieval Urban North." *Northern History* 22 (1986): 18–38.
——. "Marriage, Migration, Servanthood, and Life-Cycle in Yorkshire Towns of the Later Middle Ages: Some York Cause Paper Evidence." *Continuity and Change* 1 (1986): 141–69.
——. *Women, Work, and Life Cycle in a Medieval Economy: Women in York and Yorkshire, c 1300–1520*. Oxford, 1992.
——, ed. *Woman Is a Worthy Wight: Women in English Society, c. 1200–1500*. Stroud, England, 1992.
——. *Women in England c. 1275–1525: Documentary Sources*. Manchester, England, 1995.
Goldenberg, Joseph A. "Virginia Ports in the American Revolution." In Eller, *Chesapeake Bay in the American Revolution*, 310–40.
Goldenberg, Joseph A., Eddie D. Nelson, and Rita Y. Fletcher. "Revolutionary Ranks: An Analysis of the Chesterfield Supplement." *VMHB* 87 (1979): 182–90.
Goodfriend, Joyce D. *Before the Melting Pot: Society and Culture in Colonial New York City, 1664–1730*. Princeton, N.J., 1992.
Goodkin, Daniel. *Historical Collections of the Indians in New England, of the Several Nations, Numbers, Customs, Manners, Religion and Government before the English Planted There . . . .* 1674. *Massachusetts Historical Society, Collections* 1, ser. 1 (1792): 139–227.
"Good Newes from Virginia, 1623." Reprint. *WMQ* 5 (1948): 350–58.
Goodwin, Gary C. *Cherokees in Transition: A Study of Changing Culture and Environment prior to 1775*. University of Chicago Department of Geography Research Paper, no. 181, 1977.

Goody, Jack, Joan Thirsk, and E. P. Thompson, eds. *Family and Inheritance: Rural Society in Western Europe, 1200–1800*. Cambridge, 1976.
Gottfried, Marion H. "The First Depression in Massachusetts." *NEQ* 9 (1936): 655–78.
Gouge, William. *Of Domesticall Duties: Eight Treatises*. London, 1622.
Gough, Deborah Mathias. "A Further Look at Widows in Early Southeastern Pennsylvania." *WMQ* 44 (1987): 829–35.
Gough, Robert. "The Myth of the 'Middle Colonies': An Analysis of Regionalization in Early America." *Pennsylvania Magazine of History and Biography* 107 (1983): 393–419.
Gould, Clarence P. "The Economic Causes of the Rise of Baltimore." In *Essays in Colonial History*, 225–51.
Gould, Mark. *Revolution in the Development of Capitalism: The Coming of the English Revolution*. Berkeley, Calif., 1987.
Gowing, Laura. "Secret Births and Infanticide in Seventeenth-Century England." *P&P*, no. 156 (1997): 87–115.
Grab, Alexander. "Enlightened Despotism and Commonlands Enclosure: The Case of Austrian Lombardy." *AgH* 63 (1989): 49–72.
Grabbe, Hans-Jürgen. "European Immigration to the United States in the Early National Period, 1783–1820." *Proceedings of the American Philosophical Society* 133 (1989): 190–214.
Gragg, Larry Dale. *Migration in Early America: The Virginia Quaker Experience*. Ann Arbor, Mich., 1980.
——. "Puritans in Paradise: The New England Migration to Barbados." *Journal of Caribbean History* 21 (1988): 154–67.
Graham, Helena. " 'A Women's Work . . .': Labour and Gender in the Late Medieval Countryside." In Peter Jeremy Piers Goldberg, *Woman Is a Worthy Wight*, 126–48.
Graham, Ian Charles Cargill. *Colonists from Scotland: Emigration to North America, 1707–1783*. Ithaca, N.Y., 1956.
Graham, John. "Meetinghouse and Chapel: Religion and Community in Seventeenth-Century Maryland." In Carr, Morgan, and Russo, *Colonial Chesapeake Society*, 242–74.
Grant, Charles S. *Democracy in the Connecticut Frontier Town of Kent*. New York, 1961.
Graunt, John. *Natural and Political Observations Mentioned in a Following Index and Made Upon the Bills of Mortality*. London, 1662.
Gray, Lewis Cecil. *History of Agriculture in the Southern United States to 1860*. 2 vols. Washington, D.C., 1932.
——. "The Market Surplus Problems of Colonial Tobacco." *AgH* 2 (1928): 1–34.
Gray, Malcolm. "Farm Workers in North-East Scotland." In Devine, *Farm Servants and Labour*, 10–28.
——. *The Highland Economy, 1750–1850*. Edinburgh, 1957.
——. "Migration in the Rural Lowlands of Scotland, 1750–1850." In Devine and Dickson, *Ireland and Scotland*, 104–17.
——. "The Social Impact of Agrarian Change in the Rural Lowlands." In Devine and Mitchison, *People and Society in Scotland*, 53–69.
Gray, Ralph, and Betty Wood. "The Transition from Indentured to Involuntary Servitude in Colonial Georgia." *Explorations in Economic History* 13 (1976): 353–70.
Graymont, Barbara. *The Iroquois in the American Revolution*. Syracuse, N.Y., 1972.
Greaves, Richard L. *Society and Religion in Elizabethan England*. Minneapolis, 1981.
Green, E. R. R. "Queensborough Township: Scotch-Irish Emigration and the Expansion of Georgia, 1763–1776." *WMQ* 17 (1960): 183–99.

———. "The 'Strange Humours' That Drove the Scotch-Irish to America, 1729." *WMQ* 12 (1955): 113–23.

———. "Ulster Emigrants' Letters." In *Essays in Scotch-Irish History*, edited by E. R. R. Green, 87–103. London, 1969.

Green, Eugene. "Naming and Mapping the Environment in Early Massachusetts, 1620–1676." *Names* 30 (1982): 77–92.

Green, Eugene, and Rosemary M. Green. "Place-Name Dialects in Massachusetts: Some Complementary Patterns." *Names* 19 (1971): 240–51.

Greene, Evarts B., and Virginia D. Harrington, eds. *American Population before the Federal Census of 1790*. New York, 1932.

Greene, Jack P. *Imperatives, Behaviors, and Identities: Essays in Early American Cultural History*. Charlottesville, Va., 1992.

———. *The Intellectual Construction of America: Exceptionalism and Identity from 1492 to 1800*. Chapel Hill, N.C., 1993.

———. *Pursuits of Happiness: The Social Development of Early Modern British Colonies and the Formation of American Culture*. Chapel Hill, N.C., 1988.

———, ed. *The American Revolution: Its Character and Limits*. New York, 1987.

———. *Colonies to Nation, 1763–1789: A Documentary History of the American Revolution*. New York, 1975.

Greene, Jack P., and J. R. Pole, eds. *Colonial British America: Essays in the New History of the Early Modern Era*. Baltimore, 1984.

Greene, Lorenzo Johnson. *The Negro in Colonial New England*. New York, 1942.

Greenwood, Michael J. "Research in Internal Migration in the United States: A Survey." *Journal of Economic Literature* 13 (1975): 397–433.

Gregg, Alan C. "The Land Policy and System of the Penn Family in Early Pennsylvania." *Western Pennsylvania Historical Magazine* 6 (1923): 151–64.

Gregson, Nicky. "Tawney Revisited: Custom and the Emergence of Capitalist Class Relations in North-East Cumbria, 1600–1830." *EcHR* 42 (1989): 18–42.

Grettler, David J. "Environmental Change and Conflict over Hogs in Early-Nineteenth-Century Delaware." *Journal of the Early Republic* 19 (1999): 197–220.

Greven, Philip J., Jr. "Average Size of Families and Households in the Province of Massachusetts in 1764 and in the United States in 1790: An Overview." In Laslett and Wall, *Household and Family in Past Time*, 545–60.

———. *Four Generations: Population, Land, and Family in Colonial Andover, Massachusetts*. Ithaca, N.Y., 1970.

Gribbon, H. D. "The Irish Linen Board, 1711–1828." In Cullen and Smout, *Comparative Aspects of Scottish and Irish Economic and Social History*, 77–87.

Grigg, Susan. "Toward a Theory of Remarriage: A Case Study of Newburyport at the Beginning of the Nineteenth Century." *JIH* 8 (1977): 183–220.

Grinde, Donald, Jr. "Native American Slavery in the Southern Colonies." *Indian Historian* 10 (spring 1977): 38–42.

Grosart, Alexander B., ed. *The Poems of Sir John Beaumont, Bart., for the First Time Collected and Edited . . . .* Blackburn, England, 1869.

Gross, Robert A. "Culture and Cultivation: Agriculture and Society in Thoreau's Concord." *Journal of American History* 69 (1982): 42–61.

———. *The Minutemen and Their World*. New York, 1976.

———, ed. *In Debt to Shays: The Bicentennial of an Agrarian Rebellion*. Charlottesville, Va., 1993.

Grubb, Farley. "Educational Choice in the Era before Free Public Schooling: Evidence

from German Immigrant Children in Pennsylvania, 1771–1817." *JEH* 52 (1992): 363–76.
——. "Fatherless and Friendless: Factors Influencing the Flow of English Emigrant Servants." *JEH* 52 (1992): 85–108.
——. "German Immigration to Pennsylvania, 1709 to 1820." *JIH* 20 (1990): 417–36.
——. "Immigrant Servant Labor: Their Occupational and Geographic Distribution in the Late Eighteenth-Century Mid-Atlantic." *Social Science History* 9 (1985): 249–76.
——. "The Market Structure of Shipping German Immigrants to Colonial America." *Pennsylvania Magazine of History and Biography* 111 (1987): 27–48.
——. "Morbidity and Mortality on the North Atlantic Passage: Eighteenth-Century German Immigration." *JIH* 17 (1987): 565–85.
——. "Redemptioner Immigration to Pennsylvania: Evidence on Contract Choice and Profitability." *JEH* 46 (1986): 407–18.
——. "Servant Auction Records and Immigration into the Delaware Valley, 1745–1831: The Proportion of Females among Immigrant Servants." *Proceedings of the American Philosophical Society* 133 (1989): 154–69.
Grubb, Farley, and Tony Stitt. "The Liverpool Emigrant Servant Trade and the Transition to Slave Labor in the Chesapeake, 1697–1707: Market Adjustments to War." *Explorations in Economic History* 31 (1994): 376–405.
Gruber, Ira D. *The Howe Brothers and the American Revolution*. New York, 1972.
Grumet, Robert Steven. *Historic Contact: Indian People and Colonists in Today's Northeastern United States in the Sixteenth through the Eighteenth Centuries*. Norman, Okla., 1995.
——. "Sunksquaws, Shamans, and Tradeswomen: Coastal Algonkian Women during the Seventeenth and Eighteenth Centuries." In *Women and Colonization: Anthropological Perspectives*, edited by Mona Etienne and Eleanor Leacock, 43–62. New York, 1980.
——. "Suscaneman and the Matinecock Lands, 1653–1703." In Grumet, *Northeastern Indian Lives*, 116–39.
——, ed. *Northeastern Indian Lives, 1632–1816*. Amherst, Mass., 1996.
Gundersen, Joan R. "The Double Bonds of Race and Sex: Black and White Women in a Colonial Virginia Parish." *Journal of Southern History* 52 (1986): 351–72.
——. *To Be Useful to the World: Women in Revolutionary America, 1740–1790*. New York, 1996.
Gundersen, Joan R., and Gwen Victor Gampel. "Married Women's Legal Status in Eighteenth-Century New York and Virginia." *WMQ* 39 (1982): 114–34.
Gutman, Herbert G. *The Black Family in Slavery and Freedom, 1750–1925*. New York, 1976.
Gutmann, Myron. *Toward the Modern Economy: Early Industry in Europe, 1500–1800*. Philadelphia, 1988.
Gwyn, Julian. "The Impact of British Military Spending on the Colonial American Money Markets, 1760–1783." *Historical Papers of the Canadian Historical Association* 7 (1980): 77–99.
——. "Private Credit in Colonial New York: The Warren Portfolio, 1731–1795." *New York History* 54 (1973): 269–93.
Gwynn, Robin D. *Huguenot Heritage: The History and Contribution of the Huguenots in Britain*. London, 1985.
Gyles, John. *Memoirs of Odd Adventures, Strange Deliverances, etc.* 1736. In Vaughan and Clark, *Puritans among the Indians*, 91–131.

Haan, Richard L. "The 'Trade Do's Not Flourish as Formerly': The Ecological Origins of the Yamassee War of 1715." *Ethnohistory* 28 (1982): 341–58.
Häberlein, Mark. "German Migrants in Colonial Pennsylvania: Resources, Opportunities, and Experience." *WMQ* 50 (1993): 555–74.
Haefeli, Evan, and Kevin Sweeney. "Revisiting *the Redeemed Captive*: New Perspectives on the 1704 Attack on Deerfield." *WMQ* 52 (1995): 3–46.
Hagedorn, Nancy L. "Brokers of Understanding: Interpreters as Agents of Cultural Exchange in Colonial New York." *New York History* 76 (1995): 379–408.
Hajnal, J. "European Marriage Patterns in Perspective." In Glass and Eversley, *Population in History*, 100–143.
——. "Two Kinds of Pre-Industrial Household Formation System." In Wall, Robin, and Laslett, *Family Forms in Historic Europe*, 65–104.
Hakluyt, Richard the Younger. *Discourse of Western Planting*. 1584. In Quinn, *New American World*, 71–125.
Hall, Clayton Colman, ed. *Narratives of Early Maryland*. New York, 1910.
Hall, David D., and David Grayson Allen, eds. *Seventeenth-Century New England: A Conference Held by the Colonial Society of Massachusetts, June 18 and 19, 1982*. Boston, 1984.
Hall, David D., John M. Murrin, and Thad W. Tate, eds. *Saints and Revolutionaries: Essays on Early American History*. New York, 1984.
Hall, John Raymond. "The Three Rank System of Land Distribution in Colonial Swansea, Massachusetts." *Rhode Island History* 43 (1984): 3–17.
Hamell, George R. "Wampum, White, Bright, and Light Things Are Good to Think." In Van Dongen et al., *One Man's Trash*, 41–51.
Hamilton, Henry. *An Economic History of Scotland in the Eighteenth Century*. Oxford, 1963.
——, ed. *Life and Labour on an Aberdeenshire Estate, 1735–1750 (Being Selections from the Monymusk Papers)*. Aberdeen, Scotland, 1946.
Hammond, Isaac W., comp. *State of New Hampshire: Rolls of the Soldiers in the Revolutionary War* . . . . 4 vols. Vols. 14–17 of *New Hampshire Provincial and State Papers*. Concord, N.H., 1885–89.
Hammond, J. L., and Barbara Hammond. *The Village Labourer, 1760–1832: A Study in the Government of England before the Reform Bill*. 1911. Edited by Eric Hobsbawm. New York, 1970.
Hammond, John. *Leah and Rachel; or, the Two Fruitfull Sisters Virginia and Mary-land*. 1656. In Clayton Colman Hall, *Narratives of Early Maryland*, 281–308.
Hampsher-Monk, Iain. "The Political Theory of the Levellers: Putney, Property, and Professor McPherson." *Political Studies* 25 (1976): 397–422.
Hanawalt, Barbara A. *The Ties That Bound: Peasant Families in Medieval England*. New York, 1986.
Handley, James E. *The Agricultural Revolution in Scotland*. Glasgow, 1963.
——. *Scottish Farming in the Eighteenth Century*. London, 1953.
Hansen, Ann Natalie. *The Dorchester Group: Puritanism or Revolution*. Columbus, Ohio, 1987.
Hariot, Thomas. *A Briefe and True Report of the NewFound Land of Virginia*. 1588. In *The Roanoke Voyages, 1584–1590*, edited by David Berrs Quinn, 1:317–87. 2 vols. London, 1955.
Harley, J. B. "Meaning and Ambiguity in Tudor Cartography." In Sarah Tyacke, *English Map-Making*, 22–45.

Harnish, Hartmut. "Peasants and Markets: The Background to the Agrarian Reforms in Feudal Prussia East of the Elbe, 1760–1807." In *The German Peasantry: Conflict and Community in Rural Society from the Eighteenth to the Twentieth Centuries*, edited by Richard J. Evans and W. Robert Lee, 37–70. London, 1986.
Harper, R. Eugene. *The Transformation of Western Pennsylvania, 1770–1800*. Pittsburgh, Pa., 1991.
Harper, Roland M. "Changes in the Forest Area of New England in Three Centuries." *Journal of Forestry* 16 (1918): 442–52.
Harrell, Isaac Samuel. *Loyalism in Virginia: Chapters in the Economic History of the Revolution*. Durham, N.C., 1926.
——. "North Carolina Loyalists." *North Carolina Historical Review* 3 (1926): 575–90.
Harrington, Virginia D. *The New York Merchant on the Eve of the Revolution*. New York, 1935.
Harris, Marc. "The People of Concord: A Demographic History, 1750–1850." *Chronos*, no. 2 (1983): 65–138.
Harris, Marshall. *Origins of the Land Tenure System in the United States*. Ames, Iowa, 1953.
Harris, P. M. G. "Inflation and Deflation in Early America, 1634–1860: Patterns of Change in the British American Economy." *Social Science History* 20 (1996): 469–506.
——. "The Social Origins of American Leaders: The Demographic Foundations." *Perspectives in American History* 3 (1969): 159–344.
Harris, Ruth-Ann. "Seasonal Migration between Ireland and England prior to the Famine." *Canadian Papers in Rural History* 7 (1989): 363–86.
Harrison, William. *A Description of England; or, a Briefe Rehersall of the Nature and Quality of the People of England*. 1577–87. 2d and 3d books. Edited by F. J. Furnivall. *New Shakspere Society Publications*, ser. 6, no. 1. London, 1877.
Hart, John F. "Eminent Domain and Constitutionalism in Eighteenth Century America: Land Use and the Selective Protection of Private Property." Paper, McNeil Center for Early American Studies, February 1994.
——. "The Maryland Mill Act, 1669–1766: Economic Policy and the Confiscatory Redistribution of Private Property." *MdHM* 37 (1995): 1–27.
Harte, Thomas J. "Social Origins of the Brandywine Population." *Phylon* 24 (1963): 369–78.
Hast, Adele. *Loyalism in Revolutionary Virginia: The Norfolk Area and the Eastern Shore*. Ann Arbor, Mich., 1982.
Harvey, P. D. A. Conclusion to Harvey, *Peasant Land Market*, 328–56.
——. "English Estate Maps: Their Early History and Their Use as Historical Evidence." In *Rural Images: Estate Maps in the Old and New Worlds*, edited by David Buisseret, 27–62. Chicago, 1996.
——, ed. *The Peasant Land Market in Medieval England*. Oxford, 1984.
Haskins, George Lee. *Law and Authority in Early Massachusetts: A Study in Tradition and Design*. New York, 1960.
Hatcher, John. "England in the Aftermath of the Black Death." *P&P*, no. 144 (1994): 3–35.
——. "Labour, Leisure, and Economic Thought before the Nineteenth Century." *P&P*, no. 160 (1998): 64–115.
——. *Plague, Population, and the English Economy, 1348–1530*. London, 1977.
Hatfield, April Lee. "Reconceiving Virginia: Seventeenth-Century Intercolonial Interactions." Ph.D. diss., Johns Hopkins University, 1997.
Hatley, Tom. *The Dividing Paths: Cherokees and South Carolinians through the Revolutionary Era*. Oxford, 1995.

Hauptman, Lawrence M. "Refugee Havens: The Iroquois Villages of the Eighteenth Century." In *American Indian Environments: Ecological Issues in Native American History*, edited by Christopher Vecsey and Robert W. Venables, 128–39, 203–6. Syracuse, N.Y., 1980.

Hauptman, Lawrence M., and James D. Wherry, eds. *The Pequots in Southern New England: The Rise and Fall of an Indian Nation*. Norman, Okla., 1990.

Hawes, Austin F. "New England's Forests in Retrospect." *Journal of Forestry* 21 (1923): 209–24.

Head, Anne-Lise. "Quelques Remarques sur L'Émigration des Régions Préalpines." *Revue Suisse d'Histoire* 29 (1979): 181–93.

Heaton, Herbert. *The Yorkshire Woollen and Worsted Industries from the Earliest Times up to the Industrial Revolution*. Oxford, 1965.

Hecht, Irene W. D. "The Virginia Muster of 1624/5 as a Source for Demographic History." *WMQ* 30 (1973): 65–92.

Hedrick, Ulysses Prentiss. *A History of Agriculture in the State of New York*. 1933. New York, 1966.

Heimert, Alan. "Puritanism, the Wilderness, and the Frontier." *NEQ* 26 (1953): 361–82.

Heinton, Louise Joyner. *Prince George's Heritage: Sidelights on the Early History of Prince George's Maryland from 1696 to 1800*. Baltimore, 1972.

Henderson, Katherine Usher, and Barbara F. McManus, eds. *Half Humankind: Contexts and Texts of the Controversy about Women in England, 1540–1640*. Urbana, Ill., 1985.

Henderson, Roger C. "Demographic Patterns and Family Structure in Eighteenth-Century Lancaster County, Pennsylvania." *Pennsylvania Magazine of History and Biography* 104 (1990): 349–83.

Henderson, W. O. *Studies in the Economic Policy of Frederick the Great*. London, 1963.

Hening, William Waller. *The Statutes at Large; Being a Collection of all the Laws of Virginia, from the First Session of the Legislature, in the Year 1619 . . . .* 13 vols. Richmond, Va., 1819–23.

Henretta, James A. "Families and Farms: Mentalité in Pre-Industrial America." *WMQ* 35 (1978): 3–32.

——. "The War for Independence and American Economic Development." In Hoffman et al., *Economy of Early America*, 45–87.

Hensley, Paul B. "Time, Work, and Social Context in New England." *NEQ* 65 (1992): 531–59.

Herman, Bernard L. *Architecture and Rural Life in Central Delaware, 1700–1900*. Knoxville, Tenn., 1987.

——. "Home and Hearth: The British Colonies." In Jacob Ernest Cooke, *Encyclopedia of the North American Colonies*, 2:567–78.

Herman, Mary W. "Wampum as a Money in Northeastern North America." *Ethnohistory* 3 (1956): 21–33.

Herndon, G. Melvin. "Forest Products of Colonial Georgia." *Journal of Forest History* 23 (1979): 130–35.

——. "Indian Agriculture in the Southern Colonies." *North Carolina Historical Review* 44 (1967): 283–97.

Herndon, Ruth Wallis. "Research Note: Literacy among New England's Transient Poor, 1750–1800." *Journal of Social History* 29 (1996): 963–65.

——. "The Right to a Name: The Narragansett People and Rhode Island Officials in the Revolutionary Era." *Ethnohistory* 44 (1997): 434–62.

——. "Warned Out in New England: Eighteenth-Century Tales of Trouble." Paper, McNeil Center for Early American Studies, September 1995.

Hey, David G. *An English Rural Community: Myddle under the Tudors and Stuarts*. Leicester, England, 1974.

Heyrman, Christine Leigh. *Commerce and Culture: The Maritime Communities of Colonial Massachusetts, 1690–1750*. New York, 1984.

Hibbard, Benjamin Horace. *A History of Public Land Policies*. New York, 1924.

Higginbotham, Don. *The War of American Independence: Military Attitudes, Policies, and Practice, 1763–1789*. New York, 1971.

Higgins, Ruth Loving. *Expansion in New York, with Especial Reference to the Eighteenth Century*. Columbus, Ohio, 1931.

Higgs, Robert, and H. Louis Stettler III. "Colonial New England Demography: A Sampling Approach." *WMQ* 27 (1970): 282–94.

Hill, Christopher. *Change and Continuity in Seventeenth Century England*. Cambridge, Mass., 1975.

——. *Society and Puritanism in Pre-Revolutionary England*. New York, 1964.

Hilton, Rodney H. *Bond Men Made Free: Medieval Peasant Movements and the English Rising of 1381*. London, 1973.

——. *Class Conflict and the Crisis of Feudalism: Essays in Medieval Social History*. Rev. ed. London, 1990.

——. "A Crisis of Feudalism." In Aston and Philpin, *Brenner Debate*, 119–37.

——. *The Decline of Serfdom in Medieval England*. London, 1969.

——. *The English Peasantry in the Later Middle Ages: The Ford Lectures for 1973 and Related Studies*. Oxford, 1975.

——, ed. *The Transition from Feudalism to Capitalism*. London, 1976.

Hinderaker, Eric. *Elusive Empires: Constructing Colonialism in the Ohio Valley, 1673–1800*. Cambridge, 1997.

Hindle, Brooke, ed. *Material Culture of the Wooden Age*. Tarrytown, N.Y., 1981.

Hindle, Steve. "Persuasion and Protest in the Caddington Commons Enclosure Dispute, 1635–1639." *P&P*, no. 158 (1998): 37–78.

Hiner, N. Ray. "The Cry of Sodom Enquired Into: Educational Analysis in Seventeenth-Century New England." *History of Education Quarterly* 13 (1973): 3–22.

Hirst, Derek. *Authority and Conflict: England, 1603–1658*. Cambridge, 1986.

Hitz, Alex M. "Georgia Bounty Land Grants." *Georgia Historical Quarterly* 38 (1954): 337–48.

Hobbes, John E. "The Beginnings of Lumbering as an Industry in the New World, and First Efforts at Forest Protection: A Historical Study." *Forest Quarterly* 4 (1906): 14–23.

Hochstadt, Steve. "Migration in Preindustrial Germany." *Central European History* 16 (1983): 195–223.

Hodes, Martha, ed. *Sex, Love, Race: Crossing Boundaries in North American History*. New York, 1999.

Hodges, Graham Russell. *Slavery and Freedom in the Rural North: African-Americans in Monmouth County, New Jersey, 1665–1865*. Madison, Wis., 1996.

——, ed. *The Black Loyalist Directory: African Americans in Exile after the American Revolution*. New York, 1996.

Hodges, Graham Russell, and Alan Edward Brown, eds. *"Pretends to Be Free": Runaway Slave Advertisements from Colonial and Revolutionary New York and New Jersey*. New York, 1994.

Hoffer, Peter Charles. *Law and People in Colonial America*. Baltimore, 1992.

Hoffer, Peter Charles, and N. E. H. Hull. *Murdering Mothers: Infanticide in England and New England, 1558–1803*. New York, 1984.

Hoffman, Bernard G. "Ancient Tribes Revisited: A Summary of Indian Distribution and Movement in the Northeastern United States from 1534 to 1779. Parts I–III." *Ethnohistory* 14 (1967): 1–46.
Hoffman, Ronald. "The 'Disaffected' in the Revolutionary South." In *The American Revolution: Explorations in the History of American Radicalism*, edited by Alfred F. Young, 273–318. DeKalb, Ill., 1976.
——. "Popularizing the Revolution: Internal Conflict and Economic Sacrifice in Maryland, 1774–1781." *MdHM* 68 (1973): 125–39.
Hoffman, Ronald, and Peter J. Albert, eds. *Arms and Independence: The Military Character of the American Revolution*. Charlottesville, Va., 1984.
——. *Women in the Age of the American Revolution*. Charlottesville, Va., 1989.
Hoffman, Ronald, Thad W. Tate, and Peter J. Albert, eds. *An Uncivil War: The Southern Backcountry during the American Revolution*. Charlottesville, Va., 1985.
Hoffman, Ronald, John J. McCusker, Russell R. Menard, and Peter J. Albert, eds. *The Economy of Early America: The Revolutionary Period, 1763–1790*. Charlottesville, Va., 1988.
Hofstra, Warren R. "Ethnicity and Community Formation on the Shenandoah Valley Frontier, 1730–1800." In Puglisi, *Diversity and Accommodation*, 59–81.
——. " 'The Extension of His Majesties' Dominions': The Virginia Backcountry and the Reconfiguration of Imperial Frontiers." *Journal of American History* 84 (1998): 1281–1312.
——. "Land, Ethnicity, and Community at the Opequon Settlement, Virginia, 1730–1800." *VMHB* 98 (1990): 423–48.
——. "Land Policy and Settlement in the Northern Shenandoah Valley." In Robert D. Mitchell, *Appalachian Frontiers*, 105–26.
Hofstra, Warren R., and Robert D. Mitchell. "Town and Country in Backcountry Virginia: Winchester and the Shenandoah Valley, 1730–1800." *Journal of Southern History* 59 (1993): 619–46.
Holderness, B. A. "Credit in English Rural Society before the Nineteenth Century, with Special Reference to the Period 1650–1720." *Agricultural History Review* 24 (1976): 97–109.
——. " 'Open' and 'Close' Parishes in England in the Eighteenth and Nineteenth Centuries." *Agricultural History Review* 20 (1972): 126–39.
——. "Personal Mobility in Some Rural Parishes of Yorkshire, 1777–1822." *Yorkshire Archaeological Journal* 42 (1971): 444–54.
Holmes, Richard. *Communities in Transition: Bedford and Lincoln, Massachusetts, 1729–1850*. Ann Arbor, Mich., 1980.
Holton, Woody. "The Ohio Indians and the Coming of the American Revolution in Virginia." *Journal of Southern History* 60 (1994): 453–78.
——. " 'Rebel against Rebel': Enslaved Virginians and the Coming of the Revolution." *VMHB* 105 (1997): 157–92.
Homans, George Caspar. *English Villagers of the Thirteenth Century*. 1941. New York, 1970.
——. "The Puritans and the Clothing Industry in England." *NEQ* 13 (1940): 519–29.
Hood, Adrienne Dora. "The Gender Division of Labor in the Production of Textiles in Eighteenth-Century Rural Pennsylvania (Rethinking the New England Model)." *Journal of Social History* 27 (1994): 537–62.
——. "Industrial Opportunism: From Handweaving to Mill Production, 1700–1830." In *Textiles in Early New England: Design, Production, and Consumption*, edited by Peter Benes, 135–51. Boston, 1999.

——. "The Material World of Cloth: Production and Use in Eighteenth-Century Rural Pennsylvania." *WMQ* 53 (1996): 43–66.
——. "Organization and Extent of Textile Manufacture in Eighteenth-Century Rural Pennsylvania: A Case Study of Chester County." Ph.D. diss., University of California, San Diego, 1988.
Horle, Craig W. *The Quakers and the English Legal System, 1660–1688*. Philadelphia, 1988.
Horn, James. "Adapting to a New World: A Comparative Study of Local Society in England and Maryland, 1650–1700." In Carr, Morgan, and Russo, *Colonial Chesapeake Society*, 133–75.
——. *Adapting to a New World: English Society in the Seventeenth-Century Chesapeake*. Chapel Hill, N.C., 1994.
——. "British Diaspora: Emigration from Britain, 1680–1815." In *The Eighteenth Century*, edited by P. J. Marshall and Alaine Low, 28–52. Vol. 2 of *The Oxford History of the British Empire*. Edited by William Roger Louis. Oxford, 1998.
——. "Cavalier Culture? The Social Development of Colonial Virginia." *WMQ* 48 (1991): 238–45.
——. "Moving On in the New World: Migration and Out-migration in the Seventeenth Century Chesapeake." In Clark and Souden, *Migration and Society*, 172–212.
——. "Servant Emigration to the Chesapeake in the Seventeenth Century." In Tate and Ammerman, *Chesapeake in the Seventeenth Century*, 51–95.
——. "Tobacco Colonies: The Shaping of English Society in the Seventeenth-Century Chesapeake." In Canny and Low, *Origins of Empire*, 170–92.
——. " 'To Parts beyond the Seas': Free Emigration to the Chesapeake in the Seventeenth Century." In Altman and Horn, *"To Make America,"* 85–129.
Horsman, Reginald. "American Indian Policy in the Old Northwest, 1783–1812." *WMQ* 18 (1961): 35–53.
——. "The Image of the Indian in the Age of the American Revolution." In *The American Indian and the American Revolution*, edited by Francis P. Jennings, 1–11. The Newberry Library Center for the History of the American Indian Occasional Papers, no. 6., 1983.
Hoskins, W. G. *The Age of Plunder: King Henry's England, 1500–1547*. London, 1976.
——. *The Making of the English Landscape*. 1955. Harmondsworth, England, 1970.
——. *The Midland Peasant: The Economic and Social History of a Leicestershire Village*. London, 1957.
——. *Provincial England: Essays in Social and Economic History*. New York, 1965.
Hoskins, W. G., and L. Dudley Stamp. *The Common Lands of England and Wales*. London, 1963.
Houlbrooke, Ralph A. *The English Family, 1450–1700*. London, 1984.
——. "The Making of Marriage in Mid-Tudor England: Evidence from the Records of Matrimonial Contract Litigation." *Journal of Family History* 10 (1981): 339–52.
Houston, R. A. "Age at Marriage of Scottish Women, circa 1660–1770." *Local Population Studies*, no. 43 (1989): 63–66.
——. "The Demographic Regime." In Devine and Mitchison, *People and Society in Scotland*, 9–26.
——. "Geographical Mobility in Scotland, 1652–1811: The Evidence of Testimonials." *Journal of Historical Geography* 11 (1985): 379–94.
——. *The Population History of Britain and Ireland, 1500–1750*. Houndsmill, England, 1992.
——. *Social Change in the Age of Enlightenment: Edinburgh, 1660–1760*. Oxford, 1994.

———. "Women in the Economy and Society of Scotland, 1500–1800." In Houston and Whyte, *Scottish Society*, 118–47.

Houston, R. A., and Charles W. J. Withers. "Population Mobility in Scotland and Europe, 1600–1900: A Comparative Perspective." *Annales de Demographie Historique* 27 (1990): 285–308.

Houston, R. A., and Ian D. Whyte, eds. *Scottish Society, 1500–1800*. Cambridge, 1989.

Howard, Ronald William. "Old Age and Death: The British Colonies." In Jacob Ernest Cooke, *Encyclopedia of the North American Colonies*, 2:761–75.

Howatson, William. "Grain Harvesting and Harvesters." In Devine, *Farm Servants and Labour*, 124–42.

Howe, Mark DeWolf. "The Sources and Nature of Law in Colonial Massachusetts." In *Selected Essays: Law and Authority in Colonial America*, edited by George Athan Billias, 1–16. Barre, Mass., 1965.

Howell, Cicely. *Land, Family, and Inheritance in Transition: Kibworth Harcourt, 1280–1700*. Cambridge, 1983.

Hoyle, R. W. "An Ancient and Laudable Custom: The Definition and Development of Tenant Right in North-Western England in the Sixteenth Century." *P&P*, no. 116 (1987): 22–55.

———. "Debate: The Land-Family Bond in England." *P&P*, no. 146 (1995): 151–73.

———. "Tenure and the Land Market in Early Modern England; or, a Late Contribution to the Brenner Debate." *EcHR* 43 (1990): 1–20.

Hoyt, William D., Jr. "Colonel William Fleming in Dunmore's War, 1774." *West Virginia History* 3 (1942): 99–119.

Hsiung, David C. "Death on the Juniata: Delawares, Iroquois, and Pennsylvanians in a Colonial Whodunit." *Pennsylvania History* 65 (1998): 445–77.

Hufeland, Otto. *Westchester County during the American Revolution, 1775–1783*. Vol. 3 of *Publications of the Westchester County Historical Society*. White Plains, N.Y., 1926.

Hugget, Frank E. *The Land Question and European Society since 1650*. London, 1975.

Hughes, Jonathan R. T. *Social Control in the Colonial Economy*. Charlottesville, Va., 1976.

Hughes, Kaylene. "Populating the Back Country: The Demographic and Social Characteristics of the Colonial South Carolina Frontier, 1730–1760." Ph.D. diss., Florida State University, 1985.

Hughes, Sarah S. "Slaves for Hire: The Allocation of Black Labor in Elizabeth City County, Virginia, 1782 to 1810." *WMQ* 25 (1978): 260–86.

———. *Surveyors and Statesmen: Land Measuring in Colonial Virginia*. Richmond, Va., 1979.

Hulton, Paul. *America 1585: The Complete Drawings of John White*. Chapel Hill, N.C., 1984.

Humphrey, Thomas. "Agrarian Rioting in Albany County, New York: Tenants, Markets, and Revolution in the Hudson Valley, 1751–1801." Ph.D. diss., Northern Illinois University, 1996.

Humphries, Jane. "Enclosures, Common Rights, and Women: The Proletarianization of Families in the Late Eighteenth and Early Nineteenth Centuries." *JEH* 50 (1990): 17–42.

Hunt, William. *The Puritan Moment: The Coming of Revolution to an English County*. Cambridge, Mass., 1983.

Hurt, R. Douglas. *Indian Agriculture in America: Prehistory to the Present*. Lawrence, Kans., 1987.

Huston, James A. *Logistics of Liberty: American Services of Supply in the Revolutionary War and After*. Newark, Del., 1991.

Huzel, J. P. "The Labourer and the Poor Law, 1750–1850." In Mingay, *Agrarian History*, 755–809.

Imhof, Arthur E. "Structure of Reproduction in a West German Village, 1690–1900." In *Chance and Change: Social and Economic Studies in Historical Demography in the Baltic Area*, edited by Sune Äkerman, Hans Chr. Johansen, and David Gaunt, 23–32. Odense, 1978.

Ingram, Martin. " 'Scolding Women Cucked or Washed': A Crisis in Gender Relations in Early Modern England?" In Kermode and Walker, *Women, Crime, and the Courts*, 48–80.

Innes, Stephen. *Creating the Commonwealth: The Economic Culture of Puritan New England*. New York, 1995.

——. *Labor in a New Land: Economy and Society in Seventeenth-Century Springfield*. Princeton, N.J., 1983.

——, ed. *Work and Labor in Early America*. Chapel Hill, N.C., 1988.

Jack, Sybil M. *Trade and Industry in Tudor and Stuart England*. London, 1977.

Jackson, Harvey H., and Phinizy Spalding, eds. *Forty Years of Diversity: Essays on Colonial Georgia*. Athens, Ga., 1984.

Jackson, John W. *Valley Forge: Pinnacle of Courage*. Gettysburg, Pa., 1992.

Jacobs, Jaap, and Martha Dickinson Shattuck. "Beavers for Drink, Land for Arms: Some Aspects of the Dutch Indian Trade in New Netherland." In Van Dongen et al., *One Man's Trash*, 95–113.

James, Alfred P. *The Ohio Company: Its Inner History*. Pittsburgh, Pa., 1959.

James, Francis G. "Irish Colonial Trade in the Eighteenth Century." *WMQ* 20 (1963): 574–84.

Jameson, Hugh. "Subsistence for Middle States Militia." *Military Affairs* 30 (1966): 121–34.

Janeway, Elizabeth. *Powers of the Weak*. New York, 1981.

Jedry, Christopher M. *The World of John Cleaveland: Family and Community in Eighteenth-Century New England*. New York, 1979.

Jefferson, Thomas. *The Papers of Thomas Jefferson*. Edited by Julian K. Boyd, Charles T. Cullen, and John Catanzariti, 24 vols. Princeton, N.J., 1950– .

Jennings, Francis. *The Ambiguous Iroquois Empire: The Covenant Chain Confederation of Indian Tribes with English Colonies from Its Beginnings to the Lancaster Treaty of 1744*. New York, 1984.

——. "Brother Miquon: Good Lord!" In Dunn and Dunn, *World of William Penn*, 195–214.

——. *Empire of Fortune: Crowns, Colonies, and Tribes in the Seven Years' War in America*. New York, 1988.

——. "The Indian Fur Trade of the Susquehanna Valley." *Proceedings of the American Philosophical Society* 110 (1966): 406–24.

——. *The Invasion of America: Indians, Colonialism, and the Cant of Conquest*. Chapel Hill, N.C., 1975.

——. "The Scandalous Indian Policy of William Penn's Sons: Deeds and Documents of the Walking Purchase." *Pennsylvania History* 37 (1970): 19–39.

Jensen, Arthur. *The Maritime Commerce of Colonial Philadelphia*. Madison, Wis., 1963.

Jensen, Joan M. *Loosening the Bonds: Mid-Atlantic Farm Women, 1750–1850*. New Haven, Conn., 1986.

Jensen, Merrill. *The New Nation: A History of the United States during the Confederation, 1781–1789*. New York, 1950.

John, A. H. "Agricultural Productivity and Economic Growth in England, 1700–1760." *JEH* 25 (1965): 19–34.

Johnson, Amandus, ed. and trans. *The Journal and Biography of Nicholas Collin, 1746–1831*. Philadelphia, 1936.
Johnson, Edward. *Johnson's Wonder-Working Providence, 1628–1651*. 1653. Edited by J. Franklin Jameson. New York, 1910.
Johnson, Paul E. "The Modernization of Mayo Greenleaf Patch: Land, Family, and Marginality in New England, 1766–1818." *NEQ* 55 (1982): 488–516.
Johnson, Richard R. "The Search for a Usable Indian: An Aspect of the Defense of Colonial New England." *Journal of American History* 64 (1977): 623–51.
Johnson, Robert. *Nova Britannia: Offering Most Excellent Fruites by Planting in Virginia Exciting all Such as be Well Affected to Further the Same*. London, 1609. Reprinted in Force, *Tracts and Other Papers*, vol. 1.
Johnson, Robert C. "The Lotteries of the Virginia Company." *VMHB* 74 (1966): 259–91.
——. "The Transportation of Vagrant Children from London to Virginia, 1618–1622." In *Early Stuart Studies: Essays in Honor of David Harris Willson*, edited by Howard S. Reinmuth Jr., 137–51. Minneapolis, 1970.
Jones, Alice Hanson. *American Colonial Wealth: Documents and Methods*. 2d ed. 3 vols. New York, 1978.
——. *Wealth of a Nation to Be: The American Colonies on the Eve of the American Revolution*. New York, 1980.
——. "The Wealth of Women, 1774." In *Strategic Factors in Nineteenth-Century American Economic History: A Volume to Honor Robert W. Fogel*, edited by Claudia Goldin and Hugh Rockoff, 243–63. Chicago, 1992.
Jones, Andrew. "Land and People at Leighton Buzzard in the Later Fifteenth Century." *EcHR* 25 (1972): 18–27.
Jones, Dorothy V. "British Colonial Indian Treaties." In Washburn, *History of Indian-White Relations*, 185–94.
——. *License for Empire: Colonialism by Treaty in Early America*. Chicago, 1982.
Jones, Douglas Lamar. "Poverty and Vagabondage: The Process of Survival in Eighteenth-Century Massachusetts." *New England Historical and Genealogical Register* 133 (1979): 243–54.
——. "The Strolling Poor: Transiency in Eighteenth-Century Massachusetts." *Journal of Social History* 8 (1975): 28–54.
——. "The Transformation of the Law of Poverty in Eighteenth-Century Massachusetts." In *Law in Colonial Massachusetts, 1630–1800: A Conference Held 6 and 7 November 1981 by the Colonial Society of Massachusetts*, edited by Daniel R. Coquillette, 153–90. Boston, 1984.
——. *Village and Seaport: Migration and Society in Eighteenth-Century Massachusetts*. Hanover, N.H., 1981.
Jones, George Fenwick. *The Georgia Dutch from the Rhine and Danube to the Savannah, 1733–1783*. Athens, Ga., 1992.
Jones, Henry Z., Jr. *The Palatine Families of New York: A Study of the German Immigrants Who Arrived in Colonial New York in 1710*. Universal City, Calif., 1985.
Jones, Howard Mumford. "The Colonial Impulse: An Analysis of the 'Promotion' Literature of Colonization." *Proceedings of the American Philosophical Society* 90 (1946): 131–61.
Jones, Robert Leslie. *History of Agriculture in Ohio to 1880*. Kent, Ohio, 1983.
Jonson, Ben, George Chapman, and John Marston. *Eastward Ho!* 1605. Edited by R. W. Van Fossen. Manchester, England, 1979.
Jordan, David W. *Foundations of Representative Government in Maryland, 1632–1715*. Cambridge, 1987.

Jordan, Donald E. *Land and Popular Politics in Ireland: County Mayo from the Plantation to the Land War*. Cambridge, 1994.
Jordan, Terry G. *North American Cattle-Ranching Frontiers: Origins, Diffusion, and Differentiation*. Albuquerque, N.M., 1993.
Jordan, Terry G., and Matti Kaups. *The American Backwoods Frontier: An Ethnohistoric and Ecological Interpretation*. Baltimore, 1989.
Joyce, William L., David D. Hall, Richard D. Brown, and John B. Hench, eds. *Printing and Society in Early America*. Worcester, Mass., 1983.
Juricek, John T. "American Usage of the Word 'Frontier' from Colonial Times to Frederick Jackson Turner." *Proceedings of the American Philosophical Society* 110 (1966): 10–34.
Jütte, Robert. *Poverty and Deviance in Early Modern Europe*. Cambridge, 1994.
Kain, Roger J. P., and Elizabeth Baigent. *The Cadastral Map in the Service of the State: A History of Property Mapping*. Chicago, 1992.
Kamen, Henry. "The Economic and Social Consequences of the Thirty Years' War." *P&P*, no. 39 (1968): 44–61.
Kammen, Michael G. "Maryland in 1699: A Letter from the Reverend Hugh Jones." *Journal of Southern History* 29 (1963): 362–72.
——. "The Problem of American Exceptionalism: A Reconsideration." *American Quarterly* 45 (1993): 1–43.
Kane, Hope Frances. "Notes on Early Pennsylvania Promotion Literature." *Pennsylvania Magazine of History and Biography* 63 (1939): 144–68.
Karr, Ronald Dale. "The Transformation of Agriculture in Brookline, 1770–1885." *Historical Journal of Massachusetts* 15 (1987): 33–49.
——. " 'Why Should You Be So Furious?': The Violence of the Pequot War." *Journal of American History* 95 (1998): 876–909.
Kashatus, William C., III. "A Mutual Sufferance: Citizenry, Soldiery, and Necessity during the Valley Forge Encampment, 1777–78." *Valley Forge Journal* 3 (1987): 169–80.
——. "The Wyoming Massacre: The Surpassing Horror of the American Revolution, July 3, 1778." *Valley Forge Journal* 4 (1988): 107–22.
Katcher, Philip R. N. *Encyclopedia of British, Provincial, and German Army Units, 1775–1783*. Harrisburg, Pa., 1973.
Katz, Steven T. "The Pequot War Reconsidered." *NEQ* 64 (1991): 206–24.
Kawashima, Yasuhide. "Indian Servitude in the Northeast." In Washburn, *History of Indian-White Relations*, 404–6.
——. "Legal Origins of the Indian Reservation in Colonial Massachusetts." *American Journal of Legal History* 13 (1969): 45–56.
——. *Puritan Justice and the Indian: White Man's Law in Massachusetts, 1630–1763*. Middletown, Conn., 1986.
Kay, Marvin L. Michael, and Lorin Lee Cary. *Slavery in North Carolina, 1748–1775*. Chapel Hill, N.C., 1995.
Keary, Anne. "Retelling the History of the Settlement of Providence: Speech, Writing, and Cultural Interaction on Narragansett Bay." *NEQ* 59 (1996): 250–86.
Keen, Maurice. *English Society in the Later Middle Ages, 1348–1500*. London, 1990.
Keim, C. Ray. "Primogeniture and Entail in Virginia." *WMQ* 25 (1968): 545–86.
Kellenbenz, Hermann. "Rural Industries in the West from the End of the Middle Ages to the Eighteenth Century." In *Essays in European Economic History, 1500–1800*, edited by Peter Earle, 45–88. Oxford, 1974.
Kellenbenz, Hermann, with J. Schawacht, J. Schneider, and L. Peters. "Germany." In *An Introduction to the Sources of European Economic History, 1500–1800*, edited by

Charles Wilson and Geoffrey Parker. Vol. 1, *Western Europe*, 190–222, 241–45. London, 1977.
Keller, Kenneth W. "From the Rhineland to the Virginia Frontier: Flax Production as a Commercial Enterprise." *VMHB* 98 (1990): 487–511.
——. "The Origins of Ulster Scots Emigration to America: A Survey of Recent Research." *American Presbyterians* 70 (1992): 71–80.
——. "The Outlook of Rhinelanders on the Virginia Frontier." In Puglisi, *Diversity and Accommodation*, 99–126.
Kellogg, R. S. "Forests." In *Report of the National Conservation Commission*, edited by Henry Gannett, 1:51–74. 3 vols. Washington, D.C., 1909.
Kelly, James. "The Resumption of Emigration from Ireland after the American War of Independence, 1783–1787." *Studia Hibernica* 24 (1984–88): 61–88.
Kelly, Kevin. *Economic and Social Development of Seventeenth-Century Surry County, Virginia*. New York, 1989.
——. " 'In Dispers'd Country Plantations': Settlement Patterns in Seventeenth-Century Surry County, Virginia." In Tate and Ammerman, *Chesapeake in the Seventeenth Century*, 183–205.
Kelso, William. *Jamestown Rediscovery I: Search for the 1607 James Fort*. Jamestown, Va., 1995.
——. *Jamestown Rediscovery II: Search for the 1607 James Fort*. Jamestown, Va., 1996.
——. *Kingsmill Plantations, 1619–1800: Archaeology of Country Life in Colonial Virginia*. Orlando, Fla., 1984.
Kelso, William, Nicholas M. Luccketti, and Beverly A. Straube. *Jamestown Rediscovery*. Vol. 4. Jamestown, Va., 1998.
Kemmerer, Donald L. "The Colonial Loan-Office System in New Jersey." *Journal of Political Economy* 47 (1939): 867–74.
Kennett, Lee. *The French Forces in America, 1780–1783*. Westport, Conn., 1977.
Kent, Joan R. "Population Mobility and Alms: Poor Migrants in the Midlands during the Early Eighteenth Century." *Local Population Studies*, no. 27 (1981): 35–51.
Kerber, Linda K. *Toward an Intellectual History of Women: Essays by Linda K. Kerber*. Chapel Hill, N.C., 1997.
——. *Women of the Republic: Intellect and Ideology in Revolutionary America*. Chapel Hill, N.C., 1980.
Kermode, Jenny, and Garthine Walker, eds. *Women, Crime, and the Courts in Early Modern England*. London, 1994.
Kerridge, Eric. *The Agricultural Revolution*. New York, 1968.
Kershaw, Gordon E. *The Kennebeck Proprietors, 1749–1775: "Gentlemen of Property and Judicious Men."* Portland, Maine, 1975.
——. "A Question of Orthodoxy: Religious Controversy in a Speculative Land Company." *NEQ* 46 (1975): 205–35.
Kessel, Elizabeth A. "Germans in the Making of Frederick County, Maryland, 1730–1800." In Robert D. Mitchell, *Appalachian Frontiers*, 87–104.
——. " 'A Mighty Fortress Is Our God': Educational Organizations on the Maryland Frontier, 1734–1800." *MdHM* 77 (1982): 370–87.
Ketcham, Ralph L. "Conscience, War, and Politics in Pennsylvania, 1755–1757." *WMQ* 20 (1963): 416–39.
Kettner, James H. *The Development of American Citizenship, 1608–1870*. Chapel Hill, N.C., 1978.
Keyssar, Alexander. "Widowhood in Eighteenth-Century Massachusetts: A Problem in the History of the Family." *Perspectives in American History* 8 (1974): 83–119.

Kierner, Cynthia A. *Beyond the Household: Women's Place in the Early South, 1700–1835*. Ithaca, N.Y., 1998.
——. "Hospitality, Sociability, and Gender in the Southern Colonies." *Journal of Southern History* 62 (1996): 449–80.
——. *Traders and Gentlefolk: The Livingstons of New York, 1675–1790*. Ithaca, N.Y., 1992.
Kießling, Rolf. "Markets and Marketing, Town and Country." In Scribner, *Germany*, 145–79.
Kim, Sung Bok. "The Continental Army and the American People: A Review Essay." *New York History* 62 (1982): 460–69.
——. *Landlord and Tenant in Colonial New York: Manorial Society, 1664–1775*. Chapel Hill, N.C., 1978.
——. "The Limits of Politicization in the American Revolution: The Experience of Westchester County, New York." *Journal of American History* 80 (1993): 868–89.
——. "A New Look at the Great Landlords of Eighteenth-Century New York." *WMQ* 27 (1970): 581–614.
King, Gregory. *Two Tracts*. 1688, 1696. Edited by George E. Barnett. Baltimore, 1936.
King, J. H. C. *First Peoples, First Contacts: Native Peoples of North America*. Cambridge, Mass., 1999.
King, Peter. "Customary Rights and Women's Earnings: The Importance of Gleaning to the Rural Labouring Poor, 1750–1850." *EcHR* 44 (1991): 461–76.
——. "Gleaners, Farmers, and the Failure of Legal Sanctions in England, 1750–1850." *P&P*, no. 125 (1989): 116–50.
King, Stephen. "Migrants on the Margin? Mobility, Integration, and Occupations in the West Riding, 1650–1820." *Journal of Historical Geography* 23 (1997): 284–303.
Kingsbury, Susan Myra, ed. *The Records of the Virginia Company of London*. 4 vols. Washington, D.C., 1906–35.
Kirchner, Walther. *Commercial Relations between Russia and Europe, 1400 to 1800: Collected Essays*. Bloomington, Ind., 1966.
Kirkham, Graeme. "Economic Diversification in a Marginal Economy: A Case Study." In Roebuck, *Plantation to Partition*, 64–81, 259–61.
——. "Ulster Emigration to North America, 1680–1720." In Blethen and Wood, *Ulster and North America*, 76–117, 239–47.
Kisch, Herbert. *From Domestic Manufacture to Industrial Revolution: The Case of the Rhineland Textile Districts*. New York, 1989.
——. "The Textile Industries in Silesia and the Rhineland: A Comparative Study of Industrialization." *JEH* 19 (1959): 541–64.
Kitch, Malcolm. "Population Movement and Migration in Pre-Industrial Rural England." In Short, *English Rural Community*, 61–84.
Klein, Milton M. *The Politics of Diversity: Essays in the History of Colonial New York*. Port Washington, N.Y., 1974.
Klein, Rachel N. *Unification of a Slave State: The Rise of the Planter Class in the South Carolina Backcountry, 1760–1808*. Chapel Hill, N.C., 1990.
Klepp, Susan E. "Demography in Early Philadelphia." *Proceedings of the American Philosophical Society* 133 (1989): 85–111.
——. "Five Early Pennsylvania Censuses." *Pennsylvania Magazine of History and Biography* 106 (1982): 483–514.
Klepp, Susan E., and Billy G. Smith, eds. *The Infortunate: The Voyage and Adventures of William Moraley, an Indentured Servant*. 1743. University Park, Pa., 1992.
Klingaman, David. "Food Surpluses and Deficits in the American Colonies, 1768–1772." *JEH* 31 (1971): 553–69.

———. "The Significance of Grain in the Development of the Tobacco Colonies." *JEH* 29 (1969): 268–78.
Knapp, Jeffrey. "Elizabethan Tobacco." *Representations*, no. 21 (1988): 27–66.
Knittle, W. A. *Early-Eighteenth-Century Palatine Emigration*. Philadelphia, 1936.
Knodel, John E. *Demographic Behavior in the Past: A Study of Fourteen German Village Populations in the Eighteenth and Nineteenth Centuries*. Cambridge, 1988.
Knouff, Gregory T. " 'An Arduous Service': The Pennsylvania Backcountry Soldiers' Revolution." *Pennsylvania History* 61 (1994): 45–74.
———. "Common Soldiers in Total War: Pennsylvania Enlisted Men and the Meaning of Frontier Violence on the Revolutionary Frontier." Paper, McNeil Center for Early American Studies, October 1994.
Koehler, Lyle. *A Search for Power: The "Weaker Sex" in Seventeenth-Century New England*. Urbana, Ill., 1980.
Koellmann, W. "The Population of Barmen before and during the Period of Industrialization." In Glass and Eversley, *Population in History*, 100–143.
Komlos, John. "A Malthusian Episode Revisited: The Height of British and Irish Servants in Colonial America." *EcHR* 46 (1993): 768–82.
———. "The Secular Trend in the Biological Standard of Living in the United Kingdom, 1730–1860." *EcHR* 46 (1993): 115–44.
Konig, David Thomas. "Community Custom and the Common Law: Social Change and the Development of Land Law in Seventeenth-Century Massachusetts." *American Journal of Legal History* 17 (1974): 137–77.
———. " 'Dale's Laws' and the Non-Common Law Origins of Criminal Justice in Virginia." *American Journal of Legal History* 26 (1982): 354–75.
Kopperman, Paul E. "The British High Command and Soldiers' Wives in America, 1755–1783." *Journal of the Society for Army Historical Research*, no. 241 (1982): 14–34.
Kozy, Charlene Johnson. "Tories Transplanted: The Caribbean Exile and Plantation Settlement of Southern Loyalists." *Georgia Historical Quarterly* 75 (1991): 18–42.
Kraus, Michael. *The Atlantic Civilization: Eighteenth-Century Origins*. Ithaca, N.Y., 1949.
Kriedte, Peter. "Demographic and Economic Rhythms: The Rise of the Silk Industry in Krefeld in the Eighteenth Century." *Journal of European Economic History* 15 (1986): 259–89.
———. *Peasants, Landlords, and Merchant Capitalists: Europe and the World Economy, 1500–1800*. Translated by V. R. Berghahn. Cambridge, 1983.
Kriedte, Peter, Hans Medick, and Jürgen Schlumbohm. "Prot-industrialization Revisited: Demography, Social Structure, and Modern Domestic Industry." *Continuity and Change* 8 (1993): 217–52.
Krim, Arthur J. "Acculturation of the New England Landscape: Native and English Toponymy of Eastern Massachusetts." In Benes, *New England Prospect*, 69–88.
Kroeber, A. L. "Nature of the Land-Holding Group." *Ethnohistory* 2 (1955): 303–14.
Kross, Jessica. *The Evolution of an American Town: Newtown, New York, 1642–1775*. Philadelphia, 1983.
Kulikoff, Allan. *The Agrarian Origins of American Capitalism*. Charlottesville, Va., 1992.
———. "The Economic Growth of the Eighteenth-Century Chesapeake Colonies." *JEH* 39 (1979): 275–88.
———. "Growth and Welfare in Early America." *WMQ* 39 (1982): 359–65.
———. "Households and Markets: Toward a New Synthesis of American Agrarian History." *WMQ* 50 (1993): 340–55.
———. "Internal Migration: The British and Dutch Colonies." In Jacob Ernest Cooke, *Encyclopedia of the North American Colonies*, 2:329–44.

——. *The Making of the American Yeoman Class*. Forthcoming.
——. "The Progress of Inequality in Revolutionary Boston." *WMQ* 28 (1971): 375–414.
——. *Tobacco and Slaves: The Development of Southern Cultures in the Chesapeake, 1680–1800*. Chapel Hill, N.C., 1986.
Kupperman, Karen Ordahl. "Apathy and Death in Early Jamestown." *Journal of American History* 66 (1979): 24–40.
——. "The Beehive as a Model for Colonial Design." In Kupperman, *America in European Consciousness*, 272–92.
——. "Climate and Mastery of the Wilderness in Seventeenth-Century New England." In Hall and Allen, *Seventeenth-Century New England*, 3–38.
——. "Fear of Hot Climates in the Anglo-American Colonial Experience." *WMQ* 41 (1984): 213–40.
——. "The Founding Years of Virginia—And the United States of America." *VMHB* 104 (1996): 103–12.
——. "Presentment of Civility: English Readings of American Self-Presentation in the Early Years of Colonization." *WMQ* 54 (1997): 193–228.
——. "The Puzzle of American Climate in the Early Colonial Period." *American Historical Review* 87 (1982): 1262–89.
——. *Settling with the Indians: The Meeting of English and Indian Cultures in America, 1580–1640*. Totowa, N.J., 1980.
——, ed. *America in European Consciousness, 1493–1750*. Chapel Hill, N.C., 1995.
Kussmaul, Ann. *A General View of the Rural Economy of England, 1538–1840*. Cambridge, 1990.
——. *Servants in Husbandry in Early Modern England*. Cambridge, 1981.
Kwasny, Mark V. *Washington's Partisan War, 1775–1783*. Kent, Ohio, 1996.
Lacey, Barbara E. "Women in the Era of the American Revolution: The Case of Norwich, Connecticut." *NEQ* 53 (1980): 527–43.
——. "The World of Hannah Heaton: The Autobiography of an Eighteenth-Century Connecticut Farm Woman." *WMQ* 45 (1988): 280–304.
Lachmann, Richard. *From Manor to Market: Structural Change in England, 1536–1640*. Madison, Wis., 1987.
LaFantasie, Glenn W., ed. *The Correspondence of Roger Williams*. 2 vols. Providence, R.I., 1988.
Lamb, H. H. *The Changing Climate: Selected Papers*. London, 1966.
——. *Climate: Present, Past, and Future*. Vol. 2, *Climatic History and the Future*. London, 1977.
Lambert, Robert S. "The Confiscation of Loyalist Property in Georgia, 1782–1786." *WMQ* 20 (1963): 80–94.
Land, Aubrey C. *The Dulanys of Maryland: A Biographical Study of Daniel Dulany, the Elder (1685–1753), and Daniel Dulany, the Younger (1722–1797)*. Baltimore, 1955.
——, ed. *Bases of the Plantation Economy*. New York, 1969.
Land, Aubrey C., Lois Green Carr, and Edward C. Papenfuse, eds. *Law, Society, and Politics in Early Maryland*. Baltimore, 1977.
Landsberg, H. E., C. S. Yu, and Louise Huang. "Preliminary Reconstruction of a Long Time Series of Climatic Data for the Eastern United States." Technical Note BN-571, University of Maryland Institute for Fluid Dynamics and Applied Mathematics, College Park, Md., 1968.
Landsman, Ned C. *Scotland and Its First American Colony, 1683–1765*. Princeton, N.J., 1985.
Landy, David. "Tuscarora among the Iroquois." In Trigger, *Northeast*, 518–24.

Lane, Wheaton J. *From Indian Trail to Iron Horse: Travel and Transportation in New Jersey, 1620–1860*. Princeton, N.J., 1939.

Lanier, Gabrielle M. "Ethnic Perceptions, Ethnic Landscapes: Material and Cultural Identity in a Region of Regions." Paper, McNeil Center for Early American Studies, December 1995.

La Potin, Armand. "The Minisink Grant: Partnerships, Patents, and Processing Fees in Eighteenth-Century New York." *New York History* 56 (1975): 29–50.

Larsen, Clark Spencer, and George R. Milner, eds. *In the Wake of Contact: Biological Responses to Conquest*. New York, 1994.

Laslett, Peter. *Family Life and Illicit Love in Earlier Generations: Essays in Historical Sociology*. Cambridge, 1977.

——. "Introduction: The History of the Family." In Laslett and Wall, *Household and Family in Past Time*, 1–90.

——. "The Stem Family Hypothesis and Its Privileged Position." In *Statistical Studies of Historical Social Structure*, edited by Kenneth W. Wachter, Eugune Hammel, and Peter Laslett, 89–112. New York, 1978.

——. *The World We Have Lost: England before the Industrial Age*. 2d ed. New York, 1971.

Laslett, Peter, and Richard Wall, eds. *Household and Family in Past Time*. Cambridge, 1972.

Laslett, Peter, Karla Oosterveen, and Richard M. Smith, eds. *Bastardy and Its Comparative History: Studies in the History of Illegitimacy and Marital Nonconformism in Britain, France, Germany, North America, Jamaica, and Japan*. Cambridge, Mass., 1980.

Lauber, Almon Wheeler. *Indian Slavery in Colonial Times within the Present Limits of the United States*. Columbia University Studies in the Social Sciences, no. 134. New York, 1913.

Lawson, John. *A New Voyage to Carolina Containing the Exact Description and Natural History of That Country Together with the Present State Thereof. . . .* London, 1709.

Lazonick, William. "Karl Marx and Enclosures in England." *Review of Radical Political Economics* 6 (summer 1974): 1–32.

Leach, Douglas Edward. "Colonial Indian Wars." In Washburn, *History of Indian-White Relations*, 128–43.

——. *Flintlock and Tomahawk: New England in King Philip's War*. New York, 1958.

——. "The Military System of Plymouth Colony." *NEQ* 24 (1951): 342–64.

——. *The Northern Colonial Frontier, 1607–1763*. New York, 1966.

Leamon, James S. *Revolution Downeast: The War for American Independence in Maine*. Amherst, Mass., 1993.

Leder, Lawrence H. *Robert Livingston and the Politics of Colonial New York, 1654–1728*. Chapel Hill, N.C., 1961.

Lee, Everett S., and Michael Lalli. "Population." In Gilchrist, *Growth of the Seaport Cities*, 25–37.

Lee, James, and Jon Gjerde. "Comparative Household Morphology of Stem, Joint, and Nuclear Household Systems: Norway, China, and the United States." *Continuity and Change* 1 (1986): 89–111.

Lee, Jean Butendoff. "Land and Labor: Parental Bequest Practices in Charles County, Maryland, 1732–1783." In Carr, Morgan, and Russo, *Colonial Chesapeake Society*, 306–41.

——. *The Price of Nationhood: The American Revolution in Charles County*. New York, 1994.

Lee, William Robert. "Bastardy and the Socioeconomic Structure of South Germany." *JIH* 7 (1977): 403–25.
——. "Bastardy in South Germany: Reply." *JIH* 8 (1978): 471–76.
——. "Germany." In *European Demography and Economic Growth*, edited by William Robert Lee, 144–95. London, 1979.
——. *Population Growth, Economic Development, and Social Change in Bavaria, 1750–1850*. New York, 1977.
Leechman, Douglas, and Robert A. Hall. "American Indian Pidgin English: Attestations and Grammatical Peculiarities." *American Speech* 30 (1955): 163–71.
Lefler, Hugh T. "Promotional Literature of the Southern Colonies." *Journal of Southern History* 33 (1967): 3–25.
Lefler, Hugh T., and William S. Powell. *Colonial North Carolina: A History*. New York, 1973.
Leiby, Adrian C. *The Revolutionary War in the Hackensack Valley*. New Brunswick, N.J., 1962.
Leighly, John. "New England Town Names Derived from Personal Names." *Names* 18 (1970): 154–74.
Lemay, J. A. Leo. *The American Dream of Captain John Smith*. Charlottesville, Va., 1991.
——. *Men of Letters in Colonial Maryland*. Knoxville, Tenn., 1972.
——. *"New England's Annoyances": America's First Folk Song*. Newark, Del., 1985.
Lemon, James T. *The Best Poor Man's Country: A Geographical Study of Early Southeastern Pennsylvania*. Baltimore, 1972.
——. "Colonial America in the Eighteenth Century." In Mitchell and Groves, *North America*, 121–46.
——. "Household Consumption in Eighteenth-Century America and Its Relationship to Production and Trade: The Situation among Farmers in Southeastern Pennsylvania." *AgH* 41 (1967): 59–70.
Lemon, James T., and Gary B. Nash. "The Distribution of Wealth in Eighteenth Century America: A Century of Changes in Chester County, Pennsylvania." *Journal of Social History* 2 (1968): 1–24.
Lender, Mark Edward. "The Enlisted Line: The Continental Soldiers of New Jersey." Ph.D. diss., Rutgers University, 1975.
——. "The Social Structure of the New Jersey Brigade: The Continental Line as an American Standing Army." In *The Military in America: From the Colonial Era to the Present*, edited by Peter Karsten, 27–44. New York, 1980.
Lender, Mark Edward, and James Kirby Martin, eds. *Citizen Soldier: The Revolutionary War Journal of Joseph Bloomfield*. Newark, N.J., 1982.
Lenman, Bruce. *An Economic History of Modern Scotland, 1660–1976*. London, 1977.
Le Patourel, H. E. J. "Rural Building in England and Wales: England." In Edward Miller, *Agrarian History of England and Wales*, 820–93.
Lepore, Jill. *The Name of War: King Philip's War and the Origins of American Identity*. New York, 1998.
Lerner, Gerda, ed. *The Female Experience: An American Documentary*. Indianapolis, 1977.
Lesser, Charles H., ed. *The Sinews of Independence: Monthly Strength Reports of the Continental Army*. Chicago, 1976.
Levine, David. *Reproducing Families: The Political Economy of English Population History*. Cambridge, 1987.
Levine, David, and Keith Wrightson. *The Making of an Industrial Society: Whickham, 1560–1765*. Oxford, 1991.

———. "The Social Context of Illegitimacy in Early Modern England." In Laslett, Oosterveen, and Smith, *Bastardy and Its Comparative History*, 158–75.
Levy, Babette M. *Early Puritanism in the Southern and Island Colonies*. Worcester, Mass., 1960.
Levy, Barry. "Girls and Boys: Poor Children and the Labor Market in Colonial Massachusetts." In Canny et al., *Empire, Society, and Labor*, 287–307.
———. *Quakers and the American Family: British Settlement in the Delaware Valley*. New York, 1988.
———. "Quakers, the Delaware Valley, and North Midlands Emigration to America." *WMQ* 48 (1991): 246–52.
Levy, Daniel S. "The Life Expectancies of Colonial Maryland Legislators." *Historical Methods* 20 (1987): 17–28.
Lewis, George E. *The Indiana Company, 1763–1798: A Study in Eighteenth-Century Frontier Land Speculation and Business Venture*. Glendale, Calif., 1941.
Lewis, Johanna Miller. *Artisans in the North Carolina Backcountry*. Lexington, Ky., 1995.
Lewis, Kenneth E. *The American Frontier: An Archaeological Study of Settlement Pattern and Process*. Orlando, Fla., 1984.
Lewis, Peirce. "Learning from Looking: Geographic and Other Writing about the American Cultural Landscape." *American Quarterly* 35 (1983): 242–61.
Lewis, Theodore. "Land Speculation and the Dudley Council." *WMQ* 31 (1974): 255–72.
Lewis, Thomas Reed, Jr. "From Suffield to Saybrook: An Historical Geography of the Connecticut River Valley in Connecticut before 1800." Ph.D. diss., Rutgers University, 1978.
Leyburn, James G. *The Scotch-Irish: A Social History*. Chapel Hill, N.C., 1962.
Leyser, Henrietta. *Medieval Women: A Social History of Women in England, 450–1500*. London, 1995.
Libecap, Gary D. "Property Rights in Economic History: Implications for Research." *Explorations in Economic History* 23 (1986): 227–52.
Lindert, Peter H. "An Algorithm for Probate Sampling." *JIH* 11 (1981): 649–68.
Lindert, Peter H., and Jeffrey G. Williamson. "Revising England's Social Tables 1688–1812." *Explorations in Economic History* 19 (1982): 385–408.
Lindsay, Jean. *The Scottish Poor Law: Its Operation in the North-East from 1745 to 1845*. Elms Court, England, 1975.
Linebaugh, Donald W. "'All the Annoyances and Inconveniences of the Country': Environmental Factors in the Development of Outbuildings in the Colonial Chesapeake." *Winterthur Portfolio* 29 (1994): 1–17.
Linebaugh, Peter. "All the Atlantic Mountains Shook." In Eley and Hunt, *Reviving the English Revolution*, 193–219.
Lining, John. "Extract of Two Letters from Dr. John Lining, Physician at Charles-Town in South Carolina . . . Giving an Account of Statical Experiments . . . , Accompanied with Meteorological Observations." *Transactions of the Philosophical Society* 42 (1742–43): 490–507.
Linn, John Blair. "Indian Land and Its Fair-Play Settlers, 1773–1785." *Pennsylvania Magazine of History and Biography* 7 (1883): 420–25.
Linzer, Beth. "Population and Society: A Demographic History of Brookline, 1710–1850." *Chronos*, no. 3 (1984): 7–48.
Little, Ann M. "Men on Top?: The Farmer, the Minister, and Marriage in Early New England." In Canny et al. *Empire, Society, and Labor*, 123–50.
Livermore, Shaw. *Early American Land Companies: Their Influence on Corporate Development*. Cambridge, Mass., 1939.

Lobdell, Jared C. "Six Generals Gather Forage: The Engagement at Quibbletown, 1777." *New Jersey History* 102 (1984): 34–49.
Lockhart, Audrey. "The Quakers and Emigration from Ireland to the North American Colonies." *Quaker History* 77 (1988): 67–92.
Lockridge, Kenneth A. "Land, Population, and the Evolution of New England Society." *P&P*, no. 39 (1968): 62–80.
——. *Literacy in Colonial New England: An Enquiry into the Social Context of Literacy in the Early Modern West*. New York, 1974.
——. *A New England Town: The First Hundred Years, Dedham, Massachusetts, 1636–1736*. New York, 1970.
Loehr, Rodney C. "Self-Sufficiency on the Farm." *AgH* 26 (1952): 37–41.
Lois Green Carr: The Chesapeake and Beyond: A Celebration. Crownsville, Md., 1992.
Lomas, Tim. "South-East Durham: Late Fourteenth and Fifteenth Centuries." In Harvey, *Peasant Land Market*, 253–327.
Looney, John F. "Benning Wentworth's Land Grant Policy: A Reappraisal." *Historical New Hampshire* 13 (1968): 3–13.
Lord, Philip, Jr. *War over Walloomscoick: Land Use and Settlement Pattern on the Bennington Battlefield, 1777*. New York State Museum Bulletin No. 473. Albany, N.Y., 1989.
Lounsbury, Carl. "The Development of Domestic Architecture in the Albemarle Region." In *Towards Preservation of Place: In Celebration of the North Carolina Vernacular Landscape*, edited by Doug Swaim, 46–61. Raleigh, N.C., 1978.
Lovett, A. A., Ian D. Whyte, and Kathleen A. Whyte. "Poisson Regression Analysis and Migration Fields: The Example of Apprenticeship Records of Edinburgh in the Seventeenth and Eighteenth Centuries." *Institute of British Geographers, Transactions*, n.s., 10 (1985): 317–32.
Lower, Arthur R. M. *Colony to Nation: A History of Canada*. 4th ed. London, Canada, 1964.
Lucassen, Jan. *Migrant Labour in Europe, 1600–1900*. London, 1987.
——. "The Netherlands, the Dutch, and Long-Distance Migration in the Late Sixteenth to Early Nineteenth Centuries." In Canny, *Europeans on the Move*, 153–91.
Lucassen, Leo. "A Blind Spot: Migratory and Traveling Groups in Western European Historiography." *International Review of Social History* 38 (1993): 209–35.
Ludlum, David M. *Early American Winters, 1604–1820*. Boston, 1966.
Lurie, Nancy Oestreich. "Indian Cultural Adjustment to European Civilization." In James Morton Smith, *Seventeenth-Century America*, 33–60.
Lutz, Paul V. "Land Grants for Service in the Revolution." *New York Historical Society Quarterly* 48 (1964): 221–35.
——. "A State's Concern for Soldiers' Welfare: How North Carolina Provided for Her Troops during the Revolution." *North Carolina Historical Review* 42 (1965): 315–18.
Lynd, Staughton. *Anti-Federalism in Dutchess County, New York: A Study of Democracy and Class Conflict in the Revolutionary Era*. Chicago, 1962.
——. *Class Conflict, Slavery, and the United States Constitution*. Indianapolis, 1967.
Lyne, Gerald J. "Land Tenure in Kenmare, Bonane, and Tuosist, 1720–1770." *Kerry Archaeological and Historical Society Journal* 11 (1978): 25–55.
Lystad, Mary. *At Home in America as Seen through Its Books for Children*. Cambridge, Mass., 1984.
Macafee, William. "The Demographic History of Ulster, 1750–1841." In Blethen and Wood, *Ulster and North America*, 41–60, 233–36.

Macafee, William, and V. Morgan. "Population in Ulster, 1660–1760." In Roebuck, *Plantation to Partition*, 46–63, 257–59.
McAnear, Beverly. "Mr. Robert Livingston's Reasons against a Land Tax." *Journal of Political Economy* 48 (1940): 63–90.
McBride, John D. "The Virginia War Effort, 1775–1783: Manpower Policies and Practices." Ph.D. diss., University of Virginia, 1977.
McBride, Kevin A. " 'Ancient and Crazie': Pequot Lifeways during the Historic Period." In Benes, *Algonkians*, 63–75.
——. "Historical Archeology of the Mashantucket Pequots, 1637–1790." In Hauptman and Wherry, *Pequots in Southern New England*, 96–116.
——. "The Source and Mother of the Fur Trade: Native-Dutch Relations in Eastern New Netherland." In *Enduring Traditions: The Native Peoples of New England*, edited by Laurie Weinstein, 31–51. Westport, Conn., 1994.
McCarthy, Finbarr. "The Influence of 'Legal Habit' on English-Indian Relations in Jamestown, 1606–1612." *Continuity and Change* 5 (1990): 39–64.
MacCarthy-Morrogh, Michael. *The Munster Plantation: English Migration to Southern Ireland, 1583–1641*. Oxford, 1986.
McCary, Ben C. *Indians in Seventeenth-Century Virginia*. Williamsburg, Va., 1957.
McCleskey, Turk. "Rich Land, Poor Prospects: Real Estate and the Formation of a Social Elite in Augusta County, Virginia, 1730–1770." *VMHB* 98 (1990): 449–86.
——. "Shadow Land: Provisional Real Estate Claims and Anglo-American Settlement in Southwestern Virginia." In Crass et al., *Southern Colonial Backcountry*, 56–68.
McCloskey, Donald N. "The Persistence of English Common Fields." In Parker and Jones, *European Peasants and Their Markets*, 73–119.
McConnell, Michael N. *A Country Between: The Upper Valley and Its Peoples, 1724–1774*. Lincoln, Nebr., 1992.
McConville, Brendan. *Those Daring Disturbers of the Public Peace: The Struggle for Property and Power in New Jersey, 1701–1726*. Ithaca, N.Y., 1999.
McCowen, George Smith, Jr. *The British Occupation of Charleston, 1780–82*. Columbia, S.C., 1972.
McCoy, Drew R. *The Elusive Republic: Political Economy in Jeffersonian America*. Chapel Hill, N.C., 1980.
McCracken, John Leslie. "The Ecclesiastical Structure, 1714–60." In Moody and Vaughan, *New History of Ireland*, 84–104.
——. "The Social Structure and Social Life, 1714–1760." In Moody and Vaughan, *New History of Ireland*, 31–56.
McCullough, Robert. *The Landscape of Community: A History of the Communal Forests in New England*. Hanover, N.H., 1995.
MacCurtain, Margaret, and Mary O'Dowd, eds. *Women in Early Modern Ireland*. Edinburgh, 1991.
McCusker, John J. "How Much Is That in Real Money? A Historical Price Index for Use as a Deflator of Money Values in the Economy of the United States." *American Antiquarian Society, Proceedings* 101 (1991): 297–373.
——. *Money and Exchange in Europe and America, 1600–1775: A Handbook*. Chapel Hill, N.C., 1977.
——. Review of *Enterprise and Empire: Merchant and Gentry Investments in the Expansion of England, 1575–1630*. *Historical Methods Newsletter* 2 (June 1969): 14–18.
——. "Sources of Investment Capital in the Colonial Philadelphia Shipping Industry." *JEH* 32 (1972): 146–57.

———. "The View from British North America" [Forum on Bailyn, *Voyagers to the West*]. *Business History Review* 62 (1988): 691–96.
McCusker, John J., and Russell R. Menard. *The Economy of British America, 1607–1789*. Chapel Hill, N.C., 1985.
McDevitt, Robert F. *Connecticut Attacked: A British Viewpoint, Tryon's Raid on Danbury*. Connecticut Bicentennial Series, 10. Chester, Conn., 1974.
Macdonald, Anne L. *No Idle Hands: The Social History of American Knitting*. New York, 1988.
MacDonald, John S., and Leatrice A. MacDonald. "Chain Migration and Ethnic Neighborhood Formation and Social Networks." *Milbank Memorial Fund Quarterly* 42 (1964): 82–97.
McDonald, Michael. *Mystical Bedlam: Madness, Anxiety, and Healing in Seventeenth-Century England*. Cambridge, 1981.
McDonnell, Michael. "Other Loyalists: A Reconsideration of the Black Loyalist Experience in the Revolutionary Era." *Southern Historian* 16 (1995): 5–25.
Macdougall, Hamilton C. *Early New England Psalmody: An Historical Appreciation, 1620–1820*. Brattleboro, Vt., 1940.
McDowell, R. B. "Ireland in 1800." In Moody and Vaughan, *New History of Ireland*, 657–712.
MacFarlane, Alan. *The Family Life of Ralph Josselin, a Seventeenth-Century Clergyman: An Essay in Historical Anthropology*. New York, 1970.
———. *The Origins of English Individualism: The Family, Property, and Social Transition*. Oxford, 1978.
McGiffert, Michael, ed. *God's Plot: The Paradoxes of Puritan Piety, Being the Autobiography and Journal of Thomas Shepherd*. Amherst, Mass., 1972.
McGregor, Robert Kuhn. "Cultural Adaptation in Colonial New York: The Palatine Germans of the Mohawk Valley." *New York History* 69 (1988): 5–34.
Machin, R. "The Great Rebuilding: A Reassessment." *P&P*, no. 77 (1977): 33–56.
McIntosh, Marjorie K. *A Community Transformed: The Manor and Liberty of Havering, 1500–1620*. Cambridge, 1991.
———. "Local Change and Community Control in England, 1465–1500." *Huntington Library Quarterly* 49 (1986): 219–42.
———. "Money Lending on the Periphery of London, 1300–1600." *Albion* 20 (1988): 557–71.
———. "Servants and the Household Unit in an Elizabethan English Community." *Journal of Family History* 9 (1984): 3–23.
McJimsey, George Davis. "Topographic Terms in Virginia." *American Speech* 15 (1940): 3–38, 149–79, 262–300, 381–419.
Mack, Phyllis. *Visionary Women: Ecstatic Prophecy in Seventeenth-Century England*. Berkeley, Calif., 1992.
Mackenthun, Gesa. *Metaphors of Dispossession: American Beginnings and the Translation of Empire, 1492–1637*. Norman, Okla., 1997.
Mackey, Howard. "The Operation of the English Old Poor Law in Colonial Virginia." *VMHB* 73 (1965): 29–40.
McKendrick, Neil. "Commercialization and the Economy." In *The Birth of a Consumer Society: The Commercialization of Eighteenth-Century England*, by Neil McKendrick, John Brewer, and J. H. Plumb, 9–194. Bloomington, Ind., 1982.
McKinney, Hannah J. *The Development of Local Public Services, 1650–1860: Lessons from Middletown, Connecticut*. Westport, Conn., 1995.

McLean, Marianne. *The People of Glengarry: Highlanders in Transition, 1745–1820*. Montreal, 1991.
McMahon, Sarah F. "A Comfortable Subsistence: The Changing Composition of Diet in Rural New England." *WMQ* 42 (1985): 26–65.
——. "Provisions Laid Up for the Family: Toward a History of Diet in New England, 1650–1850." *Historical Methods* 14 (1981): 4–21.
McManus, Edgar J. *Black Bondage in the North*. Syracuse, N.Y., 1973.
MacMaster, Richard K. "The Cattle Trade in Western Virginia, 1760–1830." In Robert D. Mitchell, *Appalachian Frontiers*, 127–49.
——. "Religion, Migration, and Pluralism: A Shenandoah Valley Community." In Puglisi, *Diversity and Accommodation*, 82–98.
MacMaster, Richard K., Samuel L. Horst, and Robert F. Ulle. *Conscience in Crisis: Mennonite and Other Peace Churches in America, 1739–1789, Interpretation and Documents*. Scottdale, Pa., 1979.
McNealy, Terry A. "Bristol: The Origins of a Pennsylvania Market Town." *Pennsylvania Magazine of History and Biography* 95 (1971): 484–510.
Macpherson, C. B. *Democratic Theory: Essays in Retrieval*. Oxford, 1973.
McRae, Andrew. *God Speed the Plough: The Representation of Agrarian England, 1500–1660*. Cambridge, 1996.
McWilliams, James E. "Brewing Beer in Massachusetts Bay, 1640–1690." *NEQ* 71 (1998): 543–63.
Madden, Richard C. "Catholics in Colonial South Carolina." *American Catholic Historical Society of Philadelphia, Records* 73 (1962): 10–44.
Mager, Wolfgang. "Proto-industrialization and Proto-industry: The Uses and Drawbacks of Two Concepts." *Continuity and Change* 8 (1993): 181–215.
Main, Gloria L. "The English Family in America: A Comparison of New England, Chesapeake, and Pennsylvania Quaker Families in the Colonial Period." In *Lois Green Carr*, 143–59.
——. "Gender, Work, and Wages in Colonial New England." *WMQ* 51 (1994): 39–66.
——. "Naming Children in Early New England." *JIH* 27 (1996): 1–27.
——. "The Standard of Living in Southern New England, 1640–1773." *WMQ* 45 (1988): 124–34.
——. *Tobacco Colony: Life in Early Maryland, 1650–1720*. Princeton, N.J., 1982.
——. "Widows in Rural Massachusetts on the Eve of the Revolution." In Hoffman and Albert, *Women in the Age of the American Revolution*, 67–90.
Main, Gloria L., and Jackson Turner Main. "Economic Growth and the Standard of Living in Southern New England, 1640–1774." *JEH* 48 (1988): 27–46.
——. "The Red Queen in New England?" *WMQ* 56 (1999): 121–50.
Main, Jackson Turner. "The Distribution of Property in Post-Revolutionary Virginia." *Mississippi Valley Historical Review* 41 (1954–55): 241–58.
——. *Society and Economy in Colonial Connecticut*. Princeton, N.J., 1985.
——. *The Sovereign States, 1775–1783*. New York, 1973.
——. "Standards of Living and the Life Cycle in Colonial Connecticut." *JEH* 43 (1983): 159–65.
Malone, Patrick M. "Changing Military Technology among the Indians of Southern New England, 1600–1677." *American Quarterly* 25 (1973): 48–63.
——. *The Skulking Way of War: Technology and Tactics among the New England Indians*. Lanham, Md., 1991.
Malthus, Thomas Robert. *An Essay on the Principle of Population and a Summary View*

of the Principle of Population. 1793, 1830. Edited by Anthony Flew. Harmondsworth, England, 1970.
——. *An Essay on the Principle of Population; or, A View of its Past and Present Effects on Human Happiness* . . . . 1803. Collated with 1806 to 1827 versions. 2 vols. Edited by Patricia James. Cambridge, 1989.
Mancall, Peter C. *Deadly Medicine: Indians and Alcohol in Early America*. Ithaca, N.Y., 1995.
——. "Landholding: The British Colonies." In Jacob Ernest Cooke, *Encyclopedia of the North American Colonies*, 1:653–64.
——. *Valley of Opportunity: Economic Culture along the Upper Susquehanna, 1700–1800*. Ithaca, N.Y., 1991.
Mancall, Peter C., and Thomas Weiss. "Was Economic Growth Likely in Colonial British North America?" *JEH* 59 (1999): 17–40.
Mandell, Daniel. *Behind the Frontier: Indians in Eighteenth-Century Eastern Massachusetts*. Lincoln, Nebr., 1996.
——. " 'Standing by His Father': Thomas Waban of Natick, circa 1630–1722." In Grumet, *Northeastern Indian Lives*, 166–94.
——. " 'To Live More Like My Christian English Neighbors': Natick Indians in the Eighteenth Century." *WMQ* 48 (1991): 552–79.
Manley, Gordon. "Central England Temperatures: Monthly Means, 1659 to 1973." *Quarterly Journal of the Royal Meteorological Society* 100 (1974): 389–405.
Mann, Bruce H. *Neighbors and Strangers: Law and Community in Early Connecticut*. Chapel Hill, N.C., 1987.
Mann, Julia de Lacy. *The Cloth Industry in the West of England from 1640 to 1880*. Oxford, 1971.
Manning, Brian. "The Peasantry and the English Revolution." *Journal of Peasant Studies* 2 (1975): 133–58.
Manning, Roger Burrow. *Hunters and Poachers: A Social and Cultural History of Unlawful Hunting in England, 1485–1640*. Oxford, 1993.
——. *Village Revolts: Social Protest and Popular Disturbances in England, 1509–1640*. Oxford, 1988.
Marcus, Gail Sussman. " 'Due Execution of the Generall Rules of Righteousnesse': Criminal Procedure in New Haven Town and Colony." In Hall, Murrin, and Tate, *Saints and Revolutionaries*, 99–137.
Marietta, Jack D. "The Distribution of Wealth in Eighteenth-Century America: Nine Chester County Tax Lists, 1693–1799." *Pennsylvania History* 62 (1995): 532–45.
——. *The Transformation of American Quakerism, 1748–1783*. Philadelphia, 1984.
Mark, Irving. *Agrarian Conflicts in Colonial New York, 1711–1775*. New York, 1940.
Marks, Bayly Ellen. "The Rage for Kentucky: Emigration from St. Mary's County, 1790–1810." In Mitchell and Muller, *Geographic Perspectives on Maryland's Past*, 108–28.
Marshall, Joshua Micah. " 'A Melancholy People': Anglo-Indian Relations in Early Warwick, Rhode Island, 1642–1675." *NEQ* 68 (1995): 402–28.
Martin, Ann Smart. "Common People and the Local Store: Consumerism in the Rural Virginia Backcountry." In *Common People and Their Material World: Free Men and Women in the Chesapeake, 1700–1830: Proceedings of the March 13, 1992, Conference Sponsored by the Research Division, Colonial Willliamsburg Foundation*, edited by David Harvey and Gregory Brown, 39–53. Williamsburg, Va., 1995.
——. " 'Fashionable Sugar Dishes, Latest Fashion Ware': The Creamware Revolution in the Eighteenth-Century Chesapeake." In Shackel and Little, *Historic Archaeology*, 169–87.

——. "Frontier Boys and Country Cousins: The Context for Choice in Eighteenth-Century Consumerism." In *Historical Archaeology and the Study of American Culture*, edited by Lu Ann De Cunzo and Bernard L. Herman, 71–102. Winterthur, Del., 1996.

——. "The Role of Pewter as Missing Artifact: Consumer Attitudes toward Tableware in Late Eighteenth-Century Virginia." *Historical Archaeology* 23 (1989): 1–27.

Martin, Calvin. "Fire and Forest Structure in the Aboriginal Eastern Forest." *Indian Historian* 6 (summer 1973): 23–26; (fall 1973): 38–42.

Martin, James Kirby. "A 'Most Undisciplined, Profligate Crew': Protest and Defiance in the Continental Ranks." In Hoffman and Albert, *Arms and Independence*, 119–40.

Martin, John E. *Feudalism to Capitalism: Peasant and Landlord in English Agrarian Development*. Atlantic Highlands, N.J., 1983.

——. "Sheep and Enclosure in Sixteenth-Century Northamptonshire." *Agricultural History Review* 36 (1988): 39–54.

Martin, John Frederick. *Profits in the Wilderness: Entrepreneurship and the Founding of New England Towns in the Seventeenth Century*. Chapel Hill, N.C., 1991.

Martin, Joseph Plumb. *Private Yankee Doodle, Being a Narrative of Some of the Adventures and Dangers and Sufferings of a Revolutionary Soldier*. Edited by George E. Scheer. 1830. Boston, 1962.

Martin, Wendy. "Women and the American Revolution." *Early American Literature* 11 (1976–77): 322–35.

Martinac, Paula. " 'An Unsettled Disposition': Social Structure and Geographical Mobility in Amelia County, Virginia, 1768–1794." M.A. thesis, College of William and Mary, 1979.

Marx, Karl. *Capital: A Critique of Political Economy*. 1867–94. 3 vols. Edited by Ernest Mandel. New York, 1977–81.

——. *The Eighteenth Brumaire of Louis Bonaparte*. 1849. In *Karl Marx and Frederick Engels: Selected Works*, 95–180. New York, 1968.

Maryland Court of Appeals Record, TDM#1, 1788. Maryland State Archives, Annapolis.

Mason, Bernard. "Entrepreneurial Activity in New York during the American Revolution." *Business History Review* 40 (1966): 190–212.

Mason, Keith. "Localism, Evangelicalism, and Loyalism: The Sources of Oppression in the Revolutionary Chesapeake." *Journal of Southern History* 56 (1990): 23–54.

Masschaele, James. "The Multiplicity of Medieval Markets Reconsidered." *Journal of Historical Geography* 30 (1994): 255–71.

Massey, Gregory de Van. "The British Expedition to Wilmington, January–November, 1781." *North Carolina Historical Review* 66 (1989): 387–411.

Masur, Louis P. "Slavery in Eighteenth-Century Rhode Island: Evidence from the Census of 1774." *Slavery and Abolition* 6 (1985): 139–50.

Mate, Mavis E. "The Agrarian Economy of South-East England before the Black Death: Depressed or Buoyant?" In Bruce M. S. Campbell, *Before the Black Death*, 79–109.

——. "The East Sussex Land Market and Agrarian Class Structure in the Late Middle Ages." *P&P*, no. 139 (1993): 46–65.

Mather, Frederic Gregory. *The Refugees of 1776 from Long Island to Connecticut*. Albany, N.Y., 1913.

Mather, Increase. *A Brief History of the Warr with the Indians in New-England . . . .* Boston, 1676. In Slotkin and Folsom, *So Dreadfull a Judgment*, 78–152.

Mathews, Lois Kimball. *The Expansion of New England: The Spread of New England Settlement and Institutions to the Mississippi River, 1620–1865*. Boston, 1909.

Matson, Cathy. *Merchants and Empire: Trading in Colonial New York*. Baltimore, 1998.

Matthei, Julie A. *An Economic History of Women in America: Women's Work, the Sexual Division of Labor, and the Development of Capitalism*. New York, 1982.
Mavor, James W., Jr., and Byron E. Dix. *Manitou: The Sacred Landscape of New England's Native Civilization*. Rochester, Vt., 1989.
Maxwell, Hu. "The Use and Abuse of Forests by Virginia Indians." *WMQ*, ser. 1, 19 (1910): 73–103.
Mayer, Holly A. *Belonging to the Army: Camp Followers and Community during Revolutionary America*. Columbia, S.C., 1996.
Mayhew, Alan. *Rural Settlement and Farming in Germany*. London, 1973.
Maza, Sarah. "Stories in History: Cultural Narratives in Recent Works in European History." *American Historical Review* 101 (1996): 1493–1515.
Medick, Hans. "Village Spinning Bees: Sexual Culture and Free Time among Rural Youth in Early Modern Germany." In *Interest and Emotion: Essays on the Study of Family and Kinship*, edited by Hans Medick and David Warren Sabean, 317–39. Cambridge, 1984.
Meinig, D. W. "The Beholding Eye: Ten Versions of the Same Scene." In *The Interpretation of Ordinary Landscapes: Geographical Essays*, edited by D. W. Meinig, 33–48. New York, 1979.
——. *The Shaping of America: A Geographical Perspective on 500 Years of History*. Vol. 1, *Atlantic America, 1492–1800*. New Haven, Conn., 1986.
Melton, James Van Horn. *Absolutism and the Eighteenth-Century Origins of Compulsory Schooling in Prussia and Austria*. Cambridge, 1988.
Melvoin, Richard L. *New England Outpost: War and Society in Colonial Deerfield*. New York, 1989.
Menard, Russell R. "The Africanization of the Lowcountry Labor Force." In *Race and Family in the Colonial South*, edited by Winthrop D. Jordan and Sheila L. Skemp, 81–108. Jackson, Miss., 1987.
——. "British Migration to the Chesapeake Colonies in the Seventeenth Century." In Carr, Morgan, and Russo, *Colonial Chesapeake Society*, 99–132.
——. "Comment on Paper by Ball and Walton." *JEH* 36 (1976): 118–25.
——. "Economic and Social Development of the South." In Engerman and Gallman, *Cambridge Economic History*, 249–95.
——. "Financing the Lowcountry Export Boom: Capital and Growth in Early Carolina." *WMQ* 51 (1994): 659–76.
——. "From Servants to Slaves: The Transformation of the Chesapeake Labor System." *Southern Studies* 16 (1977): 355–90.
——. "From Servant to Freeholder: Status Mobility and Property Accumulation in Seventeenth-Century Maryland." *WMQ* 30 (1973): 37–64.
——. "Immigrants and Their Increase: The Process of Population Growth in Early Colonial Maryland." In Land, Carr, and Papenfuse, *Law, Society, and Politics in Early Maryland*, 88–110.
——. "Immigration to the Chesapeake Colonies in the Seventeenth Century: A Review Essay." *MdHM* 68 (1972): 323–29.
——. "Population, Economy, and Society in Seventeenth-Century Maryland." *MdHM* 79 (1984): 71–92.
——. "Slave Demography in the Lowcountry, 1670–1740: From Frontier Society to Plantation Regime." *South Carolina Historical Magazine* 96 (1995): 280–303.
——. "The Tobacco Industry in the Chesapeake Colonies, 1617–1730: An Interpretation." *Research in Economic History* 5 (1980): 109–77.

——. "Whatever Happened to Early American Population History?" *WMQ* 50 (1993): 356–66.

Menard, Russell R., and Lois Green Carr. "The Lords Baltimore and the Colonization of Maryland." In Quinn, *Early Maryland in a Wider World*, 167–215.

Menard, Russell R., P. M. G. Harris, and Lois Green Carr. "Opportunity and Inequality: The Distribution of Wealth on the Lower Western Shore of Maryland, 1638–1705." *MdHM* 69 (1974): 169–84.

Mendels, Franklin F. "Proto-industrialization: The First Phase of the Industrialization Process." *JEH* 32 (1972): 241–61.

Mendelson, Sara Heller, ed. " 'To Shift for a Cloak': Disorderly Women in the Church Courts." In *Women and History: Voices of Early Modern England*, edited by Valerie Frith, 3–18. Toronto, 1995.

Mendelson, Sara Heller, and Patricia Crawford. *Women in Early Modern England, 1550–1720*. Oxford, 1998.

Mensch, Elizabeth V. "The Colonial Origins of Liberal Property Rights." *Buffalo Law Review* 31 (1982): 635–735.

Merchant, Carolyn. *Ecological Revolutions: Nature, Gender, and Science in New England*. Chapel Hill, N.C., 1989.

Meriwether, Robert L. *The Expansion of South Carolina, 1729–1765*. Kingsport, Tenn., 1940.

Merrell, James H. "Cultural Continuity among the Piscataway Indians of Colonial Maryland." *WMQ* 36 (1979): 548–70.

——. " 'The Customes of Our Countrey': Indians and Colonists in Early America." In Bailyn and Morgan, *Strangers within the Realm*, 117–56.

——. "Declarations of Independence: Indian-White Relations in the New Nation." In Jack P. Greene, *American Revolution*, 197–223.

——. *The Indians' New World: Catawbas and Their Neighbors from European Contact through the Era of Removal*. Chapel Hill, N.C., 1989.

——. *Into the American Woods: Negotiators on the Pennsylvania Frontier*. New York, 1999.

Merrens, Harry Roy. *Colonial North Carolina in the Eighteenth Century: A Study in Historical Geography*. Chapel Hill, N.C., 1964.

Merrens, Harry Roy, and George D. Terry. "Dying in Paradise: Malaria, Mortality, and the Perceptual Environment in Colonial South Carolina." *Journal of Southern History* 50 (1984): 533–50.

Merrill, Michael. "Cash Is Good to Eat: Self-Sufficiency and Exchange in the Rural Economy of the United States." *Radical History Review* 4 (1977): 42–71.

——. "Putting 'Capitalism' in Its Place: A Review of Recent Literature." *WMQ* 52 (1995): 315–26.

——. "A Survey of the Debate over the Nature of Exchange in Early America." Paper presented at the Social Science History Association, New Orleans, 1987.

Merrill, Michael, and Sean Wilentz, eds. *The Key of Liberty: The Life and Democratic Writings of William Manning, "A Laborer," 1747–1814*. Cambridge, Mass., 1993.

Merritt, Jane T. "The Power of Language: Cultural Meanings and the Colonial Encounter on the Pennsylvania Frontier." Paper, McNeil Center for Early American Studies, April 1994.

Merwick, Donna. *Possessing Albany, 1630–1710: The Dutch and English Experience*. Cambridge, 1990.

Metzger, Charles H. *Catholics and the American Revolution: A Study in Religious Climate*. Chicago, 1962.

——. *The Prisoner in the American Revolution*. Chicago, 1971.
Meyer, Freeman W. "The Evolution of the Interpretation of Economic Life in Colonial Connecticut." *Connecticut History* 26 (1985): 33–43.
Michel, Jack. " 'In a Manner and Fashion Suitable to Their Degree.' " *Working Papers from the Regional Economic History Center* 5, no. 1 (1981): 1–83.
Middlekauff, Robert. *The Glorious Cause: The American Revolution, 1763–1789*. New York, 1982.
Middleton, Arthur Pierce. *Tobacco Coast: A Maritime History of Chesapeake Bay in the Colonial Era*. Newport News, Va., 1953.
Middleton, Christopher. "Peasants, Patriarchy, and the Feudal Mode of Production in England: A Marxist Appraisal: Property and Patriarchal Relations within the Peasantry." *Sociological Review* 29 (1981): 105–35.
——. "The Sexual Division of Labour in Feudal England." *New Left Review*, nos. 113–14 (1979): 147–68.
——. "Women's Labor and the Transition to Pre-Industrial Capitalism." In Charles and Duffin, *Women and Work in Preindustrial England*, 181–206.
Migliazzo, Arlin C. "A Tarnished Legacy Revisited: Jean Pierre Purry and the Settlement of the Southern Frontier, 1718–1736." *South Carolina Historical Magazine* 92 (1991): 232–52.
Mikalachki, Jodi. "Women's Networks and the Female Vagrant: A Hard Case." In *Maids and Mistresses: Women's Alliances in Early Modern England*, edited by Susan Frye and Karen Robertson, 52–69. New York, 1999.
Miles, Lion G. "The Red Man Dispossessed: The Williams Family and the Alienation of Indian Land in Stockbridge, Massachusetts, 1736–1818." *NEQ* 67 (1994): 46–76.
Miller, Edward. "Introduction: Land and People." In Edward Miller, *Agrarian History of England and Wales*, 1–33.
——, ed. *The Agrarian History of England and Wales*. Vol. 3, *1348–1500*. Cambridge, 1991.
Miller, Henry M. "An Archaeological Perspective on the Evolution of Diet in the Colonial Chesapeake, 1620–1745." In Carr, Morgan, and Russo, *Colonial Chesapeake Society*, 176–99.
Miller, Kerby A. *Emigrants and Exiles: Ireland and the Irish Exodus to North America*. New York, 1985.
Miller, Mary R. "Place-Names of the Northern Neck of Virginia: A Proposal for a Theory of Place-Naming." *Names* 24 (1976): 9–23.
Miller, Perry. *Errand into the Wilderness*. Cambridge, Mass., 1956.
Miller, Randall M. "The Failure of the Colony of Georgia under the Trustees." *Georgia Historical Quarterly* 53 (1969): 1–17.
Miller, William. "The Effects of the American Revolution on Indentured Servitude." *Pennsylvania History* 7 (1940): 131–41.
Miller, William Davis. "The Narragansett Planters." *American Antiquarian Society, Proceedings* 43 (1934): 49–115.
Millward, R. "The Emergence of Wage Labor in Early Modern England." *Explorations in Economic History* 18 (1981): 21–39.
Mingay, Gordon Edward. "The Agricultural Depression, 1730–1750." *EcHR* 8 (1956): 323–38.
——. *Enclosure and the Small Farmer in the Age of the Industrial Revolution*. London, 1968.
——, ed. *The Agrarian History of England and Wales*. Vol. 6, *1750–1850*. Cambridge, 1989.

Mitchell, Peter M. "Loyalist Property and the Revolution in Virginia." Ph.D. diss., University of Colorado, 1965.
Mitchell, Robert D. "Agricultural Change and the American Revolution: A Virginia Case Study." *AgH* 47 (1973): 119–32.
——. "American Origins and Regional Institutions: The Seventeenth-Century Chesapeake." *Annals of the Association of American Geographers* 73 (1983): 404–20.
——. "Colonial Origins of Anglo-America." In Mitchell and Groves, *North America*, 93–120.
——. *Commercialism and Frontier: Perspectives on the Early Shenandoah Valley.* Charlottesville, Va., 1977.
——. "Metropolitan Chesapeake: Reflections on Town Formation in Colonial Virginia and Maryland." In *Lois Green Carr*, 105–25.
——. "The Southern Backcountry: A Geographical House Divided." In Crass et al., *Southern Colonial Backcountry*, 1–35.
——, ed. *Appalachian Frontiers: Settlement, Society, and Development in the Preindustrial Era*. Lexington, Ky., 1991.
Mitchell, Robert, and Paul A. Groves, eds. *North America: The Historical Geography of a Changing Continent*. Totowa, N.J., 1987.
Mitchell, Robert D., and Edward K. Muller, eds. *Geographic Perspectives on Maryland's Past*. University of Maryland Department of Geography Occasional Papers, no. 4, College Park, Md., 1979.
Mitchell, Stewart, ed. *Winthrop Papers*. Vols. 1–2. Boston, 1929–31.
Mitchell, Thornton W. "The Granville District and Its Land Records." *North Carolina Historical Review* 70 (1993): 103–29.
Mitchison, Rosalind. *Life in Scotland*. London, 1978.
——. "The Making of the Old Scottish Poor Law." *P&P*, no. 63 (1974): 58–93.
——. "North and South: The Development of the Gulf in Poor Law Practice." In Houston and Whyte, *Scottish Society*, 199–225.
——. "Webster Revisited: A Re-examination of the 1755 'Census' of Scotland." In *Improvement and Enlightenment: Proceedings of the Scottish Historical Studies Seminar, University of Strathclyde, 1987–88*, edited by Thomas Martin Devine, 62–77. Edinburgh, 1989.
——. "Who Were the Poor in Scotland, 1690–1830?" In Mitchison and Roebuck, *Economy and Society in Scotland and Ireland*, 140–48.
Mitchison, Rosalind, and Peter Roebuck, eds. *Economy and Society in Scotland and Ireland, 1500–1939*. Edinburgh, 1988.
Mittelberger, Gottlieb. *Journey to Pennsylvania*. 1756. Edited and translated by Oscar Handlin and John Clive. Cambridge, Mass., 1960.
Moch, Leslie Page. *Moving Europeans: Migration in Western Europe since 1650*. Bloomington, Ind., 1992.
Moch, Leslie Page, Nancy Folbre, Daniel Scott Smith, Laurel L. Cornell, and Louise A. Tilly. "Family Strategy: A Dialogue." *Historical Methods* 20 (1987): 113–25.
Mokyr, Joel, and Cormac Ó Gráda, "The Height of Irishmen and Englishmen in the 1770s: Some Evidence from East India Company Records." *Eighteenth-Century Ireland* 4 (1989): 83–92.
——. "New Developments in Irish Population History, 1700–1850." *EcHR* 37 (1984): 473–88.
Moller, Herbert. Introduction to *Population Movements in Modern European History*, edited by Herbert Moller, 1–7. New York, 1964.

——. “Sex Composition and Correlated Culture Patterns of Colonial America.” *WMQ* 2 (1945): 113–53.
Moltmann, Günter. “The Migration of German Redemptioners to North America, 1720–1820.” In Emmer, *Colonialism and Migration*, 105–22.
Mood, Fulmer, ed. “John Winthrop, Jr., on Indian Corn.” *NEQ* 10 (1937): 121–33.
Moody, Theodore William, and W. E. Vaughan, eds. *A New History of Ireland*. Vol. 4, *Eighteenth-Century Ireland, 1691–1800*. Oxford, 1986.
Moogk, Peter N. “Mannon’s Fellow Exiles: Emigration from France to North America before 1763.” In Canny, *Europeans on the Move*, 236–60.
——. “Reluctant Exiles: Emigrants from France in Canada before 1760.” *WMQ* 46 (1989): 463–505.
Moore, Christopher. *The Loyalists: Revolution, Exile, Settlement*. Toronto, 1994.
Moore, Frank, comp. *The Diary of the American Revolution, 1775–1781*. Edited by John Anthony Scott. 1874. New York, 1967.
Morgan, Edmund S. *American Slavery, American Freedom: The Ordeal of Colonial Virginia*. New York, 1975.
——. “Headrights and Head Counts: A Review Article.” *VMHB* 80 (1972): 361–71.
——. *The Puritan Dilemma: The Story of John Winthrop*. Boston, 1958.
——. *The Puritan Family: Religion and Domestic Relations in Seventeenth-Century New England*. New York, 1944.
Morgan, Kenneth. “The Organization of the Convict Trade to Maryland: Stevenson, Randolph & Cheston, 1768–1775.” *WMQ* 42 (1985): 201–28.
Morgan, Philip D. “Black Society in the Lowcountry, 1760–1810.” In *Slavery and Freedom in the Age of the American Revolution*, edited by Ira Berlin and Ronald Hoffman, 83–141. Charlottesville, Va., 1983.
——. *Slave Counterpoint: Black Culture in the Eighteenth-Century Chesapeake and Lowcountry*. Chapel Hill, N.C., 1998.
Morison, Samuel Eliot. *Builders of the Bay Colony*. 2d ed. Boston, 1958.
Morley, Edith J., ed. [Henry] *Crabb Robinson in Germany, 1800–1805*. London, 1929.
Morner, Magnus. “Spanish Migration to the New World prior to 1810: A Report on the State of Research.” In Chiappelli, *First Images of America*, 737–810.
Morris, Francis Grave, and Phyllis Mary Morris. “Economic Conditions in North Carolina about 1780.” *North Carolina Historical Review* 16 (1939): 108–33, 296–327.
Morris, Michael P. *The Bringing of Wonder: Trade and the Indians of the Southeast, 1700–1783*. Westport, Conn., 1999.
Morris, Richard B. *Government and Labor in Early America*. New York, 1946.
——. *Studies in the History of American Law with Special Reference to the Seventeenth and Eighteenth Centuries*. 2d ed. New York, 1964.
Morton, Louis. *Robert Carter of Nomini Hall: A Virginia Tobacco Planter of the Eighteenth Century*. Charlottesville, Va., 1945.
Morton, Richard L. *Colonial Virginia*. Vol. 2, *Westward Expansion and Prelude to Revolution*. Chapel Hill, N.C., 1960.
Moynihan, Ruth Barnes. “Coming of Age: Four Centuries of Connecticut Women and Their Choices.” *Connecticut Historical Society Bulletin* 53 (1988): 5–111.
——. “The Patent and the Indians: The Problem of Jurisdiction in Seventeenth-Century New England.” *American Indian Culture and Research Journal* 2 (1977): 8–18.
Muldrew, Craig. “Interpreting the Market: The Ethics of Credit and Community Relations in Early Modern England.” *Social History* 18 (1993): 163–83.
Mullin, Gerald W. *Flight and Rebellion: Slave Resistance in Eighteenth-Century Virginia*. New York, 1972.

Murrin, John M. "'Things Fearful to Name': Bestiality in Colonial America." In Pencak and Boudreau, *Explorations in Early American Culture*, 8–43.
Mutch, Robert E. "Yeoman and Merchant in Pre-Industrial America: Eighteenth-Century Massachusetts as a Case Study." *Societas* 7 (1977): 279–302.
Myers, Albert Cook. *Immigration of the Irish Quakers into Pennsylvania, 1682–1750, with Their Early History in Ireland*. Swarthmore, Pa., 1902.
——, ed. *Narratives of Early Pennsylvania, West New Jersey, and Delaware, 1630–1707*. New York, 1912.
Nabokov, Peter, and Robert Easton. *Native American Architecture*. New York, 1989.
Nadelhaft, Jerome J. *The Disorders of War: The Revolution in South Carolina*. Orono, Maine, 1981.
Nanepashemet. "It Smells Fishy to Me: An Argument Supporting the Use of Fish Fertilizer by the Native People of Southern New England." In Benes, *Algonkians*, 42–50.
Narrett, David E. "Men's Wills and Women's Property Rights in Colonial New York." In Hoffman and Albert, *Women in the Age of the American Revolution*, 91–133.
Nash, Gary B. "The Free Society of Traders and the Early Politics of Pennsylvania." *Pennsylvania Magazine of History and Biography* 89 (1965): 147–73.
——. "The Image of the Indian in the Southern Colonial Mind." *WMQ* 29 (1972): 197–230.
——. *Quakers and Politics: Pennsylvania, 1681–1726*. Princeton, N.J., 1968.
——. *Race and Revolution*. Madison, Wis., 1990.
——. *Race, Class, and Politics: Essays on American Colonial and Revolutionary Society*. Urbana, Ill., 1986.
——. *Red, White, and Black: The Peoples of Early America*. 3d ed. Englewood Cliffs, N.J., 1992.
——. *The Urban Crucible: Social Change, Political Consciousness, and the Origins of the American Revolution*. Cambridge, Mass., 1979.
Nash, Gary B., and Jean R. Soderlund. *Freedom by Degrees: Emancipation in Pennsylvania and Its Consequences*. New York, 1991.
Nash, R. C. "South Carolina and the Atlantic Economy in the Late Seventeenth and Eighteenth Centuries." *EcHR* 45 (1992): 677–702.
——. "Urbanization in the Colonial South: Charleston, South Carolina, as a Case Study." *Journal of Urban History* 19 (1992): 3–29.
Neagles, James C., comp. and ed. *Summer Soldiers: A Survey and Index of Revolutionary War Courts-Martial*. Salt Lake City, Utah, 1986.
Neeson, J. M. *Commoners: Common Right, Enclosure, and Social Change in England, 1700–1820*. Cambridge, 1993.
Nef, John U. *The Conquest of the Material World: Essays on the Coming of Industrialization*. Cleveland, 1964.
——. *The Rise of the British Coal Industry*. 2 vols. London, 1932.
Neiman, Fraser D. "Domestic Architecture at the Cliffs Plantation: The Social Context of Early Virginia Building." In Upton and Vlach, *Common Places*, 292–314.
Neimeyer, Charles Patrick. *America Goes to War: A Social History of the Continental Army*. New York, 1996.
Nelson, Paul David. "The American Soldier and the American Victory." In *The World Turned Upside Down: The American Victory in the War of Independence*, edited by John Ferling, 35–52, 206–10. New York, 1988.
——. *William Tryon and the Course of Empire: A Life in British Imperial Service*. Chapel Hill, N.C., 1990.

Nettels, Curtis. *The Emergence of a National Economy, 1775–1815*. New York, 1962.
———. *The Money Supply of the American Colonies before 1720*. Madison, Wis., 1934.
Nevins, Allan. *The American States during and after the Revolution, 1775–1789*. New York, 1924.
Newell, Margaret Ellen. *From Dependency to Independence: Economic Revolution in Colonial New England*. Ithaca, N.Y., 1998.
New-Englands Plantation; or, A Short and True Description of the Commodities of that Countrey. 1630. Reprinted in Force, *Tracts and Other Papers*, vol. 1.
Nicholls, Michael Lee. "Competition, Credit, and Crisis: Merchant-Planter Relations in Southside Virginia." In *Merchant Credit and Labour Strategies in Historical Perspective*, edited by Rosemary E. Omner, 273–89. Frederickton, Canada, 1990.
———. "Origins of the Virginia Southside, 1703–1753: A Social and Economic Study." Ph.D. diss., College of William and Mary, 1972.
Nichtweiss, Johannes. "The Second Serfdom and the So-Called 'Prussian Way': The Development of Capitalism in Eastern German Agricultural Institutions." *Review* 3 (1979): 99–140.
Noble, John, and John F. Cronin, eds. *Records of the Court of Assistants of the Colony of Massachusetts Bay, 1630–1692*. 3 vols. Boston, 1901–28.
Nobles, Gregory. "Breaking into the Backcountry: New Approaches to the Early American Frontier." *WMQ* 46 (1989): 641–70.
———. *Divisions throughout the Whole: Politics and Society in Hampshire County, Massachusetts, 1740–1775*. Cambridge, 1983.
———. "The Rise of Merchants in Rural Market Towns: A Case Study of Eighteenth-Century Northampton, Massachusetts." *Journal of Social History* 24 (1990): 5–23.
Norona, Delf, ed. "Joshua Fry's Report on the Back Settlements of Virginia, May 8, 1751." *VMHB* 56 (1948): 22–41.
North, Douglass C., and Robert Paul Thomas. *The Rise of the Western World: A New Economic History*. Cambridge, 1973.
Norton, Mary Beth. *The British-Americans: Loyalist Exiles in England, 1774–1779*. Boston, 1972.
———. "Eighteenth-Century American Women in Peace and War: The Case of the Loyalists." *WMQ* 33 (1976): 386–409.
———. "The Evolution of White Woman's Experience in Early America." *American Historical Review* 89 (1984): 593–619.
———. *Founding Mothers and Fathers: Gendered Power and the Forming of American Society*. New York, 1996.
———. *Liberty's Daughters: The Revolutionary Experience of American Women, 1750–1800*. Boston, 1980.
Norton, Susan L. "Marital Migration in Essex County, Massachusetts, in the Colonial and Early Federal Periods." *Journal of Marriage and the Family* 35 (1973): 419–28.
Norton, Thomas Elliot. *The Fur Trade in Colonial New York, 1686–1776*. Madison, Wis., 1974.
Nylander, Jane C. "Provision for Daughters: The Accounts of Samuel Lane." In Benes, *House and Home*, 11–27.
Oakes, Elinor F. "A Ticklish Business: Dairying in New England and Pennsylvania, 1750–1812." *Pennsylvania History* 47 (1980): 195–212.
Oberg, Michael L. "Indians and Englishmen at the First Roanoke Colony: A Note on Pemisapan's Conspiracy." *American Indian Culture and Research Journal* 18 (1994): 75–89.

O'Brien, Jean M. *Dispossession by Degrees: Indian Land and Identity in Natick, Massachusetts, 1650–1790*. Cambridge, 1997.
O'Callaghan, E. B., ed. *Documentary History of the State of New-York*. 4 vols. Albany, N.Y., 1849–51.
O'Callaghan, E. B., and Berthold Fernow, eds. *Documents Relative to the Colonial History of the State of New-York*. 15 vols. Albany, N.Y., 1853–87.
O'Callaghan, Jerry A. "The War Veteran and the Public Lands." *AgH* 28 (1951): 163–68.
O'Donnell, James H., III. "Joseph Brant." In *American Indian Leaders: Studies in Diversity*, edited by R. David Edmunds, 21–40. Lincoln, Nebr., 1980.
——. *Southern Indians in the American Revolution*. Knoxville, Tenn., 1973.
Oglethorpe, James Edward. *The Publications of James Edward Oglethorpe*. Edited by Rodney M. Baine and Phinizy Spalding. Athens, Ga., 1994.
——. *Some Account of the Design of the Trustees for Establishing Colonys in America*. 1732. Edited by Rodney M. Baine and Phinizy Spalding. Athens, Ga., 1990.
Ó Gráda, Cormac. *Ireland: A New Economic History, 1780–1839*. Oxford, 1994.
O'Hara, Diana. " 'Ruled by My Friends': Aspects of Marriage in the Diocese of Canterbury, c. 1540–1570." *Continuity and Change* 6 (1991): 9–42.
Ohlmeyer, Jane H. " 'Civilizinge of Those Rude Partes': Colonization within Britain and Ireland, 1480s–1630s." In Canny and Low, *Origins of Empire*, 124–47.
Olmstead, Earl P. *Blackcoats among the Delaware: David Zeisberger on the Ohio Frontier*. Kent, Ohio, 1991.
——. *David Zeisberger: A Life among the Indians*. Kent, Ohio, 1997.
Olson, Gary D. "Thomas Brown, Loyalist Partisan, and the Revolutionary War in Georgia, 1777–1782." *Georgia Historical Quarterly* 65 (1970): 1–19, 183–208.
Olson, Sherri. *A Chronicle of All That Happens: Voices from the Village Court in Medieval England*. Toronto, 1996.
Olwell, Robert. *Masters, Slaves, and Subjects: The Culture of Power in the South Carolina Lowcountry, 1740–1790*. Ithaca, N.Y., 1998.
Onuf, Peter S. "Settlers, Settlements, and New States." In Jack P. Greene, *American Revolution*, 171–96.
——. *Statehood and Union: A History of the Northwest Ordinance*. Bloomington, Ind., 1987.
O'Reilly, William. "Conceptualizing America in Early Modern Europe." In Pencak and Boudreau, *Explorations in Early American Culture*, 101–21.
Orr, Alastair. "Farm Servants and Farm Labour in the Forth Valley and South-East Lowlands." In Devine, *Farm Servants and Labour*, 29–54.
Osberg, Lars, and Fazley Siddiq. "The Inequality of Wealth in Britain's North American Colonies: The Importance of the Relatively Poor." *Review of Income and Wealth* 34 (1988): 143–63.
Otterness, Philip. "The New York Naval Stores Project and the Transformation of the Poor Palatines, 1710–1712." *New York History* 75 (1994): 133–56.
Otto, John Solomon. "Livestock-Raising in Early South Carolina, 1670–1700: Prelude to the Rice Plantation Economy." *AgH* 61 (1987): 13–24.
——. *The Southern Frontiers, 1607–1860: The Agricultural Evolution of the Colonial and Antebellum South*. Westport, Conn., 1989.
Ouellette, Susan M. "Divine Production and Collective Endeavor: Sheep Production in Early Massachusetts." *NEQ* 69 (1996): 355–80.
Ousterhout, Anne M. *A State Divided: Opposition in Pennsylvania to the American Revolution*. New York, 1987.

Outhwaite, R. B. *Dearth, Public Policy, and Social Disturbance in England, 1550–1800*. Houndmill, England, 1991.
——. *Inflation in Tudor and Early Stuart England*. London, 1969.
——. "Progress and Backwardness in English Agriculture, 1500–1650." *EcHR* 39 (1986): 1–18.
Overton, Mark. *Agricultural Revolution in England: The Transformation of the Agrarian Economy, 1500–1850*. Cambridge, 1996.
——. "English Probate Inventories and the Measurement of Agricultural Change." In *Probate Inventories: A New Source for the Historical Study of Wealth, Material Culture, and Agricultural Development*, edited by Ad Van Der Woude and Anton Schuurman, 205–16. Utretch, The Netherlands, 1980.
——. "Re-establishing the English Agricultural Revolution." *Agricultural History Review* 44 (1996): 1–20.
Paden, John. " 'Several & Many Grievances of Very Great Consequences': North Carolina's Political Factionalism in the 1720s." *North Carolina Historical Review* 71 (1994): 285–305.
Pagden, Anthony. "The Struggle for Legitimacy and the Image of Empire in the Atlantic to c 1700." In Canny and Low, *Origins of Empire*, 34–54.
Page, John. "The Economic Structure of Society in Revolutionary Bennington." *Vermont History* 49 (1981): 69–84.
Painter, Nell Irvin. "Bias and Synthesis in History." *Journal of American History* 74 (1987): 109–12.
Palmer, Robert C. "The Economic and Cultural Impact of the Origins of Property: 1180–1220." *Law and History Review* 3 (1985): 375–96.
Palmer, Robert R. *The Age of the Democratic Revolution*. 2 vols. Princeton, N.J., 1959, 1964.
Papenfuse, Edward C., Jr. "Planter Behavior and Economic Opportunity in a Staple Economy." *AgH* 46 (1972): 297–312.
Papenfuse, Edward C., Jr., and Gregory A. Stiverson. "General Smallwood's Recruits: The Peacetime Career of the Revolutionary War Private." *WMQ* 30 (1973): 117–32.
Parish, Susan Scott. "The Female Opossum and the Nature of the New World." *WMQ* 54 (1997): 475–514.
Parker, Anthony N. *Scottish Highlanders in Colonial Georgia: The Recruitment, Emigration, and Settlement at Darien, 1735–1748*. Athens, Ga., 1997.
Parker, William N., and Eric L. Jones, eds. *European Peasants and Their Markets: Essays in Agrarian Economic History*. Princeton, N.J., 1975.
Parkhill, Trevor. "Philadelphia Here I Come: A Study of the Letters of Ulster Immigrants in Philadelphia, 1750–1875." In Blethen and Wood, *Ulster and North America*, 118–45, 247–50.
Parmenter, Jon William. "Pontiac's Wars: Forging New Links in the Anglo-Iroquois Covenant Chain, 1758–1766." *Ethnohistory* 44 (1997): 617–54.
Parr, William A. "Maryland Forests." In Richard C. Davis, *Encyclopedia of Forest History*, 2:408–9.
Parry, J. H., and P. M. Sherlock. *A Short History of the West Indies*. London, 1957.
Parry, Martin L. *Climatic Change, Agriculture, and Settlement*. Folkestone, England, 1978.
Parry, Martin L., and Terence R. Slater, eds. *The Making of the Scottish Countryside*. London, 1980.
Parsons, William T. *The Pennsylvania Dutch: A Persistent Minority*. Boston, 1976.

"A Partial List of the Families Who Arrived at Philadelphia between 1682 and 1687." *Pennsylvania Magazine of History and Biography* 8 (1884): 328–40.

"A Partial List of the Families Who Resided in Bucks County, Pennsylvania, Prior to 1687, With the Date of Their Arrival." *Pennsylvania Magazine of History and Biography* 9 (1885): 223–33.

Patten, John. *English Towns, 1500–1700*. Folkestone, England, 1978.

——. "Patterns of Migration and Movement of Labour to Three Pre-Industrial East Anglian Towns." *Journal of Historical Geography* 2 (1976): 111–29.

Patten, Matthew. *The Diary of Matthew Patten of Bedford, N.H., from Seventeen Hundred Fifty-four to Seventeen Hundred Eighty-eight*. Concord, N.H., 1903.

Pawson, Eric. *The Early Industrial Revolution: Britain in the Eighteenth Century*. New York, 1979.

Peckham, Sir George. *True Reporte of the Late Discoveries . . . by Sir Humphrey Gilbert*. 1583. In Quinn, *New American World*, 34–60.

Peckham, Howard H. *Pontiac and the Indian Uprising*. Princeton, N.J., 1947.

Pelizzon, Sheila. "The Grain Flour Commodity Chain, 1590–1790." In *Commodity Chain and Global Capitalism*, edited by Gary Gereffi and Miguel Korzeniewicz, 34–47. Westport, Conn., 1994.

Pencak, William, and George W. Boudreau, eds. *Explorations in Early American Culture. Pennsylvania History* 65, special supplement (1998).

Penn, Simon A. C. "Female Wage-Earners in Late Fourteenth-Century England." *Agricultural History Review* 35 (1987): 1–14.

Penn, William. *Letter from William Penn to the Committee of the Free Society of Traders, 1683*. In Myers, *Narratives of Early Pennsylvania*, 224–44.

——. *Some Account of the Province of Pennsylvania*. In Myers, *Narratives of Early Pennsylvania*, 202–15.

Penna, Anthony N. *Nature's Bounty: Historical and Modern Environmental Perspectives*. Armonk, N.Y., 1999.

Pennypacker, Samuel W., ed. *Valley Forge Orderly Book of General George Weeden of the Continental Army under the Command of General George Washington, in the Campaign of 1777–8 . . . .* New York, 1902.

Penrose, Maryly B., ed. *Indian Affairs Papers: American Revolution*. Franklin Park, N.J., 1981.

Percy, George. *Observations Gathered out of "A Discourse of the Plantation of the Southern Colony in Virginia by the English, 1606."* Edited by David B. Quinn. 1607. Charlottesville, Va., 1967.

Perdue, Theda. *Cherokee Women: Gender and Culture Change, 1700–1835*. Lincoln, Nebr., 1998.

——. *Native Carolinians: The Indians of North Carolina*. Raleigh, N.C., 1985.

Perkins, Edwin J. *American Public Finance and Financial Services, 1700–1815*. Columbus, Ohio, 1994.

Perlmann, Joel, and Dennis Shirley. "When Did New England Women Acquire Literacy?" *WMQ* 48 (1991): 50–67.

Perrenoud, Alfred. "Les Migrations en Suisse sous l'Ancien Régime: Quelques Questions." *Annales de Demographie Historique* 7 (1970): 251–60.

Perry, James R. *The Formation of a Society on Virginia's Eastern Shore, 1615–1655*. Chapel Hill, N.C., 1990.

Peskin, Lawrence A. "A Restless Generation: Migration of Maryland Veterans in the Early Republic." *MdHM* 91 (1996): 311–28.

Peters, Christine. "Single Women in Early Modern England: Attitudes and Expectations." *Continuity and Change* 12 (1997): 325–45.
Petersen, Christian. *Bread and the British Economy, c1770–1870*. Aldershot, England, 1995.
Peterson, Merrill, ed. *Thomas Jefferson: Writings*. New York, 1984.
Pfeiffer, Susan, and Scott I. Fairgrieve. "Evidence from Ossuaries: The Effect of Contact on the Health of Iroquians." In Larsen and Milner, *In the Wake of Contact*, 47–61.
Pfister, Ulrich. "Work Roles and Family Structure in Proto-Industrial Zurich." *JIH* 20 (1989): 83–105.
Phillips, C. B. "Town and Country: Economic Change in Kendal, c. 1550–1700." In Peter A. Clark, *Transformation of English Provincial Towns*, 99–132.
Phillips, Paul Chrisler, with J. W. Smurr. *The Fur Trade*. 2 vols. Norman, Okla., 1961.
Phillips, Roderick. *Putting Asunder: A History of Divorce in Western Society*. Cambridge, 1988.
Phythian-Adams, Charles. "Urban Decay in Late Medieval England." In *Towns in Societies: Essays in Economic History and Historical Sociology*, edited by Philip Abrams and E. A. Wrigley, 159–85. Cambridge, 1978.
Picht, Douglas R. "The American Squatter and Federal Land Policy." *Journal of the West* 14 (1975): 72–83.
Pierson, William D. *Black Yankees: The Development of an Afro-American Subculture in Eighteenth-Century New England*. Amherst, Mass., 1988.
Pingeon, Frances D. *Blacks in the Revolutionary Era*. New Jersey Historical Commission, New Jersey's Revolutionary Experience, no. 14. Trenton, N.J., 1975.
Pinkett, Harold T. "Maryland as a Source of Food Supplies during the American Revolution." *MdHM* 46 (1951): 157–72.
Plane, Ann Marie. "Putting a Face on Colonization: Factionalism and Gender Politics in the Life History of Awashunkes, the 'Squaw Sachem' of Saconet." In Grumet, *Northeastern Indian Lives*, 140–65.
Pleasants, J. Hall, and Louis Dow Scisco, eds. *Proceedings of the County Court of Charles County, 1658–1666, and Manor Court of St. Clement's Manor, 1659–1672*. Vol. 52 of *Archives of Maryland*. Baltimore, 1936.
Pleck, Elizabeth. *Domestic Tyranny: The Making of American Social Policy against Family Violence from Colonial Times to the Present*. New York, 1987.
Pocock, J. G. A. "British History: A Plea for a New Subject." *Journal of Modern History* 47 (1975): 601–28.
——. "The Limits and Divisions of British History: In Search of the Unknown Subject." *American Historical Review* 87 (1982): 311–36.
Pomfret, John E. *The Province of East New Jersey, 1609–1702*. Princeton, N.J., 1962.
——. *The Province of West New Jersey, 1609–1702: The History of the Origins of an American Colony*. Princeton, N.J., 1956.
Poos, L. R., Zvi Razi, and Richard M. Smith. "The Population History of Medieval Villages: A Debate on the Use of Manor Court Records." In Razi and Smith, *Medieval Society and the Manor Court*, 298–368.
Porter, Frank W., III. "Behind the Frontier: Indian Survivals in Maryland." *MdHM* 75 (1980): 42–54.
——. "From Backcountry to County: The Delayed Settlement of Western Maryland." *MdHM* 70 (1975): 329–49.
——. "The Nanticoke Indians in a Hostile World." In Porter, *Strategies for Survival*, 139–72.
——, ed. *Strategies for Survival: American Indians in the Eastern United States*. Westport, Conn., 1986.

Post, John D. *Food Shortage, Climatic Variation, and Epidemic Disease in Preindustrial Europe: The Mortality Peak in the 1740s*. Ithaca, N.Y., 1985.
Postan, Sir Michael M. "Feudalism and Its Decline: A Semantic Exercise." In Aston et al., *Social Relations and Ideas*, 73–87.
Potter, Janice. "Patriarchy and Paternalism: The Case of the Eastern Ontario Loyalist Women." *Ontario History* 80 (1989): 3–24.
Potter, Stephen R. *Commoners, Tribute, and Chiefs: The Development of Algonquian Culture in the Potomac Valley*. Charlottesville, Va., 1993.
Potter, Stephen R., and Gregory A. Waselkov. " 'Whereby We Shall Enjoy Their Cultivated Places.' " In Shackel and Little, *Historic Archaeology*, 23–33.
Potter-MacKinnon, Janice. *While the Women Only Wept: Loyalist Refugee Women*. Montreal, 1993.
Pound, John. *Poverty and Vagrancy in Tudor England*. London, 1971.
Pounds, N. J. G. *An Historical Geography of Europe, 1500–1840*. Cambridge, 1979.
Poussou, Jean Pierre. "Les Mouvements Migratoires en France et a Partir de la France de la Fin du XV[e] Siecle au Debut du XIX[e] Siecle: Approches pour une Synthese." *Annales de Demographie Historique* 7 (1970): 10–78.
Powell, Sumner Chilton. *Puritan Village: The Formation of a New England Town*. Middletown, Conn., 1963.
Powell, William S. "Roanoke Colonists and Explorers: An Attempt at Identification." *North Carolina Historical Review* 34 (1957): 202–26.
——, ed. "Tryon's Book on North Carolina." *North Carolina Historical Review* 34 (1957): 406–15.
Power, Garrett. "Parceling Out Land in the Vicinity of Baltimore: 1632–1696, Part 1." *MdHM* 87 (1992): 453–66.
Power, Thomas P. *Land, Politics, and Society in Eighteenth-Century Tipperary*. Oxford, 1993.
Power, Thomas P., and Kevin Whelan, eds. *Endurance and Emergence: Catholics in Ireland in the Eighteenth Century*. Dublin, 1990.
Preisser, Thomas M. "Alexandria and Evolution of the Northern Virginia Economy, 1749–1766." *VMHB* 89 (1981): 282–93.
Price, Edward T. *Dividing the Land: Early American Beginnings of Our Private Property Mosaic*. University of Chicago Department of Geography Research Paper, no. 238, 1995.
Price, Jacob M. "Buchanan & Simpson, 1759–1763: A Different Kind of Glasgow Firm Trading to the Chesapeake." *WMQ* 40 (1983): 3–41.
——. *Capital and Credit in British Overseas Trade: The View from the Chesapeake, 1700–1776*. Cambridge, Mass., 1980.
——. "Economic Function and Growth of American Port Towns in the Eighteenth Century." *Perspectives in American History* 8 (1974): 123–86.
——. "The Economic Growth of the Chesapeake and the European Market, 1697–1775." *JEH* 24 (1964): 496–511.
——. *France and the Chesapeake: A History of the French Tobacco Monopoly, 1674–1791, and of Its Relationship to the British and American Tobacco Trades*. 2 vols. Ann Arbor, Mich., 1973.
——. "The Last Phase of the Virginia-London Consignment Trade: James Buchanan & CO., 1758–1768." *WMQ* 43 (1986): 64–98.
——. "Merchants and Planters: The Market Structure of the Colonial Chesapeake Reconsidered." In *Tobacco in Atlantic Trade: The Chesapeake, London, and Glasgow, 1675–1775*, by Jacob M. Price, fourth article, 1–31. Aldershot, England, 1995.

———. "New Time Series for Scotland's and Britain's Trade with the Thirteen Colonies and States, 1740 to 1791." *WMQ* 32 (1975): 307–25.
———. "The Rise of Glasgow in the Chesapeake Tobacco Trade, 1707–1775." *WMQ* 11 (1954): 179–99.
Price, Jacob M., and Paul G. E. Clemens. "A Revolution of Scale in Overseas Trade: British Firms in the Chesapeake Trade, 1675–1775." *JEH* 47 (1987): 1–43.
Prince, Carl E., Dennis P. Ryan, Brenda Parnes, Mary Lou Lustig, Pamela Schafler, and Donald W. White, eds. *The Papers of William Livingston*. 5 vols. Trenton, N.J., 1979– .
Prince George's County, Maryland, Land Records. Maryland State Archives, Annapolis.
Prior, Mary, ed. *Women in English Society, 1500–1800*. London, 1985.
Proudhon, Pierre-Joseph. *What Is Property?* 1840. Edited and translated by Donald R. Kelley and Bonnie G. Smith. Cambridge, 1993.
Pruitt, Bettye Hobbs. "Agriculture and Society in the Towns of Massachusetts, 1771: A Statistical Analysis." Ph.D. diss., Boston University, 1981.
———. "Self-Sufficiency and the Agricultural Economy of Eighteenth-Century Massachusetts." *WMQ* 41 (1984): 333–64.
———, ed. *The Massachusetts Tax Valuation List of 1771*. Boston, 1978.
Pryce, W. T. R. "Migration and the Evolution of Culture Areas: Cultural and Linguistic Frontiers in North-East Wales, 1750 and 1851." *Institute of British Geographers, Transactions*, no. 65 (1975): 79–107.
Pryde, George S. "Scottish Colonization in the Province of New York." *New York State Historical Society, Proceedings* 33 (1935): 138–57.
Puglisi, Michael J. *Puritans Besieged: The Legacies of King Philip's War in the Massachusetts Bay Colony*. Lanham, Md., 1991.
———, ed. *Diversity and Accommodation: Essays on the Cultural Composition of the Virginia Frontier*. Knoxville, Tenn., 1997.
Pulsipher, Jenny Hale. "Massacre at Hurtleberry Hill: Christian Indians and English Authority in Metacom's War." *WMQ* 53 (1996): 459–86.
Purcell, Richard J. "Irish Contribution to Colonial New York." *Studies* 29 (1940): 591–604; 30 (1941): 107–20.
Purvis, Thomas L. "Patterns of Ethnic Settlement in Late Eighteenth-Century Pennsylvania." *Western Pennsylvania Historical Magazine* 70 (1987): 107–22.
———. "The Pennsylvania Dutch and German American Diaspora in 1790." *Journal of Cultural Geography* 6 (1986): 81–99.
———. "Were Colonial American Politics Deferential? A Case Study from the Delaware Valley." Paper, McNeil Center for Early American Studies, February 1986.
Quaife, G. R. *Wanton Wenches and Wayward Wives: Peasants and Illicit Sex in Early Seventeenth Century England*. London, 1979.
Quarles, Benjamin. *The Negro in the American Revolution*. Chapel Hill, N.C., 1961.
Quimby, Ian M. G. *Apprenticeship in Colonial Philadelphia*. New York, 1985.
Quinn, David B. *Explorers and Colonies: America, 1500–1625*. London, 1990.
———. *North America from Earliest Discovery to First Settlements: The Norse Voyages to 1612*. New York, 1977.
———. *Set Fair for Roanoke: Voyages and Colonies, 1584–1606*. Chapel Hill, N.C., 1985.
———, ed. *Early Maryland in a Wider World*. Detroit, 1982.
———, ed. *New American World: A Documentary History of North America to 1612*. Vol. 3, *English Plans for North America: The Roanoke Voyages; New England Ventures*. New York, 1979.
Quitt, Martin H. "Immigrant Origins of the Virginia Gentry: A Study in Cultural Transmission and Innovation." *WMQ* 45 (1988): 629–55.

——. “Trade and Acculturation at Jamestown, 1607–1609: The Limits of Understanding.” *WMQ* 52 (1995): 227–58.
Rabb, Theodore K. “Effects of the Thirty Years’ War on the German Economy.” *Journal of Modern History* 34 (1962): 40–51.
——. *Enterprise and Empire: Merchant and Gentry Investment in the Expansion of England, 1575–1630*. Cambridge, Mass., 1967.
——. *The Struggle for Stability in Early Modern Europe*. New York, 1975.
Raftis, J. Ambrose. *Tenure and Mobility: Studies in the Social History of the Medieval English Village*. Toronto, 1964.
——. *Warboys: Two Hundred Years in the Life of an English Mediaeval Village*. Toronto, 1974.
——, ed. *Pathways to Medieval Peasants*. Toronto, 1981.
Railton, Arthur R. “The Indians and the English on Martha’s Vineyard. Pt. 5, Gay Head Neck and Farm.” *Dukes County Intelligencer* 39 (1993): 109–64.
Rainbolt, John C. “The Absence of Towns in Seventeenth-Century Virginia.” *Journal of Southern History* 35 (1969): 343–60.
——. *From Prescription to Persuasion: Manipulation of the Seventeenth-Century Virginia Economy*. Port Washington, N.Y., 1974.
——. “A New Look at Stuart ‘Tyranny’: The Crown’s Attack on the Virginia Assembly, 1676–1689.” *VMHB* 75 (1967): 387–406.
——, ed. “The Case of the Poor Planters in Virginia under the Law for Inspecting and Burning Tobacco.” *VMHB* 79 (1971): 314–21.
Ramenofsky, Ann F. *Vectors of Death: The Archaeology of European Contact*. Albuquerque, N.M., 1987.
Ramsay, David. *The History of the American Revolution*. 1789. 2 vols. Edited by Lester H. Cohen. Indianapolis, 1990.
Ramsey, Robert W. *Carolina Cradle: Settlement of the Northwest Carolina Frontier, 1747–1762*. Chapel Hill, N.C., 1964.
Ranck, George W. “The Travelling Church: An Account of the Baptist Exodus to Kentucky in 1781.” *Register of the Kentucky Historical Society* 79 (1981): 240–65.
Rankin, Hugh, ed. “An Officer out of His Time: Correspondence of Major Patrick Ferguson, 1779–1780.” In *Sources of American Independence: Selected Manuscripts from the Collections of the William L. Clements Library*, edited by Howard Peckham, 287–360. Chicago, 1978.
Ransome, David R. “ ‘Shipt for Virginia’: The Beginnings in 1619–1622 of the Great Migration to the Chesapeake.” *VMHB* 103 (1995): 443–58.
——. “Wives for Virginia, 1621.” *WMQ* 48 (1991): 3–18.
Rasmussen, Barbara. *Absentee Landowning and Exploitation in West Virginia, 1760–1920*. Lexington, Ky., 1994.
Rasmussen, Wayne D. “Wood on the Farm.” In Brooke Hindle, *Material Culture of the Wooden Age*, 15–34.
Rawlyk, George A. “Loyalist Military Settlement in Upper Canada.” In Robert S. Allen, *Loyal Americans*, 99–103.
Rawson, David A. “The Anglo-American Frontier Settlement of Virginia’s Rappahannock Frontier.” *Locus* 6 (1994): 93–117.
Razi, Zvi. “The Erosion of the Family-Land Bond in the Late Fourteenth and Fifteenth Centuries: A Methodological Note.” In Richard M. Smith, *Land, Kinship, and Life Cycle*, 295–304.
——. “Intrafamilial Ties and Relationships in the Medieval Village: A Quantitative

Approach Employing Manor Court Rolls." In Razi and Smith, *Medieval Society and the Manor Court*, 369–91.
——. *Life, Marriage, and Death in a Medieval Parish: Economy, Society, and Demography in Halesowen, 1270–1400*. Cambridge, 1980.
——. "The Myth of the Immutable English Family." *P&P*, no. 140 (1993): 3–44.
——. "The Struggles between the Abbots of Halesowen and Their Tenants in the Thirteenth and Fourteenth Centuries." In Aston et al., *Social Relations and Ideas*, 151–67.
Razi, Zvi, and Richard Smith, eds. *Medieval Society and the Manor Court*. Oxford, 1996.
Razzell, Peter. "The Growth of Population in Eighteenth-Century England: A Critical Appraisal." *JEH* 53 (1993): 743–71.
Ready, Milton L. "An Economic History of Colonial Georgia, 1732–1754." Ph.D. diss., University of Georgia, 1970.
——. "The Georgia Concept: An Eighteenth-Century Experiment in Colonization." *Georgia Historical Quarterly* 55 (1971): 157–69.
——. "The Georgia Trustees and the Malcontents: The Politics of Philanthropy." *Georgia Historical Quarterly* 60 (1976): 264–81.
——. "Land Tenure in Trusteeship Georgia." *AgH* 48 (1974): 353–68.
Reavis, William A. "The Maryland Gentry and Social Mobility." *WMQ* 14 (1957): 418–28.
Rediker, Marcus. " 'Good Hands, Stout Heart, and Fast Feet': The History and Culture of Working People in Early America." In Eley and Hunt, *Reviving the English Revolution*, 221–49.
Redlich, Fritz. "An Eighteenth-Century German Guide for Investors." *Bulletin of the Business History Society* 26 (1952): 95–104.
——. *The German Military Enterpriser and His Work Force: A Study in European Economic and Social History*. Vol. 1. Wiesbaden, Germany, 1964.
Reich, Jerome R. *Leisler's Rebellion: A Study in Democracy in New York, 1664–1720*. Chicago, 1953.
Reid, Russell M. "Church Membership, Consanguineous Marriage, and Migration in a Scotch-Irish Frontier Population." *Journal of Family History* 13 (1988): 397–414.
A Relation of Maryland, Together With a Map of the Countrey, The Conditions of Plantation . . . . In Clayton Colman Hall, *Narratives of Early Maryland*, 70–112.
Reubens, Beatrice G. "Pre-Emptive Rights in the Disposition of a Confiscated Estate: Philipsburgh Manor, New York." *WMQ* 22 (1965): 435–56.
Reynolds, R. V. *Fuel Wood Used in the United States, 1630–1930*. U.S. Department of Agriculture, circular no. 641. Washington, D.C., 1942.
Riccards, Michael P. "Patriots and Plunderers: Confiscation of Loyalist Lands in New Jersey, 1776–1786." *New Jersey History* 86 (1968): 14–28.
Rice, Otis K. *The Allegheny Frontier: West Virginia Beginnings, 1730–1830*. Lexington, Ky., 1970.
Rich, Myra. "Prudence or Tyranny? The Role of the State in Virginia's Economy during the Revolution." *Social Science Journal* 17 (1980): 55–67.
Richards, Eric. *A History of Highland Clearances*. Vol. 1, *Agrarian Transformation and the Evictions, 1746–1886*. London, 1982.
——. *A History of Highland Clearances*. Vol. 2, *Emigration, Protest, Reasons*. London, 1985.
Richards, Eric, and Monica Clough. *Cromartie: Highland Life, 1650–1914*. Aberdeen, Scotland, 1989.
Richter, Daniel K. "Onas and the Long Knives: Pennsylvania-Indian Relations, 1783–1791." Paper, McNeil Center for Early American Studies, March 1992.

———. *The Ordeal of the Longhouse: The Peoples of the Iroquois League in the Era of European Colonization*. Chapel Hill, N.C., 1992.
Ricketson, William F. "To Be Young, Poor, and Alone: The Experience of Widowhood in the Massachusetts Bay Colony, 1675–1676." *NEQ* 64 (1991): 113–27.
Rife, Clarence White. "Land Tenure in New Netherland." In *Essays in Colonial History*, 41–73.
Rindler, Edward Paul. "The Migration from the New Haven Colony to Newark, East New Jersey: A Study in Puritan Values and Behavior, 1630–1720." Ph.D. diss., University of Pennsylvania, 1977.
Rink, Oliver A. *Holland on the Hudson: An Economic and Social History of Dutch New York*. Ithaca, N.Y., 1986.
Riordan, Liam. "Identity and Revolution: Everyday Life and Crisis in Three Delaware River Towns." *Pennsylvania History* 64 (1997): 56–101.
Ripley, Peter. "Village and Town: Occupations and Wealth in the Hinterland of Gloucester, 1660–1700." *Agricultural History Review* 32 (1984): 170–78.
Risch, Erna. *Supplying Washington's Army*. Washington, D.C., 1981.
Ritchie, Robert C. *The Duke's Province: A Study of New York Politics and Society, 1664–1691*. Chapel Hill, N.C., 1977.
Rivers, William James, ed. "Colonel Robert Gray's Observations on the War in Carolina." *South Carolina Historical Magazine* 11 (1910): 139–59.
Robb, Kenneth A. "Names of Grants in Colonial Maryland." *Names* 17 (1969): 263–77.
Robbins, William G. "The Massachusetts Bay Company: An Analysis of Motives." *Historian* 32 (1969): 83–98.
Roberts, B. W. C. "Cockfighting: An Early Entertainment in North Carolina." *North Carolina Historical Review* 42 (1965): 306–14.
Roberts, Edward Graham. "The Roads of Virginia, 1607–1840." Ph.D. diss., University of Virginia, 1950.
Roberts, Michael. "Sickles and Scythes: Women's Work and Men's Work at Harvest Time." *History Workshop* 7 (1979): 1–28.
———. " 'Words They Are Women, and Deeds They Are Men': Images of Work and Gender in Early Modern England." In Charles and Duffin, *Women and Work in Pre-Industrial England*, 122–80.
Robinson, David M. "Community and Utopia in Crèvecoeur's Sketches." *American Literature* 62 (1990): 20–31.
Robinson, Paul A. "Lost Opportunities: Miantonomi and the English in Seventeenth-Century Narragansett Country." In Grumet, *Northeastern Indian Lives*, 13–28.
Robinson, W. Stitt, Jr. "The Legal Status of the Indian in Colonial Virginia." *VMHB* 61 (1953): 247–59.
———. *Mother Earth: Land Grants in Virginia, 1607–1699*. Williamsburg, Va., 1957.
———. *The Southern Colonial Frontier, 1607–1763*. Albuquerque, N.M., 1979.
Robisheaux, Thomas. *Rural Society and the Search for Order in Early Modern Germany*. Cambridge, 1989.
Robson, Michael. "The Border Farm Worker." In Devine, *Farm Servants and Labour*, 71–96.
Roeber, A. G. *Faithful Magistrates and Republican Lawyers: Creators of Virginia Legal Culture, 1680–1810*. Chapel Hill, N.C., 1981.
———. "In German Ways? Problems and Potentials of Eighteenth-Century German Social and Emigration History." *WMQ* 44 (1987): 750–74.
———. " 'The Origin of Whatever Is Not English among Us': The Dutch-Speaking and

German-Speaking Peoples of Colonial British America." In Bailyn and Morgan, *Strangers within the Realm*, 220–83.
——. "The Origins and Transfer of German-American Concepts of Property and Inheritance." *Perspectives in American History*, n.s., 3 (1987): 115–71.
——. *Palatines, Liberty, and Property: German Lutherans in Colonial British America*. Baltimore, 1993.
Roebuck, Peter. "Rent Movement, Proprietorial Incomes, and Agricultural Development, 1730–1830." In Roebuck, *Plantation to Partition*, 82–101, 262–64.
——, ed. *Plantation to Partition: Essays in Ulster History in Honour of J. L. McCracken*. Belfast, 1981.
Roetger, R. W. "Enforcing New Haven's Bylaws, 1639–1698: An Exercise in Local Social Control." *Connecticut History* 27 (1986): 15–27.
Rogers, Alan. "Industrialisation and the Local Community." In *Region and Industrialization: Studies on the Role of the Region in the Economic History of the Last Two Centuries*, edited by Sidney Pollard, 196–211. Göttingen, 1980.
Rollins, Hyder Edward, ed. *The Pepys Ballads*. Vol. 4. Cambridge, Mass., 1930.
Romani, Daniel A., Jr. " 'Our English *Clover-Grass* Sowen Thrives Very Well': The Importation of English Grasses and Forages into Seventeenth-Century New England." In Benes, *Plants and People*, 25–37.
——. "The Pettaquamscut Purchase of 1657/58 and the Establishment of a Commercial Livestock Industry in Rhode Island." In *New England's Creatures: 1400–1900*. Vol. 18 of *The Dublin Seminar for New England Folklife*, edited by Peter Benes, 45–60. Boston, 1995.
Ronda, James P. "Generations of Faith: The Christian Indians of Martha's Vineyard." *WMQ* 38 (1981): 369–94.
——. " 'We Are Well As We Are': An Indian Critique of Seventeenth-Century Christian Missions." *WMQ* 34 (1977): 66–82.
Roper, Louis H. "The Unraveling of an Anglo-American Utopia in South Carolina." *Historian* 58 (1996): 277–88.
Rorabaugh, W. J. *The Alcoholic Republic: An American Tradition*. New York, 1979.
——. *The Craft Apprentice from Franklin to the Machine Age in America*. New York, 1986.
Rosen, Deborah A. *Courts and Commerce: Gender, Law, and the Market Economy in Colonial New York*. Columbus, Ohio, 1997.
Rosenberger, Homer T. "Migrations of the Pennsylvania Germans to Western Pennsylvania." *Western Pennsylvania Historical Magazine* 53 (1970): 320–35; 54 (1971): 58–76.
Rosenwaike, Ira. *Population History of New York City*. Syracuse, N.Y., 1972.
Rosenzweig, Roy. "What *Is* the Matter with History?" *Journal of American History* 74 (1987): 117–22.
Rose-Troup, Frances. *The Massachusetts Bay Company and Its Predecessors*. New York, 1930.
Rosivach, Vincent J. "Agricultural Slavery in the Northern Colonies and in Classical Athens: Some Comparisons." *Comparative Studies in Society and History* 35 (1993): 551–67.
Ross, Dorothy. "Grand Narrative in American Historical Writing: From Romance to Uncertainty." *American Historical Review* 100 (1995): 651–77.
Rossano, Geoffrey L. "Launching Prosperity: Samuel Townsend and the Maritime Trade of Colonial Long Island, 1747–1773." *American Neptune* 48 (1988): 31–43.

Rothenberg, Winifred. *From Market-Places to a Market Economy: The Transformation of Rural Massachusetts, 1750–1850*. Chicago, 1992.

——. "The Market and Massachusetts Farmers, 1750–1855." *JEH* 41 (1981): 283–314.

——. "A Price Index for Rural Massachusetts, 1750–1855." *JEH* 39 (1979): 975–1001.

——. "The Role of Mortgage Credit in the Endogenous Growth of Early Massachusetts: 1650–1750." Paper presented at the conference "The Economy of Early British America: The Domestic Sector," Huntington Library, San Marino, Calif., October 1995.

Rothschild, Nan A. "Social Distance between Dutch Settlers and Native Americans." In Van Dongen et al., *One Man's Trash*, 189–201.

Roundtree, Helen C. "Powhatan Indian Women: The People Captain John Smith Barely Saw." *Ethnohistory* 45 (1998): 1–29.

——. "The Powhatans and Other Woodland Indians as Travelers." In Roundtree, *Powhatan Foreign Relations*, 21–52.

——. "The Powhatans and the English: A Case of Multiple Conflicting Agendas." In Roundtree, *Powhatan Foreign Relations*, 173–205.

——, ed. *Powhatan Foreign Relations, 1500–1722*. Charlottesville, Va., 1993.

Roundtree, Helen C., and Thomas E. Davidson. *Eastern Shore Indians of Virginia and Maryland*. Charlottesville, Va., 1997.

Rowlandson, Mary. *The Sovereignty and Goodness of God Together with the Faithfulness of His Promises Displayed, Being a Narrative of the Captivity and Restoration of Mrs. Mary Rowlandson and Related Documents*. 1682. Edited by Neal Salisbury. Boston, 1997.

Royster, Charles. *A Revolutionary People at War: The Continental Army and American Character, 1775–1783*. Chapel Hill, N.C., 1980.

Ruggles, Steven. "Migration, Marriage, and Mortality: Correcting Sources of Bias in English Family Reconstitutions." *Population Studies* 46 (1992): 507–22.

——. *Prolonged Connections: The Rise of the Extended Family in Nineteenth-Century England and America*. Madison, Wis., 1987.

——. "The Transformation of American Family Structure." *American Historical Review* 99 (1994): 103–28.

Rush, Benjamin. *Essays: Literary, Moral, and Philosophical*. Philadelphia, 1806.

Rushton, Peter. "Property, Power, and Family Networks: The Problem of Disputed Marriage in Early Modern England." *Journal of Family History* 11 (1986): 205–19.

Russell, Emily W. B. "Indian-Set Fires in the Forests of the Northeastern United States." *Ecology* 64 (1983): 78–88.

Russell, Howard S. *Indian New England before the Mayflower*. Hanover, N.H., 1980.

——. *A Long, Deep Furrow: Three Centuries of Farming in New England*. Hanover, N.H., 1976.

Russo, Jean B. "Self-Sufficiency and Local Exchange: Free Craftsmen in the Rural Chesapeake Economy." In Carr, Morgan, and Russo, *Colonial Chesapeake Society*, 389–432.

Rutman, Darrett B. *Husbandmen of Plymouth: Farms and Villages in the Old Colony, 1620–1692*. Boston, 1967.

——. *John Winthrop's Decision for America, 1629*. Philadelphia, 1975.

——. "The Social Web: A Prospectus for the Study of the Early American Community." In *Insights and Parallels: Problems and Issues of American Social History*, edited by William L. O'Neill, 57–89. Minneapolis, 1973.

——. *Winthrop's Boston: A Portrait of a Puritan Town, 1630–1649*. Chapel Hill, N.C., 1965.

Rutman, Darrett B., and Anita H. Rutman. "'More True and Perfect Lists': The Reconstruction of Censuses for Middlesex County, Virginia, 1668–1704." *VMHB* 88 (1980): 37–74.
——. *A Place in Time: Explicatus*. New York, 1984.
——. *A Place in Time: Middlesex County Virginia, 1650–1750*. New York, 1984.
——. *Small Worlds, Large Questions: Explorations in Early American Social History, 1600–1850*. Charlottesville, Va., 1994.
Rutman, Darrett B., Charles Wetherell, and Anita H. Rutman. "Rhythms of Life: Black and White Seasonality in the Early Chesapeake." *JIH* 11 (1980): 29–54.
Ryan, Dennis P. "Landholding, Opportunity, and Mobility in Revolutionary New Jersey." *WMQ* 36 (1979): 571–92.
Ryden, George Herbert, ed. *Letters to and from Caesar Rodney, 1756–1784*. Philadelphia, 1933.
Saalfeld, Diedrich "The Struggle to Survive." In *Our Forgotten Past: Seven Centuries on the Land*, edited by Jerome Blum, 109–32. London, 1982.
Sabean, David Warren. *Power in the Blood: Popular Culture and Village Discourse in Early Modern Germany*. Cambridge, 1984.
——. *Property, Production, and Family in Neckarhausen, 1700–1870*. Cambridge, 1990.
Sachs, William S. "Agricultural Conditions in the Northern Colonies before the Revolution." *JEH* 13 (1953): 274–90.
Sachse, William L. "The Migration of New Englanders to England, 1640–1660." *American Historical Review* 53 (1948): 251–78.
Sainsbury, John A. "Indian Labor in Early Rhode Island." *NEQ* 48 (1975): 378–93.
St. George, Robert Blair. "'Set Thine House in Order': The Domestication of the Yeomanry in Seventeenth Century New England." In Fairbanks and Trent, *New England Begins*, 2:159–87.
Salerno, Anthony. "The Social Background of Seventeenth-Century Emigration to America." *Journal of British Studies* 19 (1979): 31–52.
Salinger, Sharon V. *"To Serve Well and Faithfully": Labor and Indentured Servants in Pennsylvania, 1682–1800*. Cambridge, 1987.
Salinger, Sharon V., and Cornelia Dayton. "Mapping Migration into Pre-Revolutionary Boston: An Analysis of Robert Love's Warning Book." Paper, McNeil Center for Early American Studies, September 1999.
Salisbury, Neal. "Indians and Colonists in Southern New England after the Pequot War: An Uneasy Balance." In Hauptman and Wherry, *Pequots in Southern New England*, 81–95.
——. *Manitou and Providence: Indians, Europeans, and the Making of New England, 1500–1643*. New York, 1982.
Salley, Alexander S., Jr., ed. *Narratives of Early Carolina, 1650–1708*. New York, 1911.
Salmon, Emily J., ed. *A Hornbook of Virginia History*. 3d ed. Richmond, Va., 1983.
Salmon, Marylynn. "Equality or Submersion? Femme Covert Status in Early Pennsylvania." In *Women of America: A History*, edited by Mary Beth Norton and Carol Ruth Berkin, 92–113. Boston, 1979.
——. "Family Management and Definitions of Motherhood in Early America." In *Lois Green Carr*, 161–70.
——. "Women and Property in South Carolina: The Evidence from Marriage Settlements, 1730 to 1830." *WMQ* 39 (1982): 655–85.
——. *Women and the Law of Property in Early America*. Chapel Hill, N.C., 1986.
Saltonstall, Nathaniel. *A Continuation of the State of New-England; Being a Farther Account of the Indian Warr . . . .* London, 1676.

——. *A New and Further Narrative of the State of New-England; Being a Continued Account of the Bloudy Indian-War*. London, 1676.
——. *The Present State of New-England with Respect to the Indian PWar*. . . . London, 1676.
Salwen, Bert. "Indians of Southern New England and Long Island: Early Period." In Trigger, *Northeast*, 160–76.
Sampson, Richard. *Escape in America: The British Convention Prisoners, 1777–1783*. Chippenham, England, 1995.
Sands, Donald B. "Regionalisms and Archaisms in Current Maine Place Names." In *American Speech: 1600 to the Present*, edited by Peter Benes. Vol. 8, *The Dublin Seminar for New England Folklife*, 35–43. Boston, 1985.
Sauer, Carl O. *Selected Essays, 1963–1975*. Berkeley, Calif., 1981.
Saunders, William L., Walter Clark, and Stephen B. Weeks, eds. *The Colonial and State Records of North Carolina*. 30 vols. Raleigh, N.C., 1886–1914.
Savory, D. L. "The Huguenot-Palatine Settlements in the Counties of Limerick, Kerry, and Tipperary." *Huguenot Society of Great Britain and Ireland, Proceedings* 18 (1947–49): 111–33, 215–31.
Saye, Albert B. "The Genesis of Georgia Reviewed." *Georgia Historical Quarterly* 50 (1966): 153–61.
——. *New Viewpoints in Georgia History*. Athens, Ga., 1943.
Schein, Richard H. "Farming the Frontier: The New Military Tract Survey in Central New York." *New York History* 75 (1993): 5–28.
——. "Unofficial Proprietors in Post-Revolutionary Central New York." *Journal of Historical Geography* 17 (1991): 146–64.
Schelbert, Leo. "On Becoming an Emigrant: A Structural View of Eighteenth- and Nineteenth-Century Swiss Data." *Perspectives in American History* 7 (1973): 441–95.
Schellekens, Jona. "Courtship, the Clandestine Marriage Act, and Illegitimate Fertility in England." *JIH* 25 (1995): 433–44.
Schlebecker, John T. "Agricultural Markets and Marketing in the North, 1774–1777." *AgH* 50 (1976): 21–36.
Schlenther, Boyd Stanley. "'English Is Swallowing Up Their Language': Welsh Ethnic Ambivalence in Colonial Pennsylvania and the Experience of David Evans." *Pennsylvania Magazine of History and Biography* 114 (1990): 201–28.
Schleunes, Karl A. *Schooling and Society: The Politics of Education in Prussia and Bavaria, 1750–1900*. Oxford, 1989.
Schlotterbeck, John T. "The Internal Economy of Slavery in Rural Piedmont Virginia." In *The Slaves' Economy: Independent Production by Slaves in the Americas*, edited by Ira Berlin and Philip Morgan, 170–81. London, 1991.
Schlumbohm, Jürgen. "From Peasant Society to Class Society: Some Aspects of Family and Class in a Northwest German Protoindustrial Parish, Seventeenth–Nineteenth Centuries." *Journal of Family History* 17 (1992): 183–99.
——. "The Land-Family Bond in Peasant Practice and in Middle-Class Ideology: Evidence from the North-West German Parish of Belm, 1650–1860." *Central European History* 27 (1994): 461–77.
Schmidt, Benjamin. "Mapping and Empire: Cartographic and Colonial Rivalry in Seventeenth-Century Dutch and English North America." *WMQ* 54 (1997): 549–78.
Schochet, Gordon J. *Patriarchalism in Political Thought: The Authoritarian Family and Political Speculation and Attitudes, Especially in Seventeenth-Century England*. New York, 1975.
Schofield, Roger S. "Age-Specific Mobility in an Eighteenth Century Rural English Parish." *Annales de Demographie Historique* 7 (1970): 261–74.

Scholten, Catherine M. *Childbearing in American Society, 1650–1850*. New York, 1985.
Schultz, Constance B. "Daughters of Liberty: The History of Women in the Revolutionary War Pension Records." *Prologue* 16 (1984): 139–53.
Schumaker, Max George. "The Northern Farmer and His Markets during the Late Colonial Period." Ph.D. diss., University of California, 1948.
Schwartz, Sally. *"A Mixed Multitude": The Struggle for Toleration in Colonial Pennsylvania*. New York, 1987.
——. "Society and Culture in the Seventeenth-Century Delaware Valley." *Delaware History* 20 (1982): 98–122.
Schwarz, L. D. "The Standard of Living in the Long Run: London, 1700–1860." *EcHR* 38 (1985): 24–41.
Schwarz, Philip J. *The Jarring Interests: New York's Boundary Makers, 1664–1776*. Albany, N.Y., 1979.
——. *Twice Condemned: Slaves and the Criminal Laws of Virginia, 1705–1865*. Baton Rouge, La., 1988.
Schweitzer, Mary McKinney. *Custom and Contract: Household, Government, and the Economy in Colonial Pennsylvania*. New York, 1987.
——. "Economic Regulation and the Colonial Economy: The Maryland Tobacco Inspection Act of 1747." *JEH* 40 (1980): 551–70.
Scott, Donald, and Bernard Wishy, eds. *America's Families: A Documentary History*. New York, 1982.
Scott, Kenneth. "Price Control in New England during the Revolution." *NEQ* 19 (1946): 453–73.
Scott, Tom. "Economic Landscapes." In Scribner, *Germany*, 1–31.
Scott, William B. *In Pursuit of Happiness: American Conceptions of Property from the Seventeenth to the Twentieth Century*. Bloomington, Ind., 1977.
Scott, William Robert. *The Constitution and Finance of English, Scottish, and Irish Joint-Stock Companies to 1720*. 3 vols. Cambridge, 1912.
Scribner, Bob. "Communities and the Nature of Power." In Scribner, *Germany*, 291–325.
——, ed. *Germany: A New Social and Economic History*. Vol. 1, *1450–1630*. London, 1996.
Searcy, Martha Condray. "1779: The First Year of the British Occupation of Georgia." *Georgia Historical Quarterly* 67 (1983): 168–88.
Seaver, Paul. *Wallington's World: A Puritan Artisan in Seventeenth-Century London*. Stanford, Calif., 1985.
Seed, Patricia. *Ceremonies of Possession in Europe's Conquest of the New World, 1492–1640*. Cambridge, 1995.
——. "Taking Possession and Reading Texts: Establishing the Authority of Overseas Empires." *WMQ* 49 (1992): 183–209.
Segal, Charles M., and David C. Stineback, eds. *Puritans, Indians, and Manifest Destiny*. New York, 1977.
Seiler, William H. "Land Processioning in Colonial Virginia." *WMQ* 6 (1949): 416–36.
Selby, John E. *The Revolution in Virginia, 1775–1783*. Williamsburg, Va., 1988.
Selement, George. "The Meeting of Elite and Popular Minds at Cambridge, New England, 1638–1645." *WMQ* 41 (1984): 32–48.
Selement, George, and Bruce C. Woolley, eds. *Thomas Shephard's Confessions*. *Publications of the Colonial Society of Massachusetts, Collections* 58 (1981).
Selesky, Harold E. *A Demographic Survey of the Continental Army That Wintered at Valley Forge, Pennsylvania, 1777–1778*. New Haven, Conn., 1987.
——. *War and Society in Colonial Connecticut*. New Haven, Conn., 1990.

Selig, Robert A. "Emigration, Fraud, Humanitarianism, and the Founding of Londonderry, South Carolina, 1763–1765." *Eighteenth-Century Studies* 23 (1989): 1–23.

Sellers, John R. "The Common Soldier in the American Revolution." In *Military History of the American Revolution*, edited by Stanley J. Underdal, 151–63. Proceedings of the Sixth Military History Symposium, U.S. Air Force Academy, 1974. Washington, D.C., 1974.

Semple, Ellen Churchill. *American History and Its Geographic Conditions*. Boston, 1903.

Sen, Amartya K. "Gender and Cooperative Conflicts." In *Persistent Inequalities: Women and World Development*, edited by Irene Tinker, 123–49. New York, 1990.

Seybert, Adam. *Statistical Annals Embracing Views of the Population, Commerce, Navigation, Fisheries, Public Lands, Post-Office Establishment, Revenues, Mint, Military and Naval Establishments, Expenditures, Public Debt and Sink Fund of the United States of America* . . . . Philadelphia, 1818.

Shackel, Paul A., and Barbara J. Little, eds. *Historic Archaeology of the Chesapeake*. Washington, D.C., 1994.

Shammas, Carole. "Anglo-American Household Government in Comparative Perspective." *WMQ* 52 (1995): 104–44.

——. "Changes in English and Anglo-American Consumption from 1550 to 1800." In Brewer and Porter, *Consumption and the World of Goods*, 177–205.

——. "The Decline of Textile Prices in England and British America prior to Industrialization." *EcHR* 47 (1994): 483–507.

——. "The Determinants of Personal Wealth in Seventeenth-Century England and America." *JEH* 37 (1977): 675–89.

——. "The Domestic Environment in Early Modern England and America." *Journal of Social History* 19 (1980): 3–24.

——. "Early American Women and Control over Capital." In Hoffman and Albert, *Women in the Age of the American Revolution*, 134–54.

——. "The Elizabethan Gentleman and Western Planting." Ph.D. diss., Johns Hopkins University, 1971.

——. "English Commercial Development and American Colonization." In Andrews, Canny, and Hair, *Westward Enterprise*, 150–74.

——. "The Food Budget to English Workers: A Reply to Komlos." *JEH* 48 (1988): 673–76.

——. "Food Expenditures and Economic Well-Being in Early Modern England." *JEH* 43 (1983): 89–100.

——. "How Self-Sufficient Was Early America?" *JIH* 13 (1982): 247–72.

——. "A New Look at Long-Term Trends in Wealth Inequality in the United States." *American Historical Review* 98 (1993): 412–31.

——. *The Preindustrial Consumer in England and America*. Oxford, 1990.

——. "The World Women Knew: Women Workers in the North of England during the Late Seventeenth Century." In Dunn and Dunn, *World of William Penn*, 99–115.

Shammas, Carole, Marylynn Salmon, and Michel Dahlin. *Inheritance in America from Colonial Times to the Present*. New Brunswick, N.J., 1987.

Shanley, Mary Lyndon. "Marriage Contract and Social Contract in Seventeenth Century Political Thought." *Western Political Quarterly* 32 (1979): 79–91.

Shannon, Timothy J. "Dressing for Success on the Mohawk Frontier: Hendrick, William Johnson, and the Indian Fashion." *WMQ* 53 (1996): 13–42.

——. "The Ohio Company and the Meaning of Opportunity in the American West, 1786–1795." *NEQ* 64 (1991): 393–413.

Sharlin, Allan. "Natural Decrease in Early Modern Cities: A Reconsideration." *P&P*, no. 79 (1978): 126–38.

Sharp, Buchanan. *In Contempt of All Authority: Rural Artisans and Riot in the West of England, 1586–1660*. Berkeley, Calif., 1980.

——. "Popular Protest in Seventeenth-Century England." In *Popular Culture in Seventeenth-Century England*, edited by Barry Reay, 271–308. New York, 1985.

Sharpe, J. A. "Plebeian Marriage in Stuart England: Some Evidence from Popular Literature." *Transactions of the Royal Historical Society*, 5th ser., 36 (1986): 69–90.

Sharpe, Pamela. "Literally Spinsters: A New Interpretation of Local Economy and Demography in Colyton in the Seventeenth and Eighteenth Centuries." *EcHR* 44 (1991): 46–65.

——. "Poor Children as Apprentices in Colyton, 1598–1830." *Continuity and Change* 6 (1991): 253–70.

Shaw, Francis J. *The Northern and Western Islands of Scotland: Their Economy and Society in the Seventeenth Century*. Edinburgh, 1980.

Shaw, J. P. "The New Industries: Water Power and Textiles." In *The Making of the Scottish Countryside*, edited by M. L. Parry and T. R. Slater, 291–317. London, 1980.

Shea, William L. *The Virginia Militia in the Seventeenth Century*. Baton Rouge, La., 1983.

Sheehan, Bernard W. " 'The Famous Hair Buyer General': Henry Hamilton, George Rogers Clark, and the American Indian." *Indiana Magazine of History* 79 (1983): 1–28.

Sheehan, James J. *German History, 1770–1866*. Oxford, 1989.

Sheller, Tina H. "Artisans, Manufacturing, and the Rise of a Manufacturing Interest in Revolutionary Baltimore Town." *MdHM* 83 (1988): 3–17.

Shepherd, James F., and Gary M. Walton. "Economic Change after the American Revolution: Pre- and Post-War Comparisons of Maritime Shipping and Trade." *Explorations in Economic History* 13 (1976): 397–422.

——. *Shipping, Maritime Trade, and the Economic Development of Colonial North America*. Cambridge, 1972.

Sheridan, Richard B. "The British Credit Crisis of 1772 and the American Colonies." *JEH* 20 (1960): 161–86.

Shipps, Kenneth W. "The Puritan Emigration to New England: A New Source on Motivation." *New England Historical and Genealogical Register* 135 (1981): 83–97.

Short, Brian, ed. *The English Rural Community: Image and Analysis*. Cambridge, 1992.

Shorter, Edward. "Bastardy in South Germany: Comment." *JIH* 8 (1978): 459–69.

Showman, Richard K., Dennis M. Conrad, et al., eds. *The Papers of General Nathanael Greene*. 9 vols. to date. Chapel Hill, N.C., 1976–97.

Shy, John W. "Logistical Crisis and the American Revolution: A Hypothesis." In *Feeding Mars: Logistics in Western Warfare from the Middle Ages to the Present*, edited by John A. Lynn, 161–79. Boulder, Colo., 1993.

——. *A People Numerous and Armed: Reflections on the Military Struggle for American Independence*. Rev. ed. Ann Arbor, Mich., 1990.

Siebert, Wilbur H. "The Dispersion of the American Tories." *Mississippi Valley Historical Review* 1 (1914): 185–97.

——. *The Legacy of the American Revolution to the British West Indies and Bahamas: A Chapter out of the History of the American Loyalists*. Vol. 17 of *Ohio State University Bulletin*. Columbus, Ohio, 1913.

——. "The Loyalist and Six Nation Indians in the Niagara Peninsula." *Royal Society of Canada, Transactions* 9 (1915): 79–128.

Siener, William H. "Charles Yates, the Grain Trade, and Economic Development in Fredericksburg, Virginia, 1750–1810." *VMHB* 93 (1985): 409–26.

Siles, William H. "Pioneering in the Genessee Country: Entrepreneurial Strategy and the Concept of a Central Place." In *New Opportunities in a New Nation: The Development of New York after the Revolution*, edited by Manfred Jonas and Robert V. Wells, 35–68. Schenectady, N.Y., 1982.

Silver, Timothy. *A New Face on the Countryside: Indians, Colonists, and Slaves in South Atlantic Forests, 1500–1800*. Cambridge, 1990.

Simler, Lucy. "Hired Labor." In Jacob Ernest Cooke, *Encyclopedia of the North American Colonies*, 2:3–15.

——. "The Landless Worker: An Index of Economic and Social Change in Chester County, Pennsylvania, 1750–1820." *Pennsylvania Magazine of History and Biography* 114 (1990): 163–99.

——. "She Came to Work/She Went to Work: The Development of a Female Rural Proletariat in Southeastern Pennsylvania, 1760–1820." Paper presented at the conference "Women and the Transition to Capitalism in Rural America, 1760–1940," DeKalb, Ill., March–April 1989.

——. "Tenancy in Colonial Pennsylvania: The Case of Chester County." *WMQ* 43 (1986): 542–69.

Simler, Lucy, and Paul G. E. Clemens. "The 'Best Poor Man's Country' in 1783: The Population Structure of Rural Society in Late-Eighteenth-Century Southeastern Pennsylvania." *Proceedings of the American Philosophical Society* 133 (1989): 234–61.

——. "Rural Labor and the Farm Household in Chester County, Pennsylvania, 1750–1820." In Innes, *Work and Labor in Early America*, 106–43.

Simmons, Richard C. "Americana in British Books, 1621–1760." In Kupperman, *America in European Consciousness*, 360–87.

Sirmans, M. Eugene. *Colonial South Carolina: A Political History, 1663–1763*. Chapel Hill, N.C., 1966.

Slack, Paul. *The English Poor Law, 1531–1782*. Houndsmill, England, 1990.

——. *Poverty and Policy in Tudor and Stuart England*. London, 1988.

——. "Vagrants and Vagrancy in England, 1598–1664." *EcHR* 27 (1974): 360–79.

Slaughter, Thomas P. *The Whiskey Rebellion: Frontier Epilogue to the American Revolution*. New York, 1986.

Slicher van Bath, B. H. *The Agrarian History of Western Europe, 500–1850*. London, 1963.

——. "Agriculture in the Vital Revolution." In *The Cambridge Economic History of Europe*, edited by E. E. Rich and Charles H. Wilson. Vol. 5, *The Economic Organization of Early Modern Europe*, 42–133. Cambridge, 1977.

Slotkin, Richard. *Regeneration through Violence: The Mythology of the American Frontier, 1600–1860*. Middletown, Conn., 1973.

Slotkin, Richard, and James K. Folsom, eds. *So Dreadfull a Judgment: Puritan Responses to King Philip's War, 1676–1677*. Middletown, Conn., 1978.

Smaby, Beverly. *The Transformation of Moravian Bethlehem: From Communal Mission to Family Economy*. Philadelphia, 1988.

Smith, Abbot Emerson. *Colonists in Bondage: White Servitude and Convict Labor in America, 1607–1776*. Chapel Hill, N.C., 1947.

——. "The Indentured Servant and Land Speculation in Seventeenth-Century Maryland." *American Historical Review* 40 (1935): 467–72.

Smith, A. Hassell. "Labourers in Late Sixteenth-Century England: A Case Study from North Norfolk." *Continuity and Change* 4 (1989): 11–52, 367–94.

Smith, Barbara Clark. *After the Revolution: The Smithsonian History of Everyday Life in the Eighteenth Century*. New York, 1985.

——. "Food Rioters and the American Revolution." *WMQ* 51 (1994): 3–38.

Smith, Betty Anderson. "Distribution of Eighteenth-Century Cherokee Settlements." In *The Cherokee Indian Nation: A Troubled History*, edited by Duane H. King, 46–60. Knoxville, Tenn., 1979.

Smith, Billy G. "Comment." *WMQ* 45 (1988): 163–66.

——. *The "Lower Sort": Philadelphia's Laboring People, 1750–1800*. Ithaca, N.Y., 1990.

Smith, C. T. *An Historical Geography of Western Europe before 1800*. Rev. ed. London, 1978.

Smith, Daniel Blake. *Inside the Great House: Planter Family Life in Eighteenth-Century Chesapeake Society*. Ithaca, N.Y., 1980.

Smith, Daniel Scott. " 'All in Some Degree Related to Each Other': A Graphic and Comparative Resolution of the Anomaly of New England Kinship." *American Historical Review* 94 (1989): 44–79.

——. "American Family and Demographic Patterns and the Northwest European Model." *Continuity and Change* 8 (1993): 389–415.

——. "Child-Naming Practices, Kinship Ties, and Change in Family Attitudes in Hingham, Massachusetts, 1641 to 1880." *Journal of Social History* 18 (1984): 541–66.

——. "The Curious History of Theorizing about the History of the Western Nuclear Family." *Social Science History* 17 (1993): 325–54.

——. "The Demographic History of Colonial New England." *JEH* 32 (1972): 165–83.

——. "Female Householding in Late Eighteenth-Century America and the Problem of Poverty." *Journal of Social History* 28 (1994): 83–107.

——. "Inheritance and the Social History of Early American Women." In Hoffman and Albert, *Women in the Age of the American Revolution*, 45–66.

——. "The Long Cycle in American Illegitimacy and Prenuptial Pregnancy." In Laslett, Oosterveen, and Smith, *Bastardy and Its Comparative History*, 362–78.

——. "A Malthusian-Frontier Interpretation of United States Graphic History before c. 1815." In *Urbanization in the Americas: The Background in Comparative Perspective*, edited by Woodrow Borah, Jorge Hardoy, and Gilbert A. Stelter, 15–24. Ottawa, Canada, 1980.

——. "Parental Power and Marriage Patterns: An Analysis of Historical Trends in Hingham, Massachusetts." *Journal of Marriage and the Family* 35 (1973): 419–28.

——. "A Perspective on Demographic Methods and Effects in Social History." *WMQ* 39 (1982): 442–68.

——. "Population, Family, and Society in Hingham, Massachusetts, 1635–1880." Ph.D. diss., University of California, Berkeley, 1973.

Smith, Daniel Scott, and Michael S. Hindus. "Premarital Pregnancy in America, 1640–1971: An Overview and Interpretation." *JIH* 5 (1975): 537–70.

Smith, James Morton, ed. *Seventeenth-Century America: Essays in Colonial History*. New York, 1959.

Smith, Jonathan. "How Massachusetts Raised Her Troops in the Revolution." *Massachusetts Historical Society, Proceedings* 55 (1923): 345–70.

——. "Toryism in Worcester County during the War for Independence." *Massachusetts Historical Society, Proceedings* 48 (1915): 15–35.

Smith, John. *The Complete Works of Captain John Smith, 1580–1631*. Edited by Philip L. Barbour. 3 vols. Chapel Hill, N.C., 1986.

Smith, Joseph H., ed. *Colonial Justice in Western Massachusetts, 1639–1702: The Pynchon Court Record, an Original Judges' Diary of the Administration of Justice in the Springfield Courts in the Massachusetts Bay Colony*. Cambridge, Mass., 1961.

Smith, Margaret Supplee, and Emily Herring Wilson. *North Carolina Women Making History*. Chapel Hill, N.C., 1999.

Smith, Merrill D. *Breaking the Bonds: Marital Discord in Pennsylvania, 1730–1830*. New York, 1991.
Smith, Paul H. "The American Loyalists: Notes on Their Organization and Numerical Strength." *WMQ* 25 (1968): 259–77.
——. *Loyalists and Redcoats: A Study in British Revolutionary Policy*. Chapel Hill, N.C., 1964.
Smith, Richard M. "Coping with Uncertainty: Women's Tenure of Customary Land in England, c. 1370–1430." In *Enterprise and Individuals in Fifteenth-Century England*, edited by Jennifer Kermode, 43–67. Stroud, England, 1991.
——. "Demographic Developments in Rural England, 1300–48: A Survey." In Bruce M. S. Campbell, *Before the Black Death*, 110–48.
——. "Geographical Diversity in the Resort to Marriage in Late Medieval Europe: Work, Reputation, and Unmarried Females in the Household Formation Systems of Northern and Southern Europe." In Peter Jeremy Piers Goldberg, *Woman Is a Worthy Wight*, 16–59.
——. " 'Modernization' and the Corporate Medieval Village Community in England: Some Skeptical Reflections." In *Explorations in Historical Geography: Interpretative Essays*, edited by Alan R. H. Baker and Derek Gregory, 140–79, 234–45. Cambridge, 1984.
——. "Some Issues Concerning Families and Their Property in Rural England, 1250–1800." In Richard M. Smith, *Land, Kinship, and Life Cycle*, 1–86.
——, ed. *Land, Kinship, and the Life Cycle*. Cambridge, 1984.
Smith, S. D. "The Market for Manufactures in the Thirteen Continental Colonies, 1698–1776." *EcHR* 51 (1998): 676–708.
Smith, Sir Thomas. *A Discourse of the Commonweal of This Realm of England, Attributed to Sir Thomas Smith*. Edited by Mary Dewar. Charlottesville, Va., 1969.
Smith, Warren B. *White Servitude in Colonial South Carolina*. Columbia, S.C., 1961.
Smith, William, Jr. *The History of the Province of New York*. Edited by Michael Kammen. 2 vols. Cambridge, Mass., 1972.
Smits, David D. " 'Abominable Mixture': Toward the Repudiation of Anglo-Indian Intermarriage in Seventeenth-Century Virginia." *VMHB* 95 (1987): 157–92.
——. " 'We Are Not to Grow Wild': Seventeenth-Century New England's Repudiation of Anglo-Indian Marriage." *American Indian Culture and Research Journal* 11 (1987): 1–32.
Smout, T. Christopher. *A History of the Scottish People, 1560–1830*. New York, 1969.
——. "Where Had the Scottish Economy Got To by the Third Quarter of the Eighteenth Century?" In *Wealth and Virtue: The Shaping of Political Economy in the Scottish Enlightenment*, edited by Istvan Hont and Michael Ignatieff, 45–72. Cambridge, 1983.
Smout, T. Christopher, Ned C. Landsman, and Thomas Martin Devine. "Scottish Emigration in the Seventeenth and Eighteenth Centuries." In Canny, *Europeans on the Move*, 76–112.
Snell, K. D. M. *Annals of the Labouring Poor: Social Change and Agrarian England, 1660–1900*. Cambridge, 1985.
Snooks, Graeme Donald. "Great Waves of Economic Change: The Industrial Revolution in Historical Perspective, 1000 to 2000." In *Was the Industrial Revolution Necessary?*, edited by Graeme Donald Snooks, 43–78. London, 1994.
Snow, Dean R. "Disease and Population Decline in the Northeast." In Verano and Ubelaker, *Disease and Demography*, 177–86.
——. "Wabanaki 'Family Hunting Territories.' " *American Anthropologist* 70 (1968): 1143–51.

Snow, Dean R., and Kim M. Lanphear. "European Contact and Indian Depopulation in the Northeast: The Timing of the First Epidemics." *Ethnohistory* 35 (1988): 15–33.
Snydaker, Daniel. "Kinship and Community in Rural Pennsylvania." *JIH* 13 (1982): 41–62.
Sobel, Mechal. *The World They Made Together: Black and White Values in Eighteenth-Century Virginia*. Princeton, N.J., 1987.
Soderlund, Jean R. "Black Importation and Migration into Southeastern Pennsylvania, 1682–1810." *Proceedings of the American Philosophical Society* 113 (1989): 144–53.
——. *Quakers and Slavery: A Divided Spirit*. Princeton, N.J., 1985.
Solar, Peter M. "Poor Relief and English Economic Development before the Industrial Revolution." *EcHR* 48 (1995): 1–20.
Soltow, J. H. "Scottish Traders in Virginia, 1750–1773." *EcHR* 12 (1959): 83–98.
Soltow, Lee. *Distribution of Wealth and Income in the United States in 1798*. Pittsburgh, Pa., 1989.
——. "Housing Characteristics on the Pennsylvania Frontier: Mifflin County Dwellings in 1798." *Pennsylvania History* 47 (1980): 57–70.
——. "Inequality amidst Abundance: Land Ownership in Early Nineteenth Century Ohio." *Ohio History* 88 (1979): 133–47.
——. "Kentucky Wealth at the End of the Eighteenth Century." *JEH* 43 (1983): 617–34.
——. "Land Inequality on the Frontier: The Distribution of Land in East Tennessee at the Beginning of the Nineteenth Century." *Social Science History* 5 (1981): 275–92.
Soltow, Lee, and Kenneth W. Keller. "Rural Pennsylvania in 1800: A Portrait from the Septennial Census." *Pennsylvania History* 49 (1982): 25–47.
——. "Tenancy and Asset-Holding in Late Eighteenth-Century Washington County, Pennsylvania." *Western Pennsylvania Historical Magazine* 65 (1982): 1–15.
Sosin, Jack M. *Whitehall and the Wilderness: The Middle West in British Colonial Policy, 1760–1775*. Lincoln, Nebr., 1961.
——, ed. *The Opening of the West*. New York, 1969.
Souden, David. "Migrants and the Population Structure of Later Seventeenth-Century Provincial and Market Towns." In Peter A. Clark, *Transformation of English Provincial Towns*, 132–68.
——. "Movers and Stayers in Family Reconstitution Populations." *Local Population Studies*, no. 33 (1984): 11–28.
——. " 'Rogues, Whores and Vagabonds'? Indentured Servant Emigrants to North America, and the Case of Mid-Seventeenth-Century Bristol." *Social History* 3 (1978): 23–41.
Spalding, Phinizy. *Oglethorpe in America*. Chicago, 1977.
——. "Oglethorpe's Quest for an American Zion." In Jackson and Spalding, *Forty Years of Diversity*, 60–79.
Spalding, Thomas W. "The Maryland Catholic Diaspora." *U.S. Catholic Historian* 8 (1989): 163–72.
Spear, Jennifer M. " 'They Need Wives': Méttissage and the Regulation of Sexuality in French Louisiana, 1699–1730." In Hodes, *Sex, Love, Race*, 35–59.
Speck, Frank G. "The Functions of Wampum among the Eastern Algonkian." *Memoirs of the American Anthropological Association* 6 (1919): 3–74.
Springer, James Warren. "American Indians and the Law of Real Property." *American Journal of Legal History* 30 (1986): 25–58.
Spruill, Julia Cherry. *Women's Life and Work in the Southern Colonies*. Introduction by Anne Firor Scott. 1938. New York, 1972.

Spufford, Margaret. *Contrasting Communities: English Villagers in the Sixteenth and Seventeenth Centuries*. Cambridge, 1974.
——. *The Great Reclothing of Rural England: Petty Chapmen and Their Wares in the Seventeenth Century*. London, 1984.
——. "The Importance of Religion in the Sixteenth and Seventeenth Centuries." In Margaret Spufford, *World of Rural Dissenters*, 1–102.
——. "The Limitations of the Probate Inventory." In *English Rural Society, 1500–1800: Essays in Honour of Joan Thirsk*, edited by John Chartres and David Hey, 139–74. Cambridge, 1990.
——. "Peasant Inheritance Customs and Land Distribution in Cambridgeshire from the Sixteenth to the Eighteenth Centuries." In Goody, Thirsk, and Thompson, *Family and Inheritance*, 156–76.
——. *Small Books and Pleasant Histories: Popular Fiction and Its Readership in Seventeenth-Century England*. Athens, Ga., 1981.
——, ed. *The World of Rural Dissenters, 1520–1725*. Cambridge, 1995.
Spufford, Margaret, and Motoyasu Takahashi. "Families, Will Witnesses, and Economic Structure in the Fens and on the Chalk: Sixteenth-and-Seventeenth-Century Willingham and Chippenham." *Albion* 28 (1996): 379–411.
Spufford, Peter. "The Comparative Mobility and Immobility of Lollard Descendants in Early Modern England." In Margaret Spufford, *World of Rural Dissenters*, 309–31.
Spurr, Stephen H., and Burton V. Barnes. *Forest Ecology*. 3d ed. New York, 1980.
Sreenivasan, Govind. "The Land-Family Bond at Earls Colne (Essex), 1550–1650." *P&P*, no. 131 (1991): 3–37.
——. "Reply." *P&P*, no. 146 (1995): 174–87.
Stahle, David W., and Malcolm K. Cleveland. "Reconstruction and Analysis of Spring Rainfall over the Southeastern U.S. for the Past 1,000 Years." *American Meteorological Society, Bulletin* 73 (1992): 1947–61.
Stahle, David W., Malcolm K. Cleveland, Dennis B. Blanton, Matthew D. Therrell, and David A. Gay. "The Lost Colony and Jamestown Droughts." *Science* 280 (1998): 564–67.
Standing, Guy. "Migration and Modes of Explanation: Social Origins of Immobility and Mobility." *Journal of Peasant Studies* 8 (1980–81): 173–211.
Stapleton, Barry. "Inherited Poverty and Life-Cycle Poverty: Odiham, Hampshire, 1650–1850." *Social History* 18 (1993): 337–55.
Stark, James H. *The Loyalists of Massachusetts and the Other Side of the American Revolution*. Boston, 1910.
Starna, William A. "The Pequots in the Early Seventeenth Century." In Hauptman and Wherry, *Pequots in Southern New England*, 33–47.
Starr, J. Barton. *Tories, Dons, and Rebels: The American Revolution in British West Florida*. Gainesville, Fla., 1976.
Statom, Thomas Ralph, Jr. "Negro Slavery in Eighteenth-Century Georgia." Ph.D. diss., University of Alabama, 1982.
Statt, Daniel. *Foreigners and Englishmen: The Controversy over Emigration and Population, 1660–1760*. Newark, Del., 1995.
Steckel, Richard H. "Nutritional Status in the Colonial American Economy." *WMQ* 46 (1999): 31–52.
Steele, Ian K. *Warpaths: Invasions of North America*. New York, 1994.
Steffen, Charles G. *From Gentlemen to Townsmen: The Gentry of Baltimore County, Maryland, 1660–1776*. Lexington, Ky., 1993.

Steinfeld, Robert J. *The Invention of Free Labor: The Employment Relation in English and American Law and Culture, 1350–1870*. Chapel Hill, N.C., 1991.

Steinitz, Michael. "Rethinking Geographical Approaches to the Common House: The Evidence from Eighteenth-Century Massachusetts." In *Perspectives in Vernacular Architecture, III*, edited by Thomas Carter and Bernard L. Herman, 16–26. Columbia, Mo., 1989.

Stevenson, David, ed. *From Lairds to Louns: Country and Burgh Life in Aberdeen, 1600–1800*. Aberdeen, Scotland, 1986.

Stevenson, W. Iain. "Some Aspects of the Geography of the Clyde Tobacco Trade in the Eighteenth Century." *Scottish Geographical Magazine* 89 (1973): 19–35.

Stewart, Mart. " 'Whether Wast, Deodant, or Stray': Cattle, Culture, and the Environment in Early Georgia." *AgH* 65 (1991): 1–28.

Stewart, Peter C. "Man and the Swamp: The Historical Dimension." In *The Great Dismal Swamp*, edited by Paul W. Kirk, 57–73. Charlottesville, Va., 1979.

Stewart, Richard W. "The 'Irish Road': Military Supply and Arms for Elizabeth's Army during the O'Neill Rebellion in Ireland, 1598–1601." In Fissel, *War and Government*, 16–37.

Stilgoe, John R. *Common Landscape of America, 1580 to 1845*. New Haven, Conn., 1982.

Stilwell, Lewis D. *Migration from Vermont*. Montpelier, Vt., 1948. Reprinted from *Vermont Historical Society Proceedings*, n.s., 5 (1937): 64–246.

Stiverson, Gregory A. *Poverty in a Land of Plenty: Tenancy in Eighteenth-Century Maryland*. Baltimore, 1977.

Stoebuck, William B. "A General Theory of Eminent Domain." *Washington Law Review* 47 (1972): 553–608.

Stone, Gary Wheller. "Artifacts Are Not Enough." In Beaudry, *Documentary Archaeology*, 68–77.

Stone, Lawrence. *The Family, Sex, and Marriage in England, 1500–1800*. New York, 1977.

——. "The Revival of the Narrative: Reflections on a New Old History." *P&P*, no. 85 (1979): 3–24.

——. *The Road to Divorce: England, 1550–1987*. Oxford, 1990.

Stout, Harry S. "The Morphology of Remigration: New England University Men and Their Return to England, 1640–1660." *Journal of American Studies* 10 (1976): 151–72.

Stretton, Tim. *Women Waging Law in Elizabethan England*. Cambridge, 1998.

Strong, John A. "Wyandanch: Sachen of the Montauks." In Grumet, *Northeastern Indian Lives*, 48–73.

Sugrue, Thomas J. "The Peopling and Depeopling of Early Pennsylvania: Indians and Colonists, 1680–1720." *Pennsylvania Magazine of History and Biography* 116 (1992): 3–31.

Supple, B. E. *Commercial Crisis and Change in England, 1600–1642*. Cambridge, 1964.

Sutter, Ruth E. *The Next Place You Come To: A Historical Introduction to Communities in North America*. Englewood Cliffs, N.J., 1973.

Sutton, Imre. *Indian Land Tenure: Bibliographical Essays and a Guide to the Literature*. New York, 1975.

Swan, Marshall W. S. "Cape Ann at the Nadir (1780)." *Essex Institute Historical Collections* 119 (1983): 252–59.

Sweeney, Kevin M. "Gentleman Farmers and Inland Merchants: The Williams Family and Commercial Agriculture in Pre-Revolutionary Western Massachusetts." In *The Farm*, edited by Peter Benes. Vol. 9, *Dublin Seminar for New England Folklife Annual Proceedings*, 60–73. Boston, 1988.

Syrett, David. *Shipping and the American War, 1775@2D83: A Study of British Transport Organization*. London, 1970.
Tagney, Ronald N. "Essex County Looks West: The Northwest Ordinance of 1787 and the Settlement of the Ohio Territory." *Essex Institute Historical Collections* 124 (1988): 86–101.
Takahashi, Motoyasu. "The Number of Wills Proved in the Sixteenth and Seventeenth Centuries: Graphs, with Tables and Commentary." In *The Records of the Nation: The Public Record Office, 1838–1988: The British Record Society, 1888–1988*, edited by G. H. Martin and Peter Spufford, 187–213. Woodbridge, England, 1990.
Tanner, Helen Hornbeck. *Atlas of Great Lakes Indian History*. Norman, Okla., 1987.
——. "The Land and Water Communication Systems of the Southeastern Indians." In Wood, Waselkov, and Hatley, *Powhatan's Mantle*, 6–20.
Tarleton, [Banastre]. *A History of the Campaigns of 1780 and 1781, in the Southern Provinces of North America*. London, 1787.
Tate, Thad W., and David L. Ammerman, eds. *The Chesapeake in the Seventeenth Century: Essays on Anglo-American Society*. Chapel Hill, N.C., 1979.
Tate, William Edward. *A Domesday of English Enclosure Acts and Awards*. Edited by M. E. Turner. Reading, England, 1978.
——. *The English Community and the Enclosure Movements*. London, 1967.
Tatter, Henry. "State and Federal Land Policy during the Confederation." *AgH* 9 (1935): 176–86.
Tawney, A. J., and R. H. Tawney. "An Occupational Census of the Seventeenth Century." *EcHR* 5 (1934): 25–64.
Tawney, R. H. *The Agrarian Problem in the Sixteenth Century*. Edited by Lawrence Stone. 1912. New York, 1967.
Taylor, Alan. *Liberty Men and Great Proprietors: The Revolutionary Settlement on the Maine Frontier, 1760–1820*. Chapel Hill, N.C., 1990.
Taylor, Arthur J., ed. *The Standard of Living in Britain in the Industrial Revolution*. London, 1975.
Taylor, George Rogers. "Comment [on paper by Lee and Lalli]." In Gilchrist, *Growth of the Seaport Cities*, 38–46.
Taylor, James David. " 'Base Commoditie': Natural Resource and Natural History in Smith's *Generall Historie*." *Environmental History Review* 17 (1993): 73–89.
Taylor, Paul S. *Georgia Plan, 1732–1752*. Berkeley, Calif., 1972.
Taylor, Peter Kei. *Indentured to Liberty: Peasant Life and the Hessian Military State, 1688–1815*. Ithaca, N.Y., 1994.
Taylor, Robert M., Jr., and Ralph J. Crandall, eds. *Generations and Change: Genealogical Perspectives in Social History*. Macon, Ga., 1986.
Tebbenhoff, Edward H. "Tacit Rules and Hidden Family Structures: Naming Practices and Godparentage in Schenectady, New York 1680–1800." *Journal of Social History* 18 (1985): 567–85.
Thayer, Theodore G. "The Land Bank System in the American Colonies." *JEH* 13 (1953): 145–59.
Theibault, John. *German Villages in Crisis: Rural Life in Hesse-Kassel and the Thirty Years' War, 1580–1720*. Atlantic Highlands, N.J., 1995.
Thibaut, Jacqueline. *This Fatal Crisis: Logistics, Supply, and the Continental Army at Valley Forge, 1777–1778*. Vol. 2 of *The Valley Forge Historical Report*. Valley Forge, Pa., 1980.
Third, Betty M. W. "Changing Landscape and Social Structure in Scottish Lowlands as

Revealed by Eighteenth-Century Estate Plans." *Scottish Geographical Magazine* 71 (1955): 83–93.

Thirsk, Joan. *Economic Policy and Projects: The Development of a Consumer Society in Early Modern England*. Oxford, 1978.

——. "English Rural Communities: Structures, Regularities, and Change in the Sixteenth and Seventeenth Centuries." In Short, *English Rural Community*, 44–61.

——. "The Farming Regions of England." In Thirsk, *Agrarian History*, 1–112.

——. Foreword to Prior, *Women in English Society*, 1–21.

——. *The Rural Economy of England: Collected Essays*. London, 1984.

——, ed. *The Agrarian History of England and Wales*. Vol. 4, *1500–1640*. Cambridge, 1967.

Thirsk, Joan, and J. P. Cooper, eds. *Seventeenth-Century Economic Documents*. Oxford, 1972.

Thomas, Brinley. "The Rhythm of Growth in the Atlantic Economy of the Eighteenth Century." *Research in Economic History* 3 (1978): 1–46.

Thomas, Peter A. "Contrastive Subsistence Strategies and Land Use as Factors for Understanding Indian-White Relations in New England." *Ethnohistory* 23 (1976): 1–18.

——. "Cultural Change on the Southern New England Frontier, 1630–1665." In Fitzhugh, *Cultures in Contact*, 131–62.

——. "In the Maelstrom of Change: The Indian Trade and Cultural Processes in the Middle Connecticut River Valley, 1635–1665." Ph.D. diss., University of Massachusetts, Amherst, 1979.

Thompson, Daniel P., and Ralph H. Smith. "The Forest Primeval in the Northeast: A Great Myth?" *Annual Tall Timbers Ecology Conference*, no. 10 (1970): 255–65.

Thompson, E. P. *Customs in Common: Studies in Traditional Popular Culture*. New York, 1993.

——. "Happy Families." *Radical History Review*, no. 20 (1979): 42–50.

——. *The Making of the English Working Class*. New York, 1963.

Thompson, Kenneth. "Forests and Climate Change in America: Some Early Views." *Climatic Change* 3 (1980): 47–64.

Thompson, Roger. *Mobility and Migration: East Anglian Founders of New England, 1629–40*. Amherst, Mass., 1994.

——. *Sex in Middlesex: Popular Mores in a Massachusetts County, 1649–1699*. Amherst, Mass., 1986.

——. "Social Cohesion in Early New England." *New England Historical and Genealogical Register* 146 (1992): 235–53.

Thompson, Thomas C. "The Life Course and Labor of a Colonial Farmer." *Historical New Hampshire* 40 (1985): 135–55.

Thompson, Tommy R. "Debtors, Creditors, and the General Assembly in Maryland." *MdHM* 72 (1977): 59–77.

——. "Personal Indebtedness and the American Revolution in Maryland." *MdHM* 73 (1978): 13–29.

Thorndale, William. "The Virginia Census of 1619." *Magazine of Virginia Genealogy* 33 (1995): 155–70.

Thornton, Russell, Tim Miller, and Jonathan Warren. "American Indian Population Recovery following Smallpox Epidemics." *American Anthropologist* 93 (1991): 28–45.

Thorp, Daniel B. "Doing Business in the Backcountry: Retail Trade in Colonial Rowan County, North Carolina." *WMQ* 48 (1991): 387–408.

——. *The Moravian Community in Colonial North Carolina: Pluralism on the Southern Frontier*. Knoxville, Tenn., 1990.

Thwaites, Reuben Gold, and Louise Phelps Kellogg, eds. *Documentary History of Dunmore's War, 1774*. Wisconsin Historical Society, Draper Series, vol. 1. Madison, Wis., 1905.

Tiedmann, Joseph S. "Patriots by Default: Queens County, New York, and the British Army, 1776–1783." *WMQ* 43 (1986): 35–63.

——. *Reluctant Revolutionaries: New York City and the Road to Independence, 1763–1776*. Ithaca, N.Y., 1997.

Tilley, Morris Palmer. *A Dictionary of the Proverbs in England in the Sixteenth and Seventeenth Centuries*. Ann Arbor, Mich., 1950.

Tillson, Albert H., Jr. *Gentry and Common Folk: Political Culture on a Virginia Frontier, 1740–1789*. Lexington, Ky., 1991.

——. "The Southern Backcountry: A Survey of Current Research." *VMHB* 98 (1990): 387–421.

Tilly, Charles, ed. *Historical Studies of Changing Fertility*. Princeton, N.J., 1978.

Tilly, Louise A., and Joan W. Scott. *Women, Work, and Family*. New York, 1978.

Timmins, Geoffrey. *Made in Lancashire: A History of Regional Specialization*. Manchester, England, 1998.

Titow, J. Z. "Medieval England and the Open-Field System." In *Peasants, Knights, and Heretics: Studies in Medieval English Social History*, ed. Rodney H. Hilton, 33–50. Cambridge, 1976.

Tittler, Robert. "Money-Lending Activities in the West Midlands: The Activities of Joyce Jeffries, 1638–49." *Historical Research* 67 (1994): 249–63.

Titus, James. *The Old Dominion at War: Society, Politics, and Warfare in Late Colonial Virginia*. Columbia, S.C., 1991.

"Tobacco Exports from Octo 1782 to Octo 1799." Auditor's Item 49, Virginia State Library, Richmond.

Todd, Barbara J. "The Remarrying Widow: A Stereotype Reconsidered." In Prior, *Women in English Society*, 54–92.

Todd, Margo. "Humanists, Puritans, and the Spiritualized Household." *Church History* 49 (1980): 18–34.

Tooker, Elisabeth. "The League of the Iroquois: Its History, Politics, and Ritual." In Trigger, *Northeast*, 418–41.

Torrence, Clayton. *Old Somerset on the Eastern Shore of Maryland: A Study in Foundations and Founders*. Richmond, Va., 1935.

Towner, Lawrence W. *Past Imperfect: Essays on History, Libraries, and the Humanities*. Edited by Robert W. Karrow Jr. and Alfred F. Young. Chicago, 1993.

Trace, Keith. "The West Jersey Society, 1736–1819." *Proceedings of the New Jersey Historical Society* 80 (1963): 267–82.

Tranter, N. L. "Population and Social Structure in a Bedfordshire Parish: The Cardington Listing of Inhabitants, 1782." *Population Studies* 21 (1967): 261–82.

Trautman, Patricia. "Dress in Seventeenth-Century Cambridge, Massachusetts: An Inventory-Based Reconstruction." In Benes, *Early American Probate Inventories*, 51–73.

Treble, J. H. "The Standard of Living of the Working Class." In Devine and Mitchison, *People and Society in Scotland*, 188–226.

Treckel, Paula A. "Breast Feeding and Maternal Sexuality in Colonial America." *JIH* 20 (1989): 25–52.

——. " 'To Comfort the Heart': English Women and Families in the Settlement of Colonial Virginia." In *Looking South: Chapters in the History of an American Region*, edited by Winfred B. Moore and Joseph F. Tripp, 133–52. New York, 1989.

Trigger, Bruce G. "Early Native American Responses to European Contact: Romantic versus Rationalistic Interpretations." *Journal of American History* 77 (1991): 1195–1215.
——, ed. *Northeast*. Vol. 15 of *Handbook of North American Indians*. Edited by William G. Sturtevant. Washington, D.C., 1978.
Trigger, Bruce G., and William R. Swagerty. "Entertaining Strangers: North America in the Sixteenth Century." In *The Cambridge History of the Native Peoples of the Americas*, edited by Bruce G. Trigger and Wilcomb E. Washburn. Vol. 1, pt., 1, *North America*, 325–98. Cambridge, 1996.
Troxler, Carole W. "A Loyalist Life: John Bond of South Carolina and Nova Scotia." *Acadiensis* 19 (1990): 72–91.
——. "Origins of the Rawdon Loyalist Settlement." *Nova Scotia Historical Review* 8 (1988): 62–76.
——. "Refuge, Resistance, and Reward: The Southern Loyalists' Claim on East Florida." *Journal of Southern History* 55 (1989): 563–97.
True, Ransom. "Land Transactions in Louisa County, Virginia, 1765–1812: A Quantitative Analysis." Ph.D. diss., University of Virginia, 1976.
Trussell, John B. B., Jr. *Birthplace of an Army: A Study of the Valley Forge Encampment*. Harrisburg, Pa., 1976.
Truxes, Thomas M. *Irish-American Trade, 1660–1783*. Cambridge, 1988.
Tryon, Rolla Milton. *Household Manufactures in the United States, 1640–1860*. Chicago, 1917.
Tucker, R. S. "Real Wages of Artisans in London, 1729–1935." In Arthur J. Taylor, *Standard of Living in Britain*, 21–35.
Tully, Alan. "Literacy Levels and Educational Development in Rural Pennsylvania." *Pennsylvania History* 39 (1972): 301–12.
——. "Patterns of Slaveholding in Colonial Pennsylvania: Chester and Lancaster Counties, 1729–1758." *Journal of Social History* 6 (1973): 284–305.
——. *William Penn's Legacy: Politics and Social Structure in Provincial Pennsylvania, 1726–1755*. Baltimore, 1977.
Tully, James. "Aboriginal Property and Western Theory: Recovering a Middle Ground." *Social Philosophy and Policy* 11 (1994): 153–80.
Turnbaugh, William A. "Assessing the Significance of European Goods in Seventeenth-Century Narragansett Society." In *Ethnohistory and Archaeology: Approaches to Postcontact Change in the Americas*, edited by J. Daniel Rogers and Samuel M. Wilson, 133–60. New York, 1993.
Turner, Michael. "Common Property and Property in Common." *Agricultural History Review* 42 (1994): 158–62.
——. *Enclosures in Britain, 1750–1830*. London, 1984.
——. *English Parliamentary Enclosure: Its Historical Geography and Economic History*. Folkestone, England, 1980.
Turnock, David. *The Historical Geography of Scotland since 1707: Geographical Aspects of Modernisation*. Cambridge, 1982.
Tusser, Thomas. *Five Hundred Points of Good Husbandry*. 1580. Edited by Geoffrey Grigson. Oxford, 1984.
Tyack, Norman C. P. "The Humbler Puritans of East Anglia and the New England Movement: Evidence from the Court Records of the 1630s." *New England Historical and Genealogical Register* 138 (1984): 79–106.
Tyacke, Sarah, ed. *English Map-Mapping, 1500–1650*. London, 1983.
Tyler, Clarence E. "Topographical Terms in the Seventeenth-Century Records of Connecticut and Rhode Island." *NEQ* 2 (1929): 382–401.

Tyler, John W. "Foster Cunliffe and Sons: Liverpool Merchants in the Maryland Tobacco Trade, 1738–1765." *MdHM* 73 (1978): 246–79.
Tyson, Robert E. "Famine in Aberdeenshire, 1695–1699: Anatomy of a Crisis." In David Stevenson, *From Lairds to Louns*, 32–51.
Ubelaker, Douglas H. "Human Biology of Virginia Indians." In Roundtree, *Powhatan Foreign Relations*, 53–75.
——. "North American Indian Population Size." In Verano and Ubelaker, *Disease and Demography*, 169–76.
Ulbricht, Otto. "The World of a Beggar around 1775: Johann Gottfried Kästner." *Central European History* 27 (1994): 153–84.
Ulrich, Laurel Thatcher. *Good Wives: Image and Reality in the Lives of Women in Northern New England, 1650–1750*. New York, 1982.
——. "Housewife and Gadder: Themes of Self-Sufficiency and Community in Eighteenth-Century New England." In *"To Toil the Livelong Day": America's Women at Work, 1780–1980*, edited by Carol Groneman and Mary Beth Norton, 21–34. Ithaca, N.Y., 1987.
——. "Martha Ballard and Her Girls: Women's Work in Eighteenth-Century Maine." In Innes, *Work and Labor in Early America*, 70–105.
——. *A Midwife's Tale: The Life of Martha Ballard, Based on Her Diary, 1785–1812*. New York, 1990.
——. "Wheels, Looms, and the Gender Division of Labor in Eighteenth-Century New England." *WMQ* 55 (1998): 3–38.
Ulrich, Laurel Thatcher, and Lois K. Stabler. " 'Girling of It' in an Eighteenth-Century New England Household." In Benes, *Families and Children*, 24–36.
Underdown, David. "The Taming of the Scold: The Enforcement of Patriarchal Authority in Early Modern England." In *Order and Disorder in Early Modern England*, ed. Anthony Fletcher and John Stevenson, 116–36. Cambridge, 1985.
Underwood, Wynn. "Indian and Tory Raids on the Otter Valley, 1777–1782." *Vermont Quarterly*, n.s., 15 (1947): 195–221.
Upton, Dell. "The Power of Things: Recent Studies in American Vernacular Architecture." *American Quarterly* 35 (1983): 262–79.
——. "Traditional Timber Framing." In Brooke Hindle, *Material Culture of the Wooden Age*, 35–93.
——. "Vernacular Domestic Architecture in Eighteenth-Century Virginia." In Upton and Vlach, *Common Places*, 315–35.
Upton, Dell, and John Michael Vlach, eds. *Common Places: Readings in American Vernacular Architecture*. Athens, Ga., 1986.
U.S. Bureau of the Census. *A Century of Population Growth*. Washington, D.C., 1909.
——. *Historical Statistics of the United States, Colonial Times to 1970, Bicentennial Edition*. 2 vols. Washington, D.C., 1975.
——. *Returns of the Whole Numbers of Persons Within the Several Districts of the United Sates* . . . . Washington, D.C., 1802.
——. *Statistical Abstract of the United States, 1995: The National Data Book*. Washington, D.C., 1995.
Valenze, Deborah. "The Art of Women and the Business of Men: Women's Work and the Dairy Industry, c. 1740–1840." *P&P*, no. 130 (1991): 142–69.
Van den Boogaart, "The Servant Migration to New Netherland, 1624–1664." In Emmer, *Colonialism and Migration*, 55–81.
Van Deventer, David E. *The Emergence of Provincial New Hampshire, 1623–1741*. Baltimore, 1976.

Van Dongen, Alexandra. "The Inexhaustible Kettle: The Metamorphosis of a European Utensil in the World of North American Indians." In Van Dongen et al., *One Man's Trash*, 115–70.
Van Dongen, Alexandra, et al. *One Man's Trash Is Another Man's Treasure: The Metamorphosis of the European Utensil in the New World*. Rotterdam, 1996.
Van Dusen, Albert E. "Connecticut: The 'Provisions State' of the American Revolution." *New England Social Studies Bulletin* 10 (October 1952): 16–23.
Van Leeuwen, Marco H. D. "Logic of Charity: Poor Relief in Preindustrial Europe." *JIH* 24 (1994): 589–614.
Vann, Richard T. "Quakerism: Made in America." In Dunn and Dunn, *World of William Penn*, 157–70.
Van Tyne, Claude H. *Loyalists in the American Revolution*. New York, 1902.
Vaughan, Alden T. *New England Frontier: Puritans and Indians, 1620–1675*. 3d ed. Norman, Okla., 1995.
——. *Roots of American Racism: Essays on the Colonial Experience*. New York, 1995.
Vaughan, Alden T., and Edward W. Clark, eds. *Puritans among the Indians: Accounts of Captivity and Redemption, 1676–1724*. Cambridge, Mass., 1981.
Vaughan, Alden T., and Daniel K. Richter. "Crossing the Cultural Divide: Indians and New Englanders, 1605–1763." *American Antiquarian Society, Proceedings* 90 (1980): 23–99.
Verano, John W., and Douglas H. Ubelaker, eds. *Disease and Demography in the Americas*. Washington, D.C., 1992.
Ver Steeg, Clarence L. *Origins of a Southern Mosaic: Studies of Early Carolina and Georgia*. Mercer University Lamar Memorial Lectures, no. 17. Athens, Ga., 1975.
Viazzo, Pier Paolo. *Upland Communities: Environment, Population, and Social Structure in the Alps since the Sixteenth Century*. Cambridge, 1989.
Vickers, Daniel. "Competency and Competition: Economic Culture in Early America." *WMQ* 47 (1990): 3–29.
——. *Farmers and Fishermen: Two Centuries of Work in Essex County, Massachusetts, 1630–1830*. Chapel Hill, N.C., 1994.
——. "The Northern Colonies: Economy and Society, 1660–1775." In Engerman and Gallman, *Cambridge Economic History*, 209–48.
——. "Work and Life on the Fishing Periphery of Essex County, Massachusetts, 1630–1675." In Hall and Allen, *Seventeenth-Century New England*, 83–117.
——. "Working the Fields in a Developing Economy: Essex County, Massachusetts, 1630–1675." In Innes, *Work and Labor in Early America*, 49–69.
Villaflor, Georgia C., and Kenneth L. Sokoloff. "Migration in Colonial America: Evidence from the Militia Rolls." *Social Science History* 6 (1982): 539–70.
Vinovskis, Maris A. "Family and Schooling in Colonial and Nineteenth-Century America." *Journal of Family History* 12 (1987): 19–38.
——. "Mortality Rates and Trends in Massachusetts before 1860." *JEH* 32 (1972): 184–213.
Vivian, Jean. "Military Land Bounties during the Revolutionary and Confederation Periods." *MdHM* 61 (1966): 231–56.
Volwiler, Albert T. *George Croghan and the Westward Movement, 1741–1782*. Cleveland, 1926.
Wacker, Peter O. "The Dutch Culture Area in the Northeast, 1609–1800." *New Jersey History* 104 (1986): 1–21.
——. "Human Exploitation of the New Jersey Pine Barrens before 1900." In *Pine Barrens: Ecosystem and Landscape*, edited by Richard T. T. Foreman, 3–23. New York, 1979.

——. *Land and People: A Cultural Geography of Preindustrial New Jersey: Origins and Settlement Patterns*. New Brunswick, N.J., 1975.
——. *The Musconetcong Valley of New Jersey*. New Brunswick, N.J., 1968.
——. "New Jersey Forests." In Richard C. Davis, *Encyclopedia of Forest History*, 2:485–87.
——. "The New Jersey Tax-Ratable List of 1751." *New Jersey History* 107 (1989): 23–47.
Wacker, Peter O., and Paul G. E. Clemens. *Land Use in Early New Jersey: A Historical Geography*. Newark, N.J., 1995.
Wadsworth, Alfred P., and Julia de Lacy Mann. *The Cotton Trade and Industrial Lancashire, 1600–1780*. Manchester, England, 1931.
Wainwright, Nicholas B. *George Croghan: Wilderness Diplomat*. Chapel Hill, N.C., 1959.
Walker, Garthine. "Women, Theft, and the World of Stolen Goods." In Kermode and Walker, *Women, Crime, and the Courts*, 81–105.
Walker, James W. St. G. "Blacks as American Loyalists: The Slaves' War for Independence." *Historical Reflections* 2 (1975): 51–67.
Wall, Maureen. *Catholic Ireland in the Eighteenth Century: Collected Essays of Maureen Wall*. Edited by Gerard O'Brien and Tom Dunne. Dublin, 1989.
Wall, Richard. "The Age at Leaving Home." *Journal of Family History* 3 (1978): 181–202.
——. "The Household: Demographic and Economic Change in England, 1650–1970." In Wall, Robin, and Laslett, *Family Forms in Historic Europe*, 493–512.
——. "Women Alone in English Society." *Annales de Demographie Historique* 18 (1981): 303–17.
Wall, Richard, Jean Robin, and Peter Laslett, eds. *Family Forms in Historic Europe*. Cambridge, 1983.
Wallace, Anthony F. C. *The Death and Rebirth of the Seneca*. New York, 1969.
——. "Political Organization and Land Tenure among the Northeastern Indians, 1600–1830." *Southwestern Journal of Anthropology* 13 (1957): 301–21.
Wallerstein, Immanuel. *Historical Capitalism*. London, 1983.
——. *The Modern World-System*. Vol. 1, *Capitalist Agriculture and the Origins of the European World Economy in the Sixteenth Century*. New York, 1974.
Walsh, Lorena S. "Community Networks in the Early Chesapeake." In Carr, Morgan, and Russo, *Colonial Chesapeake Society*, 200–241.
——. "Consumer Behavior, Diet, and the Standard of Living in Late Colonial and Early Antebellum America, 1770–1840." In Gallman and Wallis, *American Economic Growth and Standards of Living*, 217–61.
——. "The Experiences and Status of Women in the Chesapeake, 1750–1775." In *The Web of Southern Social Relations: Women, Family, and Education*, edited by Walter J. Fraser Jr., R. Frank Sanders Jr., and Jon L. Wakelyn, 1–18. Athens, Ga., 1985.
——. *From Calabar to Carter's Grove: The History of a Virginia Slave Community*. Charlottesville, Va., 1997.
——. "The Historian as Census Taker: Individual Reconstitution and the Reconstruction of Censuses for a Colonial Chesapeake County." *WMQ* 38 (1981): 242–60.
——. "Land, Landlord, and Leaseholder: Estate Management and Tenant Fortunes in Southern Maryland, 1642–1820." *AgH* 59 (1985): 373–96.
——. "Plantation Management in the Chesapeake, 1620–1820." *JEH* 39 (1989): 393–406.
——. "Provisioning Tidewater Towns." Paper, McNeil Center for Early American Studies, March 1999.
——. "Rural African-Americans in the Constitutional Era in Maryland, 1776–1810." *MdHM* 84 (1989) 327–41.

———. "Servitude and Opportunity in Charles County, Maryland, 1638–1705." In Land, Carr, and Papenfuse, *Law, Society, and Politics in Early Maryland*, 111–33.
———. "Slave Life, Slave Society, and Tobacco Production in the Tidewater Chesapeake, 1620–1820." In *Cultivation and Culture: Labor and the Shaping of Slave Life in the Americas*, edited by Ira Berlin and Philip D. Morgan, 170–99. Charlottesville, Va., 1993.
———. "Staying Put or Getting Out: Findings for Charles County, Maryland, 1650–1720." *WMQ* 44 (1987): 89–103.
———. "Summing the Parts: Implications for Estimating Chesapeake Output and Income Subregionally." *WMQ* 56 (1999): 53–94.
———. " 'Till Death Us Do Part': Marriage and Family in the Chesapeake in the Seventeenth Century." In Tate and Ammerman, *Chesapeake in the Seventeenth Century*, 126–52.
Walsh, Lorena S., and Russell R. Menard. "Death in the Chesapeake: Two Life Tables for Men in Early Colonial Maryland." *MdHM* 69 (1974): 211–27.
Walsh, Margaret. "Women's Place on the American Frontier." *Journal of American Studies* 29 (1995): 241–55.
Walter, John. "The Social Economy of Dearth in Early Modern England." In Walter and Schofield, *Famine, Disease, and the Social Order in Early Modern Society*, 75–128.
Walter, John, and Roger S. Schofield, eds. *Famine, Disease, and the Social Order in Early Modern Society*. Cambridge, 1989.
Walter, John, and Keith Wrightson. "Dearth and the Social Order in Early Modern England." *P&P*, no. 71 (1976): 22–42.
Walzer, John F. "A Period of Ambivalence: Eighteenth-Century American Childhood." In *The History of Childhood*, edited by Lloyd deMause, 351–82. New York, 1974.
Ward, Barbara McLean. "Women's Property and Family Continuity in Eighteenth-Century Connecticut." In Benes, *Early American Probate Inventories*, 74–85.
Ward, Christopher. *The War of the Revolution*. 2 vols. Edited by John Richard Alden. New York, 1952.
Ward, Matthew C. "An Army of Servants: The Pennsylvania Regiment during the Seven Years' War." *Pennsylvania Magazine of History and Biography* 119 (1995): 75–93.
———. "Fighting the 'Old Women': Indian Strategy on the Virginia and Pennsylvania Frontier, 1754–1758." *VMHB* 103 (1995): 297–320.
Warden, G. B. "Law Reform in England and New England, 1620 to 1660." *WMQ* 35 (1978): 668–90.
Warden, Rosemary S. " 'The Infamous Fitch': The Tory Bandit, James Fitzpatrick of Chester County." *Pennsylvania History* 62 (1995): 376–87.
Wareing, John. "Changes in the Geographical Distribution of the Recruitment of Apprentices to the London Companies, 1486–1750." *Journal of Historical Geography* 6 (1980): 241–49.
———. "Migration to London and Transatlantic Emigration to Indentured Servants, 1683–1775." *Journal of Historical Geography* 7 (1981): 356–78.
Waselkov, Gregory A. "Indian Maps of the Colonial Southeast." In Wood, Waselkov, and Hatley, *Powhatan's Mantle*, 292–343.
Washburn, Wilcomb E. *The Governor and the Rebel: A History of Bacon's Rebellion in Virginia*. Chapel Hill, N.C., 1957.
———. "The Moral and Legal Justification for Dispossessing the Indians." In James Morton Smith, *Seventeenth-Century America*, 15–32.
———. "Seventeenth-Century Indian Wars." In Trigger, *Northeast*, 89–100.

———. "Stephen Saunders Webb's Interpretation of Bacon's Rebellion." *VMHB* 95 (1987): 339–52.

———, ed. *History of Indian-White Relations*. Vol. 4 of *Handbook of North American Indians*. Edited by William C. Sturtevant. Washington, D.C., 1988.

Waterhouse, Edward. *A Declaration of the State of the Colony and Affaires in Virginia. With a Relation of the Barbarous Massacre in time of peace and League* . . . . London, 1622. In Kingsbury, *Records of the Virginia Company*, 3:541–79.

Waterhouse, Richard. "Economic Growth and Changing Patterns of Wealth Distribution in Colonial Lowcountry South Carolina." *South Carolina Historical Magazine* 89 (1988): 203–17.

———. "England, the Caribbean, and the Settlement of Carolina." *Journal of American Studies* 9 (1975): 259–81.

———. "Reluctant Emigrants: The English Background of the First Generation of the New England Puritan Clergy." *Historical Magazine of the Protestant Episcopal Church* 44 (1975): 473–88.

Waters, John J. "Family, Inheritance, and Migration in New England: The Evidence from Guilford, Connecticut." *WMQ* 39 (1982): 64–86.

———. "Naming and Kinship in New England: Guilford Patterns and Usage, 1693–1750." *New England Historical and Genealogical Register* 138 (1984): 161–81.

———. "Patrimony, Succession, and Social Stability: Guilford, Connecticut, in the Eighteenth Century." *Perspectives in American History* 10 (1976): 131–60.

Watson, Alan D. "Orphanage in Colonial North Carolina: Edgecombe County as a Case Study." *North Carolina Historical Review* 52 (1975): 105–19.

———. "The Quitrent System in Royal South Carolina." *WMQ* 33 (1976): 183–211.

———. "Regulation and Administration of Roads and Bridges in Colonial Eastern North Carolina." *North Carolina Historical Review* 45 (1968): 399–417.

———. "Women in Colonial North Carolina: Overlooked and Underestimated." *North Carolina Historical Review* 58 (1981): 1–22.

Wawrzyczek, Irmina. "The Women of Accomack versus Henry Smith: Gender, Legal Recourse, and the Social Order of Seventeenth-Century Virginia." *VMHB* 105 (1997): 4–26.

Webb, Stephen Saunders. *1676: The End of American Independence*. New York, 1984.

Wedgwood, C. V. *The Thirty Years War*. London, 1938.

Weeden, William. *Indian Money as a Factor in New England Civilization*. Vol. 2, nos. 8–9, *Johns Hopkins University Studies in Historical and Political Science*. Baltimore, 1884.

Weiman, David F. "Families, Farms, and Rural Society in Pre-Industrial America." In *Agrarian Organization in the Century of Industrialization: Europe, Russia, and North America*, edited by George Grantham and Carol S. Leonard, 255–77. Supplement 5 of *Research in Economic History*. Greenwich, Conn., 1989.

Weir, Robert M. *Colonial South Carolina: A History*. Millpond, N.Y., 1983.

———. *"The Last of the American Freemen": Studies in the Political Culture of the Colonial and Revolutionary South*. Macon, Ga., 1986.

Weiss, Rona Stephanie. "The Development of the Market Economy in Colonial Massachusetts." Ph.D. diss., University of Massachusetts, 1981.

Weiss, Thomas. "U.S. Labor Force Estimates and Economic Growth, 1800–1860." In Gallman and Wallis, *American Economic Growth and Standards of Living*, 19–75.

Wellenreuther, Hermann. "Urbanization in the Colonial South: A Critique." *WMQ* 31 (1974): 653–68.

Wells, Camille. "The Planter's Prospect: Houses, Outbuildings, and Rural Landscape in Eighteenth-Century Virginia." *Winterthur Portfolio* 28 (1993): 1–32.

Wells, Robert V. "Demographic Change and the Life Cycle in America." *JIH* 2 (1971): 273–82.

——. "Family Size and Fertility Control in Eighteenth-Century America: A Study of Quaker Families." *Population Studies* 25 (1971): 73–82.

——. "Illegitimacy and Bridal Pregnancy in Colonial America." In Laslett, Oosterveen, and Smith, *Bastardy and Its Comparative History*, 349–61.

——. "Marriage Seasonals in Early America: Comparisons and Comments." *JIH* 18 (1987): 299–308.

——. "The Population of England's Colonies in America: Old English or New American?" *Population Studies* 46 (1992): 85–102.

——. *The Population of the British Colonies in America before 1776: A Survey of Census Data*. Princeton, N.J., 1975.

——. "Quaker Marriage Patterns in Colonial Perspective." *WMQ* 29 (1972): 415–42.

Wennerstein, John R. "Soil Miners Redux: The Chesapeake Environment, 1680–1810." *MdHM* 91 (1996): 157–79.

Wermuth, Thomas Sylvester. " 'To Market, To Market': Yeoman Farmers, Merchant Capitalists, and the Transition to Capitalism in the Hudson River Valley, Ulster County, 1760–1840." Ph.D. diss., State University of New York, Binghamton, 1991.

Wertenbaker, Thomas Jefferson. *The Planters of Colonial Virginia*. Princeton, N.J., 1922.

Weslager, C. A. *The Nanticoke Indians: Past and Present*. Newark, Del., 1983.

West, Robert Craig. "Money in the Colonial American Economy." *Economic Inquiry* 16 (1978): 1–15.

Wetherell, Charles, and Robert W. Roetger. "Another Look at the Loyalists of Shelburne, NS, 1783–95." *Canadian Historical Review* 70 (1989): 76–91.

Whelan, Kevin. "The Catholic Community in County Wexford." In Power and Whelan, *Endurance and Emergence*, 129–70.

Whitaker, Arthur Preston. *The Mississippi Question, 1795–1803: A Study in Trade, Politics, and Diplomacy*. New York, 1934.

White, David O. *Connecticut's Black Soldiers, 1775–1783*. Connecticut Bicentennial Series, no. 4. Chester, Conn., 1973.

White, John. *The Planters Plea; or, The Grounds of Plantations Examined, . . . Together with a Manifestation of the Causes Moving Such as Have Lately Undertaken a Plantation in NEW-ENGLAND*. London, 1630. Reprinted in Force, *Tracts and Other Papers*, vol. 2.

White, Lawrence J. "Enclosures and Population Movements in England, 1700–1830." *Explorations in Entrepreneurial History* 6 (1969): 175–86.

White, Richard. *The Middle Ground: Indians, Empires, and Republics in the Great Lakes Region, 1615–1815*. Cambridge, 1991.

Whitfield, Faye W. "The Initial Settling of Niagara-on-the-Lake, 1778–1784." *Ontario History* 83 (1991): 3–21.

Whitney, Gordon G. *From Coastal Wilderness to Fruited Plain: A History of Environmental Change in Temperate North America, 1500 to the Present*. Cambridge, 1994.

Whittenburg, James Penn. "Planters, Merchants, and Lawyers: Social Change and the Origins of the North Carolina Regulation." *WMQ* 34 (1977): 215–38.

Whittington, G., and D. U. Brett. "Locational Decision-Making on a Scottish Estate prior to Enclosure." *Journal of Historical Geography* 5 (1979): 33–43.

Whittle, Jane. "Individualism and the Family-Land Bond: A Reassessment of Land

Transfer Patterns among the English Peasantry, c. 1270–1580." *P&P*, no. 160 (1998): 25–63.
——. "Inheritance, Marriage, Widowhood, and Remarriage: A Comparative Perspective on Women and Landholding in North-East Norfolk, 1440–1580." *Continuity and Change* 13 (1998): 33–72.
Whyte, Ian D. *Agriculture and Society in Seventeenth-Century Scotland*. Edinburgh, 1979.
——. "Agriculture in Aberdeenshire in the Seventeenth and Early Eighteenth Centuries: Continuity and Change." In David Stevenson, *From Lairds to Louns*, 10–31.
——. "Population Mobility in Early-Modern Scotland." In Houston and Whyte, *Scottish Society*, 37–58.
Whyte, Ian D., and Kathleen A. Whyte. "Continuity and Change in a Seventeenth-Century Scottish Farming Community." *Agricultural History Review* 32 (1984): 159–69.
——. "The Geographical Mobility of Women in Early Modern Scotland." In *Perspectives in Scottish Social History: Essays in Honour of Rosalind Mitchison*, edited by Lean Leneman, 83–106. Aberdeen, Scotland, 1988.
Wilk, Richard R., and Robert McC. Netting. "Households: Changing Forms and Functions." In *Households: Comparative and Historical Studies of the Domestic Group*, edited by Robert McC. Netting, Richard R. Wilk, and Eric J. Arnould, 1–28. Berkeley, Calif., 1984.
Willard, Margaret Wheeler, ed. *Letters of the American Revolution, 1774–1776*. Boston, 1925.
Willen, Diane. "Women in the Public Sphere in Early Modern England: The Case of the Urban Working Poor." *Sixteenth-Century Journal* 19 (1988): 559–75.
Williams, James Homer. "Great Doggs and Mischievous Cattle: Domesticated Animals and Indian-European Relations in New Netherland and New York." *New York History* 76 (1995): 245–64.
——. "Religion and Culture: Dutch-Indian Contact in New Netherland." Paper, McNeil Center for Early American Studies, April 1992.
Williams, Michael. *Americans and Their Forests: A Historical Geography*. Cambridge, 1989.
Williams, Raymond. *The Country and the City*. New York, 1973.
Williams, Roger. *A Key into the Language of America*. 1643. Edited by John J. Teunissen and Evelyn J. Hinz. Detroit, 1973.
Williams, Samuel Cole. *Tennessee during the Revolutionary War*. Edited by Frank B. Williams. 1944. New ed. Knoxville, Tenn., 1974.
Williamson, Jeffrey G., and Peter H. Lindert. *American Inequality: A Macroeconomic History*. New York, 1980.
Willoughby, Charles C. "Houses and Gardens of the New England Indians." *American Anthropologist* 8 (1906): 115–32.
Wilson, Charles. *England's Apprenticeship, 1603–1763*. 2d ed. London, 1984.
Wilson, Ellen F. "Gaining Title to the Land: The Case of the Virginia Military Tract." *Old Northwest* 12 (1986): 65–82.
Wilson, Ellen Gibson. *The Loyal Blacks*. New York, 1976.
Wilson, Lisa. *Life after Death: Widows in Pennsylvania, 1750–1850*. Philadelphia, 1992.
——. *Ye Heart of a Man: The Domestic Life of Men in Colonial New England*. New Haven, Conn., 1999.
Wilson, Sir Thomas. *The State of England: Anno Dom., 1600*. Edited by F. A. Fisher. Camden 3d ser. 52, Camden Miscellany 16. 1936.

Wiltenburg, Joy. *Disorderly Women and Female Power in the Street Literature of Early Modern England and Germany*. Charlottesville, Va., 1992.
Winchester, Angus J. L. "Ministers, Merchants, and Migrants: Cumberland Friends in North America in the Eighteenth Century." *Quaker History* 80 (1991): 85–99.
Winkel, Peter. "Naturalization in Colonial New Jersey." *New Jersey History* 109 (1991): 27–53.
Withers, Charles W. J. *Gaelic in Scotland, 1698–1981: The Geographical History of a Language*. Edinburgh, 1984.
——. "Highland, Lowland Migration and the Making of the Clotting Community, 1755–1891." *Scottish Geographic Magazine* 103 (1987): 75–83.
Withey, Lynne E. "Household Structure in Rhode Island, 1774–1800." *Journal of Family History* 3 (1978): 37–50.
Witthoft, John. *Green Corn Ceremonialism in the Eastern Woodlands. Occasional Contributions from the Museum of Anthropology of the University of Michigan*, no. 13. Ann Arbor, Mich., 1949.
Wokeck, Marianne S. "Harnessing the Lure of the 'Best Poor Man's Country': The Dynamics of German-Speaking Immigration to British North America, 1683–1783." In Altman and Horn, *"To Make America,"* 204–43.
——. *Trade in Strangers: The Beginnings of Mass Migration to North America*. University Park, Pa., 1999.
Wolf, Eric R. "The Inheritance of Land among Bavarian and Tyrolese Peasants." *Anthropologica* 12 (1970): 99–114.
Wolf, George D. "The Politics of Fair Play." *Pennsylvania History* 32 (1965): 8–24.
Wolf, Jacquelyn H. "Patents and Tithables in Proprietary North Carolina, 1663–1729." *North Carolina Historical Review* 56 (1979): 262–77.
Wolf, Stephanie Grauman. *As Various as Their Land: The Everyday Life of Eighteenth-Century Americans*. New York, 1993.
——. *Urban Village: Population, Community, and Family Structure in Germantown, Pennsylvania, 1683–1800*. Princeton, N.J., 1976.
Wood, Betty. "James Edward Oglethorpe, Race, and Slavery: A Reassessment." In *Oglethorpe in Perspective: Georgia's Founder after Two Hundred Years*, edited by Fanes Spalding and Harvey H. Jackson, 66–79, 206–9. Tuscaloosa, Ala., 1989.
——. "A Note on the Georgia Malcontents." *Georgia Historical Quarterly* 63 (1979): 264–78.
——. *Women's Work, Men's Work: The Informal Slave Economies of Lowcountry Georgia*. Athens, Ga., 1995.
Wood, Gordon. *The Radicalism of the American Revolution*. New York, 1992.
Wood, Jerome H. *Conestoga Crossroads: Lancaster, Pennsylvania, 1730–1790*. Harrisburg, Pa., 1979.
Wood, Joseph Sutherlund. *The New England Village*. Baltimore, 1997.
Wood, Neal. *John Locke and Agrarian Capitalism*. Berkeley, Calif., 1984.
Wood, Peter. *Black Majority: Negroes in Colonial South Carolina from 1670 through the Stono Rebellion*. New York, 1974.
——. "The Changing Population of the Colonial South: An Overview by Race and Region, 1685–1790." In Wood, Waselkov, and Hatley, *Powhatan's Mantle*, 35–103.
——. "Indian Servitude in the Southeast." In Washburn, *History of Indian-White Relations*, 407–9.
——. " 'Liberty Is Sweet': African-American Freedom Struggles in the Years before White Independence." In *Beyond the American Revolution: Further Explorations in Radicalism*, edited by Alfred F. Young, 149–84. DeKalb, Ill., 1993.

Wood, Peter H., Gregory A. Waselkov, and M. Thomas Hatley, eds. *Powhatan's Mantle: Indians in the Colonial Southeast*. Lincoln, Nebr., 1989.
Wood, William. *New England's Prospect*. 1634. Edited by Alden T. Vaughan. Amherst, Mass., 1977.
Woodbridge, Linda. *Women and the English Renaissance: Literature and the Nature of Womanhood, 1540–1620*. Urbana, Ill., 1986.
Woodward, Donald. *Men at Work: Labourers and Building Craftsmen in the Towns of Northern England, 1450–1750*. Cambridge, 1995.
——. "Straw, Bracken, and the Wicklow Whale: The Exploitation of Natural Resources in England since 1500." *P&P*, no. 159 (1998): 43–76.
——. "Wage Rates and Living Standards in Pre-Industrial England." *P&P*, no. 91 (1981): 28–46.
Wordie, J. R. "The Chronology of English Enclosure, 1500–1914." *EcHR* 36 (1983): 483–505.
——. "Deflationary Factors in the Tudor Price Rise." *P&P*, no. 154 (1997): 32–70.
Wright, Gavin. *The Political Economy of the Cotton South: Households, Markets, and Wealth in the Nineteenth Century*. New York, 1978.
Wright, J. Leitch, Jr. *Florida in the American Revolution*. Gainesville, Fla., 1975.
Wright, Louis B. *The Dream of Prosperity in Colonial America*. New York, 1965.
——. *Middle-Class Culture in Elizabeth England*. Chapel Hill, N.C., 1935.
Wrightson, Keith. "Alehouses, Order, and Reformation in Rural England, 1590–1660." In *Popular Culture and Class Conflict, 1590–1914: Explorations in the History of Labour and Leisure*, edited by Eileen Yeo and Stephen Yeo, 1–27. Brighton, England, 1981.
——. *English Society, 1580–1680*. New Brunswick, N.J., 1982.
——. "Two Concepts of Order: Justices, Constables, and Jurymen in Seventeenth-Century England." In *An Ungovernable People: The English and Their Law in the Seventeenth and Eighteenth Centuries*, edited by John Brewer and John Sayles, 21–46, 299–307, 312–15. New Brunswick, N.J., 1980.
Wrightson, Keith, and David Levine. "Death in Whickham." In Walter and Schofield, *Famine, Disease, and the Social Order in Early Modern Society*, 129–66.
——. *Poverty and Piety in an English Village: Terling, 1525–1700*. New York, 1979.
Wrigley, E. Anthony. "The Changing Occupational Structure of Colyton over Two Centuries." *Local Population Studies*, no. 18 (1977): 9–21.
——. "Explaining the Rise in Marital Fertility in England in the 'Long' Eighteenth Century." *EcHR* 51 (1998): 436–64.
——. *People, Cities, and Wealth: The Transformation of Traditional Society*. Oxford, 1987.
Wrigley, E. Anthony, and R. S. Schofield. *The Population History of England, 1541–1871: A Reconstruction*. Cambridge, Mass., 1981.
Wrigley, E. Anthony, R. S. Davies, J. E. Oeppen, and R. S. Schofield. *English Population History from Family Reconstruction Data, 1580–1837*. Cambridge, 1997.
Wulf, Karin. "'As We Are All Single': Unmarried Women of Property in Philadelphia County, 1693–1774." Paper, McNeil Center for Early American Studies, March 1992.
Wunder, Heide. "Peasant Organization and Class Conflict in Eastern and Western Germany." In Aston and Philpin, *Brenner Debate*, 91–100.
Wyckoff, V. J. "Land Prices in Seventeenth-Century Maryland." *American Economic Review* 28 (1938): 82–88.
——. "The Size of Plantations in Seventeenth-Century Maryland." *MdHM* 32 (1937): 331–39.
Wyckoff, William. *The Developer's Frontier: The Making of the Western New York Landscape*. New Haven, Conn, 1988.

Yelling, J. A. *Common Field and Enclosure in England, 1450–1850*. London, 1977.
Young, Alfred F. *The Democratic Republicans of New York: The Origins, 1763–1797*. Chapel Hill, N.C., 1967.
——. "The Women of Boston: 'Persons of Consequence' in the Making of the American Revolution, 1765–76." In *Women and Politics in the Age of the Democratic Revolution*, edited by Harriet B. Applewhite and Darline G. Levy, 181–226. Ann Arbor, Mich., 1990.
Young, Arthur. *Arthur Young's Tour of Ireland, 1776–1779*. Edited by Arthur Wollaston Hutton. 2 vols. London, 1892.
——. *Travels in France during the Years 1787, 1788, and 1789*. Edited and abridged by Jeffery Chapleau. Garden City, N.Y., 1969.
Young, Chester Raymond. "The Stress of War upon the Civilian Population of Virginia, 1739–1760." *West Virginia History* 27 (1966): 251–77.
Young, Christine Alice. *From "Good Order" to Glorious Revolution: Salem, Massachusetts, 1628–1689*. Ann Arbor, Mich., 1980.
Youngson, A. J. *After the Forty-five: The Economic Impact on the Scottish Highlands*. Edinburgh, 1973.
Zeichner, Oscar. "The Loyalist Problem in New York after the Revolution." *New York History* 21 (1940): 294–302.
Zell, Michael. *Industry in the Countryside: Wealden Society in the Sixteenth Century*. Cambridge, 1994.
Zemsky, Wilbur. "Nationalism in the American Place-Name Cover." *Names* 31 (1983): 1–21.
Zinkin, Vivian. "Names of Estates in the Province of West Jersey." *Names* 24 (1976): 237–47.
——. "Surviving Indian Town Names in West Jersey." *Names* 26 (1978): 209–19.
Zuckerman, Michael. *Almost Chosen People: Oblique Biographies in the American Grain*. Berkeley, Calif., 1993.

index

Abortion, 5
Accomac County (Va.) court, 234
Acculturation of colonists, 120–23, 322 (n. 114)
Adventurers, 42, 55, 85
Age: at marriage, 129, 173–75, 182, 227–29, 233, 240; of migrants, 149
Agricultural exports, 206, 209, 210–12, 282; during Revolutionary War, 256–59, 267
Agriculture: commercial, 20, 68–69, 120, 135, 204–5, 207–8, 281–82; surplus, 74, 205–6, 207–12, 215–16, 235; in Ireland, 172, 175–76; in Scotland, 173–74, 177–78; reform of, 178–79; in Germany, 180–81; diversified, 193, 206, 209–11, 213, 260–61. *See also* Farms; Subsistence agriculture; Truck farming
Albany, N.Y., 212
Alehouses, 32–33
Alexandria, Va., 216
Allen, David, 223
Allen, Ethan, 154
Allen, James, 260
American Husbandry, 203–4
Andover, Mass., 114, 149, 205
Anne Arundel County, Md., 135
Anti-Catholicism, 169
Antinomians, 67, 115
Anti-rent riots, 290
Apprentices, 23
Architectural styles, 121–22
Armstrong, William, 223
Artisans, 131, 137; and borrowing system, 221–22
Atlantic slave trade, 262, 281
Augusta County, Va., 143
Awashunkes, 100–101

Bacon, Nathaniel, 99
Bacon's Rebellion, 99
Baden-Durlach, 201
"Ballad of a Tyrannical Husband," 9
Ballads, 251; in England, 9, 24, 28, 30, 33–35, 38; about emigration, 47–48, 305 (n. 19)
Ballard, Martha, 221
Balltown, Maine, 161
Baltimore, Lord, 110–11, 158; land policy of, 152–53
Baltimore County, Md., 132
Bandits, 270–71
Barn-raisings, 220
Barter. *See* Borrowing system; Credit
Beanes, Christopher, 112
Beatty, Erkuries, 220–21, 223–24
Beaumont, John, 117
Beaumont, Sir John, 43
Beaver skins, 77, 97
Bedford, Mass., 131
Bedford County, Pa., 265–66
Bedford County, Va., 242
"Beggar Boy of the North" (ballad), 24
Beggars, 184
Belnap, Ruth, 244
Bennington, Vt., 261
Berkeley, William, 99
Berkeley Plantation, 53
Beverely, Mass., 131
Beverely family, 154
Beverley, Robert, 83
Bibles, 247

Birdsey, Nathan, 148–49
Black Death, 8–10, 14–15, 17–18
Blockades, 256–57
Blodgett, Samuel, 285
Bloomfield, Joseph, 276
Board of Trade, 126–27, 139
Boltzius, John, 192
Bond, Mary, 24
Book credit, 217–18, 223, 225
Book ownership, 241
Boone, Daniel, 153
Borden family, 154
Borrowing system, 4, 204, 206–7, 209–10, 216–17, 219–25, 245–46, 256–57, 282; and neighborhoods, 220–24; class and race limits of, 224; and commerce, 225; destruction of by Revolutionary War, 259, 282
Boston, Mass., 208, 210
Boston Overseers of the Poor, 248
Boundary trees, 112
Bouquet, Henry, 140
Bourgeoisie, 170, 227, 247–48; property concepts of, 17
Bradford, William, 226
Brandt, Sebastian, 71–72
Brandywine people, 93
Bread, 267, 282
Bridal pregnancy, 229
Bridewell Hospital (London), 47
Bristol, England, 67
Britannia (ship), 196
British army, 258–59, 268–70, 272–76
Brodhead, Daniel, 277
Brooke, Roger, 238–39
Brooke, Sarah, 238–39
Brumley, Elizabeth, 87
Bucks County, Pa.: inheritance in, 239–40; occupational structure in, 291–92
Burnaby, Andrew, 150
Butler, Edward, 149
By-employments, 10, 25

Calvert, Charles, 58
Calvert, George, 81
Calvert County, Md., 234
Calvert family, 49
Cambridge, Mass., 59
Campbell, John, 194
Camp followers, 264
Capitalism, 2, 170, 282; transition in England to, 8, 16–27, 31–32
Captivity narratives, 93–94
Carolina Regulators, 290
Carolinas: warfare in, 99–100, 142; headrights in, 110; agriculture in, 214–15, 281; slavery in, 250; during Revolutionary War, 274, 278–79
Carter, Landon, 158
Carter, Robert ("King"), 156
Carter, Robert (of Nomini Hall), 158
Cash sales, 223
Catholics: emigration of, 169, 186–88; in Ireland, 171. *See also* Anti-Catholicism; Jesuits
Chain migration, 149–50, 193–94, 200–201, 284
Chapman, George, 46
Charity, 176
Charles County, Md., 113, 115, 117; inheritance in, 240
Charleston, S.C., 215, 218, 224, 274
Chelsea, Mass., 269–70
Cherokee Indians, 78, 144
Cherry Valley, Pa., 277
Chesapeake Bay, 77
Chesapeake region: immigration of free persons to, 56–57; farm sizes in, 132; agriculture of, 213–14, 281; slavery in, 224, 249–50; inheritance in, 237–40; houses in, 241; during Revolutionary War, 265, 275; postwar emigration from, 284
Chester County, Pa., 134–35, 223; cottagers in, 136–37, 253; immigrants in, 193; widows in, 238; textile industry in, 246; slavery in, 249
Child labor, 233, 246–47; in England, 29–30
Children, 66, 69–70, 198; rearing of, 227, 246–48
Christian, William, 280
Christian Indians. *See* Indians: and Christianity
Clap, Roger, 58–59
Class: defined, 4–5; and households, 36–37
Class system, 241; in England, 17, 20–25, 299 (nn. 45, 46). *See also* Yeoman class
Climate, 72, 73, 75, 80–83; in England, 80–81; colonial acculturation to, 82–83

Clinton, Henry, 262
Coal industry, 26
Coastal trade, 209, 215
Colden, Cadwallader, 126–27
Colonies: as solution to English poverty, 39–40; as vent for surplus capital, 40; promoters of, 43–44, 47–48, 191; royal, 109
Colonist-Indian relations: and trade, 85–86; and warfare, 86, 97–100, 105–6, 141–44; and acquisition of Indian languages, 87–88, 139; sexual, 92–93; and gift exchange, 94–95, 96; and land conflicts, 95–96, 139–41
Commercial exchange, 216–17
Commodity chains, 211
Common property, 226
Commons, 9, 110, 125, 130; in England, 14, 17–19, 30, 70–71, 170–71; in Scotland, 173–74; in Germany, 180
Community life, 32–33, 120, 123. *See also* Borrowing system; Neighborhoods
Concord, Mass., 130–31
Congress. *See* Continental Congress
Connecticut, 113; landholding in, 130–31; towns in, 210; divorce in, 234–35
Connecticut Land Company, 288
Conquest, ideology of, 74, 76
Consumption patterns, 222–23, 240–42, 254; in England, 21, 170; in Ireland, 171–72; and class, 241–42
Continental army, 258–65, 267, 269–73; relations with farmers, 258–59, 271; enlistees in, 261–63; substitutes in, 265; and Morris Town, 271–73; occupation of western New York by, 276
Continental Congress, 256–57, 267, 286
Continental currency, 258–60
Contraband trade, 258–60
Contraception, 228, 230
Conversion. *See* Indians: and Christianity; Salvation
Convicts: emigration of, 185, 187, 196–97; labor of, 249
Corn, 86, 91–92, 205, 281
"Corn right," 160
Cornwallis, Charles, 279
Cottagers (Cottars, Cottiers), 12, 129, 136–37, 248, 253; in England, 13–14, 17–19, 21–23, 30, 171, 174, 175, 228; in Europe, 167–68; in Ireland, 171, 176; in Scotland, 179, 187; in Germany, 182–83
Courts: manorial, 14; county, 122, 234
Courtship, 32
Covenants, 122
Coxe, Tench, 222
Credit: 126, 217–19; crisis of, 225–26; in England, 15, 22, 30; and kinship, 222. *See also* Borrowing system
Creek Indians, 140
Crevècoeur, J. Hector St. John, 25–27
Croghan, George, 140–41, 155
Cromwell, Oliver, 67
Crossroads, development of, 214
Cultural mediators, 138–39
Cumberland County, Pa., 140, 143

Dairy products, 211, 216, 244
Danbury, Conn., 273–74
Dane, John, 60, 87
Davis, Alexander, 253
Dearth, Revolutionary War, 267
Debt: suits involving, 223, 225; *See also* Credit: crisis of
Dedham, Mass., 95, 103, 130, 205
Deerfield, Mass., 5, 99, 100, 118–19, 142, 223
Deer Island, 103
Deer skins, 77, 97
Deference, 289–90
Delaware Indians, 138–39
Delaware Valley, 69; landholding in, 133; cottagers in, 136–37, 253; agriculture in, 211–12
Denton, Daniel, 88
Depopulation, Revolutionary War, 279
Depressions, economic, 53–54
Desertion, 265
Diet, 211, 254; in England, 19; in Ireland, 172, 176; in Scotland, 179; in Germany, 183
Diggers, 17
Disease, 83–84
Divorce and legal separation, 37, 233–35
Dobbs, Arthur, 192
Doddridge, Joseph, 160
Domestic violence, 233–34; in England, 33–34
Donck, Andriaen van der, 92
Donegal, Ireland, 171
Donne, John, 40

Doweries, 236
Dower right, 232, 237–38
Drake, Judith, 33
Drew, Abigail Gardiner, 234
Droughts, 81
Dublin, Ireland, 172
Dulany, Daniel, 154, 156
Dulany family, 217
Dunmore, earl of (John Murray), 159, 262–64
Dunmore's Declaration, 262
Durfee, Richard, 267–68
Dutchess County, N.Y., 223
Dutch West India Company, 70, 91
Dwight, Timothy, 268

East Anglia, England, 59, 65–66
Eastern Shore (Md. and Va.), 101–2, 113, 137, 251
East India Company, 44–45
East New Jersey, 130–31
Eastward Ho! (Jonson, Chapman, Marston), 46
Eburne, Richard, 54
Economic growth, 281; in England, 8, 15, 296 (n. 22)
Edict of Nantes, 169
Edinburgh, Scotland, 187
Education, 247
Embargoes, 256, 267
Emigrants: numbers of, 52–53, 165, 169–70, 185–90, 199, 283, 338 (nn. 57, 62); recruiters of, 61, 70, 169, 191–93, 195; destinations of, 188, 191; trans-Atlantic voyage of, 195
Emigration:
—from Britain, 3, 163, 172, 184–89, 283; reasons for, 39–41, 53–61, 166; as part of British migration system, 40–41; and economic opportunity, 41, 47, 49, 53–54, 57–58, 60–61, 71, 165, 185–88, 194; and religion, 41, 49, 53, 57–60, 191–93; literature promoting, 42–43, 47–50, 71, 75, 191–94, 303–4 (n. 2); and search for riches, 42–43, 55, 71–72, 85; as solution to poverty, 43, 155, 175, 199; and kinship, 46–47, 57, 66, 193; temporary, 54; timing of, 55, 60–62, 63–64, 67, 188; of free persons, 56–57, 86–89; costs of, 56–57, 194; selectivity of, 60–62, 63–67, 185, 188; and community networks, 66; of prisoners of war, 178–79; letters home encouraging, 193–94; sponsored, 198
—from Europe, 70, 169
—from German lands, 163, 189–90, 200–201; 283; literature promoting, 91–94; and economic opportunity, 94; reasons for, 166, 190; fees for, 189, 194; timing of, 190; and religion, 191–93, 198; costs of, 194, 201; and kinship, 196; as solution to poverty, 198; sponsored, 198
Eminent domain, 126
Enclosure: in England, 7–8, 10, 14, 16, 17–20, 22, 29–30, 55, 60, 170, 174, 296–97 (n. 29); in Scotland, 172, 177–78
English Civil War, 54
Enlistees: Seven Years' War, 131; Revolutionary War, 261–64
Entail, 121, 133, 322 (n. 115), 325 (n. 21)
Epidemics. *See* Disease; Mortality; Virgin soil epidemics
Essex County, England, 16
Essex County, Mass., 113, 130, 136
Ethnic enclaves, 69
Ethnicity, 69
Europe, living conditions in, 166
Evans, David, 165
Evans, John, 220
Exeter, N.H., 130
Eyre, Adam, 27

Fairfax, George, Lord, 158, 162
Fairfax County, Va., 132
Fairfax proprietorship (Va.), 133
"Fair Play System," 160
Families: servitude of, 196; strategies of, 227, 235; size of, 228–31; life cycle of, 230; labor of, 242–47; frontier, and Revolutionary War, 265–67
Family economy: in Ireland, 171; in Scotland, 179
Family system, 38, 230
Famine: in England, 83; in Ireland, 172, 175–77
Farmers: small, 1; categories of, 226; political mobilization of, 290
Farming: techniques of, 92; tools for, 222
Farm making, 120–21, 144, 162, 202, 203–4, 217, 229–30, 235–36, 289; after Revolutionary War, 280–81

Farms: of English peasants, 12–13, 16, 297 (n. 32); rhythms of, 28–29, 243–45; households of, 71; size of, 113–14, 129–34, 137, 154, 161–62, 284; composite, 204; craft work on, 222; invasions of, 275–76. *See also* Agriculture; Labor—farm; Truck farming; Women—labor of: on farm
Federalists, 291; land policies of, 287–88
Femme sole. *See* Widows; Women
Fences, 79
Ferguson, Patrick, 269
Fertility, of women, 64, 129, 228, 230
Fertilizer, 92, 316 (n. 50)
Feudalism, 115–16; in England, 9–11, 13–14, 16, 229; property in, 125, 181; in Germany, 180–81, 190
Fields: in England, 14, 18, 75, 170; rotation of, 14, 137; in Scotland, 172–73; in Germany, 180; labor in, 244–45
Firewood, 80
Fischer, David Hackett, 6
Fish, Mary, 255
Fishing industry, 43, 60, 65, 206, 208, 209
Fleming, Thomas, 165
Flora and fauna, 77
Florida, land policy in, 151; during Revolutionary War, 278
Flour and flour milling, 207, 210–11, 216, 282
Folk memory. *See* Ballads
Food: riots over, 25; preparation of, 245; and lodging, Revolutionary War soldiers demand, 272–73; prices for (*see* Grain—prices for)
Foraging expeditions, Revolutionary War, 270–71
Forest industries, 76, 79–80, 198
Forests, 72, 75–76, 79–80; in England, 75–76
Fornication. *See* Premarital intercourse
Fort Pitt, 140
Franklin, Benjamin, 150, 228
Frederick County, Va., 143
Fredericksburg, Md., 216
Freeholders, 203–4
Freehold land tenure, 109
Free Society of Traders, 49
French army, 258
Frethorne, Richard, 47, 73
Frontier families, and Revolutionary War, 265–67
Frontier migration, 144–50, 160, 283–84; within Europe, 169, 189
Fry, Joshua, 147
Furman, Moore, 266
Fur trade, 65, 69–70, 77, 90–91, 97, 101, 138, 206, 212

Gardens, vegetable, 244
Gender relations, 232; in England, 9. *See also* Domestic violence; Women
Georgia: squatters on Indian land in, 140; Indian treaties with, 144; colonial migration to, 149; land policy and distribution in, 151, 155–56, 286; immigration to, 191, 199–200; partisan warfare in, 274; agriculture in, 281
Georgia trustees, 155
Gift exchange, 94–95, 96
Glasgow, Scotland, 173, 187; and tobacco trade, 213
Gleaning, 175
Gloucester, Mass., 65
Gooch, William, 148
Gordon, Dr. Roderick, 166–67
Gouge, William, 35
Gower, John, 11
Graffenried, Christopher von, 197–98
Grain: trade in, 208–9, 257; deficits of, 209
—prices for, 212; in England, 19, 21, 25, 170, 174; in Ireland, 171; in Scotland, 179; in Germany, 183–84
Granville proprietary (N.C.), 151
Great Wagon Road, 147, 284–85
Greenbrier (W.Va.) region, 143, 154
Greene, Jack P., 6
Greene, Nathanael, 266, 271
Grid system, 112
Großschönau, Saxony, 183
Grosvenor, Sarah, 5
Gun trade, 89
Gyles, John, 93–94

Hakluyt, Richard, 40, 43
Hammond, John, 40, 52
Hampshire County, Mass., 144
Hampshire County, Va., 143
Hapsburg Empire, immigration to, 181

Harrison, William, 20
Harvest: festivals at, 220; labor for (*see* Labor—farm: at harvest)
Headright system, 57, 109–11, 151, 155, 286
Heaton, Hannah, 234
Henrico County, Va., 113
Hessians, 266
Hingham, England, 66
Hingham, Mass., 66; family in, 233
Hite, Joseph, 162
Holland Land Company, 287
Home manufacture, 243, 245
Hooker, Richard, 115
Horton, Widow, 89
Households: and markets, 204–6; formation of, 226–27, 229–30; definition of, 227, 342 (n. 6); composition of, 230–31
House-raising, 220
Houses, 79, 121–22, 240–41; of English peasants, 12; log, 121; sizes of, 121–22; in Scotland, 178; in Germany, 184
Howe, William, 262
Hudson Valley, 70, 125; borrowing system in, 282
Hunter, Robert, 198
Hunterdon County, N.J., 59
Hunting. *See* Indians: hunting by
Husbandmen, 21, 28
Hutchinson, Anne, 67–68, 115

Illegitimacy, 229, 301 (n. 75); in England, 32
Immigration: demography of in colonies, 60–62; to Ireland, 44, 186. *See also* Emigration
Imports, 206, 209, 211, 222–23
Inckpen, Richard, 27
Indentured servants, 2, 47, 50–53, 55–56, 62–64, 67, 69–70, 115, 185, 188, 194–97, 236, 248–49, 261, 283; female, 56; and land, 109–10
Independence, economic, 1, 71, 126, 224–25, 256, 289–90
Indians: 1622 massacre of, 48, 85, 93, 102; colonial view of, 73, 102–3, 106; ecology of, 74, 77–79; relations with farmers of, 74, 78, 85–86, 91–92, 95, 139, 144, 198; land of, 74, 138; villages of, 76–77, 78; agriculture of, 76, 78, 85–86, 91–92, 315 (n. 48); female, 78, 88–89, 102; hunting by, 78, 89, 89, 97, 102; paths of, 78, 144; housing of, 78–79; population of, 86–87, 105, 140; acquire English language, 87; languages of, 87–88; words of used by colonists, 88, 119; policy for, 89–90, 94–101, 139–41; and reservations, 101; enslaved by colonists, 101, 103–4; and Christianity, 101–3; forced exile of, 102, 138–40; acculturation of, 102–3; as indentured servants, 103; as symbols, 106; place names of, 117–19; and epidemics (*see* Virgin soil epidemics)
—trade of, 86, 88; with colonists, 74, 88–89; culture of, 94–95
—warfare with: during Revolutionary War, 276–77, 279; after Revolutionary War, 283–84, 290
Inequality, 12, 236–37, 350 (n. 84)
Infanticide, 100–101, 184; in England, 32
Inflation, 259–60, 323 (n. 10)
Inheritance, 125, 129–30, 132, 135, 162, 206, 228, 229–30, 237–40; partible, 149, 239–40; restricted, 155; in Germany, 181–82; of widows, 237–38; of land, 239; impartible, 239–40; of daughters, 239–40. *See also* Entail; Primogeniture
Intercultural relations, 87–88, 91–95, 117–18, 139
Interpreters, 87–88, 138
Investors: in colonial ventures, 45–46; strategies of, 212, 235–36
Ipswich, Mass., 130
Ireland: immigration to, 44, 186; poverty in, 166
Iroquois Confederation, 141, 157, 212
Iroquois Indians, 78; and warfare during Revolution, 276–77
Irvine, William, 272

Jamestown colony, 86
Jefferson, Thomas, 1, 184
Jesuits, 135
Johnson, Edward, 48, 82, 205
Johnson, John, 277
Johnson, Robert, 48–49
Johnson, Sir William, 192
Joint-stock companies, 44–46, 49, 53, 109
Jones, Hugh (of Maryland), 80
Jones, Hugh (of Virginia), 80
Jonson, Ben, 46

Kastner, Johann, 184
Kent, Conn., 154
Kent, England, 25–26
Kent County, Md., 207
Kentucky: during Revolutionary War, 279–80; settlement of, 284
King, Gregory, 7
King Philip (Wampanoag Indian chief), 102
King Philip's men, 95
King Philip's War, 89, 93, 98–101, 103, 142; women widowed by, 238
Kin naming, 232–33
Kinship, 230; in England, 12, 15, 19–20; among immigrants, 66–67, 161, 193, 196; and migration, 149–50, 200–201, 284; and emigration, 165–66; in Scotland, 177
Kissam, Benjamin, 270
Kraichtal, Germany, 200–201

Labor: markets for, 23; and discipline, 48–50, 52, 72, 199, 254; slave, 199–200, 203, 215, 249–51, 262
—division of, 3–4, 69; in England, 9, 12–13, 27; in farm households, 205, 227, 233, 243–44, 253–54. *See also* Women—labor of
—farm, 256; in England, 4–15, 23–25; at harvest, 13, 15, 23, 220, 244–45, 252–53, 263, 265; colonial, 204, 209, 242–54; types of, 248–53; during Revolutionary War, 261–67
—supply of, 252; during Revolutionary War, 263–64, 266
Labor migration, 136–37, 165; within Europe, 167–68, 189. *See also* Labor—farm: at harvest; Migration
Labor theory of value, 125–26, 160
Lairds. *See* Landlords
Lancaster County, Pa., 241
Land: hunger for, 3, 71, 74, 129, 139–40, 148–49, 283–84; policy for, 3, 139, 151–57, 160, 191–92, 287–88; tenure of in England, 18; ideology of, 76, 96–97, 125–26, 150, 192–93, 203–4; Indian concepts of, 95–96; sales of Indian, 95–97, 101, 138–41; sovereignty of, 96, 106, 109; speculators in, 96, 111, 132, 139, 141, 144, 144, 151–59, 285–88; abundance of, 105, 129, 138, 150, 192–93, 228–29; development of, 106, 109, 120, 156; distribution of, 106, 109–11, 113–14; 120–21, 130–33, 153–54; 161, 209; fees for registering, 109, 112, 152; boundaries of, 111–12; value of, 120, 126, 236; prices of, 127, 129–30, 132–33, 138–39, 151–55, 161; acquisition of, 127, 135, 150, 152, 198, 201, 236; sales of, 134–35, 151–55, 157, 162, 285–88; companies for, 144, 153–54; costs of, 150–51; conflicts over, 159; redistribution of in Scotland, 172; banks of, 219; and bounties, 286; rent for (*see* Rents)
Landholding: in England, 9–11, 19, 21; in New Hampshire, 130–31, 151, 156; in New Jersey, 134, 161; in Germany, 182
Landlessness, 12, 129
Landlords, 134, 140–41, 144; in England, 9–10 in Hudson River Valley, 129, 156–59, 212; in Scotland, 177, 188–89
Land markets, 111; in England, 11–12, 19–20
Land ownership, 120–21, 137; levels of, 112–13, 127, 129–31, 133–34, 150, 159–61, 161, 288
Landscape, 74–75
Land scarcity, 105, 138, 228–29; in Germany, 181–82
Lane family, 222
Latimer, Hugh, 17
Laud, William, 54, 58
Law codes, 122
Lawson, William, 17
Leaming family, 156
Leases, developmental, 134, 153, 158–59
Legal separation. *See* Divorce and legal separation
Lenape Indians, 105
Lindsay, David, 165
Lindsay, Robert, 165–66
Linen Board (Ireland), 171
Liquor trade, 90
Litchfield, Conn., 142
Literacy, 247; in Germany, 190
Livestock, 97, 205, 207, 209–10, 213, 215–16, 235–36; in England, 10, 35; conflicts over, 95, 98, 102–3, 138; Indian, 102–3; in Ireland, 176; removal by Revolutionary War armies, 269–70

Livingston, William, 259
Livingston family, 158, 217
Livingston Manor, 198
Local exchange. *See* Borrowing system
Loen, J. M. von, 184
Logan, James, 159
Log houses, 121
London, England, 21–22, 24, 26, 51, 54, 62, 67, 185; population of, 168; poor people of, 199; tobacco merchants of, 208
Long Island, 70, 96, 114, 116; raids on, 274
Lord-peasant relationships, 9–11; in England, 13–14, 16–18; in Germany, 181
Lords of Trade, 126–27, 139
Lotteries, 45–46
Louisiana Purchase, 292
Lower Norfolk County, Va., 113
Loyalists, 255, 268–69, 278; militias of, 262; as émigrés, 280; confiscation of land of, 280, 285–86
Lumbering. *See* Forest industries
Lunenburg County, Va., 160, 161, 223
Lycoming County, Pa., 291

McCusker, John J., 6
McDonald, Michael, 200
McDonald, Nicholas, 200
Macomb, Alexander, 287
Maine, 153–54
Maize. *See* Corn
Malaria, 83–84
Malthus, Thomas, 166
Manning, William, 59
Manors, 109, 111, 158
Manufacturing, in England, 25
Manumission fees, 189, 194
Maps, 121, 147
Market: economy based on, 205–6, 226; production for (*see* Agriculture: commercial)
Marketplaces, 206, 208–9, 210–11, 216
Markets, 92, 193, 204–16, 226; in England, 4, 15. *See also* Market towns; Towns
Market towns, 206–7, 210–11, 216; in England, 14, 21, 75
Marple Township, Pa., 133–34
Marriage, 231–32; companionate, 28, 37, 232; proletarian, in England, 31–34, 37–38; settler-Indian, 92–93; of cousins, 222; and household formation, 227; and remarriage, 237–38; and inheritance, 240
Marriage contracts, 231–32; in England, 32, 35
Marston, John, 46
Martha's Vineyard, 103
Martin, Joseph Plumb, 259
Marx, Karl, 5
Maryland: settlement of, 49, 115; headrights in, 110; land in, 113, 132–33; servants in, 115; place names in, 119; poverty in, 137; debt cases in, 225; sale of Loyalist land in, 285
Masonian proprietors, 151, 156
Massachusetts, 64–65; land in, 114, 130–31; constitution of 1780, 126; transiency in, 136; Indian policy of, 141; land banks in, 219
Massachusetts Assembly, 141
Massachusetts Bay Company, 45–46, 53, 67
Massachusetts Court of Assistants, 112
Master narratives, 5–6
"Maydens of London, The" (ballad), 47, 305 (n. 19)
Mayflower (ship), 56
Mayflower Covenant, 65
Meat industry, 205, 210, 213, 214–15
Mecklenburg County, Va., 260
Menard, Russell R., 6
Merchants, 211–12; networks of, 207; tobacco, 207–8, 213–14
Merchet, 14
Microhistory, 5
Mid-Atlantic region: agriculture in, 203–4; inheritance in, 238; slavery in, 249; and British supplies, 258–59
Middlesex County, Mass.: land bank in, 219; craftsmanship in, 222; illegitimacy rates in, 229
Middleton, John, 112
Mifflin County, Pa., 241
Migration: within England, 2–3, 15, 22–23, 26–27, 41, 54, 185; intracolonial, 3, 107, 114–15; intercolonial/state, 50, 67–68, 107, 127, 139, 144–47, 157–58, 200, 283–84; Indian, 104–5, 138; short-distance, 114–15, 135–37, 144; direction of, 144–48, 200, 284–85; and warfare, 147; immigrant, 147, 196–97, 198, 200–

201; motivations for, 148–49; chain, 149–50, 193–94, 200–201, 284; within Europe, 167–69, 184–85, 186, 187, 189, 198; seasonal, 168
Military: principles of wartime service in, 265; raids by, 269, 278
Militia: in Indian wars, 100; during Revolutionary War, 262, 264–66, 279–80
Minisink partners (N.Y.), 157
Miscegenation, 92–93
Missionaries, 88
Mittelberger, Gottlieb, 196
Mogridge, Katherine, 24
Mohawk Valley, 279
Monacy Manor, 285
Money, 90, 223, 258–60
Montgomery County, N.Y., 160, 161
Moore, Henry, 150
Moraley, William, 165, 195
Moravians, 201, 218
Morris Town, N.J., 271–73
Mortality, 64, 66, 83–84, 198, 199, 231, 237, 313 (n. 27); in England, 83; shipboard, 195–96
Mortgages, 218–19
Morton, Thomas, 91, 92
Mosely, Joseph, 137
Munster, Ireland, 177

Naragansett Bay region, 208
Natick, Mass., 95, 103
Natural increase, 129
Neighborhoods, 220–24
Neolin, 143
New England: settlement of, 48–49, 86; immigration to, 64–65; livestock conflicts in, 95; warfare in, 98–99; land distribution and landholding in, 110, 113–14, 130, 153–54; food imports to, 130; agriculture in, 204, 207–8, 209–10; household composition in, 230; domestic violence in, 233–34; inheritance in, 237–38; child labor in, 248; servants in, 248
"New England's Annoyances" (Johnson), 48, 82
New Englands Plantation (book), 48
Newfoundland, 81
New Hampshire, 130–31, 151, 156
New Haven, Conn., 68
New Jersey, 134, 161
"Newlanders," 193
New Netherland, 70
New Sweden, 69–70
Newtown, N.Y., 249
New York: land distribution in, 156–57; landlords in, 156–59; population growth in, 157–58; mortgage and credit market in, 218–19; sale of Loyalist land in, 285–86
New York, N.Y., 211–12
Nipmuck Indians, 98
Nonatum/Newton, Mass., 89
North Carolina, 132, 192
Northern frontiers, 284
Northern Neck (Va.), 133, 158
Northumberland County, Pa., 160
Northwest Ordinance, 287–88

Occupational distribution, 131, 291–92; in Germany, 183; in England, 297–99 (nn. 37, 43, 44)
Oglethorpe, James, 191, 199
Ohio, 288
Ohio Company (first), 153
Ohio Company (second), 288
Olinda, Hilletie van, 87
Onondaga (Iroquois) council, 141
Open fields. *See* Fields
Opequon, Frederick County, Va., 161
Orange County, N.C., 155
Orphans, 231, 248
Osborne, Goodwife, 88
Oxen, 244

Palatine Germans, 157, 197–98
Palmer, Abigail, 276
Panie, Edward, 24
Parke, Robert, 193
Parkman, Ebenezer, 246–47, 251
Parliament (British), 256–57
Partisan warfare, 268–69, 273–75, 290–91
Patch, Mayo Greenleaf, 226–27
Patriarchalism, theory of, 36, 231, 233–35, 251, 349 (n. 75); in New England, 233–34
Patriarchy, 29, 34–35, 226–27, 230–31, 289
Patron-client relations, 22, 156, 217–18
Patten, Matthew, 216–17, 221
Pattison, Jane, 234–35

Pattison, Jeremiah, 234
Paxton, Pa., 144, 159, 160
Paxton Boys, 290
Payne, Robert, 30–31
Peasant agriculture: in England, 9–11; in Europe, 167–68; in Britain, 171, 172–74, 175–77; in Germany, 180–81
Peasants, 125; English, 4, 7–8, 13; images of, 11; and marriage, 13–14, 168, 180; in Europe, 167–68; in Germany, 180–82; dispossession of (*see* Enclosure); relationships with lords of (*see* Lord-peasant relationships)
Peasant uprisings, 10; ideology of, 7–8; in England, 7, 16
Peddlers, 15
Penn, William, 49, 60, 69, 111, 191
Penn family, 141; land policy of, 152–53
Pennsylvania: settlement of, 49–50, 60–61; immigration to, 69, 191; headrights in, 110; squatters on Indian land in, 140; migration in, 148; land bank in, 219; inheritance in, 239–40; houses in, 241; Revolutionary War enlistment in, 265; occupational structure in, 291–92
Pequot Indians, 86–87
Pequot War, 98, 100
Perkins, Elizabeth, 234
Persistence, 160–61; rates of, 114–15; reasons for, 135, 149, 166–67, 187
Phelps and Gorham tract (N.Y.), 287
Philadelphia, Pa., 211, 259
Philadelphia County, Pa., 238
Philip, King (Wampanoag Indian chief), 102
Phillipsburg Manor, 286
Pick, William, 260
Pickens, Andrew, 275
Piedmont (Va.), 196
Piscataway Indians, 88, 101, 105
Pisquetomen, 139–40
Pittsylvania County, Va., 225
Place names: conflicts over, 117–19; of land tracts, 119–20
Planter's Plea, The (book), 49
Plunder, during Revolutionary War, 258–59, 268–69, 272–73, 290
Plymouth Adventurers, 45
Plymouth colony, 65, 76, 100, 115, 226
Pocahontas, 93
Pond, John, 73
Pontiac's War, 143–44
Poole, Robert, 87
Poor: children of, 136, 248; auctions of, 137
Poorhouses, 137, 175, 179
Poor laws: in England, 25, 31, 122; in Ireland, 176
Poor relief, 136–37; in Scotland, 172, 179–80 in England, 175; in Germany, 184; during Revolutionary War, 267
Population, 64, 70, 129, 157–58, 161, 209, 240, 285; of England, 9–10, 17–18, 22–23, 62; origin of colonial, 51; density of, 107, 127, 129, 146, 150; pressure on land of, 129, 148; distribution of, 144–48, 285; in Europe, 166; in Scotland, 172; in Ireland, 175; in Germany, 181; exchanges of during Revolutionary War, 278, 280
Portsmouth, R.I., 103
"Possession fence," 160
Potatoes, 175, 176
Poverty, 129, 135–37, 150, 203, 242; in England, 7–9, 24, 39, 58, 174–75; in Europe, 166, 184; in Ireland, 171; in Scotland, 178, 179; in Germany, 183–84; in Britain, 199; of widows, 238
Powhatan Indians, 85–87, 94, 97–98; confederacy of, 77
Pownall, Thomas, 139
"Praying Indians." *See* Indians: and Christianity
Premarital intercourse, 229
Prenuptial agreements. *See* Marriage contracts
Presbyterians, Ulster, 171; emigration of, 186–87
Price-fixing, Revolutionary War, 260
Primogeniture, 239–40; *See also* Inheritance
Prince George's County, Md., 112; place names in, 119; landholding in, 132–33; household composition in, 230
Prisoners of war, 93–94, 255; as farm laborers, 266
Privateering, 256–57
Proclamation of 1763, 140, 144, 153
Proletarian households, in England, 33, 34
Property: feudal, 11, 125, 181; private, 17, 71, 75, 106, 111, 115–16, 126; among Indians, 101

Proprietary colonies, 49, 109
Proprietors of land, 151, 154
Protestantism, 34–35
Protests, rural, 174, 187, 291
Proverbs, in England, 33
Providence, R.I., 118
Provisions: military demand for, 257–60, 266–67, 271–72, 276; for cities (*see* Truck farming)
Prussia, 181–82
Puritans, 42, 58–60, 65–66, 68, 70; in England, 34, 54
Purry, Jean, 192
Putting-out system, 173, 182–83
Pynchon, Thomas, 91
Pynchon, William, 91
Pynchon family, 60, 90, 116–17, 217

Quakers, 50, 58, 186, 193, 219, 247; English persecution of, 60; as immigrant sponsors, 198; credit and, 222; female fertility among, 229–30; marriage rituals of, 233
Queen Anne's Manor, 285
Queens County, N.Y., 270
Quincy, Mass., 120
Quitrents, 116, 151–52

Rack rents, 175
Raids, Revolutionary War, 273–75, 278; frontier, 277; rumors of, 279
Raleigh, Sir Walter, 117
Ramsey, David, 126, 150
Rape, 276
Recruiters: of servants, 51–52, 195; of emigrants, 61, 70, 169, 191–93, 195
Redemptioners, 194–97, 282
Redwood, Joanne, 29
Refugees: Seven Years' War, 143; Revolutionary War, 268, 277–80
Rensselaer, Jeremiah van, 70, 116
Rensselaerwyck, 111
Rents, 117, 135; in Ireland, 175–76, 187. *See also* Anti-rent riots; Quitrents; Rack rents
Requisitions, of Revolutionary War armies, 259, 266–67, 269–71
Remarriage, 237–38
Revere, Paul, 5
Revolutionary War: and farm economy, 256; size of armies in, 258; and contraband trade during, 258–60
Rhode Island, 136
Rice: plantations for, 132; production of, 203, 205, 207, 215, 256–57
Rituals: of reciprocity, 220–25; of entertainment, 241
River systems, 77
Roads, 120, 144, 215–16
Roanoke colony, 42, 62, 77; Raleigh names, 117
Roanoke Indians, 86
Roberts, John, 139
Rock, John, 253
Rogers, John, 59
Rogers, Samuel, 54
Rogers, William, 162
Rolfe, John, 93
Ross, John, 192
Rotterdam, Holland, 195
Rowan County, N.C., 221
Rowlandson, Mary, 93
Roxbury, Mass., 120
Royalists, 64
Rumors, of colonial conditions, 46–47, 165

St. Eustatius, 257
St. George's Parish, S.C., 250
St. John's Berkeley Parish, S.C., 132
St. Mary's County, Md., 58, 284
Salem, Mass., 65, 117
Saltonstall, Nathaniel, 95
Salt riots, 267
Salvation, 58–59
Savage, Thomas, 87
Saxony, 182
Schooling. *See* Education; Literacy
Scotland, 172, 173, 177–78, 188–89; clan chiefs in, 178; tacksmen in, 178
Seasoning, 72, 83–84, 199
Sedgwick, Pamela, 246
Self-sufficiency, 4, 11–13, 204, 207, 256, 259. *See also* Subsistence agriculture
Selwyn-McColluh proprietary (N.C.), 152
Serfdom, 10, 181, 189, 229
Servants, 249; recruiters of, 51–52, 195; work of, 52, 56, 196–97, 251, 47 197; in Scotland, 177; families as, 196; sale of, 196–97; and conflicts with masters,

251–52; runaway, 261–62. *See also* Convicts; Indentured servants; Redemptioners; Servants-in-husbandry
Servants-in-husbandry, 13, 14, 22–23, 27–28, 32, 37, 41, 56, 58, 61, 72, 228
Servant trade, 50–52, 62–63, 196
Settlement patterns, 107–8, 210, 217–18
Seven Years' War, 131, 135, 138, 142–44, 156, 200
Shawnee Indians, 138, 139
Shenandoah Valley, 154–55; servant trade in, 196; agriculture in, 215–16, 259
Shepherd, Thomas, 59
Shipboard mortality, 195–96
Shopkeepers. *See* Storekeepers
Sibbes, Richard, 34
Silliman, Gold Selleck, 255
Silliman, Mary Fish, 255
Simm, Hugh, 253
Simm, Simon, 253
Simmonds, William, 55
Skickllamy, 138
Slash-and-burn agriculture, 78, 311 (n. 13)
Slaveholding: in Georgia, 155, 199–200; in Virginia, 223; distribution of, 249–50, 262
Slave-hunting, 104
Slaves, 249; Indian, 101, 103–4; as labor, 199–200, 203, 215, 249–51, 262; domestic economy of, 224; female, 250–51; and conflicts with masters, 251; runaway, 261–62
Smallwood, William, 261
Smith, Henry, 234
Smith, John, 43, 48, 77, 86
Smith, Rev. Thomas, 209
Smith, Sir Thomas, 7
Smith, William, Jr., 157
Snyder, Benjamin, 212, 223
Society for the Propagation of Christian Knowledge, 191
Somerset County, Md., 115
South Carolina: immigration to, 67–68; Indians in, 104; landholding in, 132; land policy in, 151; migration from, 200
Southern Appalachia, 287
Southside (Va.), 154, 214, 218
Sovereignty. *See* Land: sovereignty of
Spicer family, 156
Spinning. *See* Textile manufacturing
Spotswood, Alexander, 152, 154
Springfield, Mass., 117
Squatters, 3, 96, 126, 134, 139–41, 141, 144, 150, 153, 158–60, 198, 286, 288; in England, 23
"Squaw Sachem," 100–101
Standish, Arthur, 7
Starvation, 62, 66
Staten Island, 98
Step-parents, 237
Stevenson, Randolph, and Cheston, 196–97
Storekeepers, 211, 218–19; Scots as, 213–14, 218, 223; and borrowing system, 222–23
Stout, William, 27
Stumpel, Johann Heinrich Christian von, 192, 200
Subsistence agriculture, 3–4, 204, 208–9, 235, 260–61. *See also* Self-sufficiency
Sudbury, Mass., 114, 205
Sullivan, John, 277
Surry County, Va., 113, 117, 249
Surveying, 111–12, 121, 162
Susquehannock Indians, 90–91
Swansea, Plymouth, 113–14
Symmes, John, 288

Tacocolie, 139
Talbot County, Md., 113
Tapp, William, 102
Tenancy, 71, 111, 126, 129–31, 133–35, 137, 150, 153, 156–58, 212–13, 285–86; in England, 18; levels, 116–17; in Ireland, 171, 175, 175; in Scotland, 177–78
Tennessee, 141
Textile manufacturing, 221, 245–46; in England, 9, 16, 25–26, 58–59, 171, 174; in Europe, 167–68; in Ireland, 171, 175–76; in Scotland, 173; in Germany, 182–83
Thatcher, James, 268
Thirty Years' War, 181
Thompson, Alexander, 200
Tobacco: prices for, 63, 213, 217–18; production of, 64, 203, 205, 207–8, 213, 281; trade in, 173, 207–9, 213, 256–57, 281; Oronoko, 213; sweet-scented, 213; inspection acts for, 214; warehouses for, 214
"Tomahawk right," 160

Tories. *See* Loyalists
Torture, 275
Towns, 108, 210–11, 214, 343 (n. 17); proprietors of, 110, 154; military occupation of, 278
Trade, 68–69; coastal, 209, 215; contraband, 258–60
Trading companies, 43–45
Transiency, 115, 135–37, 144, 160; in England, 22, 24–25, 33, 58; in Germany, 184
Transylvania Company, 153
Travel, costs of, 144, 201
Treaties, 141, 144, 283
Truck farming, 205–7, 210–13, 215, 260; in Ireland, 172
Tryon, William, 150
Tsenacomoco/Tsenacommach, 117
Tukapewillin, Joseph, 103
Turner, Amos, 136
Tuscarora Indians, 198
Tusser, Thomas, 27–30, 38

Ulster County, N.Y., 223
Ulster Irish immigrants, 161
Urbanization, 291; in England, 26–27, 168, 299 (n. 45); in Europe, 168; in Scotland, 172–73 in Germany, 183

Valentine, Mary, 193
Valley Forge, Pa., 266, 273; civilian/army conflict at, 271
Vermont, 154, 156
Virginia, 81, 122; settlement of, 39–40, 48–49, 52, 55, 86, 115, 140; migration of immigrants to, 67; warfare in, 97–99; headrights in, 109–10; land distribution in, 113, 132, 151–52, 154, 161; Indian place names in, 117–18; land prices in, 133; migration patterns, 148; tobacco in, 213, 257
Virginia Assembly, 96
Virginia Company, 45–46, 48–49, 51, 62–63, 71, 86, 109
Virginia Council, statement on Indians, 141–42
Virgin soil epidemics, 77, 86–87, 105

Wage labor, 67, 130, 131, 135–37, 198, 248, 252–53; in England, 13, 22, 27, 30–31, 33, 51, 72, 228; Indian, 95, 103–4; in Ireland, 176; in Scotland, 177, 179
Wages: in England, 10–11, 18–19, 63, 170; in Scotland, 179; Revolutionary War, 266
Wagon roads. *See* Roads
"Walking purchase," 141
Wampum, 90
Ware, Ebenezer, 136
Warham, John, 59
Warnings out, 136
Warren, Sir Peter, 218
Warwick, R.I., 102–3
Washington, George, 143, 158, 259, 269–72, 277
Waterhouse, Edward, 48, 85
Watertown, Mass., 54
Wealth: accumulation of, 236; distribution of, 237
Weaving. *See* Textile manufacturing
Wenham, Mass., 131, 149
Westchester County, N.Y., 268–69
West Country, England, 65
"West-Country Man's Voyage to New England" (ballad), 47
West Indies, 68, 211, 215, 282
West Jersey: place names, 119–20; proprietors, 152
West Jersey Society, 159
Weston, Thomas, 65, 92
Wet nurses, 246
Wheat, 207, 210–12, 214–15, 251, 281
White, John, 59
White, Samuel, 220
Whitelaw, James, 194
Widows, 237–39
Wife Lapped in Morels Skin (chapbook), 34
Wigwams, 78–79, 91
Wilderness, 73–75, 78
Williams, Eunice, 5
Williams, Roger, 68, 88, 96, 98
Wilson, Thomas, 20
Windsor, Conn., 68
Wingandacola, 117
Winslow, Edward, 76
Winstanley, Gerrard, 17
Winthrop, John, 39, 46, 65–66, 73, 252
Winthrop, John, Jr., 91
Woad, 31

"Woman's Work is never done, A" (ballad), 35
Women: as scolds in England, 31–32, 33–34; rights of in Protestantism, 34–35; English bourgeois, 35; and property, 35–36; images of, 36; as immigrants, 62–63, 65, 69; fertility of, 64, 129, 228, 230; in Ireland, 175; and borrowing system, 220–21; legal position of, 230–31; roles of as wives, 232; during Revolutionary War, 275–76
—labor of, 3–4, 243–45; in England, 9, 13, 15, 27–30, 31–32, 174–75; as servants, 56, 251; in Ireland, 171; in Scotland, 173, 179; in Germany, 180; on farm, 9, 13–14, 238; 244–46; as slaves, 250–51; during Revolutionary War, 255–56, 264, 266
Wood, William, 88, 120
Worcester County, Mass., 241
Wrentham, Mass., 135
Wyatt, Sir Francis, 102
Wyoming Valley: settlement of, 141; Indian warfare in, 277

Yakatastange, 140
Yamasee War, 100, 105
Yeoman class, 20–21, 28, 203–4, 241; ideology, 291–92
York County, Maine, 233
Yorkshire, England, 186
Young, Arthur, 166, 176

Zurich, Switzerland, 183